KU-345-659

ANDERSONIAN LIBRARY
WITHDRAWN
FROM
LIBRARY
STOCK
UNIVERSITY OF STRATHCLYDE

Alluvial Mining

The geology, technology and economics of placers

Alluvial Mining

The geology, technology and economics of placers

Eoin H. Macdonald

Consulting Mining Engineer

London New York

CHAPMAN AND HALL

First published 1983 by
Chapman and Hall Ltd
11 New Fetter Lane, London EC4P 4EE

Published in the USA by
Chapman and Hall
733 Third Avenue, New York NY10017

Printed in Great Britain
at the University Press, Cambridge

ISBN 0 412 24630 9

British Library Cataloguing in Publication Data

Macdonald, Eoin H.

Alluvial mining.
1. Hydraulic mining
I. Title

622'.32 TN278

ISBN 0-412-24630-9

Library of Congress Cataloging in Publication Data

Macdonald, Eoin H.

Alluvial mining.
Bibliography: p.
Includes index.
1. Hydraulic mining. 2. Placer deposits. I. Title

TN278.M24 1983 622'.32 82-22215
ISBN 0-412-24630-9

Contents

Preface

Alluvial mining studies were once considered the sole prerogative of engineers and only rudimentary standards of geology were thought necessary for those engaged in exploration and property valuation. Practical experience was gained largely through trial and error and there was little grasp of the physical background to sedimentation or of advanced mathematical models in ore assessment.

But attitudes are changing and this book is an attempt by the author to provide in one volume, a broad and up-to-date treatment of the subject following several decades during which important new concepts have been proposed and developed. There has been, in the first place, a great addition to the knowledge of placer geology, geomorphology and sedimentology; and of geophysics and photogrammetry as supplementary tools in exploration. Impressive numbers of new research programmes have provided better means of assessing data from sampling. New and improved laboratory techniques and fresh approaches of chemistry and physics have been and are being applied to an ever greater extent.

Although the transportation of solids in pipelines is not an exact science, important new information has reduced the risk of failure in the design of slurry pipeline systems. It is emphasized that levels of performance in all sections of wet mining and processing plants are influenced greatly by how well the slurry pumps perform. Chapter 5 brings together all of the concepts and data required in the preliminary design stage of an operation. This will also enable the engineer to collaborate more effectively with the manufacturer throughout.

On markets, the basics have changed little since Pliny the Elder first observed that 'in some places the earth is dug for riches, when life demands gold, silver, silver-gold and copper, and in other places, for luxury, when gems and colours for tinting walls and beams are demanded'. But if the basics have not changed, the complexities of operating in conditions of widely fluctuating world prices and conditions of supply and demand now add significantly to all the problems of operation.

Important sections of the book describe current machines and techniques employed in mining and mineral dressing plants. Some of these are just beginning to be mentioned in the literature and reference is made to significant advances in creating modular gravity, electrostatic and electromagnetic circuitry. Many new design features now permit the mining of lower grade deposits with less usage of power and water than before and with minimum impact on the environment.

Because of the great difficulties of coping with the current flood of literature, models have been created to assist in relating the new developments to practical applications. As well as setting out the basic guidelines for researching new discoveries, such models also provide a means of testing hypotheses and explanations of current beliefs.

All of the above matters are discussed along with certain fundamental aspects of mine planning and development. Nevertheless, the net result represents only one stage in a rapidly evolving progression and there is ample scope for the application of new concepts to bring about more accurate and complete interpretations of the natural phenomena discussed. For example, in the fields of offshore exploration and mining, bold new ideas are being put to the test at hitherto unattainable depths of water.

The book is aimed primarily at professionals or those undergoing professional training in technical colleges and universities. However, it recognizes that the topics are of interest to both lay and professional operators and while there is nothing beyond the mathematical ability of any science graduate, only such equations are used as are needed to explain or develop a particular theme or to describe real situations as they occur. Some researchers may find the mathematical treatment too elementary and, for them, a comprehensive set of references will provide the opportunity to follow up matters of interest in greater depth. Metric terminology has been used throughout and a list of conversion factors is appended for those who follow the Imperial system of measurement.

Acknowledgements

Possibly the greatest benefit to an author is the willingness of friends and associates to examine critically what is written and make useful suggestions for improvement. In this I have been most fortunate and, while the facts and opinions expressed are my own as are the responsibilities for any errors and omissions, I am greatly indebted for their comments to Ian Terrill of Mineral Deposits, Southport, Q; John Grover of Peko Wallsend; Douglas Boyle of Riofinex, Geological Mission in Saudi Arabia; Graham Miller of Mitchell Cotts Projects; L. J. Noakes (ex Director of the Bureau of Mineral Resources, Canberra) and Murray Hill of Warman International.

Additional to my personal knowledge and experience, I have obtained, from many sources, information on machine performances and criteria for design. Particular thanks are due to Warman International Limited, Mineral Deposits Limited and Readings of Lismore for permission to reprint data from internal reports and brochures. Full acknowledgements of such sources are made at appropriate places in the text.

Finally I would like to acknowledge that much of the inspiration for this book arose from my association with many fine geologists and engineers whose enthusiasm for learning is a great credit to their countries and themselves. Outstanding in my memories are Kim Won Jo (South Korea), Sa-ngob Kaewbaidhoon and Dr Charan Achalbhuti (Thailand), U Tint Zaw (Burma), Sutedjo Sujitno (Indonesia), D. Santokh Singh and Mokty bin Mahmood (Malaysia), Chao-yi Meng and Ching Chao (Taiwan), Muhammad Aslam (Pakistan) and Ravi Prakesh (India). There are many others whom I remember with great affection and respect for their help in developing my opinions and improving my knowledge.

1

Background and the Environments

1.1 PLACERS, AN HISTORICAL FRAMEWORK

Primitive man sought gold and gemstones for aesthetic reasons and records show that the market for gemstones, especially the coloured ones, was well established in Babylon as far back as 4000 years BC. Diamonds are referred to in Exodus xxviii, 18 and xxxix, 11 but were apparently less well regarded at that time, perhaps because of their extreme hardness and the consequent difficulties in cutting and polishing. Tin became valuable as an essential ingredient for the making of bronze weapons (the Bronze Age lasted from about 4000 to 1000 years BC) and, according to Raeburn (1927) mention is made in the writings of Strabos and Diodorus Siculus to quotations from the lost writings of Pythias concerning the mining of tinstone in Cornwall. Further early references to tin are found in the Bible (Numbers xxi, 9) and, in Ezekial xxvii, 12, articles made from tin are listed amongst merchandize sold in the markets of Tyre. The metal for these is supposed to have come from mines in southern Spain.

There is reason to believe that gold has been known and prized for at least 12 000 years and Rose (1902) attributes early recognition of the metal to its distinctive colour and common occurrence in surface and near-surface sediments. He also noted that in the code of Menes who reigned in Egypt around 3600 BC, about 2000 years before Moses, the ratio of values of gold and silver was given as 1 part of gold equal to $1\frac{1}{2}$ parts of silver; and that among the ancient rock carvings in upper Egypt there are several illustrations of the art of washing auriferous sands by hand. The oldest of these carvings date back to about 2500 BC.

The earliest mention of mercury for amalgamation with gold occurs in the works of Theophrastus, about 300 BC, and Vitruvius, about 13 BC described how gold could be recovered, using mercury, from cloth in which gold thread had been interwoven. Theophilus, a monk in the eleventh century, described a method for washing sands from the Rhine on wooden tables, the final clean up being with mercury to recover the gold.

The gemstone crysoberyl is referred to in an ancient Ayurvedic text called the Rasaraja Tarangini (Gopinath, 1978) dated around 2000 BC and the main three varieties were recognized as:

common crysoberyl: the colour of a bamboo leaf
alexandrite: lustrous like a peacock's eye
cymophane (catseye): shining as a cat's eye.

Reference is also made by Gopinath to a description of the Yudhishtira's coronation between 2000 and 1400 BC given in the similarly ancient Mahabharata epic. Amongst the many gifts were vaiduriam (a loose Sanskrit name for crysoberyl and gemstones found with it, for example Vaiduriam-neela such as blue vaiduriam or sapphire), from the mountains on the west coast of India.

Other early descriptions of minerals are found in the works of ancient Greek scholars. The first written work on the subject is attributed to the philosopher Theophrastus (372–287 BC). A book on 'Geography' published by Strabos around 7 BC noted the use of running water in sediment transportation and referred to the existance of alluvial gold deposits in river bed sections in Spain and Iran. Pliny the Elder (*Natural History*, AD 77), who recorded the mineralogical thought of his time, observed how miners recovered gold dust from the streams and described the attritional effects of the currents on river gold. Figures given in these works suggest that, from Spain alone, the annual production was around 200 000 ounces troy.

Pliny also asserted that 'gold is found in the world in three ways to say nothing of that found in India by the ants and in Scythia by the Griffins'. This set the pattern of thought for a very long time and during the following 1300 years all publications were based on folk lore and fables with very little substance. In fact, early geological thinking was almost entirely split between the teachings of two Greek philosophers; Thales, who in 640 BC founded what later became known as the Neptunian School; and Xenon (270 BC) who was the first to propose Plutonism. The Neptunians ascribed the formation of all rocks and strata to water and explained the subsequent creation of individual land forms as being due to catastrophic floodings such as those described in the biblical legend of Noah's Ark. Plutonists, on the other hand, believed that fire as well as water was responsible for the evolution of all things living and dead and, generally, reason took second place to a strong belief in the supernatural that allowed both schools to attribute to divine providence any unpleasant facts contradicting their accepted theories. 'If the Gods have so chosen, who are we to question Their will' has been a common response to difficult questions throughout the ages.

As an example of facts as the basis for myths, Strabos in his writings reported that 'Around 4000 BC in the country of Saones . . . the winter torrents brought down gold which the barbarians collected in troughs pierced with holes and lined with fleeces.' They hung them on trees, to dry, and then beat them to recover the gold. They also laid hides, hair side up on streambeds to collect the gold settling out from sands washing over them. Amongst the hides were sheepskins, hence the legend of the Golden Fleece and Jasons Voyage when he landed near the shore of Euxine, around 1200 BC, was probably no more than a simple act of dacoity by him and the Argonauts.

Descriptions of the dangers encountered and obstacles overcome were the normal responses of a superstitious people to new experiences and perils that they did not fully understand.

Throughout ensuing times, miners continued to exploit alluvial sources for gold, tin and gemstones. Metallurgists learned to smelt metals and prepare certain alloys and chemistry as a science had its beginnings around AD 300. Chemistry (chemia, from the Greek *χημεια*) was initially defined as the art of transmuting base metals into gold and silver and, from the fourth until the fifteenth century, alchemists dominated the scene. Some information on the composition of ores, minerals and gemstones is contained in middle-eastern works of around the eleventh century, but it was not until Agricola published his descriptions of alluvial mining operations during the period 1546 to 1556 and reduced to some semblance of order the basic facts of gold metallurgy that any substantial evidence of scientific reasoning emerged to explain observed phenomena. Of the metals industry, generally, Agricola said

> Many persons hold the opinion that the metal industries are fortuitous and that the occupation is one of sordid toil, and altogether a kind of business requiring not so much skill as labour. But as for myself, when I reflect carefully upon its special points one by one, it appears to be otherwise. For the miner must have the greatest skill in his work that he may know first of all what mountain or hill, what valley or plain, can be prospected most profitably, or what he should leave alone; moreover, he must understand the vcins, stringers and seams in the rocks. Then he must be thoroughly familiar with the many and varied species of the earths juices, gems, stones, marbles, rocks, metals and compounds. . . . Lastly, there are the various systems of assaying substances and preparing them for smelting. . . . Furthermore there are the many arts and sciences of which the miner should not be ignorant. First there is Philosophy, that he may discern the origin, cause and nature of subterranian things. . . . Secondly there is Medicine, that he may be able to look after his diggers and other workmen that they do not meet with those diseases to which they are more liable than workmen in other

occupations. . . . Thirdly follows Astronomy that he may know the divisions of the heavens and from them judge the directions of the veins. Fourthly there is the art of surveying. . . . Fifthly, his knowledge of Arithmetical Science should be such that he may calculate the cost to be incurred in the machinery and the working of the mine. Sixthly his learning should comprise Architecture, that he, himself, may construct the various machines . . . or so that he may be able to explain the method and construction to others. Next he must have a knowledge of Drawing that he can draw plans of his machinery. Lastly, there is the Law, especially that dealing with the metals, that he may claim his own rights . . . that he may not take another man's property and so make trouble for himself and that he may fulfil his obligations to others according to the Law.
(Agricola, 1556, *De Re Metallica* as translated by Herbert C. Hoover).

At around Agricola's time, also, Paracelsus was expanding the fields of Chemistry and Alchemy was being relegated to the past. The main break through came later between 1779 and 1848 when Berzelius, a Swedish chemist developed, with his students, the underlying principles on which the present chemical classification of minerals depends.

1.1.1 Observations, questions and truth

There have always been men who observe, question and seek the truth and, of these, the renowned Russian scientist M. V. Lomonosov (1711–1765) developed revolutionary concepts well ahead of the views of his time. According to Smirnov (1976), Lomonosov laid scientific foundations for developing the theory of the accumulation of minerals on which the Russian and Soviet School of Geology finally arose. Lomonosov was one of the first to recognize that placers result from 'the fracturing of lodes and nowhere is it more hopeful to seek them as along rivers in the upper reaches of ore mountains'. However, much of his writings still reflected the low level of scientific thinking at that time.

Professor D. Z. Sokolov (1788–1852) then demonstrated that all of the known placers in the Ural Mountains had formed from the disintegration of gold-bearing lodes in primary source rocks. According to Tikkomirov (1953), he identified morphological features that could not be explained by current Neptunist theories. These theories proposed that gold placers in the Urals had been transported either by 'the primeval ocean that covered the whole globe, or by raging torrents from India, or by other currents'. A Neptunist in his early years, Sokolov later broke away and Smirnov notes that he spent much of the time prior to his death reviewing, critically, both his own early writings and those of his fellows at home and abroad.

It is interesting to note that although such placer minerals as wolframite, ilmenite and platinum were known to Russian placer miners at the beginning of the nineteenth century, technology had not caught up with their usefulness and they had so little application in industry that in 1828 some of the Russian coinage was actually struck from platinum. Platinum was also found in concentrates recovered from some Californian beach-sand deposits around 1860 but again it was considered virtually useless as an industrial metal. Realization of its true value and many uses came later and, by the early twentieth century, 95% of the worlds production was being won from placers in the Urals and Siberia.

1.1.2 The key to the past

Other geologists also found that the more they questioned, the more they had to lay aside dogmas that had distorted and inhibited the thinking of man for thousands of years and in Europe, James Hutton, a contemporary of the dominant Neptunist of his time, A. G. Werner, propounded his theory of 'uniformitarianism' largely from studies of sedimentary processes in streams, lakes and along foreshores. This theory, that the same processes are in operation today as in the past and that, hence, 'the key to the past is in the present' was at that time a very radical departure from the prevailing views on placer formation. Indeed, Huttons concept entirely revolutionized much contemporary thinking. In 1802, Playfair showed that 'no valley is independent of the rivers flowing in it' but, instead, develops various characteristic features governed by its geology and the environment within which it is placed. This was again almost heretical by earlier standards but, in 1830, Charles Lyell was able to state, against very little opposition, in his *Principles of Geology* that, 'given enough time, whole landscapes can be created or destroyed by the action of the slow yet relentless forces of Nature'.

In America by 1880 G. K. Gilbert had introduced the concept of grade and, from the results of a series of quantitative studies of such features as river volumes, stream velocities and gradients, had established guidelines for much of the later work in this field. Gilbert maintained that stream beds and other flow surfaces constantly adjust to changes in energy conditions by varying their profiles and that, for each set of variables, a state of dynamic equilibrium is reached that may change significantly only through appreciable changes in the controlling factors.

This was not an entirely new idea because engineers in Europe had long suggested a balance between stream erosion and deposition and, in fact, more than one hundred years earlier A. Brahms had published an equation that rated the competence of flowing streams in terms of the sixth power of their velocities (see Chapter 2, p. 81). Gilbert's great contribution was to both formalize the principles involved and to develop new avenues for study.

1.1.3 Youth, maturity and old age

With the turn of the twentieth century, W. M. Davis described a geographical cycle in which streams and landscapes progress from youth, through maturity to old age. This concept was eagerly seized upon by geologists who, at this time, were beginning to take a closer interest in the development of alluvial land forms. Although Davis offered only broad explanations and neglected the physical processes involved, his concept dominated geological thought for several decades and set the general framework for the study of geomorphology as a separate discipline. In Russia, Bogdonovich (*Ore Deposits*, Volumes 1 and 2; 1912–13) distinguished placers as separate entities amongst other conditions of ore formation. Grabau referred to the new science of 'lithogenesis' in his *Principles of Stratigraphy*, published in 1913.

1.1.4 Shelf geology

The study of sediments and sedimentation offshore began with the voyage of HMS *Challenger* (1872–76) when many samples of the sea floor were made available for geological investigations. Succeeding expeditions provided additional material for study. In 1920, Andree wrote what may have been the first text devoted to marine geology, and scientists began to realize that the ocean floor was not the generally featureless surface previously envisaged. The wartime development of sonar devices and improved methods of navigation opened up a whole new world for exploration. The continental shelf was recognized as an integral part of the adjacent land mass and miners began intensive investigations for placers offshore, recognizing that they had originally been formed on land.

Mapping in the offshore placer environment soon reached the stage of providing explorers with a broad picture of the topography and geology of large sections of the marine basins as a guide to selecting specific areas of interest. More stress was placed on developing and improving methods for surveying, detailed mapping and sampling in localized offshore terraines and regional work, because of its massive cost, became an increasing responsibility of governments. Progress has been hastened because of international cooperation, particularly in the CCOP (ESCAP)* countries of south-east Asia, and because marine scientists everywhere, are now increasing their efforts to understand an environment about which so much has still to be learned.

* The Committee for Co-ordination of Joint Prospecting for Mineral Resources in Asian Offshore Areas (abbreviated to CCOP) is an intergovernmental body established under the aegis of the United Nations Economic Commission for Asia and the Far East (ECAFE) which is now known as the United Nations Economic and Social Commission for Asia and the Pacific (ESCAP) and, since 1972, assisted

1.2 ENGINEERING AND PLACERS

Early man discovered, by observation and experience, that the more dense minerals lodged preferentially in hollows and behind obstructions. He soon learned to assist nature by using such devices as the 'golden fleece' and by constructing channels alongside the stream bed to process the auriferous gravels. Valuable heavy minerals were also recovered by panning, using vessels that were hollowed out from wood. It seems probable in fact that European miners, when they first flocked to the placer gold fields in Alaska, California and Australia, used methods and equipment that had been developed hundreds or even thousands of years earlier.

Mechanical processes may have been first applied to placer mining with the introduction and development of hydraulicing in the USSR in 1830 (Popov, 1971), and in California around 1849 according to Wolff (1976). Hydraulic elevators were used to pump cassiterite-bearing sands in Malaya prior to 1900, and similar installations found service in Nigeria, Switzerland and New Zealand around the same time. Hydraulic mining (see Chapter 6) was developed for use wherever a sufficient fall of the land allowed water to be delivered to the face at pressure heads of 30 m or more. Wood-stave piping was used at first, then light-gauge spiral or butt-welded steel piping and, with the introduction of high-pressure water pumps and centrifugal gravel pumps it became, for a time, the most common mechanical method for mining placers.

1.2.1 Dredging

Dredging for placer ores was first carried out using a 'bag and spoon' type dredger, reportedly about 1565, in the lowlands of Europe (Dekker, 1927). The art spread to other places and primitive forms of dredging have been reported along the Gold Coast of Africa since the middle of the eighteenth century.

Nevertheless, the history of dredging may be said to date from 1372 when the first description of a device for elevating fluids through the use of

---

by the United Nations Development Programme (UNDP) through a project entitled 'Technical Support for Regional Offshore Prospecting in East Asia'.

CCOP issues Technical Bulletins annually, containing technical and scientific studies on marine geology and offshore prospecting for mineral resources as well as the results of surveys undertaken through CCOPs sponsorship. Apart from contributions to scientific knowledge, these studies and survey reports have helped in arousing interest in the mineral potentials of the marine shelves of the region and have played an important role in attracting risk capital from industry to east Asia. The office of the Project Manager/Co-ordinator, CCOP, ESCAP, is located at Sala Santitham, Bangkok-2, Thailand.

rotation was recorded. In that year, M. Le Demour sent the description of a novel type of pump to the French Academy. The pump was essentially a straight tube attached in an inclined position to a vertical open-ended pipe, the bottom of which was immersed in water. When rotated at the required velocity about its longitudinal axis, water was discharged from the end of the tube.

In 1816, M. Jorge submitted an improvement by providing a means whereby only the arms were rotated while the vertical pipe remained stationary. Papin in 1705 invented the centrifugal pump in somewhat the form it is known today (Erickson, 1956).

The year 1855 saw the first recorded use of a centrifugal pump in a seagoing hopper dredger on the Charleston, S.C., Bar, South Carolina, USA (Sanford, 1904). It suffered many initial breakdowns but proved the success of the method before the vessel sank during the Civil War. The principle of hydraulic transportation of fluid–solids mixtures in pipelines gained general acceptance, and in 1867 the French engineer, Bazin, used it to excavate the Suez Canal.

The first suction cutter dredger in the USA was developed by Atkinson in 1862, improved by Duval in 1869, and subsequently adapted by Vivian for use on the Chicago drainage in Illinois. In 1871, General Q. A. Gillmore (Scheffauer, 1954) proposed the conversion of a steam ship for dredging the mouth of the St. John River, Florida, and by 1876 cutterhead dredges were in extensive use.

A 'suction dredger' for dredging onshore was invented and patented in 1874 by Barnes, Simons, and Brown (Anon, 1874). In 1889, Wallace C. Andrews filed a patent claim 'to convey solid material in an artificial or even a natural condition of requisite fine division, introducing some into a liquid, conveying the mixture through pipes by gravity or force pumps.'

The first steam engine to be applied to dredge service was constructed in England by Watt in 1795. In 1855, in Germany, Schwartzcoff and Hoffman built a pump and steam engine assembly on a barge and called it an 'hydraulic dredge'. They also developed the floating pontoon line system for disposal of the spoil using leather sleeves as flexible joints. A further development was a mechanical cutter head surrounding the suction nozzle. The dredge was subsequently used for land reclamation, canal dredging, etc., near Berlin.

Diesel electric power was first used for dredger propulsion and operation in 1922, and the long-distance pumping of fine tailings through pipelines was commenced in 1930. Barges equipped with centrifugal dredging plant were used in the New England District of New South Wales, Australia, for dredging tin placers at Skeleton Creek, the Brickwood lead near Tingha, and the Wye Water lead near Emmaville. The deposits were too small to justify the use of bucket dredgers and too wet to be mined by conventional hydraulic sluicing methods.

In 1947, a diesel electric dredger *Medernbing* was built for alluvial tin mining in Indonesia (Cooper, 1958). Its dredging capacity was about 125 $m^3 h^{-1}$ to a depth of 30 m. More recently in this application hydraulic dredgers, utilizing jet lift pumps, dredge from depths of up to 100 m in waters too deep for conventional suction dredging.

Authors differ on the timing of the introduction of bucket elevator dredgers to the placer industry. Rose (1902) refers to a report by A. Grothe (*Mineral Industry for 1899*, p. 326) in which he states that the first dredge was operated on the Clutha River, New Zealand in 1864. O'Neill (1976) lists, as first, a bucket dredger operating on the same river in 1882; followed by another at Bannack, Montana, in 1894; and a third at Oroville, California in 1898. However, according to Lord (1976), the first bucket elevator dredger was built in Otago, New Zealand, in 1867; the Bannack dredger was built in 1895; and the first closely connected bucket line unit in 1898 (presumably at Oroville). The most authoritative source, McLintoch (1966), states that the first steam bucket dredger to operate on placers in New Zealand was the *Dunedin* and that it was built in 1886. Dredging for gold in the Urals is believed to have commenced around 1916 with the introduction of three American-built dredgers and since then, many different types have been designed and built in the USSR for Russian conditions.

According to Romanowitz and Cruikshank (1968) a fundamental difference between the New Zealand- and California-type dredgers was that the former, by virtue of its natural disposal methods was able to leave the dredged areas in much the same condition as before dredging commenced. However, it was less able to deal with tough digging conditions and its application was largely restricted to formations consisting of mostly fines and wash gravels generally smaller than 1 inch (2.54 cm) diameter. The California-type dredger on the other hand was more efficient in dealing with the larger gravels but, by reversing the order of spoil deposition, i.e. by depositing the washed gravels back on the top of the restored ground, it met with serious ecological objections. An adaptation of the California Dredger, called a 'resoiling dredger' was developed to overcome these objections although it was found to cost 20% more than the original dredger.

Less uncertainty attaches to the inauguration of bucket dredging offshore and the honour for the first operating unit goes to a Captain Miles who commenced dredging for tin offshore Phuket Island, Thailand in 1908. Shown to be successful, another dredge was put into service near Singkep Island, Indonesia in 1911 and at least ten bucket dredges now operate in Indonesian waters alone.

Of the various types of dredgers, the cutterhead suction and bucket-line types are used most in placer mining. Improvements in design follow the need for operating at greater depths and in more rigorous climatic conditions. Devices that compensate for ocean swell and pumps operating closer to the dredging face are only two of the many recent innovations for increasing dredging capabilities. Improved systems have also been developed for land operations and bucket-wheel excavators are used in both wet and dry conditions. Automated underwater dredgers similar to some oil drilling platforms have already been built and operated in more than 30 m of water. However, they are not yet in general use.

Emphasis is now being placed on developing and improving methods for surveying, mapping and sampling in localized offshore terraines and the regional work is becoming more and more the responsibility of such agencies as the Commission for the Geological Map of the World and its working groups.

Improved methods of navigation have led to the more accurate positioning of survey vessels and drilling platforms and have eliminated many of the problems of co-ordinating traverses for bottom profiling and relocating mineral zones for subsequent testing. A great number of systems are known using both line of sight and remote fixing and most of the current requirements can now be met.

Some of the most important achievements in oceanographic mapping have resulted from the newly acquired ability to take measurements of the topography of the ocean floor using side-scanning sonar and echo sounding devices originally used for tracking down submarines in wartime. Seismic methods of reflection and refraction first used on shore are being adapted for determining the nature and stratigraphy of sediments on the sea floor. Computer utilization for data storage and processing has increased the density and efficiency of sampling and miniaturization has reduced package sizes and costs to the extent that some self-propelled prototypes are now capable of moving freely over the ocean floor guided only by acoustic signals from the surface.

1.2.2 Minerals processing

Early recovery methods were based upon the use of dish-shaped pans, rockers, long toms and other short sluice types for small hand operations and it is to be observed that similar devices have been developed independently throughout the world wherever placers were mined. The use of animal hides spread along the surface of the river bed has already been referred to. In Australia, according to Rose (1902), devices called 'fly catchers' were invented for catching the fine gold floating off from the riffles of sluice boxes. The fly catchers consisted of weirs constructed across the river bed on which were attached boards covered with blanketing or coarse gunny sacking. At intervals the blankets were taken up and washed to recover the fine gold.

It is not known when dish-shaped pans were first used for recovering gold but Junner (1973) refers to the writings of Bosman (1698), Barbot (1732) and Meredith (1812) to assert that the local inhabitants of the Gold Coast, West Africa, used such devices prior to the arrival of the Portugese in 1471. Barbots description of how the natives of the Lower Ankobra River obtained alluvial gold from the larger streams near falls and rapids reads:

> They plunge and dive under the most rapid streams, with a brass basin or wooden bowl on their heads into which they gather all they can reach to at the bottom; and when full return to the bank of the river with the basin on their heads again where other men and women are ready to receive and wash it, holding their basins or bowls against the stream till all the earth and dross is washed away: the gold, if there is any in the basin, by its own weight sinking to the bottom. When thus cleaned and separated they turn it into another vessel till quite clear of sand or earth. The gold comes up some in small grains, some in little lumps as big as peas or beans, or in very fine dust. This is a very tedious and toilsome way of gathering gold; for I have been assured that the most dextrous diver cannot get above the value of two ducats a day, one day with another.

According to Meredith, in 1812, the women in other parts of the Gold Coast:

> . . . put the earth into a wooden bowl, where it undergoes frequent ablutions by a circular motion until the lighter parts are washed away; the heavier parts of the earth that remain are put into another bowl; this process is repeated several times until there is nearly a bowlful collected; it then undergoes a careful examination and frequent washings, and the gold at length is perceived at the bottom of the bowl, where it is allowed to remain until the whole of the earth is washed away, when they take it out and dry it either by the sun or fire. During this process there is much dexterity and ingenuity to be seen, which are only acquired by much practice.

As Junner remarks, this description could well be a description of how the process of panning is carried out today. Figure 1.1 shows tribal Indian women panning for gold in Punna Puzha, Kerala State, India.

Conventional sluice boxes were too long for the early dredgers and tables were used instead. These were made as wide as possible to reduce the depth of slurry passing over them; they were covered in coir or cocoa matting to recover the coarse gold and with burlap to recover the fines. Table widths varied generally from 5 to 25 m.

According to Cleaveland (1976) the first jigs came into service on dredgers around 1915 but all known systems were inefficient in the finer sizings. Modern techniques are now being developed around the appli-

Fig. 1.1 Tribal Indian women panning for gold in Punna Puzha, Kerala State, India

cation of high-capacity, flowing-film, gravity separators, such as Reichert Cones and other undercurrent sluice types, in conjunction with selective hydrocyclone systems. As a result, many of the tailings from previous operations offer similar recovery rates to those achieved when the deposits were first mined.

1.3 THE PLACER ENVIRONMENTS

1.3.1 Introduction

An environment is the total of surrounding influences affecting the formation or continuance of a finite being and in this context, placer environments are distinguished from other sedimentary environments and from each other by important differences that place additional emphasis on the nature of their provenances and the geomorphological settings within which they are laid down. Environmental factors marking these differences are:

(1) Source rock geology, which determines how the individual mineral types are distributed in their parent bodies and, hence, the order in which they become exposed at the earth's surface (Chapter 3).

(2) Climate and chemistry, which combine to order the rate and form of their release (Chapter 2).
(3) Surface geometry and boundary conditions, which exert physical constraints on transportation and deposition (Chapter 2).
(4) Elements of environmental change, which modify existing patterns of mineral distribution (Chapter 1).

Sediments on slopes and in channels near the headwaters of streams have been exposed to sub-aerial forces of destruction for a comparatively short time only and, hence, may comprise all sediment types and sizes down to the finest of silts and colloids. On the other hand, the sediments in beach deposits have normally travelled great distances and may well have survived numerous cycles of weathering and erosion. They are composed largely of comparatively fine-grained and well-graded sediments from which the chemically unstable minerals and very fine particles have been removed. Varying quantities of locally derived shell fragments and materials eroded from adjacent rocks may also be present.

The rocks and minerals in placers exhibit varying degrees of chemical and mechanical breakdown and buried deposits are found in a variety of forms. They are best known in Quaternary sediments and are rare in the older compacted rocks because such superficial and ephemeral deposits along ancient streams and strandlines have been largely eroded away as a result of climatic changes and tectonism. Few of the most ancient placers are likely to be preserved in the sedimentary record, however, remnants of what may have been mineral-bearing beach deposits have been found in the Triassic Sandstones in New South Wales and in Precambrian sediments in north-Western Australia. Fossil-terrace deposits may be all that remain of early channel fill placers in mountainous regions, but many large fluvial deposits have been preserved by the drowning of valleys in lower, estuarine environments. The so called 'deep leads' (see Chapter 3) of eastern Australia, California and elsewhere are remnants of tin- or gold-bearing stream placers, that have been preserved from erosion by basalt flows which filled the valleys, generally in Tertiary times.

1.3.2 Classifying the environments

This task has exercised the minds of geologists and engineers for many years and various groupings have been proposed according to such criteria as:

(a) Past and present geomorphological cycles (Lindgren, 1911)
(b) Present location (Brooks, 1913)
(c) Conditions of deposition (Jenkins, 1946)
(d) Lithology (Wells, 1969)

(e) Links with bedrock (Smirnov, 1976)
(f) Techniques for exploration and mining (Macdonald, 1983)

Each classification represents a different viewpoint and reflects something of the attitudes of the author and his particular interests. For example, the classification based upon techniques for exploration and mining considered how the various placer types might respond to specific exploration and mining techniques in common use. On this basis, an exploration team having the skills and experience to find and exploit one type of placer should be equally capable of finding and developing any other placer type in the same group. The classification was used to stress the different levels and types of expertise required in the various environments onshore and offshore but many difficulties arose in trying to achieve clear lines of demarcation between some groupings.

Obviously a model to satisfy all criteria would be extremely complex and might tend to confuse rather than to instruct. Accordingly in this book, common ground is sought with other classifications by relying upon three major groupings; continental, transitional and marine, and within those groupings considering only those sub-divisions having distinctly different morphologies. The system is related to exploration and mining techniques appropriate to each subdivision in Table 1.1.

The continental placer environment

The main characteristics of this environment (Fig. 1.2) are summarized in Table 1.2 in terms of its principal sub-environments.

Common exploration techniques rely on the use of similar drill types and pitting to obtain samples. All are found close to their provenance and a similarly common range of methods applies to their exploitation. Small differences are due principally to differences of scale and to the availability of adequate water supplies for mining and treatment.

The transitional placer environment

This environment (Fig. 1.3) commences offshore where the wave orbits first disturb sediments on the sea floor, and extends inland to the limits of aeolean transportation. Placer accumulations are found at all levels but of particular interest are those that form at the base of frontal dunes on open beaches and in the hinterland where high-level dunes may be built up from a succession of individual sand migrations. Deposits formed in deltaic conditions sometimes reach economic proportions but are usually the most transient of all. The main sub-environments are described in Table 1.3.

The marine placer environment

The marine environment, illustrated in Fig. 1.4, is a submerged continuation of the adjacent land which provides the bedrock geology. Dissection

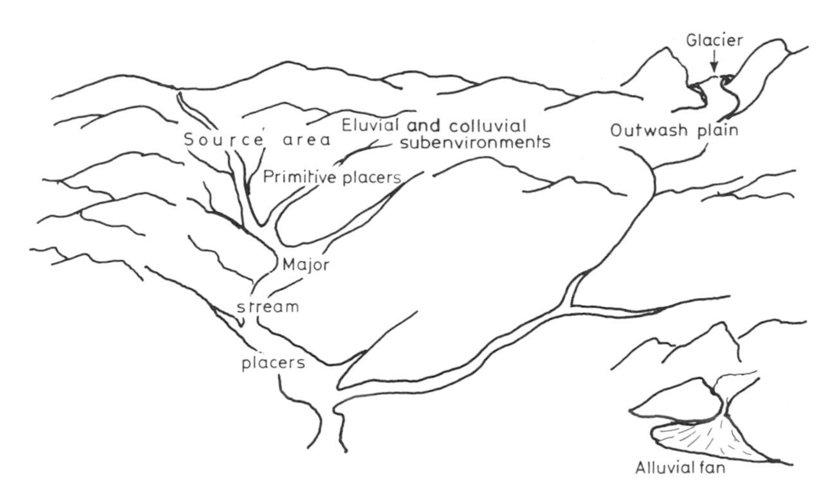

Fig. 1.2 The continental placer environment

of the bed during periods of emergence in the Caenozoic has developed a generally fluvial morphology, modified to some degree by the action of waves, tides and currents during submergence. The relict features have since been more or less masked by a flow of sediments from the land.

In most cases, primary source mineralization has been followed by continental-type processing; sometimes also by a further stage of working in the transitional environment; and finally by submergence. Drowned placers, typical of the Sunda shelf in south-east Asian waters are found in:

(1) Quaternary valleys which have acted as repositories for the deposition of valuable heavy minerals eroded from nearby host rocks.
(2) Terraces along the valley walls.
(3) Eluvial and colluvial concentrations along the valley sides and slopes.

Generally in this environment, the valuable minerals are found in the channels of ancient streams incised into the bedrock or at higher levels in a sequence of strata such as are common near the tin islands of Bangka and Belitung in Indonesian waters where the succession is: young alluvials; young marine sediments; old alluvials; old marine sediments; bedrock.

Tin placers occur in both the old and the young alluvials as well as at bedrock. The sequence is covered by a thin layer of sandy mud and coral fragments.

More complex systems result from the active reworking of sediments during submergence and the action of fast-moving currents along the sea floor. Additional complexities arise from repeated cycles of emergence and submergence and changes in conditions of supply, transportation and deposition can usually be inferred, from differences in sediment patterns

Table 1.1 The placer environments

Environment	Sub environment	Main products	Environmental process elements	Exploration techniques (Chapter 4)	Mining methods (Chapter 6)
Continental	Eluvial	Au, Pt, Sn, WO_3 Ta, Nb, gem stones (all varieties)	Percolating waters, chemical and biological reactions, heat, wind and rain	Soil sampling, shallow pitting, churn drilling	Open cast, hydraulic sluicing, hand mining
	Colluvial	Au, Pt, Sn, WO_3 Ta, Nb, gem stones (all varieties)	Surface creep, wind, rainwash, elutriation. Frost	Stream and soil sampling, shallow pitting and trenching	Hydraulic sluicing bulldozer and loader—hand mining
	Fluvial	Au, Pt, Sn, rarely Ta, Nb, diamonds and corundums	Flowing streams of water	Stream sampling, geophysics, pitting, churn auger and pit-digging drills, Banka Drills	Bucket dredging in active beds, bucket dredging, hydraulic sluicing and dozer-loader operations in old stream beds
	Desert	Au, Pt, Sn, WO_3 Ta, Nb, gem stones (all varieties)	Wind with minor stream flow. Heat and frost	Shallow pitting, churn and pit digging drills, geophysics	Various earth-moving combinations
	Glacial	Au (rarely)	Moving streams of ice and melt waters	Stream sampling and pitting	Hydraulic sluicing

Transitional	Strandline	Ti, Zr, Fe, ReO, Au, Pt, Sn	Waves, currents, wind, tides	Hand augering and sludging, sample splitting allowable	Suction cutter dredging, bulldozer and loaders, bucket-wheel dredging
	Coastal Aeolean	Ti, Zr, Fe, ReO	Wind and rain splash	Power augers (hollow), sample splitting allowable	Suction cutter dredging, buldozers and buried loaders
	Deltaic	Ti, Zr, Fe, ReO	Waves, currents, wind, tides and channel flow	Hand augering and sludging, sample splitting allowable	Specially designed shallow depth dredges having great mobility
Marine	Drowned placers	Au, Pt, Sn, diamonds, minor Ti, Zr, Fe, ReO industrial sand and gravel	Eustatic, isostatic and tectonic movements—net rise in sea level	Geophysics (seismic refraction and reflection) bottom sampling, remote sensing, hammer, jet, vibro and banka drills, positioning	Bucket line dredging, jetting, clamshell, rarely suction-cutter dredging

Table 1.2 Sub-environments of the continental placer environment

Sub-environment	Features of distinction
Eluvial	Weathering *in situ*; upgrading largely through the removal of soluble minerals and colloids; some surface material removed by sheet flow, rivulets and wind. All minerals may be represented in partially weathered sectors; end products at surface may be chemically stable.
Colluvial	Downslope movement of weathered rock controlled principally by gravity; all placer minerals may be represented but sorting poorly developed.
Fluvial	The most important sub-division of the environment characterized by wide range of depositional land forms; most deposits formed within few kilometers of source rocks; particle size reduces and sorting improves with distance from source; only chemically and physically stable particles persist. Many deposits are relict from earlier times.
Desert	Somewhat similar characteristics to eluvial and colluvial sub-environments except for wind as principal transporting agency; flash flooding in some desert areas induces fluviation in gutters and channels; all types of placer minerals may be represented but principal varieties are tin, gold and pegmetite minerals.
Glacial	Deposits unsorted and unstratified; rare economic concentrations; morraines and tills in outwash plains sometimes give rise to more important concentrations due to subsequent stream action; upgrading also occurs along shorelines where glaciers discharge out into the sea.

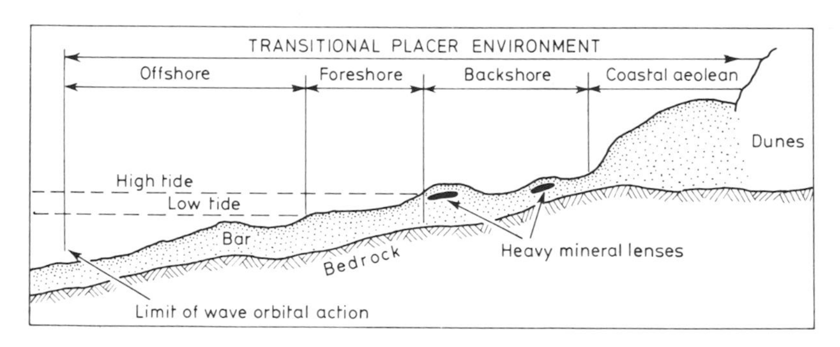

Fig. 1.3 Generalized section across transitional placer environment

and stratigraphic breaks. Davies (1979) has suggested that, along the New South Wales shelf, some deeply buried seams of heavy minerals may have escaped re-working and that evidence of terracing was clear. There is no general case.

Table 1.3 Sub-environments of the transitional placer environment

Sub-environment	Features of distinction
Strandline	Beach placers are formed by the action of waves, tides, currents and wind; principal minerals are the most resistant of the lower-density varieties such as rutile, ilmenite, zircon and monazite; sometimes gold, platinum, tin and diamonds; rarely others.
Aeolean	Placers formed in sand blown up from beaches; dunal systems develop from both stationary and transgressive dunes; mineral suites as for beaches but particle size generally finer.
Deltaic	Formed around mouths of rivers carrying large-sediment loads. Seaward margins favour reworking of sediments and repetitive vertical sequences common—lower-density varieties common.

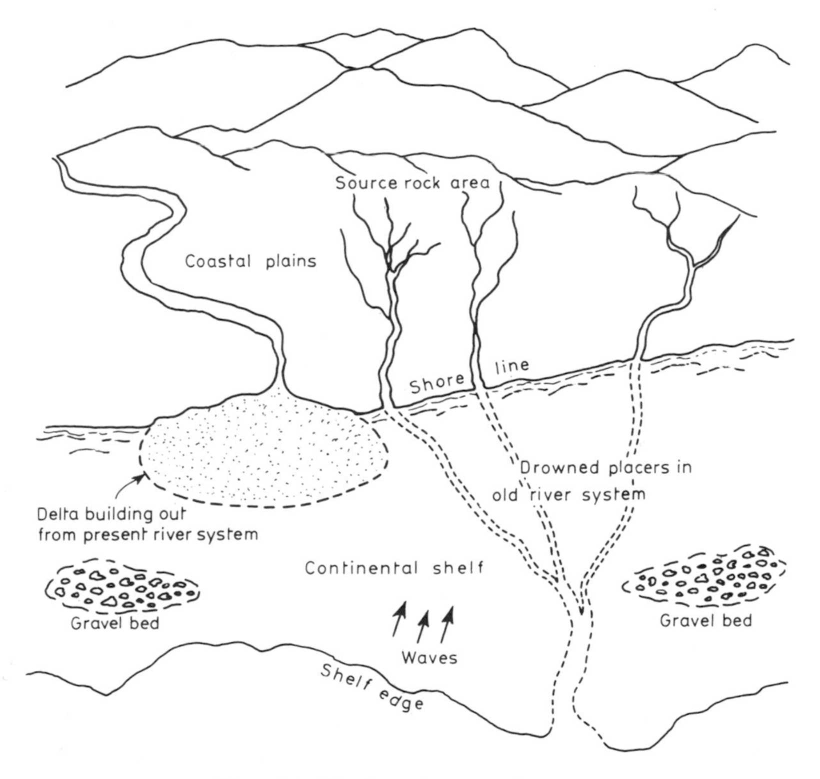

Fig. 1.4 Marine placer environment

The environment extends seawards from the breaker line at water depths below those at which the action of wave orbits is felt. Shelves range from a few kilometres to 100 km or more in width and depths to a maximum of 500 m at the break between the shelf and the continental slope.

1.3.3 Environmental change

Global conditions have never remained constant for long and, in a world that began some 4.5 billion years ago, the geography of all lands and oceans has changed many times. However, the record is far from complete and there are many gaps in man's knowledge of the events of the past and in his understanding of the ways in which the early mechanisms of placer sedimentation may have differed from those operating today. It is known that from the beginning certain intervals of time have been characterized by mountain building and other large-scale diastrophisms, and by the intrusion of igneous magmas along lines of weakness and fracturing in the earth's crust. For every uplift there has been a major environmental change, with strengthened forces of destruction seeking to flatten out the land and re-distribute its waste. Of the valuable minerals, some have remained close to their source while others have moved on, either to be lost or to take part in further sequences of multi-phase deposition (Fig. 1.5). Each group of minerals has favoured a particular environment according to its capacity to survive.

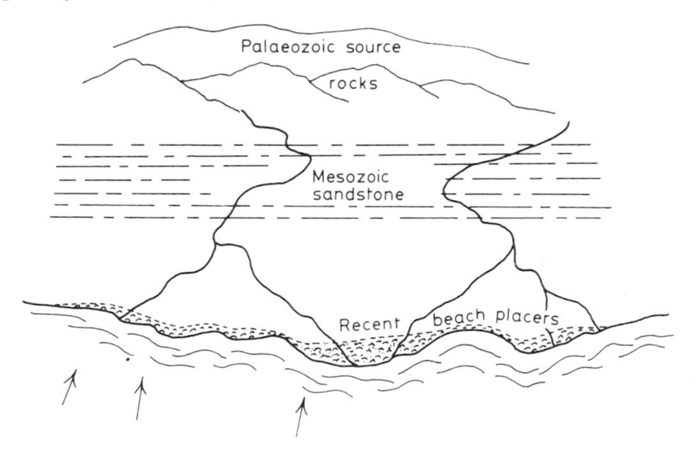

Fig. 1.5 Elements of multi-phase deposition

Some placer deposits are very old and in their present locations in respect of considerable thicknesses of overlying younger sediments and rocks, lies a clear indication that others have yet to be found. As has already been shown, and will be enlarged upon in Chapter 3, each environmental system has produced some distinctive features and each environmental change has in some way modified the existing forms. Of the survivors, some have been compacted or indurated, others have been reworked and still others have been tilted or folded, thus modifying their original features and making difficult their recognition. In some cases the early deposits were preserved from excessive alteration by a protective cover of younger sediments or other rock types, or by being in some other way isolated from active water horizons. Yeend (1976), cites the lava-capped Tertiary channels of the Sierra Nevada regions of California and the San Juan Ridge of northern Columbian diggings where some 200 m of volcanic breccia overlay sand, clay and minor amounts of fine gravel. Similar examples are found in the famous 'deep lead' gold placers of Ballarat, Victoria and in Western Australia where the raised beaches at Eneabba are covered by wind-blown sediments. The major episodes are distinguished by wide-spread basal conglomerates that mark transgressive seas when they tend towards lithological homogenity and are typically heterogeneous when related to rapid subsidence (Krumbein and Sloss, 1963).

In the beginning

Elements of environmental change affecting placers relate first to the origin of the earth and then to the total environment when placers were first formed. Arguing the case of the earth as a molten mass, cooling from the outside towards the centre, Obruchev (1959) visualized the primeval state of the world at the time of crustal hardening, as one in which the continents were represented by fields of consolidated lava with very uneven surfaces from which vapours escaped with a hiss; and by gases, including asphyxiating and poisonous varieties, that were liberated in different places from the fissures. The water was still hot in the seas which smoked like cauldrons. The atmosphere was heavy and saturated with vapours; dark clouds covered the sky and blotted out the sun.

Turekian (1976) took another view following along the lines of the Kant Laplace hypothesis of a hot and rotating gaseous nebula turning faster and faster as it cools and sloughing off rings of gas to condense into planets that orbit about the central core (the sun). With the cooling of the vapour the high-temperature minerals condensed and accumulated making up the inner parts of the terrestrial bodies. Melting and fractionation then occurred as the result of heating, primarily from gravitational energy. According to Turekian, this initial process would have taken less than 1 million years to complete and no volatiles in any form would have accumulated in that time. The planets were then veneered with carbonaceous meteorite materials. Subsequent melting and recrystallization in the outer parts of the planets released the volatiles to the surface and if any planetary body was large enough they would be retained. According to this model, a dynamic upper mantle and crustal system developed on the earth soon after the accumulation process due to the extensive energy release and heating of the interior. The earth was provided with an ocean and an atmosphere soon after the planet formed.

Other hypotheses include the Kuiper model described by Lepp (1973) which proposes that the temperature of the initial gaseous body must have been near absolute zero. All leave unanswered many questions bearing on the availability of minerals at the earth's surface and their subsequent, partial re-distribution although on one point there is general agreement: great changes have since taken place in both the composition of the atmosphere and the disposition of the oceans and these changes have affected to varying degrees, the breakdown of rocks by weathering, the subsequent removal of the debris away from its source and the preferential survival and deposition of detrital heavy minerals in placers.

There is ample evidence also that, at the time of crustal hardening, volcanic activity must have been catastrophic with eruptions of molten rock and ash and the effusion of mixtures of gases comprising such volatiles as CH_4, NH_3, HCl, carbon and sulphur compounds, and water vapour on a much larger scale than in modern times. At first, steam in the atmosphere condensing and falling as rain must immediately have been re-evaporated by the heat at the earth's surface. Subsequently, as the surface cooled, water would have begun to accumulate at lower levels and to commence flowing in streams. Presumably, at this stage, placer formation could also have begun but it is unlikely that such deposits would have survived for long. It seems more likely that because of the sheer magnitude of the forces acting, vast quantities of sediments were generated and laid down in the oceans of those times with much less local sorting than might have occurred in less violent conditions.

Atmospheric change

Assuming a progressive degassing of the earth's interior through volcanic and hydrothermal activity, any free oxygen would immediately have reacted with hydrogen or methane to form carbon dioxide and water in the original gaseous mixtures. As additional evidence of this lack of oxygen in volcanic gases reaching the surface, Lepp offers the observation that the iron in volcanic rocks is largely in the ferrous state (Fe^{2+}). If free oxygen had been present the ferrous iron would have been oxidized to ferric iron (Fe^{3+}).

Free oxygen probably first became available in small quantities in the atmosphere from the dissociation of water vapour into hydrogen and oxygen by the action of ultraviolet rays given out by the sun. At first, the effects on the composition of the atmosphere may have not been very significant because of the masking effects of thick clouds of dust and other substances. Later as the atmosphere cleared, the absence of a protective layer of ozone such as the layer that shields the earth today meant that the ultraviolet rays must have been much more intense and appreciable quantities of oxygen would have been made available to react with surface rocks. The escape of hydrogen from the earth's gravitational field along with neon and other volatile gases is assumed to have taken place because there has been no build-up with time. Possibly, as postulated by Turekian, had the earth been larger, they might have been retained in the atmosphere and the history of Earth and its life would then have been quite different.

Alone, the ultraviolet–water vapour dissociation process could not have accounted for the present high content of oxygen in the atmosphere or for its evolution into the atmosphere of today but, rather, it paved the way for the development of plant life capable of generating oxygen by photosynthesis. The process of photosynthesis manufactures carbohydrates from carbon dioxide and water taken from the atmosphere when green plants are exposed to sunlight. The catalyst is chlorophyll and oxygen is set free in the process.

The first evidence of primitive plant life capable of photosynthesis is found in rocks some 3000 million years old. However, experiments showing that amino acids can be synthesized from mixtures of methane, water vapour, hydrogen and ammonia in an electrical field suggest that earlier forms of life may have been developed in the oxygen-free environment of earlier primeval times. There is as yet no direct proof that such life forms existed but the accumulation in Archaeozoic seas of organic compounds of carbon, hydrogen and nitrogen and massive limestones derived from organic calcarious silts could have provided a starting point for all other forms of life. Organic carbon compounds may thus have played a part in precipitating gold from solutions in some Precambrian placers.

Plant and animal life has had a profound effect on sedimentary processes throughout the ages and if it can be said that 'the processes creating and destroying land forms are also those that regulate biotic activity at the earth's surface' (Douglas, 1977) the corollary is also true that processes affected by biotic activity serve also to regulate the creation and destruction of land forms. Until the first appearance of land-based plant life in the Proterozoic era to control the flow of water over the surface of the land, desert conditions prevailed and the flow of sediments was likely to have been intermittant but massive depending upon the frequency and intensity of precipitation.

A few highly metamorphosed placers of Proterozoic age are known but the Precambrian may have been more a time of sediment manufacture than one of sorting and probably it was not until the development of more prolific vegetation in the Palaeozoic era that the sedimentation processes were regulated sufficiently for placers to be formed wherever a suitable provenance became eroded at higher levels.

Life during Cambrian times was concentrated in the oceans but, with increasing supplies of oxygen, a rapid development of surface plant life in the Devonian gave rise to its proliferation during the Carboniferous period. This period, noted for its vast forests in marshy areas and various amphibians, reptiles and insects may have had an atmosphere not unlike the present-day atmosphere. The Permian period was cold and dry with intensified vulcanism but the earth's atmospheric purification systems were well established and probably as capable of dealing with increased emanations of volcanic gases as, in the modern world, they cope with industrial wastes.

From an oxygen-free beginning dry air has evolved to its present mixture of 78 % nitrogen, 21 % oxygen, 0.93 % argon, 0.03 % carbon dioxide plus trace quantities of such rare gases as neon, helium and krypton. The atmosphere contains, in addition, up to 5 % of moisture and, while the composition is relatively constant globally, the moisture content of the atmosphere varies with the climate from place to place and from season to season. It diminishes with elevation to zero at a height of about 14.5 km above sea level.

If, as is thought, this composition has been relatively constant since around the middle of the Palaeozoic era it can be expected that placers since that time have been formed in accordance with models that are predictable from present-day observations. Difficulties in establishing genetic relationships for the minerals in some of the earlier placers, such as the gold-bearing banket deposits of Africa, may stem partly from their extreme ages, partly because of the environmental changes to which they have been subjected but perhaps also because of sedimentation differences stemming from differences in the atmosphere and climates of those times.

Effects of regional climatic changes

Changes from arid to semi-arid conditions gave rise to increased rates of sediment transport; dry alluvial fans became channelled and the greater availability of water in source areas brought additional quantities of mineral-bearing sediments down into the valleys. Deposition was generally limited to close environs of the provenance, and sorting to the networks of channels formed.

Changes from semi-arid to temperate climatic conditions led to the large-scale remobilization of sediments and to the production of new sediments at a greater rate than before. Major fluvial systems were developed and, because of the massive re-working of valley fill sediments, large placers were formed from sediments emplaced in previous episodes. Such conditions tended to inhibit placer formation in some areas however, the accompanying conditions of deep weathering prepared great masses of material for easy erosion when any major uplifts or changes back to more temperate conditions provided for large-scale mobilization.

Oceanic change

Oceans were first formed when water was released as a liquid during the first 1000 million years and since that time the total volume of water appears to have remained relatively constant with new oceans forming while others have been destroyed. For example, Bullard (1969) suggests that it is virtually certain that the Atlantic Ocean is less than 150 million years old and that an earlier ocean in the Lower Palaeozoic, 650 to 400 million years ago, gave rise to the sediments now in the Caledonian, Hercynian, and Appalachian mountains of Europe and North America.

Wilson (1976), following upon Wegener's (1966) hypothesis of continental drift, has described a hypothetical mid-Mesozoic world in which all continents were joined together in one land mass, Pangaea, and in which there was only one ocean. Before that, he suggests that there had presumably been a long history of periodic assembly and disassembly of continents and fracturing and spreading of ocean floors as convection cells in the mantle proceeded to turn over in different configurations. Calder (1975) placed Britain on the equator some 80 million years earlier, sandwiched between Africa, Europe and North America. McKenzie and Sclater (1976) concluded from their studies that Australia broke away from Antarctica and began moving north around 45 million years ago. Figure 1.6 illustrates the break-up of Pangaea and the separation of continents as seen by Wilson in line with current trends of thought.

Application of theories of ocean floor spreading, plate tectonics and continental drift offer broad explanations for some of the problems of relating ancient placers to their provenances and in reconciling their present stratigraphical relationships with conditions at the time of formation. Generally, however, it is difficult to speculate about the details. For example, some geologists look to the diamond pipes of South Africa for provenances for the alluvial diamonds found in Suriname and other parts of South America. But this does not mean that the search for kimberlites in those countries should cease. An explanation, because it appears plausible, is not necessarily the correct one and an enormous amount of work is required along the fringes of the land and on continental shelves before explorers can translate present drift theories into specific practical applications.

Again, there is a division in time between such past environments that are capable of being related directly to present conditions and older environments for which there is no direct link. For the atmosphere it may have been around Mid Palaeozoic times; for the oceans it was probably not until the beginning of the Caenozoic era that the distribution of oceans and continents assumed much its present form and that the land masses became closely defined according to the limits of their platform boundaries. Many of the previous shelves and their deposits must have been destroyed when the land masses came together from time to time; new ones would have been formed when other masses broke apart.

In the modern world of marine-minerals exploration the continental shelves have been studied most because of cheaper and easier access to the

Fig. 1.6 Schematic representation of break-up of Pangaea (after Wilson, 1963)

sea floor than in the deeper parts of the ocean, and because of the already known shelf resources of detrital heavy minerals and construction materials. Such deposits were formed under sub-aerial conditions during repeated periods of emergence which date back to the Proterozoic era, however, the present shelf deposits relate mainly to the Quaternary period of the past 1.8 million years to the present.

Quaternary sea level changes

Perhaps no other series of events has more profoundly affected the environments in which placers are formed and preserved or destroyed, than the fluctuations in level between land and sea that followed the successive ice ages and interglacial intervals of the Pleistocene. Perhaps, also, in no other geological period have environmental agencies been so active, worldwide, in the weathering of land masses or have so many regions been attacked successively in such extremes of climate. Fluctuating sea levels were dominant Caenozoic features and Van de Meene (1976) has recognized in south and south-east Asia, at least eight ice ages that have resulted in separate lowerings of sea level of 100 m or more.

Other factors have also influenced the movement of shorelines in coastal and platform areas and Bowen (1978) describes the five principal mechanisms:

(1) Long-term tectonic movements
(2) Glacio-isostacy
(3) Hydro-isostacy
(4) Geoidic variations
(5) Glacio-eustacy

Tectonism

Long-term tectonic changes resulting from such deep-seated causes as ocean floor spreading and plate collisions have effected the equilibrium of shorelines, probably from when the land masses and oceans first formed. Globally, the orogenies have overlapped in both space and time and while some land masses have been rising, others have been sinking or tilting. Base levels have changed constantly and, with those changes, variations have taken place in the ability of streams to erode and transport the products of weathering away from where they were formed. Smirnov (1976) refers to 'loose sediments in upper Tertiary and Quaternary times in the Soviet Far East which were formed under the influence of vertical tectonic dislocations in various earlier epochs with varying amplitude and sign'. Holmes (1969) illustrates typical structures of the orogenic belts as described in Fig. 1.7.

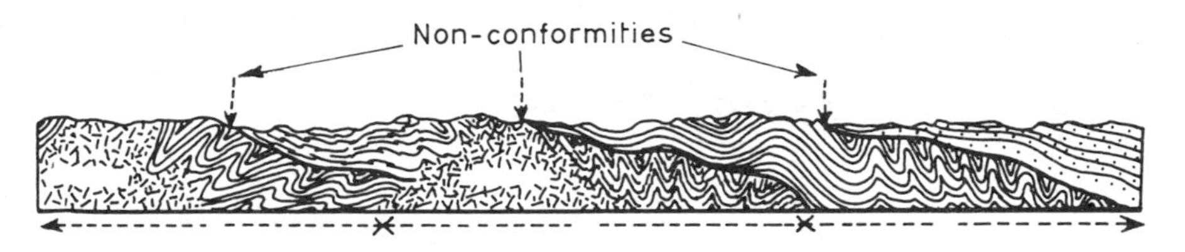

Fig. 1.7 Typical structures of the Orogenic belts

Glacio-isostacy

Isostatic adjustments follow the loading and unloading of land masses that result from the alternate freezing and thawing of ice sheets. For example, a section of the Hudson Bay coastline has risen by more than 300 m since the last ice age and is still rising. Isostatic adjustments occur because, during glaciation, the imposed stresses extend out beyond the ice sheet and when the ice melts and runs back into the sea the stresses are relieved. Figure 1.8 (after Walcott, 1970) illustrates the glacio-isostatic adjustments from an ice sheet 1.8 km thick, of radius 450 km and assumed flexural parameter of the lithosphere, 150 km.

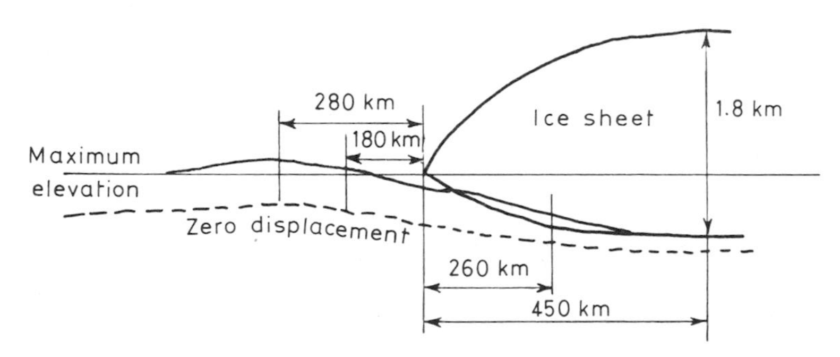

Fig. 1.8 Glacio-isostatic relations (after Walcott, 1970)

This type of uplift helps to preserve beaches and beach deposits by placing them away from active marine forces and a series of fossil deposits may be formed depending upon the number and duration of the periods of stillstand. The raised beaches and gold-bearing gravels at Nome, Alaska are believed to owe their present positions to isostatic uplift.

Hydro-isostacy

According to Bowen (1978) the dimensions of hydro-isostatic adjustments have not been adequately evaluated except on a theoretical basis. However, some workers have suggested that hydrostatic pressures resulting from the return of the water to the oceans following the last ice age do much to explain why the older submerged sea levels lay at different depths at the beginning of the Holocene Transgression, e.g. −130 to −170 m in Australia and −90 to −130 m off North East America.

Geoidic variations

Geoidic variations in sea level are due to local differences in the earth's gravitational field. Measurements have been taken from satellite data and Shimer (1969) explains a 30 m lower water level in part of the Indian Ocean balanced by a bulge of 30 m near Indonesia as being due to gravitational differences. Morner (1976) refers to geoidic maps which show a 180-m sea level difference between the 'high' of New Guinea and the 'low' of the Maldive Islands. Bowen suggests that the same phenomenum may also supply an explanation for some of the anomalous sea levels of the Holocene and the various interpretations that have been made from them.

Glacio-eustacy

Glacial eustatic changes in sea level resulting from the cyclic removal of water from the oceans during the ice ages, and its return during the warmer interglacials, have affected all oceans similarly but not all lands. Other factors have also influenced the changing shorelines and consequently many of the simple hypotheses of early investigators, such as Zeuner (1945), have now been put aside. Zeuner, from measurements of relict strandlines and marine terraces in Mediterranean regions postulated a global raising and lowering of sea level throughout the Quaternary, due almost entirely to glaciation and deglaciation. The model, as it is understood today, is much more complex because although the shorelines of the world reflect the very great influence of eustatic sea level differences, they also reflect changes due to the other processes already referred to.

In a similar mould Fairbridge (1961) outlined a model showing a descending sequence of high and low levels during the Quaternary from 100 m above, to 100 m below present mean sea level. Again, most geologists now reject simple altimetric measurements since these apply only locally and not worldwide. In fact, the existence of high Quaternary sea levels is generally rejected and, although Tija (1973) referred to a 'wide acceptance' around 30 m to 50 m above present sea level, Flint (1971) had already suggested a tentative maximum of 20 m, while emphasizing that it did not necessarily attain that level. Haile (1975) from a study of postulated late Cainozoic high sea levels concluded that 'all evidence which had been cited for Quaternary sea levels above 6 m in peninsular Malaysia is either invalid, misquoted or inconclusive'.

There is a greater readiness to accept the concepts of low eustatic sea levels of some magnitude and Haile cites a number of studies which indicate the presence of anastomizing channels at depths approaching 165 m in a valley system offshore West Sarawak. He ascribed the cause of features extending to 110 m as probably glacio-eustatic sea-level lowering. Davies (1979) identified three groups of terraces offshore a section of the coastline of south-eastern Australia at depths of less than 60 m, less than 120 m and more than 120 m respectively. He related these levels largely to eustatic lowering but recognized the uncertain effects of tectonic warping and isostatic adjustments to the pressures of the sediment load. A degree of downward tilting is ascribed to sediment loading on the shelves and it seems clear that the break between shelf and slope marks the general trend of the shoreline during the last major glaciation. Irregularities in relief along these breaks are now thought to have resulted from differences in compaction of the sediments and the effects of this on channelling by ocean currents. Submarine canyons are also a feature of most shelf areas and these probably had terrestrial beginnings.

Morner (1976) grouped the causes with which eustatic changes may be related to, (a) change in ocean volume due to epeirogenesis, orogenesis and sediment volume changes; (b) change in ocean volume through the addition of water released from magmatic processes, sediment compaction, water from the atmosphere, temperature changes, glacial expansion and retreat; (c) changes in ocean surface or geoid. Shimer (1969) suggests another cause, connected with changes in the earths rotation. This is thought to have been reduced by as much as 2% since the Cretaceous and, according to Shimer, some geologists have calculated that it should have resulted in a lowering of sea level at the equator by as much as 180 m with a corresponding rise in polar regions.

A large degree of uncertainty still prevails and, perhaps, all that is really known is that individual movements of the land have either added to or subtracted from the eustatic variations so that in some cases the differential has been positive, in others negative. There is little doubt, however, that the gross morphology of coastal plains and shelf areas has been largely determined by the events of the Pleistocene and that placer environments both on- and offshore have been greatly affected by the resulting changes in base level.

1.4 THE PLACER GEOLOGICAL TIME SCALE

The Quaternary was a favoured period for placer formation because of its high energy surface conditions, climatic extremes and relative freedom from major diastrophisms that might otherwise have destroyed them. Placer formation was also widespread during the Tertiary until interrupted by the volcanic activities and upheavals of the Alpine Orogeny. These events followed an epoch in which some of the most important gold placers were formed. The Tertiary gold gravels of California occupy a special place in Wells (1969) placer groupings.

Considerable surface activity in the Mesozoic era was followed by successive global diastrophisms and undoubtedly, many of the placers formed during that era have since been destroyed. However, some have survived and Mesozoic drainage channels, preserved under lava flows and later sediments have yielded great quantities of placer tin in the New England District of New South Wales, Australia. Some important gold placers have also been preserved in the far north-western portion of the same state buried, in part, beneath rocks of upper Cretaceous age. They are regarded, provisionally, as Middle Cretaceous.

The Paleozoic and Precambrian are noted more for primary mineralization than for placers and most of the later placers can be traced directly to Precambrian provenances. In some instances, however, very old gold bearing gravels have been consolidated into rock. The classic example is the Witwatersrand (pre-Devonian) deposit of South Africa which has contributed about one-half of all of the gold recovered in the world to date. Most other surviving placers of pre-Carboniferous age contain only the very stable minerals ilmenite, rutile, zircon and monazite. Table 1.4 describes the geological time scale for placers.

1.5 PLACERS – WHAT THE FUTURE HOLDS

As the inevitable result of increasing world demands for minerals and centuries of exploitation many resources are being overtaxed and rates of depletion must be matched by new discoveries if the needs of future generations are to be met in full. Historically, most of the commercially viable placer deposits have been found by local inhabitants attracted by the colour, brilliance or beauty of a particular stone or mineral; by prospectors with a love for solitary places and infinite patience but very little else; or by chance. Only a few deposits of note have been found by scientifically equipped parties set up specifically for the purpose and, then, only in very recent times.

1.5.1 Technological requirements

In recent times, however, tremendous advances have been made towards satisfying some of the technological requirements of modern explorers and nowhere more than in the fields of geophysics and electronics where instrument capabilities are continuing to expand in terms of resolution, penetration and delineation, at the same time becoming increasingly miniaturized electronically. Miners, for their part, have an ever widening array of machines for extracting materials from the ground, for dredging to greater depths and for recovering valuable minerals from more complex ores. This has had a marked effect upon attitudes and world-wide interest in a more scientific approach to placer exploration is evidenced by the creation of such scientific bodies as the CCOP (ESCAP) and series of courses such as those held at the Mackay School of Mining, University of Nevada, over the past few years. The communication of knowledge and experience is an essential aspect of future development. Such undertakings provide a forum for the exchange of views, policies and requirements and for the getting together of staff experts and technical advisors to consider recent developments, new techniques and progress in sponsoring and implementing training projects and programmes, and to arrange for the publishing of relevant scientific papers.

Some degree of confidentiality is understandable amongst competing commercial organizations, but this should never be carried to the point where progress is hindered substantially or the industry, as a whole, will suffer damage. There is clear evidence that geoscientists are co-operating

Table 1.4 The placer geological time scale

Eon	Era	Period	Epoch	Age (millions of years)	Geology
Phanerozoic	Cainozoic	Quarternary	Recent	0–0.015	Beginnings of Caenozoic ice ages early Tertiary and arid conditions worldwide. Gold bearing Tertiary gravels of California laid down in Eocene stream beds. Volcanic activity in Oligocene preceded Alpine Orogeny and formation of Alps, Himalayas and Pyrenees. Land masses again elevated during Alpine Orogeny and subsequently weathered. Opening of Atlantic Ocean. Holocene transgression followed last ice age about 20 000 years BP. A very favourable era for placers.
			Pleistocene	0.015–1.8	
		Tertiary	Pliocene	1.8–5.0	
			Miocene	5.0–22.5	
			Oligocene	22.5–37.5	
			Eocene and Paleocene	37.5–65.0	
	Mesozoic	Cretaceous		65–136	Primary tin deposits of great South Asian tin belt formed in the Jurassic. Much of the continental drift may have taken place at the end of the Cretaceous and continents separated from one another. A sea split Australia in two, climatic conditions deteriorated and the sequence of ice ages commenced. The Urals were weathered during the Triassic and generally the Mesozoic era featured extensive subtropical weathering and favoured the formation of placers. Very important diamond source rocks of Karoo systems (Cretaceous in South Africa and probably late lower Jurassic in Siberia).
		Jurassic		136–195	
		Triassic		195–235	

Palaeozoic		Permian Carboniferous Devonian Silurian Ordovician Cambrian	235–280 280–345 345–395 395–435 435–500 500–570	Reduced physical changes and abundance of sedimentary rocks. Periods separated on basis of intervals of continental uplift followed by submergence and intervals in which the ocean encroached on the land—Caledonian orogeny reached its peak in Silurian Period with burst of igneous activity and granitic intrusions. The Hercynian orogeny ushered in the Permian Period and the Australian continent took on its present shape and dimensions. A mountain chain was formed which stretched from Poland to Alabama. Rare placers such as diamonds in Permian of Yukutia but major gold deposition in Witwatersrand.
Precambrian	Proterozoic	Upper Middle Lower	570–1400 1400–1800 1800–2300	Evidence of glaciation and marine sedimentation—source rocks contain widespread and very important primary deposits. Precambrian important source of diamonds. Beginning of plentiful oxygen supply in atmosphere. Titanium and zircon placers in Proterozoic shales in USSR.
Archaean			2300 plus	Archaeozoic a time of igneous activity and mountain building—Surviving sedimentary rocks highly metamorphosed and intruded by granites. No evidence yet found of placer formation – however, a long period of erosion marked the end of this era. Precambrian rocks form a stable granitic shield in Western and Central Australia.

with one another to solve common technical problems as they arise: it is equally important that manufacturers should also learn from one another and that international co-operation is forthcoming so that any fresh developments in one country or organization can be used to advantage for the common good.

Basically, however, future constraints to development in the placer industry are likely to be imposed by economic factors rather than for technological reasons. Given the time and adequate funds, man's ingenuity will usually allow him to find better and cheaper ways of doing things but rarely are such funds available in the relatively small field of placer technology. Instead, most modern placer techniques have evolved from scientific discoveries in other fields and there may be little change in the future. For example, it may be expected that heavy earth-moving research programmes will be funded more for civil engineering than for mining projects and geophysical advances will be made primarily for oil, gas and hard-rock minerals exploration. Only in the further development of mineral processing equipment may a significant proportion of research capacity and funds be directed specifically to the problems of the placer industry.

Apart from the continuing search for land-based deposits much of the future is also bound up with placers on the sea floor. Technologies not available to placer geologists even a few decades ago are allowing exploration to move out into deeper and more difficult waters and future dredgers may have to dig up to three times their present digging depths in much rougher seas. None of the problems are insurmountable but geologists and engineers will have to exercise resourcefulness and imagination and not be daunted by problems that are, in reality, no more perplexing to them, today, than were many of the things they now take for granted to their forefathers.

1.5.2 Ecological considerations

Because placer operations take place either at the surface of the land or on the shallow sea bed they inevitably interfere for a time, at least, with man's enjoyment of a particular environment when it is used also for other purposes. Little thought was given in the past to preserving many features of the ecology now considered important but attitudes are changing and, just as miners onshore are learning to conduct their operations with a minimum of disturbance to others, so too must miners offshore guard against the adverse effects of mining on resources that are relied upon by other sections of the community. The fishing industry is naturally concerned with the maintenance of spawning and feeding grounds and coastal tourism depends upon the preservation of beaches and landscapes. Particularly in the marine environment, much of the future of all

Fig. 1.9 Illustrating the sequence of mining followed by land-surface contouring, the planting of rye grasses and finally the regeneration of natural species as practiced in the Australian beach-mining industry

exploiting agencies may depend on the awareness that no one action is an end in itself and all possible reactions should be evaluated in advance so that marine mining does not fall into disrepute. Governments are having to become aware of the needs for protecting the environment at all levels and an essential feature of all evaluation exercises is an environmental impact statement examining the propriety of the proposed operation. Figure 1.9 illustrates the main technique used for restoration of mined-out areas in the high-level dunes of eastern Australia. The sequence of surface contouring and preliminary stabilization with fast growing rye grasses is an essential first stage in the overall procedure.

Largely for ecological reasons bulk materials such as sand and gravel are becoming in short supply in many countries and where future construction needs are unlikely to be met onshore, exploration for such materials will be extended out into the ocean. Although many deposits are buried under considerable depths of mud and silt offshore the scope for exploration is enormous and technologies briefly referred to in Chapter 5 point the way to possible means of exploitation.

This work is an attempt to describe the fundamental aspects of placer technology and to relate present theory to practice through the eyes of one who has been both operator and investigator. It seeks to encourage the development of a synthesis covering all of the related sciences and their physical backgrounds; some important contributions are discussed, and where possible, parallels are drawn from known occurrences in the field. The data have been collected from the author's experience and from a variety of sources that are acknowledged in the text. As an introduction to the subject the opening chapter has traced the placer industry from its beginnings to the present; it discusses the various placer environments and some effects of environmental change. Chapter 2 discusses the dynamic background to sedimentation; it describes how sediments are formed, their physical properties and how they behave in fluids in motion in the various depositional placer environments. The geology of placers in Chapter 3 deals first with the provenances and then with placer formation in the three main placer environments. Chapter 4 distinguishes exploration from prospecting and describes a range of techniques and equipment for both large- and small-scale activities. Chapter 5 deals with slurry pumps and pumping. Mining methods and machines are described in Chapter 6 and, in Chapter 7, treatment processes include various forms of gravity concentration and magnetic, high tension and electrostatic separation. The final chapter is devoted to the valuation of placer deposits and an outline is given of the main stages in determining final feasibility.

REFERENCES

Agricola, G. *De Re Metallica*. Translated from Latin in 1912 by Hoover, H. and Hoover, L. H., and published in 1956 by Dover Publications Inc., New York.

Anon (1847) *Professional Papers on Indian Engineering*, Geological Survey of India, New Delhi.

Bowen, D. Q. (1978) *Quaternary Geology*, Pergamon Press, Oxford.

Brooks, A. H. (1913) The mineral deposits of Alaska, *Mineral Resources of Alaska: U.S. Geol. Surv. Bull.* 592.

Bullard, Sir Edward (1969) The origin of the oceans, *The Scientific American*, Sept. 1969, 15–26.

Calder, Nigel (1975) *The Restless Earth*, Omega Publishing Co., London.

Cleaveland, Norman (1976) Dirt, diamonds and gold dust, in *Placer Exploration and Mining Short Course*, Mackay School of Mines, Nevada.

Cooper, H. I. (1958) *Practical Dredging*, Brown Son and Ferguson Ltd, London.

Davies, P. J. (1979) Marine geology of the continental shelf off southeast Australia, *BMR Bulletin*, **195,** Canberra.

Dekker, P. M. (1927) *Dredging and Dredging Appliances*, London Technical Press Limited, London.

Douglas, Ian (1977) *Humid Landforms*, Australian National University Press, Canberra.

Erickson, O. I. E. (1956) The Hydraulic Dredge. Its History, Development and Operation. *Dock and Harbour Authority*, August Edition 133–137.

Fairbridge, R. W. (1961) Eustatic changes in sea level, *Physics and Chemistry of the Earth*, **4,** 99–185.

Flint, R. F. (1971) *Glacial and Quaternary Geology*, John Wiley and Sons, New York.

Gopinath (1978) The secret treasure of Kerala, *New Delhi Magazine*, Issue 5.

Haile, N. S. (1975) Postulated Late Cainozoic high sea levels in the Malay Peninsula, *JMBRAS*, **48,** 78–88.

Holmes, Arthur (1969) *Principles of Physical Geology*, Clarendon Press, London, Oxford.

Jenkins, Olaf P. (1946) New techniques applicable to the study of placers, *California Mineral Bulletin*, **135.**

Junner, N. R. (1973) *Gold in the Gold Coast*, Ghana Geological Survey Department, *Memoir No.* 4.

Krumbein, W. C. and Sloss, L. L. (1963) *Stratigraphy and Sedimentation*, W. H. Freeman & Company, San Francisco.

Lindgren, Waldemar (1917) *Mineral Deposits*, McGraw Hill, New York.

Lepp, Henry (1973) *Dynamic Earth*, McGraw Hill, New York.

Lord, J. R. (1976) Placer Mining Evaluation, Methods and Applications, in *Placer Exploration and Mining Short Course*, Mackay School of Mines, Nevada.

Macdonald, Eoin H. (1983) Placer Exploration and Mining, in *AGID Guidebook on Mineral Resources Development* (ed. Woakes, M.), in press, Bangkok.

McKenzie D. P. and Sclater, J. G. (1976) The Evolution of the Indian Ocean, in *Continents Adrift and Continents Aground: The Scientific America*, 138–48.

McLintoch, A. H. (ed) (1966) *Encyclopaedia of New Zealand*, New Zealand Government Printing Office, Wellington.

Morner, N. A. (1976) Eustacy and geoidic changes, *Journ. of Geol.*, **84**(2) 123–51.

Obruchev, V. (1959) *Fundamentals of Geology*, Foreign Languages Publishing House, Moscow.

O'Neil, Patrick R. (1976) Placer Mining with Bucket Ladder Dredges, in *Placer Mining and Exploration Short Course*, Mackay School of Mines, Nevada.

Popov, G. (1971) *The Working of Mineral Deposits*, MIR Publishers, Moscow.
Raeburn, C. (1927) *Alluvial Prospecting*, Thomas Murby & Co., London.
Romanowitz, C. M. and Cruickshank M. J. (1968) The evolution of floating dredges for mining operations, in *Proceedings of the Second International Surface Mining Conference Minneapolis*: Sept 18–20.
Rose, T. K. (1902) *The Metallurgy of Gold*, Chas. Griffin and Co., London.
Sanford, J. C. (1954) Trans. Amer. Soc. Civ. Eng. Vol. 54 Part C., in *The Hopper Dredge*, U.S. Corps of Engineers.
Scheffauer, F. C., (1954) *The Hopper Dredge*, Office of the Corps of Engineers, US Army, Washington DC.
Shimer, John A. (1969) *This Changing Earth*, Barnes and Noble, New York.
Smirnov, V. I. (1976) *Geology of Mineral Deposits*, MIR Publishers, Moscow.
Tikkomirov, V. V. (1953) *The Geological Sciences in Russia in the Middle of the 13th Century*, 1 Zv.an, SSSR, Moscow. pp. 78–95.
Tija, H. D. (1973) *Geomorphology in Geology of the Malay Peninsular*, (eds Gobett, D. J. and Hutchinson, C. S.) Wiley–Interscience, New York.
Turekian, Karl K. (1976) *Oceans*, Prentice Hall Inc., New Jersey.
Van De Meene, E. A. (1976) *Notes on the Development of Quaternary Geology in East and South East Asia*, UN(ESCAP) CCOP, Technical Publication No. 5, Bangkok.
Walcott, R. L. (1970) Isostatic response to loading of the crust in Canada, *Can. Journ. of Earth Sciences*, **7**, 716.
Wegener, Alfred L. (1966) *The Origin of Continents and Oceans*, Methuen, London.
Wells, John H. (1969) Placer Examination, Principles and Practice, Technical Bulletin No. 4, US Dept. Int. Bur. Land Management, Washington DC.
Wilson, J. Tuzo (1976) Continental drift, in *Scientific American, Continents Adrift and Continents Aground*, Freeman & Co, San Francisco.
Wolfe, Ernest N. (1976) Small scale placer mining methods in Alaska, in *Placer Mining and Exploration, Short Course*, Mackay School of Mines, Nevada.
Yeend (1976) The geology of placers, in *Placer Exploration and Mining, Short Course*, Mackay School of Mines, Nevada.
Zeuner, F. E. (1945) *The Pleistocene Period*, The Ray Society, London.

2

Placer Sediments and Sedimentation

2.1 THE PHYSICAL BACKGROUND

Alluvial deposits are formed as the result of natural processes that link the properties of sedimentary particles with the physical behaviour of flowing streams of air and water in the form of wind, rivers, waves, tides and ocean currents. Fluid energy is translated into solids movement when impact and viscous forces build up sufficiently to overcome the inertia of the particles at rest; deposition takes place wherever the gravitational forces prevail. Individual deposits are characteristic of the sediments of which they are composed and of the strength and nature of the forces involved in their movement and sedimentation. In this chapter, attention is directed specifically to those processes in which high energy forces mould and shape the depositional landforms known as placers.

2.1.1 The energy balance

The initial energy to lift moisture into the atmosphere is provided by heat from the sun; unequal heating of the earths surface promotes mass movements of air, and winds associated with these movements and with the earths rotation transfer energy to the waves and currents and carry water from the ocean to the land. In a cycle of events that varies from place to place according to physiographic relief, latitude and proximity to the sea, a similar amount of water is returned to the sea, thus maintaining the hydrological and, hence, the solar energy balance (Fig. 2.1).

According to Lepp (1973), the flow of solar energy amounts to 2 calories/cm^2/min of time and causes the evaporation each year of 440 000 cubic kilometres of water. Much of it falls in the highlands and is distributed between a variety of drainage networks, some is absorbed by the soil and later transpired back into the atmosphere through plants and

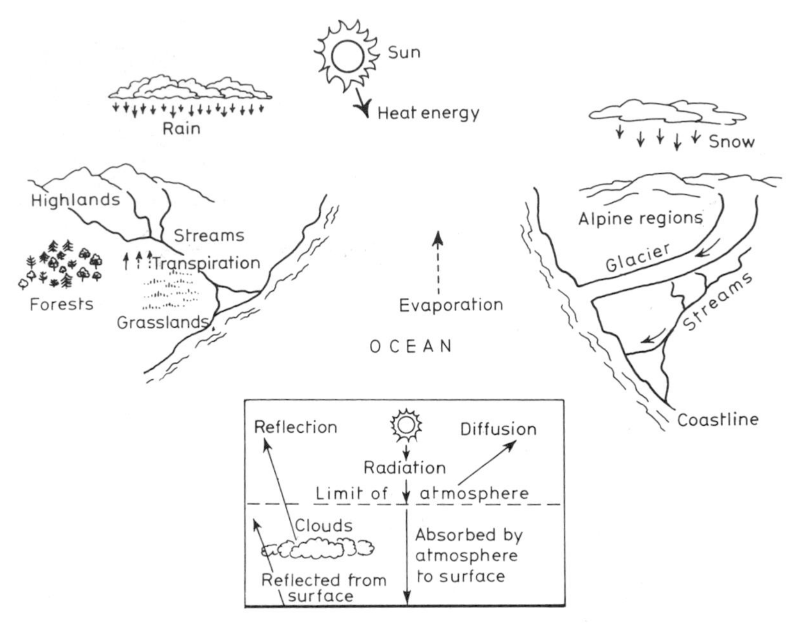

Fig. 2.1 Hydrological balance and solar energy balance (inset)

various animal forms and some takes part in the chemical and hydrolytic actions of rock weathering. Stream flow commences in areas of high relief and, in its flow back to the sea, the energy of position (potential energy) is converted to kinetic energy and gradually expended in overcoming friction and doing work. Importantly in placer environments the work includes the segregation of particles according to size, shape and density and, as in other sedimentary environments, the distribution of energy is governed by the geometry of the boundaries along which flow takes place.

The energy equation

Because the velocity of the fluid varies from point to point the kinetic energy in the flow varies also, sometimes increasing and sometimes decreasing, thus demonstrating the presence of other forms of energy and their interchangeability. This consideration led Bernoulli to his equation of energy which, for any point on a streamline, may be written:

$$\text{Total energy } (E) = \tfrac{1}{2}\rho V^2 + P + \rho g Z = \text{a constant} \qquad (2.1)$$

Z is the elevation of a point relative to some arbitrary datum, e.g. sea level. The term $\rho g Z$ represents the energy of position, i.e. the potential energy; P is the total pressure at the point and $\frac{1}{2}\rho V^2$ is the energy of motion, or kinetic energy. This expression, together with the continuity equation:

$$Q = AV \qquad (2.2)$$

is fundamental to the design of all hydraulic machines and pipelines (see Chapter 5) and applied to natural flow conditions defines the different forms of energy at every point in a stream.

To illustrate the above points: an element of water moving from rest in the headwaters of a stream contains mostly potential energy: the product of its density ρ, the acceleration due to gravity g and its elevation above sea level Z. Losses of potential energy with decreasing Z are compensated for by gains in kinetic energy some of which is expended in turbulence and changes in momentum, particularly at the foot of rapids and waterfalls where losses are high. A further interchange of energy takes place when the stream changes direction. Flowing around a bend the velocity increases along the outside and is retarded along the inside of the bend and elements of water in a horizontal line across the bend have equal quantities of potential energy and total energy but different proportions of kinetic and pressure energy. Since the kinetic energy is greatest at the outside of the bend the pressure is least and a transverse current is set up at the surface which is compensated for by a reverse current at the bottom as discussed in Chapter 3. At sea level the remaining energy of motion is finally spent in turbulence and intermolecular friction.

2.1.2 Process–response elements

The ongoing process–response model has been described by Krumbein and Sloss (1963) and is shown in Table 2.1.

Table 2.1 *A generalized sedimentary process–response model after Krumbein and Sloss (1963)*

Process elements	Response elements
Geometry of the environment	The geometry of the deposits
Materials of the environment	Properties of the sediment formed
Energy of the environment	Areal variations in the sedimentary properties
Biological elements of the environment	

Process elements

In relating the model to placers in this chapter, attention is focussed upon sediments of a particular type and in a particular size range. A review is made of the main processes involved in the various placer environments and of the means by which climate and chemistry combine to set the valuable particles free. Sediment movement is studied in order to explain how the natural processes of entrainment, transportation and deposition react to bring about the preferential deposition and accumulation of the valuable mineral types. The geometry of the environment and its deposits is touched on briefly and discussed more fully in Chapter 3. The materials of the environment are air, water, ice, rocks and minerals; biological features include burrowing animals, humic materials and bacteria; the energy of the environment has been discussed briefly in the opening paragraphs and will be referred to again many times throughout the text.

Response elements

Sedimentary responses are influenced by the natures of their environments. Forces imposed by moving streams of air and water govern deposition in the continental environment; coastal and shallow platform areas are dominated by waves, tides and currents; winds are associated with most of the movements and rivers flowing into the sea influence formative processes along some coastlines. Placers in the marine environment are modified by both sub-aerial and marine forces during alternate periods of emergence and submergence.

Individual responses are reflected in the size and shape of the placers so formed and other features evolve with differences that are characteristic of their environments. The sediments also exhibit distinctive features due to the variable effects of individual particle properties and their varying levels of resistance to the forces imposed upon them. Characteristic mineral types survive in different environments and are deposited in specific placer units within relatively small but predictable environs.

The great complexity of sedimentary response is clear from the Reinbeck–Singh model (Table 2.2) which describes the possible range of

Table 2.2 *Possible modifications in a depositional environment caused by energy of environment, availability of sediment and type of sediment (after Reinbeck-Singh, 1975)*

Availability of sediment	Type of sediment	Climate								
		Arid			Humid			Sub-Arctic		
High	Coarse									
	Medium									
	Fine									
	Mixed									
Low	Coarse									
	Medium									
	Fine									
	Mixed									
		High	Med	Low	High	Med	Low	High	Med	Low
		Energy								

modifications arising from only one small group of the many factors influencing placer deposition.

Constructed around four very general size classifications, three basic climates, three levels of energy and two levels of sediment supply the model gives rise to seventy-two possible modifications. An almost infinite number of modifications is inherent in the placer model. For example, as a process element, energy may be applied through the movement of masses of air, ice and water on land and at sea by waves, tides and currents. Its distribution is governed by boundary conditions which change constantly with the flow. Its magnitude is a function of an ever changing climate and of a variety of local features such as channel slope and size.

2.2 SEDIMENTARY PROCESSES

Sedimentary processes commence with the initial break down of the source rocks by forces of weathering and erosion in conditions that vary both spatially and with time according to the geology of the source rocks, their geographical location and the strength and nature of the sub-aerial forces of destruction. World climates vary through a full range of environments from the extreme cold of a perpetual cover of snow and ice in polar and high alpine regions, to the extreme heat of the equatorial zone. Distinctive weathering patterns are associated with each environment. Mechanical forces dominate the actions in alpine and polar regions and in deserts regardless of whether they are hot or cold. Chemical and biochemical agents are prime weathering factors in humid conditions favouring plant growth and biological proliferation. A balance is more nearly struck between the various sets of forces in temperate regions where most of the important placers are formed.

2.2.1 Mechanical agents of weathering

The initial breakdown of primary source rocks is due to a variety of internal stresses that cause rupturing during the cooling of igneous magmas; stresses that follow the release of enormous static pressures when the overlying strata are eroded away and stresses imposed by earth tremors and other tectonic movements. Such stresses cause the near surface rocks to spall and crack, thus providing openings for other weathering agents to penetrate and exposing fresh surfaces to attack. Continued mechanical breakdown at the surface of the earth is due mainly to temperature changes and extremes.

Temperature changes and extremes

Temperature variations at the earth's surface range from about $-20°C$ to $+20°C$ and give rise to repetitive stresses of various intensities. Different materials respond in a variety of ways: crystalline rocks such as granite are subjected to unequal expansion and contraction because they contain dark-coloured crystals which absorb heat and cool down faster than the light-coloured grains: crystals of contrasting size, shape and specific heat exert a variety of stresses that seek to fracture the rock along its grain boundaries and impose internal strains within the crystals themselves. Sparkes (1960) asserts that a rise in temperature of about 65°C will cause a block of granite 100 ft thick to expand an inch and contract by the same amount when it cools, thus illustrating the enormity of the forces involved in containing the movement. Scott (1932) describes 'whole mountains crumbling to arkose and sand in Turkestan . . . such are the features wrought by an arid sun and shade with a range of 80° Fahrenheit from day to night'.

However although there are suggestions of widespread destruction due to heat energy alone there is also a mass of evidence to suggest that simple expansion and contraction alone causes only minor exfoliation and splitting. For example Griggs (1936) detected no change within a rock that was alternately and rapidly heated and cooled by blasts of hot and cold dry air although tests were continued through a range of 110°C for a sufficient number of times to represent diurnal temperature changes over a period of 244 years. Similar experiments using jets of hot and cold water resulted in appreciable mechanical changes in the equivalent of $2\frac{1}{2}$ years of diurnal temperature variations. The effects of fluctuating and extreme temperatures may be better understood when more is learned about the surface rocks of the moon where there is no atmosphere and the temperature varies from more than 100 to less than minus 160°C. The present weight of evidence is against the concept of widespread destruction due to heat energy changes alone and points clearly to the need for water to be involved. On earth no desert is completely dry and the presence of even small quantities of moisture in the atmosphere ensures some chemical action, however slight.

In sub-zero temperatures water, when it freezes, expands by 9.2% in volume and pressures are exerted that rise eventually to nearly $2400\,kg\,cm^{-2}$ at -22°C. These pressures are well in excess of forces holding the rocks together and mechanical breakdown occurs wherever moisture can penetrate into cracks or pores in the rocks and become frozen. Frost in the arctic regions is most destructive in the upper few centimetres of surface material because below this level freezing occurs only once, after which no further action can take place. Even in the more temperate climates where frosting is superficial it still plays an important role. Once initiated, frost fracturing proceeds at an accelerated pace as new openings are made and additional planes of weakness are opened up.

2.2.2 Chemical agents of weathering

It is debatable whether any appreciable change can take place without the assistance of some chemical or biological action and there may be no place on earth in which one group of processes acts entirely alone. Certainly, the production of valuable placer accumulations occurs on a significant scale only where a wide range of forces are active and where they supplement each others efforts.

The principal agents of chemical weathering are oxygen, carbon dioxide and other dissolved gases and impurities present in the atmosphere and, most importantly, water. Water, as a solvent for these gases and salts is the means by which they are brought into contact with surface and sub-surface materials. By entering into the lattices of minerals such as orthoclase and exchanging H^+ ions for cations water promotes decomposition by hydrolysis; by carrying the waste products away, it provides for continuity of the reactions in fresh materials. The most important reactions take place where there is a sustained movement of ground water in a zone of aeration between the top of the water table and the ground surface. Within this zone active solutions may either dissolve a rock completely or remove some of its constituents leaving a porous mass behind. In primary source material the weathering contours generally follow the zoning of the deposit.

Chemical weathering is greatest near cities and industrial centres where sulphur compounds in the atmosphere acidify the water as H_2SO_4, and CO_2 forms carbonic acid. Rainwater in the modern environment is mostly acidic with a pH slightly less than 7.0 but this acidity increases as a result of human activity and, according to Douglas (1977) measurements as low as 4.0 have been recorded at Plymouth, UK, the centre of a large industrial area. At the other end of the scale, pH readings as high as 7.4 have been obtained from rainwater falling along the Ivory Coast of Africa where dense vegetation filters impurities out of the atmosphere and there is little man-made atmospheric contamination.

Past atmospheres, because of intermittent periods of intense volcanic activity and extremes of climatic change, must have been exposed to a similar wide range of conditions giving rise to varying degrees of weathering in different places and at different times. There must also have been similarly wide effects on rocks exposed to atmospheric and meteoric waters due to differences in the oxidation potential (eH) and, hence, certain minerals have survived in sediments preferentially. For example, quartz, one of the most stable minerals in most environments, is less likely to survive in humid tropical regions where the surface waters are alkaline and its absence, in some cases, is an indication of the character of previous climates.

2.2.3 Biological agents of weathering

All types of animal and plant life have some effect on the weathering of source rocks *in situ* acting either alone, or in conjunction with the mechanical and chemical processes already referred to, to accelerate or retard the forces of destruction. The root systems of trees exert great pressures to widen cracks as they grow and expand. On slopes the vegetation both helps to bind the soil together and to disturb it when trees sway violently or topple over. Although providing a measure of protection against erosion from winds and running water, plants also contribute organic acids that are solvents for some materials and provide sustenance for earthworms and other burrowing species that disturb the soil. Large amounts of soil are brought to the surface thus opening up the lower parts for increased aeration and percolation by ground waters. For example, Williams (1968) estimates that termites in the Brocks Creek Area, Northern Territory, Australia move almost $0.48\,m^3$ of earth annually for every hectare infested. Biological activity is also linked with rock decay at sea where benthic organisms in shallow coastal waters are similarly active on the sea bed.

2.2.4 The susceptibility of minerals to change

Minerals formed at high temperatures and pressures at an early stage of crystalization are more readily oxidized than most others. Because of their oxygen deficient beginnings they are more vulnerable when exposed to the atmosphere than are the later forming species. Olivine thus weathers faster than many other ferro-magnesium minerals, and quartz is found at the other end of the scale. Consequently the weathering sequence proposed by Goldich (1938) is in the reverse order to the Bowen series which shows the normal order of crystallization from a magma (Table 2.3).

With time, oxidation also effects changes in some of the more resistant minerals and the leucoxenization of ilmenite is a case in point. Ilmenite ($FeO \cdot TiO_2$) has a theoretical TiO_2 content of 52.6% but this gradually increases as the ferrous iron Fe^{2+} is oxidized to the soluble ferric state Fe^{3+} and is leached out of the grains by circulating waters. Standard grade leucoxenes contain 80–90% of TiO_2 and, in this form, are sometimes used as substitutes for rutile (96% TiO_2) for the manufacture of welding rods. Leucoxene is also becoming an important raw material for beneficiation in chlorine route technology in the pigment industry although it is less preferred for the sulphate route because of its poor solubility characteristics in H_2SO_4 (Macdonald, 1973). Pure oxides like rutile (TiO_2) and cassiterite (SnO_2) are amongst the most chemically stable of the valuable placer minerals. The pure carbon, diamond, is probably the most stable both physically and chemically.

Table 2.3 *Common minerals in order of susceptibility to chemical weathering*

Decreasing temperature of crystallization / Increasing susceptibility to weathering		
Olivine	Calc. Plagioclase	Gabroic
$(Mg, Fe)_2SiO_4$	$CaAl_2Si_2O_8$	
Augite	Cal. Soda Plagioclase	
$Ca(Mg, Fe, Al)(Al, Si)_2O_6$	$(Ca, Na)(Al, Si)_4O_8$	Dioritic
Hornblende	Soda Calc. Plagioclase	
$Ca_2(Mg, Al)_4(Al, Si)_8O_{22}(OH)_2$	$(Na, Ca)(Al, Si)_4O_8$	
Biotite	Soda plagioclase	
$K(Mg, Fe)_3Al, Si_3O_{10}(OH)_2$	Na, Al, Si_3O_8	
Orthoclase		
$K Al Si_3O_8$		
Muscovite		
$K Al_2Si_3O_{10}(OH)_2$		Granitic
Quartz		
Sio_2		

2.2.5 The resistance of minerals to change

Of the metallic ores, those containing sulphides weather fastest and sulphide minerals such as galena, sphalerite, chalcopyrite etc. are rarely found in placers except near the source. There is a similar unlikelihood of placers forming from the weathering of gold-bearing rocks containing abundant manganese, iron sulphides or chlorides except where precipitating agents such as calcite, pyrrhotite, chalcocite, olivine etc. are also present, or where erosion is very rapid. Solutions containing chlorides and H_2SO_4 promote natural chlorination processes which take the gold into solution and carry it away (Emmons, 1917).

Resistance to change is provided in varying degrees by the chemistry of placer minerals and by their physical properties of hardness, cleavage, toughness, size, crystal shape and textural associations. Certain minerals survive preferentially compared with many of the rock-forming minerals with which they set out. This was demonstrated by Ruhe and Cady (1967) when they measured the particle size distribution in two different soils to show that of the minerals present, zircon persisted in the lower-size fractions better than any of the others. Data showing the percentage variation of the different minerals are presented in Table 2.4.

In similar studies conducted along the River Tamar, South West England, Samad (1977) found the more chemically stable varieties to be the most persistant of the detrital minerals; mechanical action appeared to be of minor importance only. Emery and Noakes (1968), in considering the relative transportability and survival of detrital heavy minerals, suggested a maximum travel distance of 8 km for cassiterite and a median distance for gold of 15 km deducing, thereby, that placers containing these minerals are seldom far removed from their parent bodies. Gold, however, does not always conform and Wells (1969) points out that finely divided

Table 2.4

Mineral species	Sangamon soil			Loveland loesees		
	Microns 100–50	Microns 50–20	Percentage variation	Microns 100–50	Microns 50–20	Percentage variation
Epidote	24	29	+21	18	25	+39
Tourmaline	2	3	+50	4	3	−25
Zircon	4	11	+175	1	11	+1000
Garnet	10	10	Nil	5	9	+80
Hornblende	45	20	−56	54	23	−57
Titanium minerals	10	20	+100	10	18	+80
Apatite	3	5	+66	4	9	+125
Others	2	2	Nil	4	2	−50

gold travels great distances, citing a 400 mile stretch of the Snake River, Western Wyoming, where skim bars have been worked intermittantly since 1860.

Experiments conducted by Linkholm (1968) suggest that gem quality diamonds are more persistant than the industrial types. In one of these trials, an 18 in (45.72 cm) diameter × 24 in (61 cm) long ball mill was charged with 250 lb (114 kg) of 1 in (2.54 cm) steel balls and 1 litre of −6 mm + 4 mm gravels from the central treatment plant of CDM together with six industrial quality diamonds from Bakwanga in the Congo, and six gem-quality diamonds from south-west African marine terraces. The whole charge was covered with water and run at 30% of critical speed for a total milling time of 950 h.

The contents of the mill were washed over a 60 mesh screen each hour and the oversize returned to the mill. After 7 h all of the industrial diamonds has disintegrated. After 950 h the total weight loss of the terrace diamonds of gem quality was hardly measurable at 0.01%. After 950 h the total weight loss of the gravels was 40%.

The results of this experiment support observations in south-west Africa of diamond-bearing marine terraces in which less than 3% of the diamonds are of industrial quality, however, the observed frailty of the industrial diamonds in these investigations is not necessarily a norm for all varieties.

Table 2.5 lists some of the common placer minerals in order of increasing resistance to mechanical wear.

2.2.6 Erosion

As the natural adjunct to weathering, erosion is the total of all the processes that effect the movement of weathered material away from the parent rocks. Erosion begins when the first rock parts are dissolved by

Table 2.5 *General order of increasing resistance to mechanical wear of common placer minerals*

Economic minerals	Non-economic minerals
Wolframite	Pyrite
Cassiterite	Fluorite
Columbite	Enstatite
Tantalite	Augite
Gold	Apatite
Monazite	Olivine
Kyanite	Hornblende
Haematite	Garnet
Zircon	Staurolite
Rutile	Tourmaline
Ilmenite	Quartz
Diamonds (gem quality)	

It will be noted that while olivine is the most chemically unstable mineral described in Table 2.3 it is intermediate in its resistance to impact forces and abrasion.

circulating waters or, as fragments, are entrained by moving masses of ice, water or wind to be carried away and deposited elsewhere. On the slopes, where the production of waste materials is influenced by the intensity and duration of storms as well as by other climatic conditions, additional factors of importance are the geometry of the ground surfaces, the hydraulic efficiency of the drainage system, slope gradients and the density of vegetal cover.

Desert conditions pertain to tropic regions in the absence of adequate supplies of water and to arctic and alpine regions where the ground is covered by a perpetual armouring of snow and ice. In tropical deserts, high ambient temperatures are associated with hot dry winds and erosion is due mainly to the action of the wind. However, some precipitation occurs from time to time, even in the most arid conditions and both rain splash and sheet flow play a part in moving the particles from rest.

Erosive forces in alpine regions are generated from the piling up of snow in valleys and on the mountain slopes. Under increasing pressure the snow turns to ice and forms glacial masses that flow downslope under the influence of gravity, tearing and grinding off fragments of rock and incising deeply into the land. The flow is slow but continuous, the rate of flow depending largely upon seasonal variations in rates of precipitation. The spoil is deposited in streams of melt water flowing through tunnels at the base of the ice, in morrains and outwash plains on land and as accumulations on the sea floor when glacial icebergs melt at sea.

High temperatures along shorelines give rise to massive meteorological disturbances and to copeous precipitation in mountain areas. However, in the more humid climates the extent of the resulting erosion may be inhibited by profuse vegetal growth which tends to promote deep weathering of rocks with less surface erosion than in the more temperate zones. Although rain splash is still a factor in breaking up and fluidizing the soil, much of the available kinetic energy in the raindrops is dissipated by impact on leaves and branches before it hits the ground. The water either soaks into the soil or runs off as surface flow and considerable protection is given to the surface by the cover of leaf and other organic debris. Percolating waters transpire back into the atmosphere from the leaves of plants or circulate through joints, cracks and pores in the regolith until they reach and replenish the water table reservoirs.

Erosion by marine agencies

Additional erosive forces are brought to bear along the fringes of the land where variations in solar and lunar attraction affect the strength and height of tidal wave forms and accelerate the erosion and resculpturing of shore lines. Coastal physiography constantly adjusts to the forces acting and the most rapid changes take place when the ocean attacks the land successively at different levels. This is a daily occurrence because of the variability of waves, tides and winds. It is both seasonal and occasional when climatic variations and atmospheric disturbances cause storms at sea. On a geological time scale, tectonic uplifts and subsidences, isostatic adjustments and eustatic sea-level changes during a succession of ice ages and interglacial periods of thaw have altered the levels of attack along shorelines for hundreds of millions of years.

During such changes, intermittently retreating coastlines are attacked at increasingly higher levels as the sea moves in on the land. Stream energy is reduced because of decreasing gradients and sediments are deposited progressively upstream thus providing a protective cover over the recently active beds. In such conditions, strandline placers developed at various levels of stillstand are usually short-lived, being either quickly re-mobilized by wave action or blown about in the direction of the prevailing winds.

Intermittently advancing coastlines leave behind them at various levels of stillstand raised beaches and marine terraces that are more permanent although increasingly remote from the shoreline as the sea recedes (Fig. 2.2). Rejuvenated streams bring down fresh quantities of coarser-grained sediments to take part in the continuing process of marine sedimentation. Some incision of raised beaches follows the development of new water courses and, as they become uncovered, sediments deposited at higher sea levels are again subjected to reworking and redistribution by the forces of erosion.

Fig. 2.2 Relict beach sand concentration, Sonmiani, Pakistan

2.3 THE PROPERTIES OF SEDIMENTS

The physical properties of sedimentary particles are influenced by their provenance, and placer accumulations commonly reflect many of the parent qualities of grain size, shape and density, modified according to the duration and intensity of the forces applied to their fragmentation and wear. Some particles become rounded, some remain jagged and others are flattened, but most have recognizable characteristics that affect their hydraulic behaviour (see Table 2.9). Properties such as bulk density, porosity and permeability are important factors affecting the movement of ground waters and weathering *in situ*.

2.3.1 Particle size

Particle size is the most important single factor determining the behaviour of solids in fluids and clastic sediments range in size from large boulders to fine-grained sediments such as silt. A number of systems for classification have been proposed. Of these, the system of Udden (1898), subsequently modified by Wentworth (1922), found early acceptance. With 1 mm as the base, the ratio 1:2 in one direction and 2:1 in the other, forms a geometrical progression ($\frac{1}{8}$, $\frac{1}{4}$, $\frac{1}{2}$, 1, 2, 4, 8) that is generally convenient for both plotting and subsequent mathematical treatment because fine-grained sediments

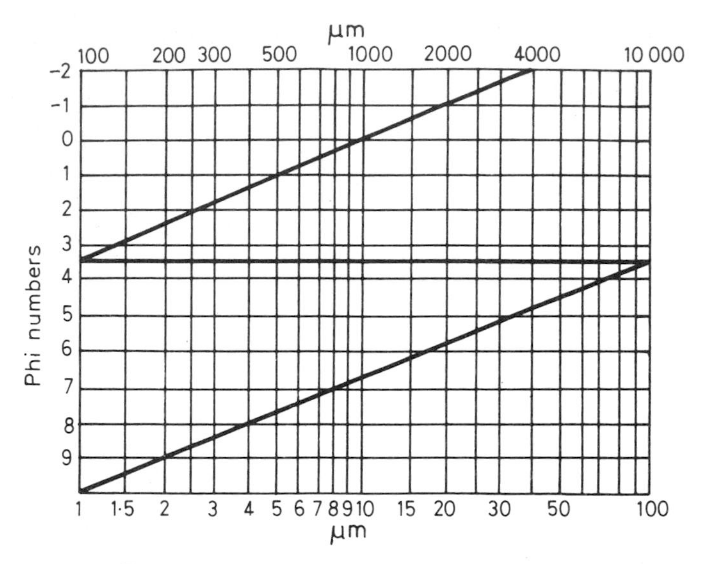

Fig. 2.3 Conversion micrometer to phi scale

generally exhibit log normal characteristics when separated into appropriate particle size fractions. In other words, if a sediment is sized in classes or grade intervals according to some standard measure and plotted logarithmically, its graph will assume normal proportions.

The need to cope with very small measurements in the finer sizings was only partly overcome by expressing them in terms of micrometres (1 μm = 0.001 mm) and Krumbein (1941) found it much more convenient to express each division of the Wentworth Scale in terms of the negative logarithm to the base 2 of the diameter d in millimetres. In doing so he introduced phi (ϕ) units of measurement. The phi scale, being simply a logarithmic transformation of the Udden–Wentworth progression, expresses each division of the progression as one ϕ unit, and retains its essential features in a simplified form. Whole numbers are substituted for the small Wentworth fractions and, because the negative logarithm is used, the ϕ sizings increase inversely with the particle size. This has the advantage of all geometric grade scales of giving equal significance to size ratios. As Krumbein points out, the difference of 1 cm in the size of a boulder is negligible whereas a difference of 1 μm in the size of a colloidal clay particle may be sufficient to double or halve it. Figure 2.3 is a chart for converting micrometres to ϕ units over the range of sediments from fine clay ($\phi = 10$) to very fine gravel ($\phi = -2$).

Particle size measurement

Large stones and boulders may be measured in place using flexible steel rules or tapes to measure the principal intercepts. A high level of accuracy is not required for individual measurements but the frequency with which each class is represented needs to be known. A few boulders can usually be bi-passed or pushed to one side without seriously interfering with mining but suitable measures for handling large quantities can be taken only if the group characteristics are understood (see Chapter 6).

The gravel class is typical of most gold, tin and gemstone placers and again the individual measurements need only be approximated while determining how much oversize material can be screened out ahead of the recovery plant. Provision can then be made for the handling and disposing of valueless reject materials and to ensure that recovery units have adequate but not excessive capacities for treating the valuable fractions. Flow-sheet design depends on accurate measurements of quantities in each sizing and these may be determined by sieving bulk samples and recovering individual fractions that can be measured volumetrically or weighed.

Sand particles in the size range 2–0.125 mm ($-1\,\varphi$ to $+3\,\varphi$) are measured satisfactorily for most purposes using nests of standard sieves and a mechanical shaker. Less confidence is placed on sieve analyses of the finer sizes 0.125–0.0625 m ($3\,\phi$–$4\,\phi$) because of dust losses, fine particle attraction and fine aperture clogging. The sand class is by far the most important for valuable placer minerals and most commercial operations are built around detailed measurements of their physical properties.

Suspensions of silts and clays, often classified broadly as slimes, rarely contain economically recoverable quantities of valuable heavy minerals although occasional exceptions are found in gold placers where particles as small as $6\,\phi$–$7\,\phi$ may be recovered by gravity methods and even smaller particles by heap leaching. Some cassiterite grains in the $5\,\phi$–$6\,\phi$ range may be recovered using more sophisticated plant arrangements as described in Chapter 7 but significant recoveries of particles smaller than $4\,\phi$ are unusual in placer operations. Measurements of size, quantity and settling rates are mainly important because very fine particles cause problems in all phases of placer operations and the harmful effects can be minimized only if the quantities and settling characteristics have been determined and remedial measures taken. Slimes problems are discussed in more detail elsewhere in this text.

The $-4\,\phi$-size material in a sample can be determined within practical limits by decantation and evaporation or through the use of a suitably calibrated density gauge such as the 'Marcy'. Calibrated cylinders of the Emery (1938) settling-tube type are useful for gauging the sedimentation properties. Two-dimensional size parameters can be measured microscopically, but this is a long and tedious task.

Table 2.6 describes the principal size classifications of the Wentworth and phi scales and how they may be measured.

Table 2.6 Sediment size classification and measurement

Type	mm Wentworth Scale	Units	Sieve mesh (BSS)	Measurement
Boulders	256	−8.0		Direct measurement
Cobbles	256 to 64	−8.0 to −6.0		
Coarse to very coarse gravel	64 to 16	−6.0 to −4.0		
Medium gravel	16 to 8	−4.0 to −3.0	2.5	Sieving
Fine gravel	8 to 4	−3.0 to −2.0	5	
Very fine gravel	4 to 2	−2.0 to −1.0	8	
Very coarse sand	2 to 1	−1.0 to 0	16	
Coarse sand	1 to 0.5	0.0 to 1.0	30	
Medium sand	0.5 to 0.25	1.0 to 2.0	60	
Fine sand	0.25 to 0.125	2.0 to 3.0	120	
Very fine sand	0.125 to 0.0625	3.0 to 4.0	150	
Coarse silt	0.0625 to 0.0312	4.0 to 5.0	300	Direct microscopic measurement or sedimentation
Medium silt	0.0312 to 0.0156	5.0 to 6.0		
Fine silt	0.0156 to 0.0078	6.0 to 7.0		
Very fine silt	0.0078 to 0.0039	7.0 to 8.0		
Very coarse to fine clay	0.0039 to 0.0002	8.0 to 11.0		

Particle size distribution

Sediment analysis, amongst other things, is concerned with particle size distributions in the form of histograms, cumulative curves, frequency curves and other forms of plotting and, as already noted, most naturally occurring sediments have distributions that approach normality if plotted to a logarithmic scale. In some cases the undersize material recovered from coarse screening also exhibits log normal characteristics (Tanner, 1959). Figure 2.4 compares the normal and skewed plots of the same distribution obtained from logarithmic and arithmetic plotting.

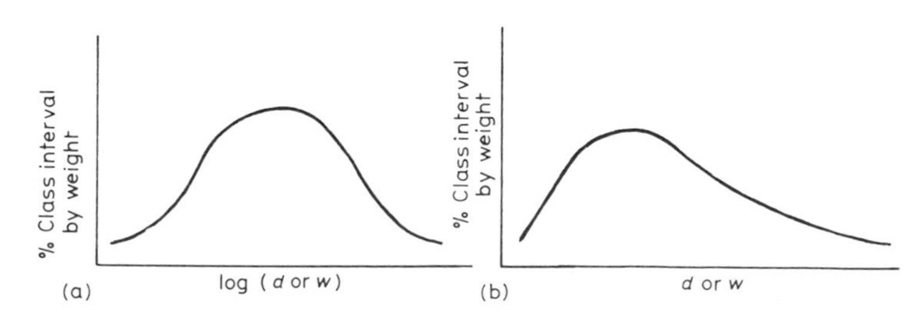

Fig. 2.4 Diagrammatic comparison of logarithmic (a) and direct plotting (b) of a log normal frequency distribution

Log normal distributions are conveniently plotted on a cumulative probability scale as cumulative percentages versus log d in which case they plot as straight lines and yield standard deviations represented as S in Fig. 2.5. Values for other sediment properties used in mathematical analyses are also calculated from these data and together provide an overall picture of the hydrodynamic conditions at the time of deposition. The subject is dealt with in various textbooks and other literature describing sedimentary petrology and geostatistics (see Milner, 1962; Blais and Carlier, 1968; Royal and Hosgit, 1974; and others).

Sands having normal distributions are less common. One example, with special applications for resin coating in foundry moulding practice was sampled by the author at Wamberal, New South Wales. Two of a number of samples, screened on a special nest of sieves, yielded the results described in Table 2.7.

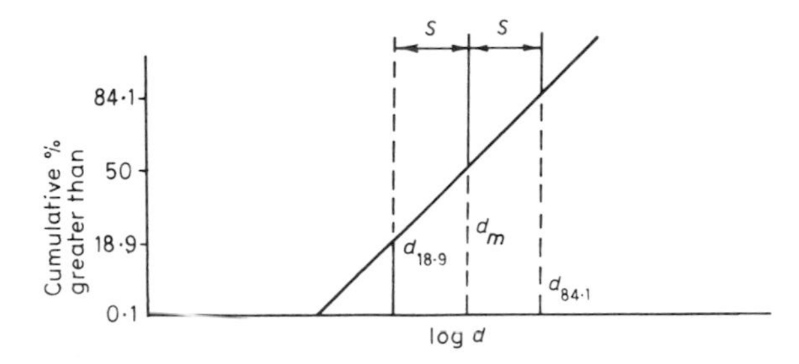

Fig. 2.5 Plot of log d versus cumulative percentage on probability scale

Table 2.7 Sand analyses showing normal size distribution: Wamberal, New South Wales, Australia

ASTM screen no.	Sieve opening (mm)	Percentage retained		
		Sample 1	Sample 2	Average
10	2.00	—	—	—
20	0.841	0.02	0.06	0.04
30	0.595	0.17	0.30	0.24
40	0.420	1.23	2.12	1.67
50	0.297	14.80	14.82	14.81
70	0.210	66.45	69.96	68.21
100	0.149	16.54	14.40	13.20
140	0.105	0.61	1.06	0.84
200	0.074	0.11	0.12	0.11
270	0.053	0.03	0.04	0.04
PAN	−0.053	—	—	—

Bi-modal distributions

Although most mineral sands accumulations are unimodal many alluvial gravels are bi-modal. Gordon (1978) has found a paucity of particles in the size range 4–8 mm. On the other hand Allen (1978) suggests a depression in modes in the 1–6 mm interval, commonly 1–2 mm and attributes the deficiencies to either the source rock grain sizes of resistant parts, or to:

(1) Hydrodynamical sorting involving preferential entrainment and transport of 1–6 mm particles, or
(2) Selective fragmentation during transport.

Minister and Toens (1970) studied the deposition of sand in gravel beds in flume experiments and noted that 'bankets' are essentially gold-bearing conglomerates in which the conglomerate particles are bi-modal and the detrital heavy minerals are hydraulically equivalent to the sand-sized mode. The high density of pebble packing in these conglomerates suggests that they were open-work gravels subsequently filled with sand and heavy minerals. In their opinion the distribution of sand in a gravel bed is controlled by the surface layer of pebbles; the sizes of sand particles are not independent of the pebble sizes and gravel beds thinner than 1.5 times the pebble diameters accumulate fine sediments much more slowly than thicker gravel beds (see Chapter 3).

2.3.2 Particle shape

Individual particles tend to assume characteristic shapes. Micas remain platy because of strong basal cleavages. Brittle minerals like wolfram and scheelite shatter easily into smaller fragments and quickly disappear from the stratigraphic column. Gold flattens into wafer-thin flakes because of its extreme ductility and gem-quality diamonds, some ninety times harder than any other stone, retain something of their octahedral form through all but the most severe conditions of mechanical wear.

Various approaches to describing the shape of sedimentary particles include comparisons between the shape of the individual particle and some recognizable standard shape. Expressions such as needle-like, tear drop and globular are used primarily in sedimentary petrography for identifying minerals according to their habits and general characteristics. For placer minerals, individual particles are compared with some geometrically exact solid bodies to give shape factors that can be used to explain differences in the sedimentation and transport of particles of equal mass and density but different shape.

Placer technologists are concerned mainly with particle shape parameters that describe various levels of roundness and sphericity; roundness as an indication of wear and the rigours of transportation; sphericity as a reflection of the sedimentation characteristics of particles. Such parameters are gauged by the relative lengths of principle axes or intercepts and, as shape factors, are usually expressed in terms of an 'equivalent' diameter derived as a length connecting particle areas and volumes. Such parameters are used as factors of proportionality to compare size determinations by different methods such as sedimentation and sieving and as conversion factors expressed in terms of a perfect sphere equal to unity.

Roundness

By definition, roundness is the relationship in two dimensions that the outline of a particle bears to its maximum inscribed circle and various suggestions have been put forward for indices of roundness as an indication or evidence of wear. The significance arises from the degree to which most placer minerals lose their sharp edges and corners wear smooth during transportation thus helping to distinguish between particles of the same mineral types derived from different source rocks in different locations.

The simplest and most useful comparisons for placer investigations are made using visual comparison charts, many of which are based upon the Waddell index ρ which relates the radius r of the circle that can be just inscribed in the particle to the mean radius of curvature of the particle R in the expression:

$$\rho = \frac{r}{R} \tag{2.3}$$

This concept was extended by Herdan (1960) to:

$$\rho = \sum \frac{(r/R)}{N} \tag{2.4}$$

where r is the radius of curvature of a corner of the particle surface, R the radius of the maximum inscribed circle in the longitudinal section of the particle and N the number of corners. As the corners wear down r approaches R and $\rho_{r \to R} = 1$.

Briggs (1977) describes a number of roundness indices according to Cailleux (1945), Cailleux and Tricant (1963) and Powers (1953). Powers

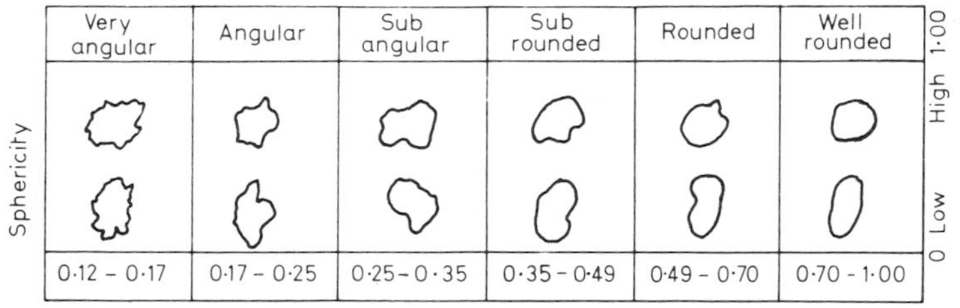

Very angular	Angular	Sub angular	Sub rounded	Rounded	Well rounded
0·12 – 0·17	0·17 – 0·25	0·25 – 0·35	0·35 – 0·49	0·49 – 0·70	0·70 - 1·00

Sphericity: High 1·00 – Low 0

Fig. 2.6 Visual comparison chart for Powers roundness index

roundness index is possibly the most widely accepted and Fig. 2.6 presents a typical visual comparison chart for assessment of Powers roundness index. Briggs suggests that when considering changes in shape within a single environment (e.g. downstream changes in sphericity) or attempting to distinguish between different environments on the basis of particle shape, it is generally advisable to use a single lithology. In doing so a rock type should be chosen that is present in sufficient quantities and which will also respond to the processes under study. King (1975) notes further that 'in order to provide valid bases for differentiating various processes or deposits, it is necessary to ascertain that the indices for the different samples differ by an amount greater than a certain value, which can be obtained by statistical analysis of many samples'. Shepard (1973) cautions that, due to other factors, roundness does not necessarily increase with distance of travel and that Russell (1939) and Kuenan (1958) have shown instances where it has decreased. Note, also, that in his studies of the sediments of the Tamar River, southwest England, Samad found a general decrease in both roundness and sphericity with decreasing particle size. It should not be overlooked that such observations may have their explanations in the grain-size characteristics of the individual source rock minerals and the subsequent differential rates of wear of large and small particles in transport. Detrital minerals in order of their resistance to abrasive rounding are listed in Table 2.8 (taken from Berkman, 1976).

Sphericity

Spheres have the smallest surface areas for any given volumes and higher settling velocities than any other shapes. Spherical particles are also the most readily transported of particles making up the traction load of a stream and, hence, sphericity is an obvious standard against which to measure the sedimentation properties of particles of other shapes.

Table 2.8 Detrital minerals in descending order of resistance to rounding

Quartz (most resistant)	
Tourmaline	Hypersthene
Microcline	Spodumene
Staurolite	Apatite
Titanite	Monazite
Magnetite	Augite
Garnet	Haematite
Ilmenite	Brookite
epidote	Kyanite
Zircon	Enstatite
Hornblende	Fluorite
Rutile	Siderite
Diallage	Barite (least resistant)

Sphericity was defined initially by Wadell (1932) as the ratio between the surface area S of a particle and the surface area S' of a sphere with the same volume as the particle so that:

$$\text{Sphericity } \psi = \frac{S'}{S} \tag{2.5}$$

where ψ has a limiting value of unity for a true sphere. He later proposed an alternative definition in terms of equivalent volumes and this was adapted by Krumbein (1941) who defined sphericity in terms of the three principle axes:

$$\psi = 3\left(\frac{bc}{a^2}\right)^{0.5} \tag{2.6}$$

where a is the longitudinal axis, b the maximum diameter at right angles to it and c, the maximum diameter along the third right-angle plane from the same point.

The geometry of a sphere is adequately described by its true diameter D and mathematically:

$$\begin{aligned} \text{Projected area} &= \pi D^2/4 \\ \text{Volume} &= \pi D^3/6 \\ \text{Surface area} &= \pi D^2 \\ \text{Specific surface} &= c/D \end{aligned}$$

The geometry of a sedimentary particle has no such unique diameter and can be described only through a series of nominal diameters of varying physical meanings and numerical values. These include:

(1) Equivalent settling diameter: diameter of a sphere having the same settling rate as the particle in the same conditions of fluid density, temperature and gravitational force.
(2) Sedimentation diameter: the diameter of a quartz sphere that settles in a given medium at the same terminal velocity as that of the particle.
(3) Equivalent volume diameter: the diameter of a sphere having the same volume as that of the particle.
(4) Surface area diameter: the diameter of a sphere having the same surface area as that of the particle.
(5) Projected area diameter: the diameter of a circle of equal projected area, either maximum or minimum.

Many other characteristic diameters have been proposed, but all are difficult to measure and in placer technology it is convenient to use only such measurements as are readily approximated. Of these, the equivalent settling diameter, being influenced by volume, shape and density is generally the most realistic although, for beach sands, the equivalent volume diameter may be a more useful measure. This is because the range of sphericities for fine sands generally starts at about $\psi = 0.75$ and average diameters can be closely approximated by screen sizing.

A common problem in determining the sedimentary properties of sediments containing both sieve and sub-sieve sizes (see Table 2.6) is how to correlate values that are determined by different methods. A convenient technique is to calculate a subsidiary shape factor by plotting each set of results on the one sheet and then equating the results to obtain the required factor. Figure 2.7 presents the results of two hypothetical sets of data to show that the two lines are brought into congruence by multiplying the sedimentation values by the factor ϕ. The result is a standard set of measurements, applicable throughout.

2.3.3 Particle density

Density is mass per unit volume and individual particle measurements can usually be determined using standard laboratory techniques. The calculation of bulk densities from weight–volume measurements is more difficult because in addition to particle size, shape and distribution, beds of particles have bulk properties that vary according to the size, shape and distribution of the voids and the degree to which these may be filled with air, gas or water. The amount of void space (porosity) is a function of how the particles are arranged and, of pressure through compaction. Changes take place when the material is disturbed. In a well-compacted bed the particles tend to arrange themselves so that the smaller particles fill the voids between larger ones and the total space occupied is a minimum. Newly disturbed sediments form aggregations, with small particles holding larger particles apart, instead of fitting in to the voids between them, thus increasing the overall volumes and decreasing the bulk densities.

In sampling, there is rarely a need to calculate bulk densities for placers containing the more valuable minerals gold, tin etc. because values for these are normally expressed in terms of weight of product per volume of ground treated, both measurements being readily obtained. Introducing

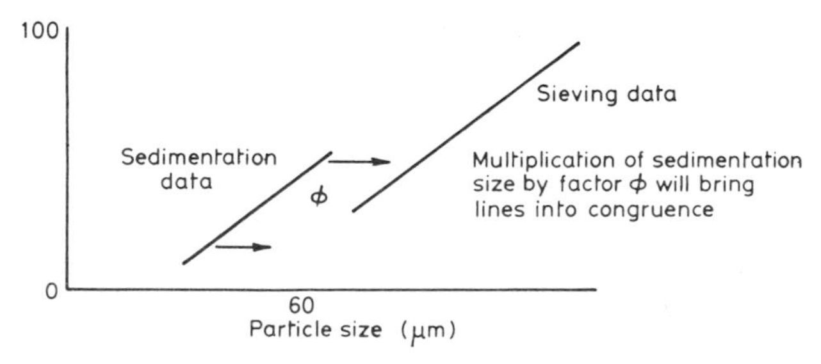

Fig. 2.7 Hypothetical sieving and sedimentation relationships

conversion figures where they are not needed creates opportunities for error and in no way assists the valuations. Mineral-sands placers, on the other hand, are evaluated by methods which yield results in terms of weight of product per weight of sample and, for them, the calculation of bulk densities is an important consideration.

Bulk densities for beach sediments may also be affected very significantly by differences in their heavy minerals content and individual areas of high and low mineral grades should be calculated separately. Preliminary calculations may be based upon conversion factors described in Fig. 2.8. Figure 2.8 is a generalized plot of bulk densities prepared experimentally from sediments containing high percentages of the light heavy rock-forming minerals (densities 2.8–3.6), quartz (2.65) and minor amounts of zircon, rutile and ilmenite (4.2–5.0). The equation to this graph is

$$BD = 1.56 + 0.0128\,C \tag{2.7}$$

approximately. Haulage measurements are normally related to measurements of various materials *in situ* through the use of 'swell' factors that take account of differences in volume between equal quantities of compacted and non-compacted materials. A swell factor of 30% is commonly used for sediments of all types on the assumption that 1 m^3 of sediments *in situ* will occupy a volume of 1.3 m^3 when dug and loaded into a truck. Such ready assumptions are seldom accurate and often lead to serious errors in estimating tonnages and grades as, also, do attempts to apply in practice the figures quoted for bulk densities in various textbooks. For example one text suggests the following range of values:

Continuously submerged silt and clay 20–60 lb ft^{-3}
Dry sand mixtures 85–100 lb ft^{-3}
Dry fine sediments up 125 lb ft^{-3}

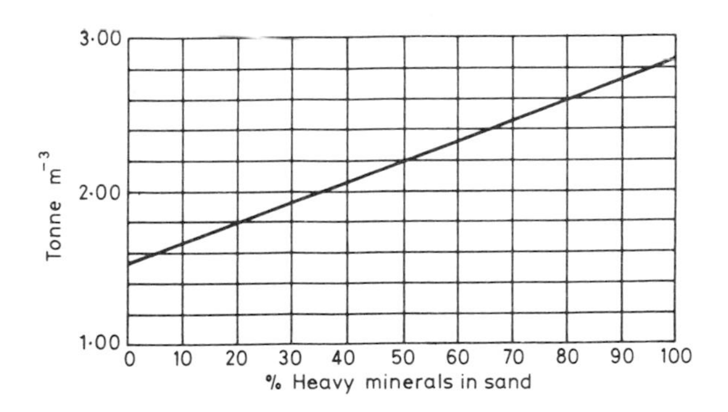

Fig. 2.8 Generalized plot of percentage heavy minerals versus tonnes m $^{-3}$ for heavy-minerals bearing sands

The distinction between dry sand mixtures and dry fine sediments is far from clear; no lower limit is given for the bulk density of the fine dry sediments; no grounds are stated for the range of 20–60 lb ft^{-3} for continuously submerged silt and clay although, if used, they leave open possibilities of errors as high as 200%.

None of these figures are of any practical use except that, by illustrating the wide range of possible variations in the bulk densities of sediments of different types and in various degrees of compaction, they emphasize the need to prepare fresh plots and to calculate fresh swell factors for all major deposits. This is particularly important in cases where the overall grades are marginal and tolerances for error must necessarily be small or where, as in some mineral-sands deposits, the percentages of valuable minerals in the heavy-mineral fraction of the sediments varies with the total content of heavy minerals. An example given in Chapter 8 illustrates this point.

Density measurements

Regardless of how the sediment samples are taken they are always disturbed and it is seldom easy to return them to reasonable approximations of their original states. Usually a considerable number of measurements is needed to strike a satisfactory mean using methods that differ according to the lithology of the materials concerned and their location above or below water level. Direct measurements may sometimes be made above the water table by boring out a small cylinder of material leaving smooth, regular sides and filling the cavity with a measurable volume of very viscous oil or a quick-setting material such as plaster of Paris. The bulk density is calculated by dividing the dry weight of the excavated material by the volume of the filler used to fill the excavation made.

Other methods employ physical processes of compaction that include vibration, tamping and compression in controlled conditions using calibrated measuring boxes with vertical walls braced firmly for strength and rigidity. A favoured technique for saturated sand-sized particles is to make the additions gradually into just sufficient water to cover the sample and to simultaneously liberate entrained air bubbles by puddling. When no more bubbles appear the sediments are then settled into place by vibration. The measured volume of the compacted sand is compared with its dry weight to calculate the bulk density of the original drowned sediments in their undisturbed state.

2.3.4 Characteristic properties of placer minerals

The characteristic properties of common placer minerals (derived from Baker, 1962) are described in Table 2.9 (page 64).

2.4 SEDIMENTATION

2.4.1 Introduction

Almost all branches of placer technology are linked with considerations of how particles of rocks and minerals behave in fluids both in nature and in subsequent man-made operations. Geologists are concerned with depositional land forms and with the preferential movement and settling of sediments containing valuable heavy minerals. Engineers measure the sedimentation properties of detrital materials in order to determine how they may best be slurried and pumped from one point to another. Mineral processers are faced with the task of segregating one or more mineral types from a host of others by either wholly or partly exploiting differences in particle sedimentation rates.

Many of the problems are complex and available knowledge is often based upon a somewhat loosely related synthesis of practical experience, scale-model technology and theoretical analysis; consequently there are still many gaps and serious difficulties in trying to bring the three together. Operators with placer experience are closest to the problems but have little time to study the mechanisms involved or to record and integrate all of the data from daily observations: model studies are directed towards solving specific problems but leave others unresolved: the theoretical approach is often far removed from practicalities and researchers are seldom given opportunities to test hypotheses in the field.

Thus, while a few of the known analytical devices developed by such workers as Hjulstrom (1935), Shields (1936) and Bagnold (1954) are useful for predicting the behaviour of certain classes of sediments in some environments, in other environments the wide range of variables and their almost infinite capacity to provide fresh combinations pose many problems that cannot yet be resolved by such means. No one, for example, has been able to assess closely the effects on sediment transport of such factors as the variable geometry of natural slopes and channels and the interlocking of particles of different sizes and shapes in stream beds; and neither in the laboratory or in the field has the natural arrangement of particles in any cross-section of a fluvial channel been reproduced exactly by artificial means.

It is not enough, therefore, for engineers and mineral processors to adopt blindly such empirical formulae and procedures as are in current favour in the hope that they will work. Nor will geologists who merely categorize structures as fluvial, transitional or marine escape from the resulting maze of misunderstandings. Only by observing closely and then seeking rational explanations in terms of the mechanisms involved can workable solutions be reasonably assured. The following treatment deals broadly with sedimentation as an introduction to placer formation in Chapter 3 and to

Table 2.9 *Characteristic properties of common placer minerals*

Mineral	Hardness (Mohs Scale)	Relative density	Crystal system	Common mesh size (BSS)	Weathered shape
Apatite	5	3.17–3.23	Hex.	100	Rounded elongated (egg shaped) grains, occasional stumpy euhedral prisms
Cassiterite	6–7	6.8–7.1	Tetrag.	100–240	Sometimes prismatic crystals, angular, rounded grains and fracture fragments
Chromite	5–6	4.3–4.6	Cubic	52–150	Sometimes octohedral crystals, sub-rounded to rounded grains, irregular fragments
Colombite	9	5.15–5.25	Orth	Various	Fracture fragments and sub-rounded grains
Corundum	9	3.95–4.15	Hex.	14–30	Irregular fracture fragments, sometimes rounded
Diamond	10	3.50–3.53	Cubic	Various	Octahedral crystals, sometimes rounded
Garnet (var)	6.5–7.5	3.42–4.27	Cubic	14–100	Commonly irregular, fractured, very variable to well rounded
Gold	2.5–3.0	19.3	Cubic	5–200 but variable	Rounded or flattened grains, rods, wire-like pellets and flakes
Hornblende	5–6	3.0–3.3	Mono	60–199	Platy, cleavage flakes, some with ragged ends, rounded stumpy prisms
Hypersthene	5–6	3.4–3.5	Orth	60–200	Rugged prismatic crystals, sometimes sub-rounded, rarely well rounded, prismatic forms
Ilmenite	5–6	4.5–5.0	Hex	52–150	Irregular sub-rounded to rounded grains usually equidimensional

Kyanite	4–7	3.56–3.68	Tri	52	Sub-angular prismatic, sometimes stumpy rounded grains
Leucoxene	Variable	3.5–4.5	Amorphous	14–200	Rounded to sub-rounded and angular pitted irregular grains, coated
Magnetite	5.5–6.5	5.17–5.18	Cubic	14–100	Occasional octahedral crystals angular and well rounded, irregular equidimensional
Monazite	5.0–5.5	4.6–5.4	Mono	30–150	Rare crystals, well-rounded grains, usually ellipsoidal, occasionally sub-euhedral tablets
Olivine	6.5–7.0	3.27–3.37	Orth	60	Usually irregular and much fractured rarely euhedral
Osmiridium	6–7	19–21	Hex	Various	Small flattened grains
Rutile	6.0–6.7	4.18–4.25	Tetrag.	14–200	Prismatic crystals, sometimes broken, rounded anhedral grains, fracture fragments, occasional twins
Sphene	5.0–5.5	3.54	Mono	Various	Occasionally diamond shaped, commonly irregular or sub-rounded
Spinel (Var)	7.5–4.6	3.6–4.6	Cubic	14–100	Rounded, octahedral grains
Staurolite	7.0–7.5	3.65–3.67	Orth	30–100	Usually irregular to platy
Tantalite	6.0–6.5	7.9–8.0	Orth	Various	Fractured, irregular, sub-rounded
Tourmaline	7.0–7.5	2.98–3.20	Hex	60–150	Well-rounded, angular fractures
Wolframite	5.0–5.5	7.10–7.90	Mono	Various	Submetallic cleavage fragments
Xenotime	4.0–5.0	4.59	Tetrag.	30–150	Commonly rounded, sub-angular, rectangular flakes, equidimensional
Zircon	7.5	4.20–4.86	Tetrag.	14–200	Prismatic and stumpy pyramidal crystals, common, rounded grains and fracture fragments

the mechanics of solids–fluid flow in Chapters 5, 6 and 7. References are made, where applicable, to both commercial and natural examples.

2.4.2 The flow of solids through fluids

Solids and fluids in motion behave variously according to individual differences in the solids characteristics and because of dynamic forces that are governed by the density and viscosity of the fluids and by the velocity of flow. The forces involved are called *body forces* when they act from a distance and *surface forces* when they are in direct physical contact with a solid surface or element of fluid. Depending upon how the forces interact particle movement is effected, (a) without losing contact with other particles or the bed (sliding, rolling and creep), (b) by bouncing along in intermittent contact with other particles in motion and with the stationary bed (saltation), and (c) with or without intermittent contact with other particles away from the bed (suspension).

The velocity of the particles moving in accordance with (a) and (b) is below that of the fluid and such particles form the traction load of a stream. The velocity of particles moving in accordance with (c) is the same as that of the fluid and such particles form the suspension load of the stream. Moss (1962–3) has pointed out that since no one particle is independent of its neighbours or of local changes in the velocity distribution resulting from their movement, all three methods of solids–fluid flow may be represented in the one sample.

Body forces

The main body forces with which placer technology is involved are:

(1) Forces of electrostatic attraction, repulsion and ionization, both natural and artificially induced. (See Chapter 4, geophysical prospecting and Chapter 7, the separation of conductor from non-conductor minerals).
(2) Magnetic forces due to the earths magnetic field and artificially induced magnetism (See Chapter 4, geophysical prospecting and Chapter 7, the separation of magnetic from non-magnetic minerals).
(3) Centrifugal forces arising from the rotation of the earth–moon and planetary systems (See Chapter 1, the environments).
(4) Gravitational forces exerted by the sun, moon and earth and artificially induced gravity.

Prime consideration is given in this chapter and in Chapters 3, 5, 6 and 7 to gravitational forces, particularly those of the earth's gravitational field. International units are adopted that express mass in terms of grams (g); weight as kilograms (kg) or tonnes (1000 kg); and the unit of force as the weight of 1 kg (dyne).

The gravitational force

Mass as the quantity of matter in a body does not change; however, the acceleration due to gravity, g, varies with distance from the earth's centre of mass and bodies of equal mass vary similarly in weight from place to place on the earth's surface. The variations, although small, have important implications for geophysicists who are concerned with very slight differences in the weights of bodies of equal mass at various locations and elevations. Engineers, on the other hand, deal in more robust units and, while recognizing the distinction are usually content to assume a standard value of $9.81\,\mathrm{m\,s^{-2}}$ for g and hence constant body weights regardless of position. Where there is a need for more precise measurements and to recognize the variations in gravity in different latitudes, approximations may be made from the calculated values for g in Table 2.10.

It follows that the value for g is greatest at sea level at the poles and least on high mountains on the equator. For greater accuracy using the base value $g = 9.781\,236\,\mathrm{m\,s^{-2}}$ the numerical value for g at any desired place can be calculated from the formula:

$$g = 9.781\,236\,(1 + 0.005\,243 \sin^2 l)\,(1 - 0.000\,000\,097e) \qquad (2.8)$$

where l is the latitude of the place, north or south of the equator, and e is the elevation in metres above mean sea level.

Example: To find the value of g at a location of 28°39′N, latitude, 213 m above sea level

$$g = 9.781\,236\,(1.001\,21)\,(0.999\,979\,62)$$
$$= 9.793\,\mathrm{m\,s^{-2}}$$

Certain other units used by placer hydraulics engineers relate to the density of liquids and, again, average values are accepted for most calculations. Thus, for water:

W = weight density or weight per unit volume ($9.81\,\mathrm{dyne\,cm^{-3}}$)
ρ = mass density or mass per unit volume ($1.00\,\mathrm{g\,cm^{-3}}$)

and the relationship between the weight density and the mass density is invariably given by:

$$W = \rho g \qquad (2.9)$$

Table 2.10 *Values of coefficient g at different latitude*

Latitude	0°	10°	20°	30°	40°	50°	60°	70°	80°	90°
g ($\mathrm{m\,s^{-2}}$)	9.781	9.783	9.787	9.784	9.802	9.811	9.819	9.826	9.830	9.831

All other solids and fluids have their densities represented as ratios that compare their weight densities with the weight density of pure water. The relationship is the relative density (RD).

The effect of the earths gravitational field is to attract all matter towards its centre of mass with a force that is expressed by Newton's Second Law of Motion:

$$F = Mg \tag{2.10}$$

where F is the force in dynes.

A particle in free fall in a vacuum continues to fall with constant acceleration until brought to rest on impact with a solid boundary. The same particle falling freely in a fluid meets with a resistance to motion that is due to the fluid properties of density and viscosity; the gravitational force does not change but the motion of the particle is affected. Two main premises govern all aspects of wet and dry gravity concentration:

(1) High density particles settle faster in a fluid than do lower density particles.
(2) Particles of equal density settle faster in fluids of lower densities than in fluids of higher densities.

A solid in air thus appears to be heavier than when it is in water and, from Archimedes' Principle, the apparent difference in weight is equal to the weight of the fluid displaced. The weight of air displaced by a solid is negligible; the weight of water displaced is both appreciable and predictable if the volume of the solid is known.

Effectively therefore, two opposing forces act on a solid at rest in a fluid, the force of gravity acting vertically downwards and a buoyancy force acting vertically upwards. The resultant of the two forces is positive or negative depending upon the density of the solid relative to the density of the fluid. If the fluid is water and the relative density of the solid is greater than unity it will sink; if less than unity the solid will float, only partially submerged; if equal to unity the solid will remain suspended in the water at whatever level it is brought to rest.

Buoyancy

The principle of buoyancy can be used to explain much of the behaviour of solids in gravity treatment processes and why, with the exception of certain pneumatic processes referred to in Chapter 7, particles of different densities are separated more readily in wet than in dry gravity treatment plant. Consider, for example, two equal spheres, one of rutile ($\rho' = 4.25$) and the other of silica ($\rho' = 2.65$). The resultant gravitational force Fg is, in each case:

$$Fg = \frac{\pi d^3}{6}(\rho' - \rho)g \tag{2.11}$$

where $\pi d^3/6$ is the volume of the sphere and ρ' and ρ are the densities of the solids and fluid respectively.

Since the relative density of air is negligible compared with that of water, for which $\rho = 1$, the apparent densities of the two spheres in air and water are:

Mineral	*In air*	*In water*
Rutile	4.25	$(4.25 - 1) = 3.25$
Silica	2.65	$(2.65 - 1) = 1.65$

therefore the ratio of apparent densities:

$$\rho_{\text{air}} = 4.25 \div 2.65 = 1.60:1$$
$$\rho_{\text{water}} = 3.25 \div 1.65 = 1.97:1$$

All gravity treatment processes work more efficiently when the ratios of apparent densities of individual particles of different minerals are large than when they are small and, obviously, the differences become more pronounced with increasing fluid densities. Fluids such as bromoform (RD 2.8–2.9) provide a means of separating the light minerals from the heavies in the laboratory; artificially densified fluid media allow such separations to be made on a commercial scale as described in Chapter 7.

Surface forces

Surface forces act in direct contact with a fluid element or solid body. Forces that act normal to the surface are called static or dynamic forces depending upon whether the fluid and the body are at rest or in relative motion.

Static pressures

The static pressure exerted against a surface of area A under a fluid column of height h is given by the expression:

$$P = \rho g h \qquad (2.12)$$

At sea level with air as the fluid; P becomes P_{atm}, the atmospheric pressure on an element of matter at sea level. Hence, the same element of matter submerged beneath the sea has a column of water and a column of air above it and the total static pressure is given by:

$$P = P_{\text{w}} + P_{\text{atm}} \qquad (2.13)$$

The three main fluids in which sedimentation takes place in placer environments are:

fresh water: $\rho = 1.00\,\text{g cm}^{-3}$
sea water: $\rho = 1.025\,\text{g cm}^{-3}$
air: $\rho = 1.250\,\text{kg m}^{-3}$ (dry air at sea level)

Impact forces

Solids and fluids in relative motion are subjected to dynamic pressures and impact forces resulting from changes in momentum. By definition, the momentum or impact I is the product of mass and velocity:

$$I = MV \qquad (2.14)$$

and the force changing the momentum is its time derivative:

$$F = M\frac{\mathrm{d}V}{\mathrm{d}t} \qquad (2.15)$$

As applied to the theory of centrifugal pumps (see Chapter 5), equation 2.15 states that when a body in steady motion undergoes a change in angular motion, the resultant of the external forces acting on the body is a torque equal to the time rate of change of the angular motion. As applied to hydraulic sluicing operations which rely heavily on the impact of high-velocity jets of water, the impact force illustrated in Fig. 2.9 is first exerted as a force of destruction to break down the bank of material to be slurried. In the horizontal plane and for a vertical bank face, since force equals momentum:

$$F = \rho(AV)V = \rho AV^2 \qquad (2.16)$$

where A is the cross-sectional area of the jet (m^2).

Viscous forces

Part of the kinetic energy of a solids–fluid mixture in motion is expended in overcoming viscous forces arising from intermolecular friction between adjacent fluid elements and fluid drag along solid surfaces in contact and relative motion with the fluid. Viscous forces act tangentially, the resulting shear stresses generated at the surface being governed by the fact that air and water are similarly Newtonian in character in that their rates of deformation remain constant for any given temperature and

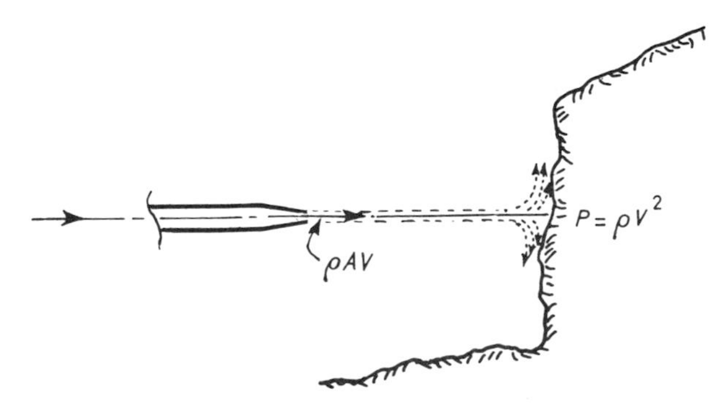

Fig. 2.9 Dynamic pressure exerted by free jet of water on bank of alluvium

pressure, regardless of the duration of the forces acting. Viscous shearing forces are defined by Newton's Law of Viscosity:

$$\tau = \mu \frac{dv}{dy} \tag{2.17}$$

which relates the shear stress τ to the rate of angular deformation dv/dy through a proportionality factor μ representing the viscosity of the fluid.

It is important to note that, while pressure variations have little effect on the viscosities of real fluids such as air and water, such viscosities are sensitive to changes in temperature and, whereas the viscosities of water and other liquids decrease with increased temperatures, those of air and other gases increase with increased temperature. Some typical relationships for air and water at different temperatures are listed in Table 2.11.

To account for the additional effects of turbulence, an eddy viscosity η has been introduced into equation 2.17 and for turbulent flow the shear stress is given as

$$\tau = (\mu + \eta) \frac{dv}{dy} \tag{2.18}$$

The eddy viscosity η is generally several orders of magnitude greater than the dynamic viscosity μ and, hence, the shear stresses in turbulent flow for similar velocity gradients are also many times greater than in laminar flow.

The boundary layer

Shear stresses are transmitted to a solid surface through a thin layer of fluid surrounding it. Within this narrow region of stress the velocity is zero at the interface where a thin film of fluid adheres to the solid surface without slip; it increases parabolically to reach its maximum value asymptotically at some distance into the stream and, since the rate of angular momentum and the velocity gradient dy/dx measured normal to

Table 2.11 Viscosities of air and water at atmospheric pressure

Temperature °F	Dynamic viscosity (lb s ft^{-2})		Kinetic viscosity (ft^2 s^{-1})	
	Air ($\times 10^{-7}$)	Water ($\times 10^{-5}$)	Air ($\times 10^{-4}$)	Water ($\times 10^{-5}$)
40	3.62	3.24	1.46	1.67
60	3.74	2.34	1.58	1.21
80	3.85	1.80	1.69	0.930
100	3.96	1.42	1.80	0.736
120	4.07	1.17	1.89	0.610
150	4.23	0.906	2.07	0.476

the direction of flow are related to the velocity at every point, τ is a maximum value at the interface and zero at the point of maximum velocity in the body of the flow. In other words, the means for transferring stress is contained wholly within the boundary layer and no external forces have any effects upon the underlying surface.

The development of a boundary layer is described in Fig. 2.10 for low viscosity flow in an open channel with ideal entrant conditions. As illustrated, the boundary layer is laminar and very thin at the upstream end but, as the layer moves down the channel in the direction of flow, the build up of shearing stresses slows down additional fluid elements and the thickness of the layer increases. A corresponding increase in inertial effects promotes turbulence and leads to the formation of a turbulent boundary layer. Such a layer, when fully developed, is composed of a thin laminar sublayer in contact with the solid surface; a buffer zone in which a transition takes place from the laminar to the turbulent state and, beyond this, the fully turbulent zone of the boundary layer.

An understanding of boundary layer theory is a pre-requisite to understanding many aspects of sedimentation both in nature and in the process plant and various applications of the theory are referred to elsewhere in the text. One example in the fluvial placer environment, is found in the separation of a boundary layer from its surface of attachment wherever an adverse pressure gradient develops along a boundary. Typically this may occur through a sudden deceleration of the flow or where the surface curvature is pronounced. Figure 2.11 illustrates the formation of eddies due to separation at the downstream side of a rock bar, with consequent scouring and erosion of the bed. Further examples are found in the plant, where separation occurs due to such causes as non-streamlined flow through orifices and sudden expansions and contractions in pipelines. Boundary layers developed in sluices and spirals provide the means whereby heavy particles settling out from the main flow are collected and recovered preferentially (see Chapter 7).

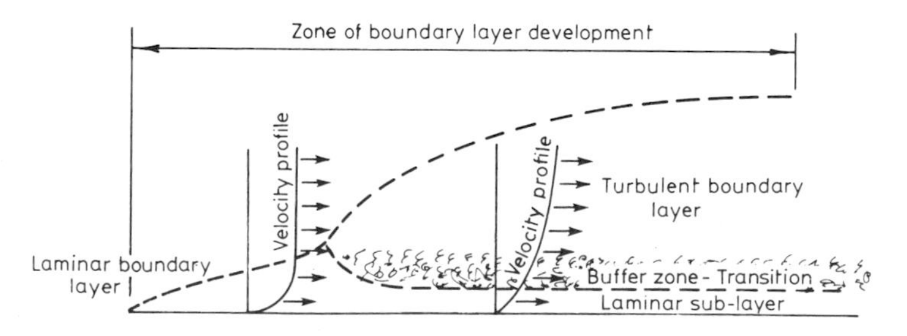

Fig. 2.10 Development of boundary layer in open channel with ideal entry conditions

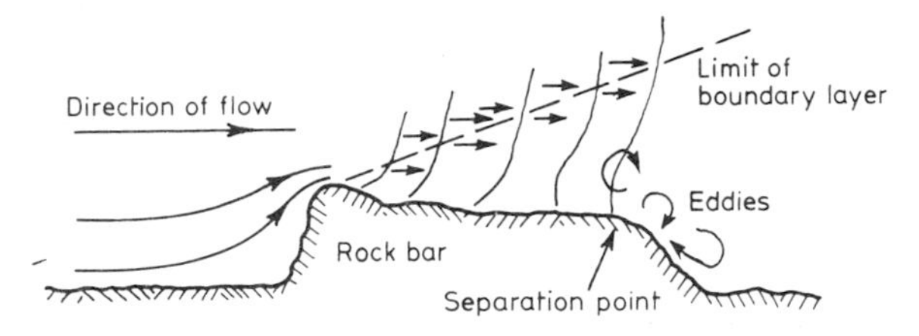

Fig. 2.11 Adverse pressure gradient causing separation in stream flow passing over a rock bar

Patterns of flow

Fluid flow may be laminar, turbulent or transitional:

(1) Laminar flow is flow in which the fluid moves in layers with only a molecular change in momentum. Solid particles in suspension are enveloped in layers of fluid that remain in contact and move along with them without disturbing the pattern of flow.
(2) Turbulent flow is flow at high velocities and is characterized by the erratic motion of particles along and across the direction of flow. Inertial forces predominate and, because of violent transverse changes in momentum, the pattern of flow is complicated by eddies and other secondary motions that are impressed upon the stream.
(3) Transitional flow describes the character of flow when viscous forces and inertial forces vie with one another for control while flow passes from the laminar to the turbulent state.

Although all types of flow are common, placer technology is concerned mainly with flow in the turbulent state. With few exceptions, solids movement is insignificant in laminar flow and, for the most part, particles are entrained, transported and deposited in conditions of turbulence. Safe conditions for the transportation of solids in pipelines obtain only above a certain critical velocity V_{cr} that ensures continued turbulence (see Chapter 5) and transitional flow velocities do not provide a safe margin for transport.

The Reynolds Number (NR)

The different natures of laminar and turbulent flow were demonstrated originally by Osborne Reynolds who found, in a series of experiments dealing with the flow of water through glass tubes, that fluids behave differently in motion depending upon the interaction between the inertial and viscous forces. By equating the two sets of forces he arrived at a dimensionless number which is now known as the Reynolds Number (NR) and is defined as:

$$\mathrm{NR} = \frac{\rho l V}{\mu} \tag{2.19}$$

where l is a characteristic length and V is a characteristic velocity.

Reynolds found that, regardless of pipe size or fluid viscosity, flow commenced changing from the laminar to the turbulent state at around NR = 12 000 and that, returning from the turbulent to the laminar state, the inertial effects persisted and the full transition took place only when the NR had dropped to 2000. The same effect is demonstrated by the observation that higher velocities are required for particle entrainment than for particle transportation, all other conditions being the same.

Applied to flow in an open channel, the Reynolds Number may be found by substituting the hydraulic radius of the channel (area of cross-section of channel divided by the wetted perimeter) for the characteristic length l. According to Allen (1965) a critical value of about NR = 500 is generally accepted as marking the transition from laminar to turbulent flow in rough open channels such as natural streams possess.

Lift forces

Lift forces derive from differences in pressure between the top and bottom of a particle in a flowing stream and act in the direction of the reduced pressure. Opinions differ upon the values and effects of such forces, but theoretically they exist and probably, in some circumstances, they have a direct influence on particle entrainment and its subsequent behaviour.

Solids and fluids in relative motion

Mechanisms of free and hindered settling that determine the behaviour of solids in fluids depend upon the resistance exerted by all fluids to the movement of solids and particle properties of size, shape and density. Exact determinations are possible only under completely homogeneous conditions wherein the resistance to movement is a function of the velocity of free fall (terminal velocity) of a particle in an infinite fluid.

The terminal velocity

The terminal velocity of a particle in free fall is reached when the resistance forces are equal in amplitude and opposite in sign to the gravitational forces. At this stage the acceleration due to gravity is zero and the particle continues to fall at constant velocity. The corollary can be stated that a body at rest or in straight-line motion at constant velocity is in equilibrium with the flow when the total of all of the forces and their moments in any direction is constant. Terminal velocities are of particular importance in all applications of solids–fluid movement both in nature and in minerals processing. In all cases, there are fundamental differences in the settling of fine and coarse particles (Fig. 2.12).

Fine particle settling

Under the limiting conditions of free fall in a fluid, very small particles fall slowly at low values of NR and if the fall velocity is low enough for flow

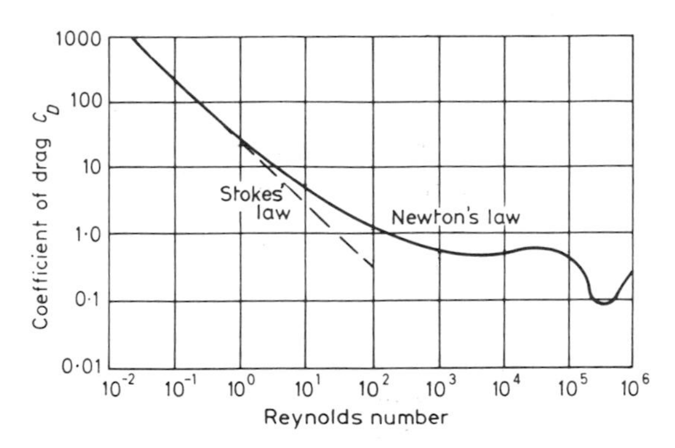

Fig. 2.12 Drag coefficient C_D *versus NR for sphere settling in water*

conditions to be laminar the resistance to motion is defined by Stokes' Law which states that, for a perfect sphere, the resistance to motion R is:

$$R = 6\pi\mu r V \tag{2.20}$$

where r is the radius of the sphere. It must be emphasized in this regard that Stokes' Law holds good only for very small particles and experimentally it has been shown that for perfect spheres of quartz (RD = 2.65) settling freely in clear water the critical upper size limit is a diameter of about 62.5 μm. The terminal velocity is given by the expression for laminar settling:

$$V_{\text{lam}} = C_{\text{D}} \frac{(\rho' - \rho) g d^2}{\mu} \tag{2.21}$$

where C_{D} is the coefficient of drag, ρ' and ρ the densities of the solid and fluid and d the diameter of the sphere.

Stokes' Law applies closely to slimes settling and to the thickening of fine particles in plant residues etc. up to a critical size of about 50 μm for normal sediments (i.e. not perfect spheres.) It applies approximately to most sediments up to 100 μm in size.

Coarse particle settling

Particles larger than about 100 μm in size settle in conditions of turbulence and are accountable to Newton's Law which states that the resistance to motion R of a sphere in a fluid is:

$$R = \frac{\pi}{2} \rho' r^2 V^2 \tag{2.22}$$

The terminal velocity is given by the expression for turbulent settling:

$$V_{\text{turb}} = \left[\frac{4}{3}\left(\frac{\rho' - \rho}{\rho}\right) \cdot \frac{gd}{C_{\text{D}}}\right]^{0.5} \tag{2.23}$$

According to Gaudin (1957) the experimental verification of Newtons Law expressed in equation 2.22 is not satisfactory and the insertion of a coefficient of resistance is needed to make the facts fit a relationship in which the resistance varies as the velocity squared. This coefficient which is also referred to as a coefficient of drag, varies according to differences in particle shape and the velocity of flow.

Shape effects

Shape effects are more important for large than for small particles settling in a fluid because of their higher fall velocities. Sedimentary particles are irregular in shape and at higher velocities tend to orientate themselves with their maximum cross-sections increasingly normal to the line of motion so as to cause the greatest resistance. Eddying effects are dominant and C_D is likewise increasingly a function of NR. Lack of an axis of symmetry results in a wobbling effect that adds to the difficulties of evaluating the effects of shape.

Experimental work seeking to explain shape effects has dealt mainly with solids having axes of symmetry. Brown (1949) calculated the fall velocity of an ellipsoidal particle with axes in the ratios 1:2:4 to be 72% of that for a sphere of the same volume if the shortest axis is parallel to the direction of fall within the Stokes range commencing at NR = 0.1; and 86% and 100% respectively, with the intermediate and longest axes parallel to the direction of fall. Such a particle would tend to fall at high values of NR with the shortest axis parallel to the direction of motion and have a fall velocity of 50% of that for the corresponding sphere at NR = 100 and only 20% at NR = 1000.

Brown (1956) recognized the relationship between shape, as represented by an average volume diameter, and sphericity, and sought a means of comparing the behaviour of a series of geometric shapes in terms of the average screen aperture size d_{avg}, the equivalent diameter d_s and the ratio of specific surfaces n. Thus, in consistant units:

$$\tau = \frac{d_{avg}}{d_s}\left(\frac{1}{n}\right) \qquad (2.24)$$

where $d_s = d$ for a perfect sphere.

Although the relationship was determined only for geometrically true solids it has some application wherever the geometry of irregular particles can be deduced statistically from the results of sizing analyses based upon sieving or sedimentation. The geometrical shapes selected by Brown were those for which there are similarities in nature and hence reasonable approximations for d_s/d_{avg} can usually be interpolated using the data presented in Table 2.12. Care must be taken to select the correct shape for comparison because of the wide range of variables.

Table 2.12 *Sphericity and the value of d_s/d_{avg} related to screen size (after Brown 1956)*

Shape	Sphericity	d_s/d_{avg}
Sphere	1.00	1.00
Octahedron	0.847	0.965
Cube	0.806	1.24
Prisms		
$a \times a \times 2a$	0.767	1.564
$a \times 2a \times 2a$	0.761	0.985
$a \times 2a \times 3a$	0.725	1.127
Cylinders		
$h = 2r$	0.874	1.135
$h = 3r$	0.860	1.31
$h = 10r$	0.691	1.96
$h = 20r$	0.580	2.592
$h = 1.33r$	0.858	1.00
$h = r$	0.827	0.909
Discs		
$h = r/3$	0.594	0.630
$h = r/10$	0.323	0.422
$h = r/15$	0.254	0.368

Size and time effects

The relationship between particle size and time for solids in motion in a fluid is a determining factor in such gravity concentration processes as jigging and air tabling (see Chapter 7). In a fluidized and rapidly pulsating jig bed, particles having higher settling rates have a net downward movement that is enhanced by the repeated reversals of flow along the distances travelled, while the lighter particles have a net upwards movement towards the surface of the bed and are progressively discharged as waste. Small differences in behaviour are accentuated and swarms of mineral particles are stratified roughly according to their individual fall velocities.

The efficiency of sorting is also a function of particle size and time in natural surroundings. Heavy mineral concentrations on beaches and in active fluvial lag gravels (see Chapter 3) are similarly built up in grade over extended periods. However, there is much less control over the operating variables and sorting is much less complete than in the plant.

Leliavsky (1966) offers the following time periods (Table 2.13) for sedimentary particles settling freely in a lake or river, 6.1 m deep, to demonstrate the great variation in settling times between fine and coarse particles.

Table 2.13 Time periods for settling of sedimentary particles in still water

Coarse sand	$d = 1.00$ mm	7 seconds
Fine sand	$d = 0.1$ mm	12 minutes
Silt	$d = 0.01$ mm	20 hours
Clay particle	$d = 0.001$ mm	80 days
Colloids	$d = 0.0001$ mm	20 years

Hydraulic equivalence

It was thought initially that the phenomenon of free fall was an end in itself and Rubey (1933) in describing the size distribution of heavy minerals in a water-laid sandstone ascribed the distribution of individual grains to differences in their fall velocities. He concluded that it was natural for small, dense particles to report with larger, less-dense particles and, with others, developed the concept of hydraulic equivalence or equi-settling of particles having different sizes, densities and shapes.

Pettijohn (1949) measured the hydraulic equivalence of some placer minerals according to the number of Udden size grades between the size of a given mineral particle and the size of the quartz grain with which it was deposited. Some anomalies were noted and the results listed in Table 2.14 showed that, in some cases, the concept of hydraulic equivalence fell short of explaining the resulting pattern of sedimentation.

Hindered settling

In addition to free fall, Rittenhouse (1943) had already recognized the effects of 'varying hydraulic equivalence at the time of deposition and certain unknown factors' amongst which, presumably, were differences in solids concentration (crowding), various impact forces variously directed

Table 2.14 Hydraulic equivalence of some detrital minerals (after Pettijohn, 1949)

Mineral	Density	Hydraulic equivalence
Tourmaline	3.1	0.2
Hornblende	3.2 (variable)	0.2
Apatite	3.2	0.4
Pyroxene	3.4 (variable)	0.3
Hypersthene	3.4	0.4
Sphene	3.5	0.5
Kyanite	3.8	0.3
Garnet	3.8 (variable)	0.6
Zircon	4.6	0.9
Ilmenite	4.7	1.0
Magnetite	5.2	1.0

and the reactions of particles, both singly and in groups, to the different types and states of flow, i.e. hindered settling. Later, Brady and Jobson (1973) were to conclude from their observations of sedimentation along shorelines that sorting variations were normal responses to local hydraulic conditions; and to suggest that fall velocities alone may have little or no effect on the local segregation of heavy detrital minerals in some circumstances. Disturbing influences are mainly due to crowding and to wall effects.

Crowding

It has been demonstrated that the motion of a particle in a fluid is influenced by the presence of other particles through impact and fluid drag and because each particle carries with it some elements of fluid and displaces others whatever path it takes. Worster and Denny (1955) studied the behaviour of concentrations of particles falling in Newtonian conditions and found a general linear reduction in settling rates with increased concentrations to a limiting value of zero. For the test conditions, a rule-of-thumb generalization reduced the free settling velocity for individual particles by about 15% for every 10% of solids content by volume. On this basis both solids and fluids come to rest at around 67% of solids by volume.

Again, empirically, Allen (1970) calculated the terminal velocity of one sphere in a medium containing other settling spheres from the equation:

$$V_s = V_0 (1 - C)^n \tag{2.25}$$

where V_s is the falling velocity in the swarm of spheres, V_0 is its free settling velocity, C is the volumetric concentration of the spheres (occupied space ÷ available space) and n is an exponent that varies from 4.6 in the Stokes range to 2.4 for flow at large Reynolds Numbers. Experimentally and in carefully controlled conditions it has also been shown that, for an average spacing S between the particles, the terminal velocity of a sphere of diameter d in a concentration of similar spheres approximates:

80% of the Stokesian value for $d/S = 0.1$
65% of the Stokesian value for $d/S = 0.16$

Examples such as these offer quantitative data for estimating the effects of crowding using perfect spheres, closely sized and the knowledge gained has been put to good use in the design of process plant dealing with slurries of both sized and unsized particles (see Chapters 5, 6 and 7). Applied to studies of sedimentation in placer environments it has led to a much greater understanding of the mechanics of placer formation and enrichment, thus assisting explorers and prospectors in their tasks of valuation and, miners, in the subsequent exploitation of the deposits found.

Wall effects

Most theoretical analysis considers the behaviour of individual particles in an infinite fluid, i.e. one in which the particles settle freely away from any outside influence. Such analysis presupposes a particle boundary layer that does not make contact with any other boundary layer attached to other particles in the flow or to the solid boundaries, or walls, along which the flow takes place.

Wall effects have been studied experimentally and, according to Brown (1949), a particle falling freely in a cylinder of water has a settling velocity that is reduced by about 2.5% of its Stokesian value for a particle diameter 0.01 times that of the cylinder and by as much as 28% if the ratio is reduced to 0.1 times that of the cylinder.

There are no equally precise data from which to estimate wall effects in natural stream channels and other surroundings. It can only be supposed that such effects exist and that certain features resulting from sedimentation in streams are due to the proximity of some particles to the solid walls, boulders and other flow boundaries during settling. Wall effects are probably of greater interest to the hydraulic process engineer who may find, in them, possible explanations for some apparently anomalous settling in bins.

2.4.3 Sediment transportation and settling

Turning now to natural processes of sedimentation it has already been shown that the finest particles in flowing streams are primarily clay minerals derived from the decomposition of certain feldspars; sands are predominantly silicates and quartz; gravels begin to be recognized as fragments of their parent rocks and boulders have all of the source materials represented. Each class of material consists of a great number of particles differing in size, shape and density; each has a characteristic motion relative to other particles and the surrounding fluid and, for each particle, there is a certain critical velocity below which no movement can take place. Fluid velocity is undoubtedly the main dynamic force but the mechanism through which it acts to move a particle from rest differs according to the types and states of flow, the boundaries over which flow takes place, and swirl.

Fundamental differences in the characteristics of process-response elements in the various conditions of placer formation are reflected in the different natures of the deposits formed. Processes in the continental placer environment are related largely to flow in open channels; aeolean processes are governed by the flow of air across open surfaces; processes in the transitional placer environment are controlled by the action of waves, tides, currents and wind in the shallow waters of the continental shelf and on open beaches. Responses to these processes are characterized by the entrainment, transportation and deposition of sediments under separate fluvial, aeolean and marine conditions. The following treatment deals first with the fluvial and aeolean placer environments; processes in the marine sector of the transitional environment are then dealt with separately.

Fluvial particle entrainment

The relationship between particle size and the velocity of flow is fundamental and a general rule often applied in fluvial environments is to rate the competence of a stream in terms of the largest particle it can move. Stream power has been acknowledged from antiquity and primitive man probably realized that the ability of a stream to move solid objects was in some way related to its velocity. However, according to Leliavsky (1966), it was probably Brahms who, in 1753, first proposed a mathematical solution and his formula equating the critical velocity V_{cr} to the submerged weight W of a particle through a proportionality constant K has survived in a number of forms to this day. The same equation:

$$V_{cr} = KW^{1/6} \tag{2.26}$$

was apparently developed independently by Airy in 1834 and is now generally referred to as the Brahms–Airy Equation.

It seems clear that the law has some validity when applied to large particles free to move but when Lacey (1930) attempted to describe general bed load movement in terms of the sixth power of a critical non-scouring, non-silting velocity and the second power of a nominal size coefficient, Rubey (1937) was to point out that since crowding and other mechanical factors had been ignored the formula was too simplistic for common use. It was Rubey's opinion, also, that the relationship could not apply to particles with radii less than the thickness of the boundary layer. For particles smaller than this, the absolute velocity is less important than the velocity gradient and the resulting pressure difference between the top and bottom of the grains. Machin (1948) suggested that the solid load of a stream varies with the third power of the velocity if all of the sizes are represented; with a higher power if they are all fine grained and a lower power if they are all coarse.

Du Buats (1816) proposed a range of critical entrainment velocities for various sediments that recognized differences in entrainment mechanisms in the smaller sizes (Table 2.15). It is interesting to note that his observations of similar velocities for the erosion of individual beds of sand and potters clay were recorded more than 100 years before Hjulstrom (1935) published critical erosion curves describing similar phenomena, although in greater detail.

Hjulstrom's graph relating critical entrainment velocities to particle size in fluvial environments is illustrated in Fig. 2.13. It shows that progressively higher velocities are needed to entrain particles larger and

Table 2.15 Critical scour velocities (after du Buats, 1816)

Sediment	V_{cr} (cm s^{-1})
Potters clay	10.6
Sand and gravel	
Size of aniseed	10.8
Size of peas	18.9
Size of marsh plant beans	32.5
Coarse sand	21.6
Sea pebbles: about 1″ diam. or more	65.0
Egg-shaped pebbles: in shape of large, sharp-edged flint	120.0

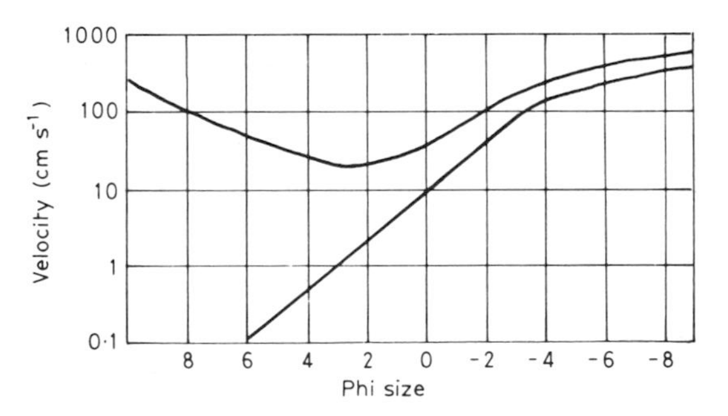

Fig. 2.13 Critical entrainment and erosion velocities in fluvial environments (adapted from Hjustrom, 1935)

smaller than about 125 μm (3 ϕ) and that similar critical velocities will just erode beds composed of 10 ϕ grains and $-5\,\phi$ particles respectively. However, these values ignore cohesion through cementation and the mechanical interlocking of grains; Allen (1970) has noted that such beds continue to resist erosion until velocities are high enough for separation to occur. Fine-grained sediments may then be torn away from the bed in clusters for which the average sizes entrained are governed by factors which defy analytical treatment.

Particles larger than about 2 mm project high enough above the boundary layer to initiate vortices that affect the smaller particles around them. Fine particles are moved and scouring occurs in such circumstances even though the average stream velocities are well below their critical entrainment levels. Swirl is the net result of rotational momentum in the fluid and an eddy so produced has both linear and rotational velocities. The linear velocity is the resultant of components of velocity acting along the axis of the eddy and the horizontal velocity of the stream. The rotational movement is around the axis of the eddy and the resultant of all

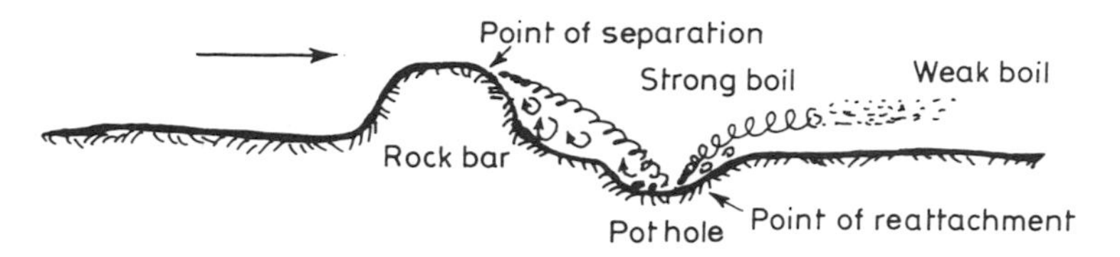

Fig. 2.14 Separation and potholing in stream beds

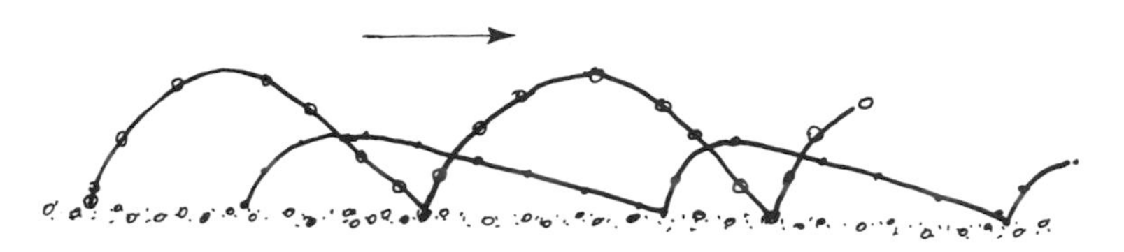

Fig. 2.15 Typical trajectories for large and small particles of wind blown sand

of the forces determines the direction of local movement. The instantaneous velocities resulting from free-shear turbulence are thus greatly in excess of the temporal mean. They result in the formation of spirals and vortices which disturb the pattern of flow and exert very large forces against points of attachment when contact is made with a solid surface. Typical examples illustrated in Fig 2.14 offer partial explanations for scouring in stream beds and for the preferential accumulation of heavy minerals in some 'pothole' situations.

Aeolean particle entrainment

Wind sweeping along the ground surface is turbulent and the entrainment of sand commences when particles roll along the bed, strike other particles and bounce into the air to be carried forward in a flat trajectory only to fall and strike other particles which, in turn, are dislodged from the bed and carried forward. This is the process of saltation and typical trajectories are described in Fig. 2.15.

Aeolean entrainment mechanisms differ from fluvial entrainment mechanisms largely because the two fluids concerned, air and water, have widely different densities and viscosities. The effects of the differences are twofold: because the viscous resistance to movement is lower, the solid particles move faster in air than in water; because the apparent densities of particles are higher and drag forces are lower the maximum size of particles that can be entrained by moving streams of air is smaller. In fact, particles larger than 2 mm diameter are seldom affected by wind pressure or drag and such particles are unlikely to be set in motion by impact forces because only quite small particles are moved in saltation.

Figure 2.16 describes the resultant fluid forces acting on a single particle at the threshold of movement. F_l is a lift force normal to the direction of

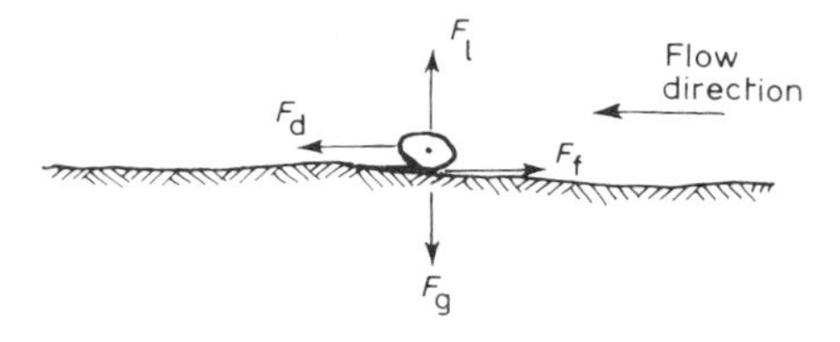

Fig. 2.16 Resultant fluid forces on single particle at threshold of movement opposed by gravitational and frictional forces

flow. F_d is a drag force in the direction of flow, F_f is the frictional force resisting movement and F_g is the gravitational force acting downwards.

Fluvial transportation

The solid load of a stream is the total quantity of sediments transported through a given cross section per unit of time. It comprises a bed load of coarser particles that are pushed and rolled along the bed of the stream or travel in saltation, and a suspended load of the finer particles having settling rates lower than the vertical components of velocity opposing their deposition.

Referring again to Hjulstrom's graph (Fig. 2.14) it is seen that, particularly in the finer sizings, much higher velocities are required for entrainment than for transportation. For example a particle of size 2 ϕ for which the critical velocity of entrainment is around 16 cm s^{-1}, is not deposited until the velocity has been reduced to 1.1 cm s^{-1}.

Goncharov (1938) derived sediment transport equations based on the horizontal forces generated by the currents and critical drag forces due to eddies in the flow. Smirnov (1976) listed critical flow velocities for various sized particles after Goncharov and Velikanov as shown in Table 2.16

Efforts such as this, to develop functions analytically for single particles that might be equally applicable to the movement of populations of particles lead to some difficulties. Important amongst these is that, while

Table 2.16 Critical flow velocities

Average particle diameter (mm)	Velocity (m s $^{-1}$)	Average particle diameter (mm)	Velocity (m s $^{-1}$)
0.10	0.27	15	1.10
0.25	0.31	25	1.20
0.50	0.36	50	1.50
1.00	0.45	75	1.75
2.50	0.65	100	2.00
5.00	0.85	150	2.20
10.00	1.00	200	2.40

the average velocities across a plane can be measured quite readily, the fluids–solids ratios cannot be measured accurately by sampling except for very fine silts in homogeneous suspension. However, Leliavsky (1966), using data from seven different laboratories has derived an empirical equation for critical drag that relates the particle diameter to the critical shear stress as follows:

$$V_{cr} = \frac{500}{3} \times 2a = 166d \qquad (2.27)$$

where d is in millimetres, $a = d/2$ and V_{cr} is in $g\,m^{-2}$. This equation appears to have a general application for the Stokes range of sediments.

According to Bagnold (1971) the physics of the transportation of sand by running water is still not understood because of difficulties of observation and because of the difficulty in separating the effect of the motion of the surrounding water from that of the solid particles. The velocity of the stream is controlled largely by the geometry of the stationary surface and this introduces serious complications because the degree of unevenness itself is a function of the movement of the grains of which the surface is built.

Shape effects

Fall velocities alone do not determine the forward motion of a particle in a flowing stream for which the general order of transportability is in terms of decreasing sphericity. Platy particles and discs are usually the most transportable and these tend to move in erratic saltation because of their high maximum cross-section: weight ratios. Spherical particles roll more easily than other particles and, since their settling velocities are higher, become more closely associated with the bed load than with the more rapidly moving particles in saltation and suspension. Rods and other elongated particles usually outrun spheres in the bed load because, during transport, they both roll and turn end for end, in the process coming into contact with the higher velocity components of flow at a distance from the bed. The flow of sediments in suspension is commonly between 1 and 10% of the total load and the coarsest grains carried in suspension range generally from 45 to 125 μm depending largely on the velocity of flow.

The principal flow regimes are classified broadly in Table 2.17 for most average conditions.

Only those particles for which the flow velocities are above critical values at their points of attachment are moved directly by the stream; not all of the particles are moved at once and some particles move only because of impact from other particles or because they become undermined and topple over. Boulders may shift their positions fractionally but will not move far until broken down by chemical and mechanical forces into finer fragments that are more readily carried away.

Table 2.17 Fluvial sedimentary flow regimes

Particle size	Sediment class	Flow regime
1–16 μm (6ϕ plus)	Very fine clay to very fine silt	Homogeneous flow in suspension
16–250 μm (5ϕ–2ϕ)	Fine silt to fine sand	Homogeneous flow in suspension during flooding. Long-range saltation in normal flow
250–2000 μm (1ϕ– −1ϕ)	Medium to very coarse sand	Particles move largely by rolling and saltation
2000 μm–256 mm (−1ϕ– −8ϕ)	Very coarse sand, gravel and cobbles	Particles move by sliding, rolling and saltation
256 mm plus (−8ϕ)	Boulders	Particles move only by sliding or rolling

Aeolean transportation

Most movement takes place within a few centimetres from the surface and photographs show that grains rise initially with only a small forward velocity compared with what they will attain later (Bagnold, 1971). This is because, regardless of the absolute velocity of the wind, the velocity at a height of 3 mm is always the same and the velocity gradient near the surface is comparatively flat. The bulk of the transport is in the saltation load and movement takes place, for the most part, within a few centimetres of the surface. Mabbutt (1977) has observed that the rate of travel is about half the wind velocity measured at 1 m above the ground and involves, mainly, the fractions between 0.15 and 0.25 mm with an upper limit of 0.5 mm.

Bagnold (1954) used closely sized materials to show that the lowest critical velocities are associated with a particle size of around 80 μm if wind is the only dynamic force (Fig. 2.17). Higher velocities are needed for eroding beds of unconsolidated sediments both smaller and larger than the critical size. However, when impact is also taken into account, the sand grains are entrained at much lower velocities and for impact alone, the relationship between grain size and critical wind velocities is linear throughout.

Fluvial deposition

The forward motion of bed-load sediments is cyclical, particle entrainment being followed by transportation, deposition and re-entrainment when velocities once more approach critical values. But these values will not always be the same, even for particles of the same size and shape, because

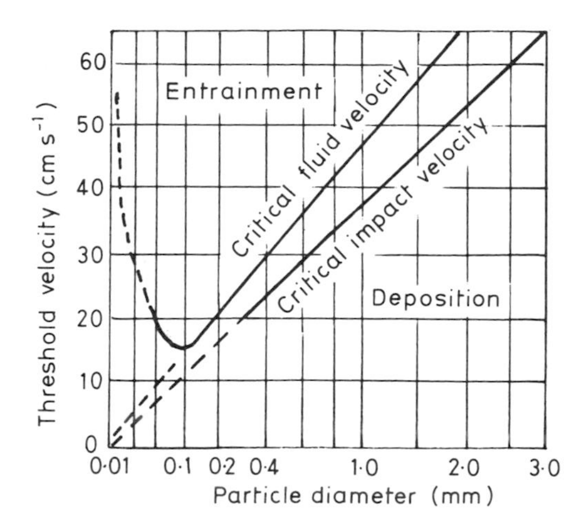

Fig. 2.17 Relationship between particle diameter and critical wind velocities

of differences in orientation that result from differences in stream conditions at the time of deposition. In stable flow, large particles entrained from positions of minimum resistance to movement come to rest in positions of maximum resistance. In flood times particles may be deposited with no preferred orientation as soon as the stream loses its competence. Different parts of the same deposit may also be expected to exhibit different depositional features because of the wide variations in flow conditions, even over the same cross-sections. There are few data from which to predict the relationship between flow conditions and deposition and most formulae and diagrams imply a simplicity for which the only parallels in placer environments may be in those locations where the particles are relatively closely sized.

Sundborg (1968) carried out mineralogical investigations on relative proportions of magnetite, hematite and ilmenite in river sands near the iron-ore mines in Sweden. He concluded that although the ratio was invariably magnetite, hematite, ilmenite; this was not because of any hydrodynamical difference between the three minerals but resulted from differences in their primary ratios in the original rocks and their respective resistances to wear. However, in the relationship between magnetite (5.20 g cm^{-3}), amphiboles (3.30 g cm^{-3}) and quartz (2.65 g cm^{-3}) a displacement towards the finer sizes for denser material was evident in most of the samples, see Fig 2.18. Clearly, however, other environmental factors were also involved and, in Sundborg's investigations, the influence of shape was well recognized; flaky particles of micas and sericite schists having a considerably coarser grain size than the quartz and felspars with which they were associated.

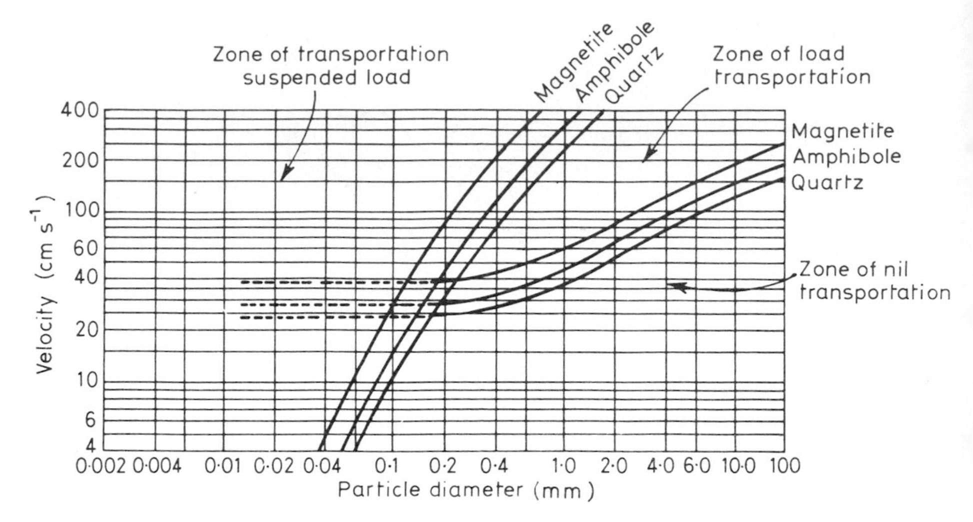

Fig. 2.18 Relationship between water velocity, particle size and state of sediment movement for particles of different densities (quartz and feldspars 2.65 g cm^{-3}; amphiboles 3.30 g cm^{-3}; magnetite 5.20 g cm^{-3}).

Sorting

Obviously, if all sedimentary particles were equally resistant and equally dense, sorting would be by size and shape alone and the relatively small amounts of the valuable minerals released would soon become widely dispersed and pass unnoticed amongst the enormous quantities of waste. In the event, the more granular particles of gold, platinum and cassiterite sink quickly towards bedrock and form concentrations in natural sedimentary traps, or accumulate in the slow-moving lag gravels close to their source.

The lighter heavy minerals, having lower sedimentation rates but higher stabilities, occur as subsidiary minerals in tin and gold placers and form economic concentrations only when subjected to high-energy conditions in an environment in which all of the particles are fine grained and already closely sized. In these circumstances even small density differences may be significant factors in sorting.

The efficiency of sorting improves generally in a downstream direction because of closer particle size relationships, but rich concentrations may be formed wherever the local environment provides high-energy conditions and some form of natural lodgement for the heavier particles.

Aeolean deposition

Particles smaller than 0.05 mm have Stokesian settling characteristics and are carried away in suspension. Particles between 0.25 and 2 mm move by surface creep, largely as the result of impact, and have a forward movement generally less than 1 $cm\,s^{-1}$ compared with a few metres per second in saltation. Such particles come to rest as soon as the wind velocity drops below its critical value for entrainment.

Sorting

Sand grains proceeding by surface creep and saltation undergo a continual gravity sorting operation that acts largely in the horizontal plane in the direction of the wind and depends for the most part on the settling characteristics of the particles in conditions of hindered settling. Since there is no dilatation of the bed there is no tendency for the heavier particles to work their way downward and form concentrations at bedrock. Consequently, in aeolean placers of this type, the heavy mineral concentrates take the form of thin seams or only partly defined beds of mineralized material sandwiched between layers of impoverished sand particles.

2.4.4 Transitional-marine sedimentation

The physics of solids–fluid movement in fluvial and aeolean conditions is equally applicable to studies of sedimentary behaviour in shallow waters along the shorelines. However, there are additional complexities and a knowledge of physical oceanography is needed in order to understand them. Factors of importance are discussed in the following brief notes on waves, tides and currents.

Waves

Waves are generated when any form of energy is transferred to a mass of water. The frictional drag of the wind creates ripples that grow to waves as increasing amounts of energy are received and stored. Gravitational forces exerted by the sun and moon produce the wave forms known as tides. Regional changes in atmospheric pressure create additional wave forms by selectively raising and lowering the surface. Impulsively generated waves (Tsunamis) occur during and after submarine earthquakes, landslides and volcanic eruptions offshore. The most common and also the most important waves involved in placer formation are wind waves and tides.

Wind waves

Wind wave generation is a function of the response of the free surface of water to turbulent shear stresses and to the inter-relationship between wave growth and changing wind conditions due to changing wave configurations. It is expressed physically by pressure differences between the windward and leeward aspects of the waves, relationships that are characteristically complex because of the gusty nature of the winds and the influence of other waves. Energy is also transferred from the impact of flying spray from 'white caps', particularly in stormy weather.

Wind waves reach maximum values governed largely by the strength and duration of the winds and the distance (fetch) over which they act. Stewart (1969) concluded from his observations that the tangential stress on surfaces increases at least with the square of the wind velocity and that, for example, 5 h of 60 knot winds will put more momentum into the waves than a week of 5 knot winds. Darbyshire (1956) found that waves reach their peaks in fetches of little more than 100 miles for a wide range of wind velocities and do not increase above a certain maximum value, no matter how long the winds blow.

Waves of oscillation

The passage of a wave imparts an orbital motion to particles of water so that they move forward and down from the crest and then backward in a nearly circular orbit without appreciable change of position. Called waves of oscillation (Fig. 2.19) they have a height H measured as the vertical distance from trough to crest, a horizontal length L from one crest to the next and a period T which is the time taken for two successive crests to pass a given point. It follows that the velocity V is given by:

$$V = LT^{-1} \tag{2.28}$$

and, in deep water, where the depth d is large compared with L it can be shown that wave lengths and velocities are connected by the expression:

$$V = \left(\frac{gL}{2\pi}\right)^{0.5} \tag{2.29}$$

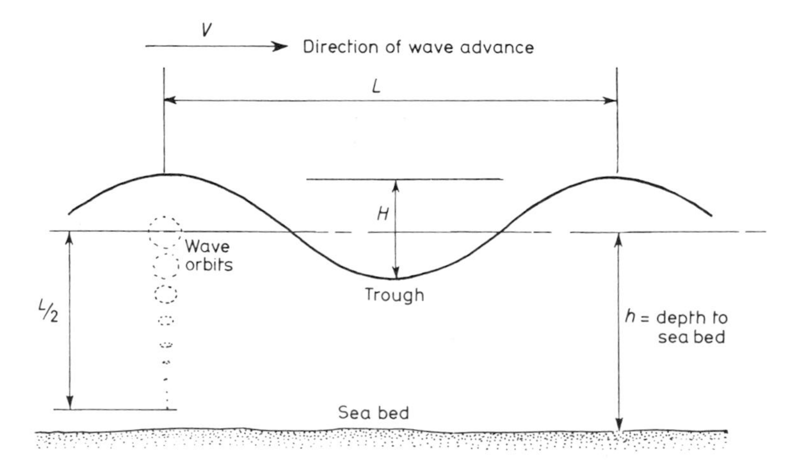

Fig. 2.19 Wave characteristics – waves of oscillation in deep water

Wave heights, although governed by the dimensions L, V and T cannot be related to air flow by any known formula although Holmes (1969) refers to a rule of thumb estimate by mariners in which limiting height in feet is equal to about one-half the wind velocity in miles per hour.

Wave forms collapse, either by plunging or spilling when the ratio $h:d$ is approximately 1.3:1 for long-period waves at the outer edge of the surf zone, or 1.7:1 approximately for short-period steep waves. At break point, the water at the crest of the wave has the same velocity as the wave and a massive transformation of energy takes place as waves of oscillation become waves of translation and potential energy is converted to energy of motion.

Plunging waves

Plunging waves discharge their energy violently on steeply sloping bottoms or in shallows which suddenly impede the movement at the interface and cause the front of the wave to curl over and collapse. One form of plunging wave described by Shepard (1973) has a hollow concave upward front and results from long swells moving onto a gently sloping beach. Plunging waves entrain sand abruptly into short-lived zones of turbulence and deposit it as quickly to form sand bars on either side of the breaker line. They tend to act only on the lower part of the foreshore and are the most erosive of the wave types.

Spilling waves

Depending upon the ratio between open-sea steepness and wave length, spilling waves may be produced either on steep or gently sloping beaches but are most common on the latter. When surging takes place on steeply sloping beaches the wave does not actually break but simply rushes forward. Spilling waves are generally responsible for beach building, particularly when combined with strong offshore winds that promote bottom currents towards the land. However, when the waves strike at an acute angle to the beach, they may also produce a scouring effect.

Changes in form take place in shallow waters as soon as the wave orbits come into contact with the bottom and their motion is impeded. The components of velocity opposite to the direction of the waves are retarded most. This has the effect of flattening the orbits and changing their form from open circular to open eliptical in the horizontal plane, thereby increasing the net flow of water particles towards the shore. Wave velocity becomes increasingly dependent upon depth and at depths less than 25 m, according to Turekian (1976), equation 2.29 no longer applies and, instead, V is given by:

$$V = (2gh)^{0.5} \tag{2.30}$$

where h is the depth of water.

Observations at the Scripps Institute Pier have suggested that orbital velocities follow closely the form of the wave and are nearly always greater under the wave crest where they are directed towards the land than are the seawards velocities under the trough.

Wave reflection and refraction
Some waves strike directly against cliff faces without breaking and these are reflected at their angles of incidence. However, most waves are subject to some degree of refraction as soon as the bottom is felt. Friction is the retarding force and wave energy is concentrated against any obstructions in more or less direct proportion to the frictional resistance. The consequences of this are twofold, (a) the wave crests move more rapidly and tend to swing around parallel to the bottom contours, and (b) longshore currents are developed away from the headlands towards the bay. Figure 2.20 describes a number of typical wave refraction patterns. The orthogonals represent the distribution of energy along the coastline and explain how the various currents are formed.

Although the effects are similar, waves generated by winds in coastal waters have somewhat different characteristics to those in deep water because their fetch is shorter and additional influences result from the geometry of the seabed and the physiography of the coastline. The waves

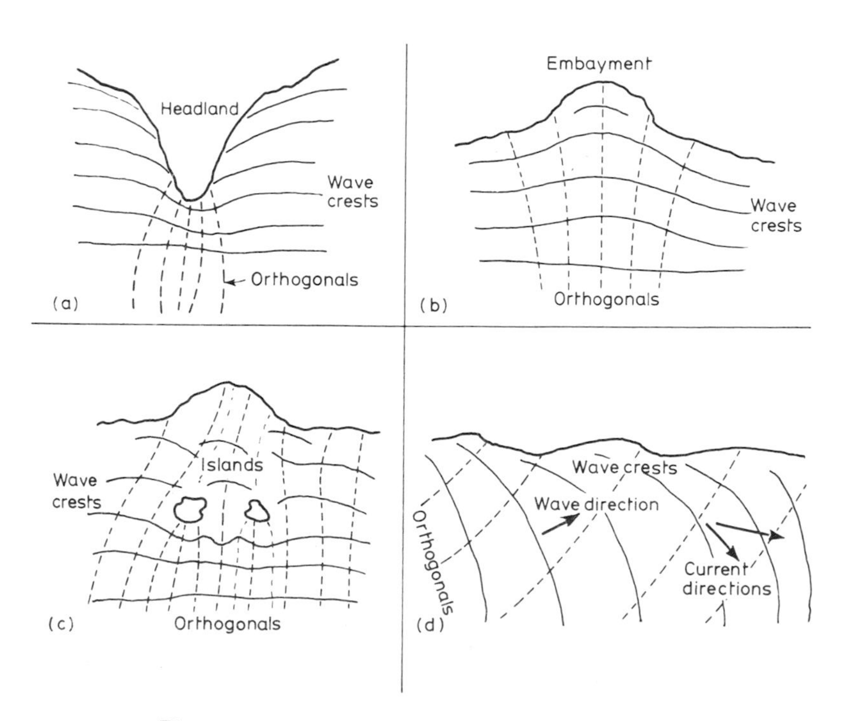

Fig. 2.20 Diagrammatic wave refraction patterns

are shorter, steeper and less regular than the ocean swells on which they become superimposed. They either damp the swell down if the winds are blowing offshore or add to its dimensions if they are from the sea.

The main effects of wind waves in the transitional placer environment are:

(1) To supply the energy for promoting and sustaining the currents that move sediments about in the shallow coastal waters and to provide a constant interchange of sediments between the land and the sea.
(2) To actively sort the beach materials according to differences in size and density. Shape differences are relatively unimportant in this environment.

Tides

Tides are very long waves of length about one-half the circumference of the earth and a semi-diurnal tidal cycle of 24 h and 50 min. Two high and two low tides occur each day and, because the solar day is 24 h and the moon circles the earth every 24 h and 50 min, the tides occur 50 min later each day.

Tides depend upon changing interplanetary configurations but are complicated by the presence of irregular land masses, shoals and shallow platforms. The tide-generating forces are due to the gravitational effects of the sun and moon and, of these, the moon is dominant having more than twice the attractive power of the sun. Spring tides are generated when the sun and moon are aligned on the same side of the earth; 'neap' tides occur when they are on opposite sides. Enormous quantities of water are set into motion both towards and away from the land, but in ever changing amounts because of such complexities as the rotation of the earth and sun about their respective axes and because of the elliptical orbits of the earth and the moon.

Tidal amplitudes are also affected by strong winds that pile water up in one direction or another, thus causing tides that are higher or lower than would have resulted from astronomical forces alone. The irregular shape of some ocean basins is another disrupting factor and Rogers and Adams (1966) refer to broad, flat coasts on the open sea where tidal ranges are only 1–2 ft; whereas the funnel-like structure of the Bay of Fundy in Nova Scotia causes water to pile up towards its head and permits tidal movements of up to 20 m.

Currents

Ocean currents affecting the movement of solids in shallow offshore waters may result from windstress on the surface of the water, wave reflections and refraction, tidal movements and density differences due to the differential entrainment of sediments, particularly in the surf zone.

The principal currents so formed are, (a) currents of mass transport acting towards the shore, (b) longshore currents which act parallel to the shore, and (c) rip and density currents acting away from the shore.

Currents of mass transport

Currents of mass transport result from the slight but positive advance of open-ended wave orbits along the sea bed and are the principal means by which sediments beyond the breaker line are moved towards the shore. In relatively calm shallow depths, their motion is determined by the wave characteristics, the gradient of the sea bed and its texture, rippled or smooth.

Studies of orbital velocities and their effects on sediments along the shallow ocean bed show that currents of mass transport act seawards for particles in suspension and towards the land for particles in contact with the bed. Oscillating currents of $10\,\mathrm{cm\,s^{-1}}$ are sufficient to move fine particles along the bed and while substantial movement appears limited to depths less than 9 m, very long waves may create currents of mass transport at depths of as much as 30 m. Much has still to be learned about the onshore movement of sand grains from beyond the breaker line but wave heights are important factors as shown by King (1966) in experiments carried out in a wave tank at Cambridge University (Fig. 2.21).

The onshore movement of sediments is also clearly related to the gradient of the sea bed. Along any coastline sector, the onshore movement of sand ceases above a certain angle although flatter slopes allow freer movement because the horizontal components of velocity are greater. In field observations in the Geographe Bay System of Western Australia, the influence of both particle size and density are clear and the richest heavy mineral accumulations onshore appear to be associated with shelf gradients around 1:200. Along the same section of the coastline, gradients steeper than 1:50 have apparently inhibited the onshore movement of even the lighter heavy minerals although large supplies of the larger and less-dense sand particles have been moved on shore.

General movement commences, and smooth beds of sand on the shallow sea floor become rippled, when orbital velocities achieve the critical values at which particles of sand are first set into motion by rolling. With time, the ripples increase in height as they move towards the land up to a maximum size governed by particle size and water depth. Ripples can only form when the flow is turbulent but once formed they tend to promote turbulence. At twice the critical speed vortexing occurs and increased amounts of sand are carried forward in saltation.

Onshore movement is generally favoured by flat waves acting over rippled flat gradients at shallow depths. The form of the wave governs both the rate and direction of the current, and above a certain critical wave steepness the direction of maximum transport is away from the shore.

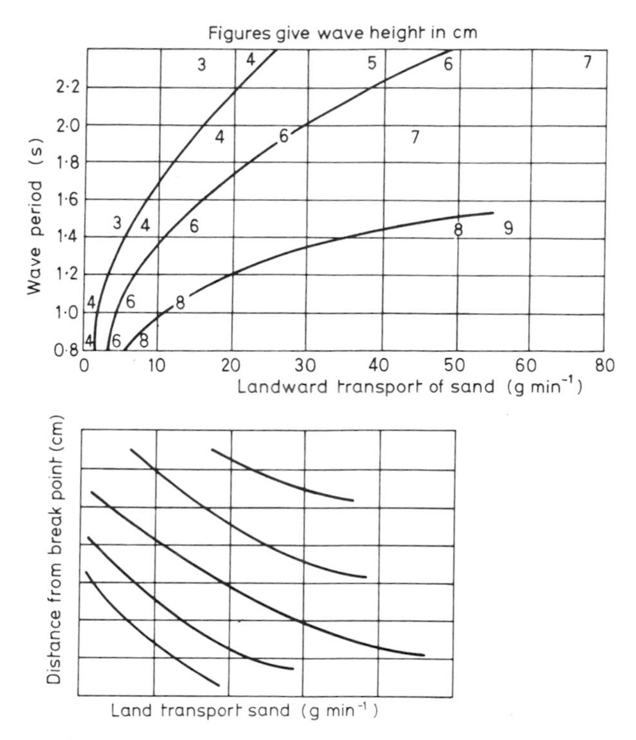

Fig. 2.21 Relationship between wave characteristics and volume of sand transported towards the land

Rip currents

Rip currents are common features of most coastlines and when concentrated into a few channels may contain a great deal of energy. They form, usually, from a build up of water between an offshore bar and the land. This may result from a heavy swell or strong onshore winds and must be compensated for by an equal volume flowing seawards. For example, if a 20-m wide channel drains a 200-m wide beach sector the rip flow obviously attains a high velocity and, as a current, is uni-directional from the surface to the bottom. Shepard points out however, that the bottom current does not extend much beyond the surf zone, even though the surface components might persist for as much as a kilometer from the shore. Consequently, rip currents carrying sediments away from the beaches usually dump most of them just beyond the breaker line from whence they may once more be moved towards the shore by currents of mass transport or along the coastline by longshore currents. Rip currents are generally destructive to shorelines but are an integral part of the mechanism of longshore movement (Fig. 2.22).

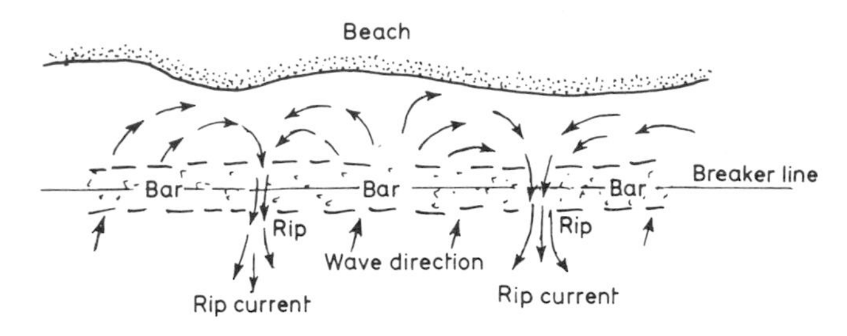

Fig. 2.22 Rip currents developed at low points along offshore bar

Density currents
Density currents are caused by the entrainment of sediments in the surf zone into slurries of sand and water having higher densities than that of clear salt water. Since the pressure gradient is seawards from the breaker line the direction of current flow is also seawards. However, this does not necessarily result in a loss of sand because, as King has observed, in 98 % of the experiments carried out in the wave tank at Cambridge University, the net movement of sand outside of the breaker line is towards the shore. The few exceptions were involved with very steep waves and in his experiments only small quantities of sediments were affected.

On the other hand, the same mechanism is responsible for the large-scale transportation of sand away from the sections of the continental shelf where, under the force of gravity, steep gradients such as those of submarine canyons, may promote very rapid movement. In this form they are known as 'turbidity' currents.

Longshore currents
Waves acting at an angle to the beach set up currents inside the breaker line with components of force in the longshore direction. Longshore currents also develop from the refraction which follows wave convergence between canyons; divergence at canyon heads (Shepard and Inman, 1950) and from the circulation of major eddies offshore. The cumulative effects of these processes together with the modifications from rip and other bottom currents produce longshore currents and foster littoral transportation along coastlines.

Along some coastlines, where the prevailing winds blow from opposite quarters from one season to the next, the direction of the longshore currents may change accordingly. Nevertheless, each coastline is normally influenced by one predominant flow pattern as shown by the development of sand spits in the direction of flow and from accumulations of heavy minerals at the down current ends of beaches as described in Chapter 3.

Sediment movement predominates in the swash zone of the foreshore when flat waves are acting. Movement follows a zig-zag pattern along the axis of the foreshore and optimum conditions for transportation are found when the waves strike at an angle of 30° to the beach. With steep waves most of the sand movement is in the surf zone, partly in suspension and partly in traction.

Table 2.18 Principal variables affecting the transitional placer environment

	Beach profile		
Variable	Eroding	Building up	Remarks
Waves	Steep and acting at an angle to the beach	Flat acting normal to the beach	
Tides	High – particularly in conjunction with steep waves and longshore currents	Low to normal	
Currents	Rip currents: These act perpendicular to and flow seawards from the shore Of currents due to wind: Onshore winds cause surface currents which move towards the shore and equilibrium is maintained by return currents (undertow) near the sea bottom which flow away from the shore Density currents: Differences in sediment concentration between adjacent layers of water cause a pressure drop seawards of the breaker line.	Currents of mass transport: Near the bottom these are always in the sense of propagation of the wave Offshore currents due to wind: Surface currents acting away from the shore are balanced by currents near the bottom acting towards the shore	Longshore currents: These may help in the erosion of one beach by transporting material away from and contribute to the build-up of another by supplying material to that beach.
Winds	Onshore winds cause waves to steepen and become destructive	Offshore winds flatten seas and lead to build up	Both offshore and onshore winds may erode beaches by the removal of the sand in the direction of the wind.

2.4.5 Summary

It is difficult to generalize on the effect of any one variable in the transitional placer environment and present theory is inadequate to explain all of the observed effects. In both ocean and atmosphere, process and response elements interact and every action affects some other action so that the two are indivisible parts of a whole. Nevertheless the first step towards understanding the complexities is to study each individual process separately. Table 2.18 summarizes the principal factors affecting shorelines and hence the formation and stability of transitional placer deposits.

REFERENCES

Airy, W. (1834) *Proc. Inst. Civ. Eng.*, p. 227.
Allen, J. R. L. (1965) A review of the origin and characteristics of recent alluvial sediments, *Sedimentology*, **5**, 89–191.
Allen, J. R. L. (1970) *Physical Processes of Sedimentation*, George Allen and Unwin, London.
Allen, J. R. L. (1978) Studies in Fluvial Sedimentation; a comparison of fining upwards cyclotherms, with special reference to coarse-member composition and interpretation, *J. Sed. Pet.*, **40,** 298–323.
Bagnold, R. A. (1953, reprinted 1973) *The Physics of Blown Sands and Desert Dunes*, Methuen, London.
Baker, George (1962) *Detrital Heavy Minerals in Natural Accumulates*, AIMM Monogram No. 1, Melbourne.
Berkman, D. A. (1976) *Field Geologists Manual*, AIMM Monograph No. 9, Melbourne.
Blaise, R. A. and Carlier, P. A. (1968) Applications of Geostatistics in Ore Evaluation, in *Ore Reserve Estimation and Grade Control*, (eds Gill, J. E., Blaise, R. A. and Haw, V. A.) Can. Inst. Min. Met. Special Vol No. 9, pp. 41–48.
Brady, L. L. and Jobson H. E. (1973) *An Experimental Study of Heavy-Mineral Segregation Under Alluvial Flow Conditions*, U.S. Government Printing Office, Washington.
Brahms, A. (1753) *Anfangsgrunde der Deich – U Wasserbaukunst.*
Briggs, David (1977) *Sediments*, Butterworth, London.
Brown, C. B. (1949) Sediment transportation, in *Engineering Hydraulics*, (ed. Hunter Rouse), John Wiley & Sons, New York pp. 769–857.
Brown, G. G. (1956) *Unit Operations*, John Wiley, New York.
Cailleux, A. (1945) Distinction des galets marines et fluviatiles, *Bull. Soc. Geol. France* xv, 5, 375–404.
Cailleux, A. and Tricant, J. (1963) *Initiation a L'etude des Sables et les Galets*, Centre de Documentation, Universitaire, Paris.
Darbyshire, J. (1956) An investigation into the generation of waves when the fetch of the wind is less than 100 miles, *QFR Met. Soc.* **82,** 461–8.
Douglas, I. (1977) *Humid Landforms. An Introduction to Systematic Geomorphology*, Vol. 1. Australian National University Press, Canberra.
Du Buats, L. G. (1816) *Principles d'Hydraulique*, Paris.

Emery, K. O. (1938) Rapid method of mechanical analysis of sands, *Journ. Sed. Petrology*, **8,** 105–111.
Emery, K. O. and Noakes, L. J. (1968) *Economic Placer Deposits of the Continental Shelf*, ECAFE Technical Bulletin Vol. 1, CCOP, Bangkok.
Emmons, W. H. (1917) The enrichment of ore deposits, *US Geol. Surv. Bull*, **625,** 305–324.
Gaudin, A. M. (1957) *Principles of Mineral Dressing*, Tate-McGraw Hill, New Delhi.
Goldich, S. S. (1938) A study in rock weathering, *Journ. Geol.* **46,** 17–58.
Goncharov, V. N. (1938) *Movement of Sediments in a Steady Stream*, ONTI, Moscow.
Gordon, Wolman M. (1978) Alluvium, in *Encyclopaedia of Earth Sciences Series: The Encyclopaedia of Sedimentology* (eds Rhodes W. Fairbridge and Joanne Bourgeois), Dowden, Hutchinson and Ross, Pennsylvania.
Griggs, D. T. (1936) The factor of fatigue in rock exfoliation, *Journ. Geol.* **44,** 781–6.
Hjulstrom, F. B. (1935) Studies of the morphological activities of rivers as illustrated by the River Fyris, *Bull of Geol. Inst*, Upsalla, **25,** 227–527.
Herdan (1960) *Small Particle Statistics*, Elsevier, New York.
Holmes, Arthur (1969) *Principles of Physical Geology*, Thomas Nelson & Sons, London.
King, Cocklaine A. M. (1966) *Beaches and Coasts*, Edward Arnold, London.
King, Cocklaine A. M. (1975) *Techniques in Geomorphology*, Edward Arnold, London.
King, H. W., Wisler, C. D. and Woodburn, J. G. (1948) *Hydraulics*, John Wiley, New York.
Krumbein, W. C. (1941) Measurement and geological significance of shape and roundness of sedimentary particles, *Journ. Sed. Pet.* **11,** 64–72.
Krumbein, W. C. and Sloss L. L. (1963) *Stratigraphy and Sedimentation*, W. H. Freeman, San Francisco.
Kuenan, P. H. (1958) Some Experiments on fluvial rounding, *Koninkle/Ned. Ahad, Wetenck, Proc. Sev. B.*, Amsterdam, **61** 1.
Lacey, G. (1930) Stable channels in alluvium, *Min. Proc. Inst. Civ. Eng.*, **229,** 271.
Leliavsky, Serge (1966) *An Introduction to Fluvial Hydraulics*, Dover Publications, New York.
Lepp, Henry (1973) *Dynamic Earth*, McGraw Hill, New York.
Linkholm, A. A. (1968) Occurrence mining and recovery of diamonds, *Mining and Minerals Engineering*, August.
Mabbutt, J. A. (1977) Desert landforms, in *An Introduction to Systematic Geomorphology*, Vol. 2, MIT Press, Cambridge, Mass.
Macdonald, Eoin H. (1973) *Manual of Beach Mining Practice*, 2nd edn, Department of Foreign Affairs, Australian Government Publishing Service, Canberra.
Machin, J. H. (1948) The concept of the graded stream, *Bull. Geol. Soc. Amer.* **59,** 463–512.
Milner, H. B. (1962) *Sedimentary Petrography*, Allen and Unwin, London.
Minter, W. E. L. and Toens, P. D. (1970) Experimental simulation of gold deposition in gravel beds, *Trans. Geol. Soc. South Africa*, 89–98.
Moss, A. J. (1962–3) The physical nature of common sandy and pebbly deposits. Parts 1 and 2, *Am. Journ. Science*, **260,** and **261,** 1962–3.
Pettijohn, F. J. (1949) *Sedimentary Rocks*, Harper & Bros, New York.
Powers, M. C. (1953) A New Roundness Scale for Sedimentary Particles, *Journ. Sed. Petrol.* **23,** 117–19.
Reinbech, H. E. and Singh, I. B. (1975) *Depositional Sedimentary Environments*, Springer, Berlin.

Rittenhouse, G. (1943) The transportation and deposition of heavy minerals, *Bull. Geol. Soc. Amer.*, **54,** 1725–80.
Rogers, John J. W. and Adams, John A. G. (1966) *Fundamentals in Geology*, Harper International, New York.
Royal, A. G. and Hosgit, E. (1974) Local estimation of sand and gravel reserves by geostatistical methods, *Trans. Inst. Min. Met.* (*Sect A*) *Min Industry* **83,** A53–62.
Rubey, W. W. (1933) The size distributions of heavy minerals within a waterlaid sandstone, *Journ. Sed. Petrol*, **3,** 3–29.
Rubey, W. W. (1937) The force required to move particles in a stream bed, *US Geol. Surv. Prof Paper* 189E, 121–40.
Ruhe, R. V. and Cady, J. G. (1967) The relation of Pleistocene geology and soils between Bently and Adair in southwestern Iowa, *US Dept Agric. Tech. Bull.* **1349.**
Russel, R. J. (1939) Effects of transportation on sedimentary particles, in *Recent Marine Sediments* (ed. P. D. Trask). *Amer. Assoc. Petrol. Geol. Bull. No. 8*, 32–47.
Samad, Abdus (1977) *Sediments of the River Tamar*, PhD thesis.
Scott, W. B. (1932) *An Introduction to Geology*, Macmillan, New York.
Shepard, Francis P. (1973) *Submarine Geology*, Harper & Row, New York.
Shepard, F. P. and Inman, D. L. (1950) Nearshore circulation related to bottom topography and wave refraction, *Trans. Amer. Geol. Un.* **31** 196–212.
Shields, A. (1936) *Anwendung Der Achalichheitsmechanik und der Turbulenzforschung auf die Geschiebebewegung*, Mitteilungen der preuss, Versuchsanst. F. Wasserbau. U. Schiffbau, Berlin.
Smirnov, V. I. (1976) *Geology of Mineral Deposits*, MIR Publishers, Moscow.
Sparkes, M. A. (1960) *Geomorphology*, Longmans, Green & Co., London.
Stewart, R. W. (1969) The atmosphere and the ocean, *Scientific American* **221,** 3.
Sundborg, A. K. E. (1968) Some aspects of fluvial sediments and fluvial morphology, *Geographiska Annaler*, **50,** A.3, 121–35.
Tanner, W. F. (1959) Sample components obtained by the method of differences, *Journ. Sed. Pet.* **29,** 408–11.
Turekian, Karl, K. (1976) *Oceans*, Prentice Hall, Yale, New Jersey.
Udden, J. A. (1898) *Mechanical Composition of Wind Deposits*, Public No. 1, Augustana Library.
Waddell, H. (1932) Volume, shape and roundness of quartz particles, *Journ. Geol.* **43,** 250–80.
Wells, John H. (1969) Placer examination, principles and practice, *Tech. Bull. No. 4*, US Dept Int. Bur. Land Management, Washington DC.
Wentworth, C. (1922) A scale of grade and class terms for clastic sediments, *Journ. Geol.*, **30,** 377–92.
Williams, M. A. J. (1968) Termites and soil development near Brocks Creek, Northern Territory, *Aust. Journ. Science*, **31,** 153–4.
Worster, R. C. and Denny, D. F. (1955) Hydraulic transport of solid material in pipes, *Proc. Inst. Mech. Eng.* **169,** 32, 563.

3

The Geology of Placers and Their Formation

3.1 INTRODUCTION

Placer deposits are specific types of alluvial land forms containing concentrations of sand, gravel, metallic minerals and gemstones. They result from sedimentary processes that have persisted for long periods of time and a close reconstruction of their geomorphic histories is a prerequisite to understanding them in their present forms. Placers are best known in Recent to Pleistocene sediments and are less known in the older compacted series due to lack of knowledge and to the greater difficulties of seeking them out. Attention has already been drawn to the relevant process-response elements in Chapters 1 and 2 and to differences in the physical and chemical properties of the various placer minerals: these factors are now related to the provenances of placers and are used in describing the various placer types.

3.2 THE PROVENANCES AND RELATED PLACER MINERALS

The word 'provenance' is derived from the French 'provenir' meaning to originate and come forth. In this chapter the provenances encompass the source areas from which the sediments derive, the actual source rocks being distinguished broadly between two classes of primary mineralization. In one, intrusive granitic massifs mobilize crustal elements associated with continental land masses; in the other, the thinner oceanic crust is intruded and covered by outpourings of basic lavas and other volcanics. Further subdivisions are concerned with differences in the composition and characteristics of rocks of the same general type, differences in the levels of emplacement or present erosional levels of the intrusive bodies and differences in the ages of the rocks concerned.

Placer gold is derived mainly from the weathering of quartz or calcite reefs and stringers within igneous and metamorphic rocks, although rich concentrations are not necessarily indicative of a central rich source. For example, the Raigarh Goldfield of Madhya Pradesh, India, covers an area of about 1000 km^2; it has yielded large quantities of placer gold for many hundreds of years with no historical record of underground mining and recent investigations to discover economically viable reef deposits have been no more successful. Rich alluvial gold concentrations at Edie Creek and Bulolo in New Guinea were also probably derived from widespread stringers rather than from single primary bodies although none have been found. On the other hand, in other placer mining districts such as the famous old gold diggings at Ballarat, Victoria, Australia, many of the richest deposits were directly traceable to distinct lines of lode. Generally the most important deposits of gold and other placer minerals have been formed from weathering over a large area of provenance and an abundance of mineralized veinlets, widely dispersed (Fig. 3.1).

Evans (1981) has proposed an alternative provenance for some placer gold deposits suggesting that gold, in small amounts but evenly distributed throughout ultrabasic rocks, may be chemically dissolved and reprecipitated in possibly commercial placer concentrations by the normal

Fig. 3.1 Cluster of auriferous veinlets worked by tribal Indians in Nilambur Valley Goldfield, Kerala, India

process of lateritization. He cites as possible examples a French Guiana gold placer where the only possible source appeared to be the thick, red, lateritic soil cut by the stream channel; and some of the gold deposits in the higher Tertiary channels of California's Mother Lode country and on the west slope of the Sierra Nevada. A similar environment to that of French Guiana, in Suriname, provides another possible example at Royal Hill where the auriferous laterites, totalling some 5 million tonnes, have given rise to extensive placer workings. Evans makes reference to an article by Dr Landsweert (1869) who wrote that

> Everyone concurs in the belief that alluvial gold has been derived at some time or other from lodes: but seeing that the largest piece of gold ever found in the matrix is insignificant when compared with the nuggets that have sometimes been found in the alluvium, it has been a difficult matter to reconcile belief with experience . . . The occurrence of larger nuggets in gravel deposits than have been found in quartz lodes with the fact that alluvial gold almost universally has a higher standard of fineness, would seem to imply a different origin for the two.

The platinoids, chromite, ilmenite and magnetite occur in basic and ultrabasic intrusives and lavas, chiefly in peridotites, gabbros, pyroxenes and norites. Zircon, monazite and rutile are common minor constituents of the more acid igneous rocks. Tin and tungsten minerals are developed in a variety of deposits in granodiorites and granites as well as in association with gold of hydrothermal origin, high temperature greissen and pegmatite minerals. Tin derives from the more acid varieties of these source rocks as does wolframite, ferberite and hubnerite of the tungsten minerals. Such rocks normally carry some bi-product ilmenite, zircon and monazite. Scheelite, another tungsten mineral, is associated with the less acid granodiorites and is found also in contact metamorphic zones. Titaniferous magnetite is a notable constituent of andesitic rocks of Caenozoic island arcs; the purer magnetites are characteristic of magmatic segregations or of metamorphic facies which also constitute the main source rocks for other durable minerals such as garnets, quartz, staurolite, kyanite, rutile and some gemstones.

Pegmatites are source rocks for gemstones, tin, tungsten, tantalite–columbite minerals, rare earth minerals and a host of radioactive compounds. Typically associated with late phases of granitic intrusives, they are normally emplaced in the roof of the intrusive or in the adjacent metamorphic aureoles and may be mineralogically simple or highly complex. The most favourable types for tantalum group minerals are pegmatites with a wide development of albite-muscovite and albite-lepidolite assemblages to which the highest Ta:Nb ratios are normally related. Mineral zoning is revealed in some pegmatite bodies with different zones varying widely in mineral composition. This is typical of many of the rare metal pegmatites; Rao (1975) has described one such pegmatite in Bihar, India (see Fig. 3.2).

Carbonatite deposits are distinct from other groups in terms of both the geological and physico-chemical environments of their formation. Their closest link is with the diamond deposits in kimberlites; they are usually developed within ancient stable continental blocks or cratons and probably derive from injections from the upper mantle. Carbonatites have been recognized as important source rocks for certain valuable detrital minerals only in the past few decades, possibly because they are comparatively rare. The valuable mineral carbonatite types are divided into three main classes; (a) apatite–magnetite, (b) rare metal–rare earth, and (c) phlogopite types which are distinguished by concentrations of mica in the outer margins of carbonatite stocks. The most commonly known of the derived placer minerals are rutile, anatase, ilmenite, columbite, magnetite, apatite and several niobium minerals.

Table 3.1 lists some of the more common provenances and the economic minerals that derive from them.

3.2.1 Time-related provenances

Noakes (1976) considered provenance on a time scale and pointed out that, in south-east Asia, large-scale concentrations of rutile, zircon, monazite

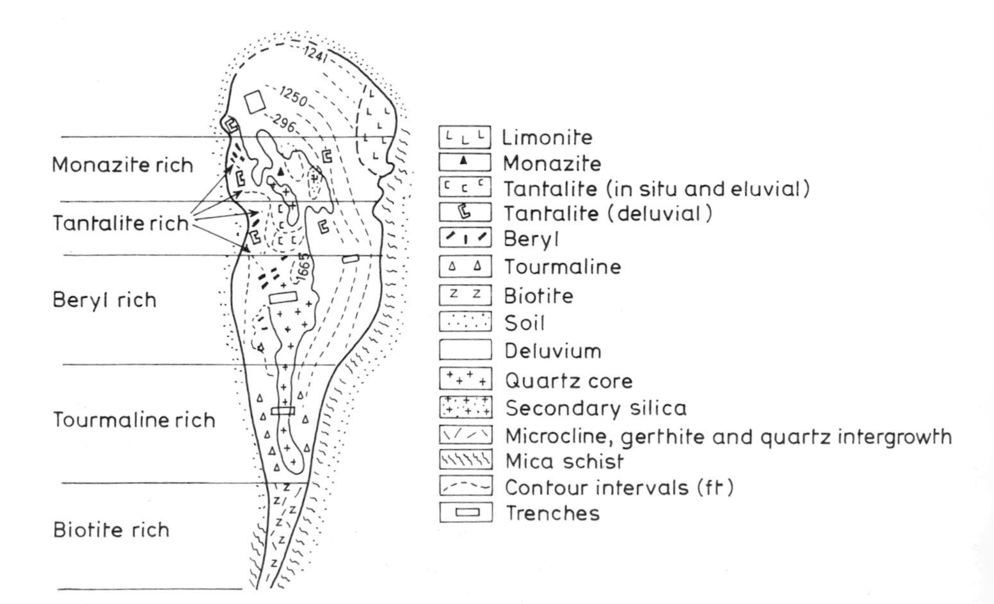

Fig. 3.2 Geological map of the Tantalite Pegmatite, Kharridih, Bihar, India, illustrating the pattern of zoning (adapted from Rao, 1975)

Table 3.1 Provenances and related minerals

Provenance	Economic mineral	Associated minerals
Ultramafic and mafic terrains including pyroxenites and norites	The platinoids	Olivine, enstatite, basic plagioclase, chromite, titano-magnetite, ilmenite, chrome spinel, diopsidic augite
Granitoid terrains and related pegmatites and greisens	Cassiterite, monazite, zircon, rutile, gold	Wolframite, potash feldspar, quartz, topaz, beryl, spodumene, petalite, tourmaline, tantalite, columbite, monazite, fluorite, sphene
Plateau basalts	Magnetite, ilmenite	Pyroboles, basic plagioclase, apatite
Syenitic rocks and related pegmatites	Zircon, rare earth minerals including uranium- and thorium-bearing minerals	Ilmenite, magnetite, fluorite, pyroboles, potash feldspar, apatite, feldspathoids, zircon
Contact metamorphic aureoles – skarns	Scheelite, rutile, occasionally corundum	Diopside, grossularite, wollastonite, calcite, basic plagioclase, epidote
Kimberlites	Diamonds	Ilmenite, magnetite, pyrope garnet, pyroxene including diopside, kyanite, sphene, apatite
High-grade metamorphic terrain	Gold, rutile, zircon, gemstones	Kyanite, pyroboles, quartz, sillimanite, almandine garnet, feldspars, apatite
Serpentine belts	Platinoids, chromite, magnetite	Chrome garnet, pyroxenes, olivine
Carbonatites including associated rare basic igneous rocks	Rutile, ilmenite, magnetite, rare-earth minerals, uranium, niobium, thorium and zirconium minerals	Potash feldspars, calcite, pyroboles, garnets, apatite

and high titania ilmenite derive principally from pre-Mesozoic rocks, mainly Precambrian complexes. Cassiterite and its associated minerals have for their provenance granite rocks principally of Mesozoic age, while deposits of titaniferous magnetite are provided mainly by andesitic rocks associated with island arcs of the Tertiary and Quaternary eras. He related the age of source rocks in south-east Asia to broad elements of provenance as shown in Table 3.2.

3.2.2 Tectonically-related provenances

The provenances are also related to the earth's deformational movements. For example, it is becoming increasingly evident that the geological setting for kimberlite pipe emplacements was primarily a period relatively unaffected by ancient orogenic activity, particularly where there was very little horizontal folding. The pipes generally penetrated very old metamorphic rocks and the environment is typified by the kimberlite zone of South Africa which is confined between folded Permian rocks to the south and the Lebombo folded rocks to the northeast. Similar cratonic environments contain provenances for gold in greenstone belts and for platinum in basic and ultrabasic intrusives.

Ophiolites indicate older island arc emplacement. Those of Papua New Guinea shed osmiridium, platinum and chromite but only minor gold whereas the adjacent metamorphic terrain of the Owen Stanley Range sheds gold over wide areas, some placers being of economic grade.

It is noteworthy that all of the major strandline concentrations along the coasts of eastern and western Australia, and in India, Sri Lanka, South Africa and Florida contain valuable heavy minerals derived from mainly Precambrian shield areas and that the recycling of more stable minerals through younger sediments has played an important role in transport and final concentration. For such minerals the provenances must be exceedingly large because only small proportions of the original minerals content of source rocks find their way eventually onto the beaches where they can be concentrated in conditions of high energy. Precambrian rocks are not necessarily better source rocks for rutile, ilmenite, zircon and other beach minerals than are many other rock types. However, the stability of the Precambrian shield areas over long periods of time ensured favourable combinations of sufficiently large provenances and sustained periods of weathering and transport, such conditions being essential for major concentrations to be developed from source rocks in which the valuable minerals occur only as minor constituents. By contrast, large concentrations of titaniferous magnetite in New Zealand and in parts of south-east Asia have been derived from relatively small provenances during short periods of time; titaniferous magnetite is found in dominant concentrations in Cainozoic areas.

Plate margins are associated with lines of active volcanoes along island arcs and with volcanic activity in the continental areas. Noakes (1976) refers to studies in which the content of TiO_2 is used to distinguish between oceanic and continental volcanics or between circum-oceanic volcanics, i.e. those lying landward of the troughs, and oceanic volcanics situated seawards. According to Chayes (1964), oceanic volcanics have a considerably higher content of TiO_2 than their landward equivalents. Hess and Poldervaart (1967) assert that oceanic rocks of basaltic composition average 2.6% TiO_2 against 1.6% TiO_2 for those of continental origin. In discussing these and similar studies, Noakes suggested that it was tempting to link the prevalence of titaniferous magnetite in Cainozoic rocks with processes of plate tectonics in comparatively recent times, and to accept that tectonic styles have changed and evolved over geological time, but that much more research was needed.

3.2.3 Regional aspects of provenance and placer minerals' distribution

In placers, only the more stable minerals survive and, of these, the valuable types persist to varying degrees. Any of the valuable minerals may be found in eluvial or colluvial concentrations but more chemically vulnerable or brittle minerals such as wolframite, and scheelite seldom persist beyond the slopes except where they are wholly or largely enclosed in a protective gangue. Cassiterite and tantalite survive in fluvial conditions for a few kilometers and then they, too, disappear. Highly stable and ductile minerals tend to flatten under impact rather than shatter, and particles of gold or platinum are more transportable in that form than many other less dense but more equant minerals. Because of their great resistance to weathering and abrasion, the lighter heavy minerals rutile, ilmenite, zircon and monazite form economic concentrations on beaches and back-dune areas hundreds of kilometers from their source. Of the gemstones, diamonds may be found in any placer sub-environment because of their extreme hardness and durability. Rubies and sapphires are slightly less resistant but they, too, are found in most fluvial conditions as well as on the slopes. Emeralds disintegrate rapidly because of their many flaws and internal structure. Figure 3.3 illustrates the general order of disappearance of the various continental placer minerals away from their source rocks, and the appearance along shorelines of concentrations of some of the more resistant types.

None of these placer minerals is distributed uniformly throughout the world although there is some evidence to suggest a pattern of distribution in Metallogenic Provinces. Schulling (1967) plotted all known major and minor tin occurrences around the Atlantic Ocean to describe a series of tin belts that may have been joined together in a virtually unbroken line

Table 3.2 *Broad elements of provenance of detrital heavy minerals in south east asia* (*derived from Noakes 1976*)

Age	Source Rocks	Minerals	Location	Remarks
Precambrian	Complexes including granitic rocks, gneisses and schists	Ilmenite, zircon, monazite and minor rutile	Korea and Vietnam	Precambrian province extends northwards into China
Precambrian to Palaeozoic	Complexes including granite rocks, gneisses and schists	Zircon, ilmenite, monazite and rutile	Prachaub, Kiri khan, Gulf of Thailand	Predominantly zircon
Palaeozoic (principally)	Including lower Palaeozoic and Devonian low-grade metamorphics intrude by Hercynian (lower Carboniferous) granite rocks	Possibly contributes to mineral suites in South Vietnam – see above	Vietnam north of Hue and south of Nga Tranz	Provenance needs further investigation
Principally Mesozoic	Granitic intrusives	No record of heavy mineral occurrences in primary rocks	Coastline of south China extending down into Vietnam	Perhaps a recycling through Mesozoic sediments

Late Carboniferous to Triassic	(a) Coarse grained biotite granites (b) Fine to medium grained, two mica granites (c) Adamelites	Cassiterite with subordinate ilmenite, leucoxene, zircon, monazite, xenotime and tanto-columbite	Tin belt of south-east Asia from north Burma to tin islands of Indonesia	For cassiterite, (a) most favourable (b) less favourable (c) least favourable, regardless of age
Palaeozoic	Andesites and basalts to meta-sediments and granites	Titaniferous magnetite, ilmenite, leucoxene and traces of zircon and rutile	Western Sumatra	
Mesozoic Tertiary rocks and Quaternary volcanics	Caenozoic sediments, volcanics and intermediate intrusives with Mesozoic sediments, granitic and basic rocks in places. Quaternary volcanics range from acid to	Titano-magnetites	Philippines Japan Indonesia Taiwan	Quarternary volcanics generally higher Ti content than older rocks. Prevalence of deposits in south-east Asia region linked with island arc development
	basic ultra-basics (Mesozoic)	Chromite	Southern Sulawesi	

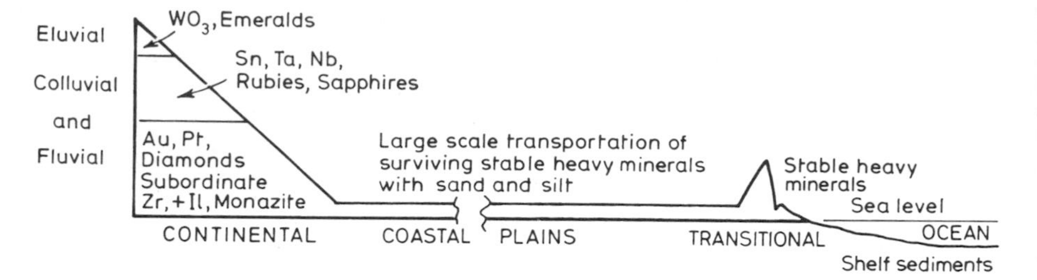

Fig. 3.3 Order of disappearance of characteristic placer minerals with distance from source and reappearance of concentrations of most stable types in transitional setting

before the continents broke apart. He warned, however, that with the perspectives of smaller scales some of the linear trends are less apparent and require careful analysis before assuming theoretical significance.

Liddy (1972) related all known mineralization in the South Pacific to a structural pattern governed primarily by the continental mass of Asia and its Sunda Shelf; the Australian continent and its Sahul Shelf; and island arcs, the locations of which were determined by the two continents. However, Taylor (1976) noted that although the major south-east Asian tin zone is normally regarded as a single province extending some 3000 km from Thailand to Indonesia, it could equally be regarded as a series of adjacent provinces, particularly in Malaysia where eastern and western tin zones are geographically related but display different characteristics.

Emery (1966) described the regional distribution of exposed igneous and associated metamorphic rocks (Fig. 3.4) to show that, with a few excep-

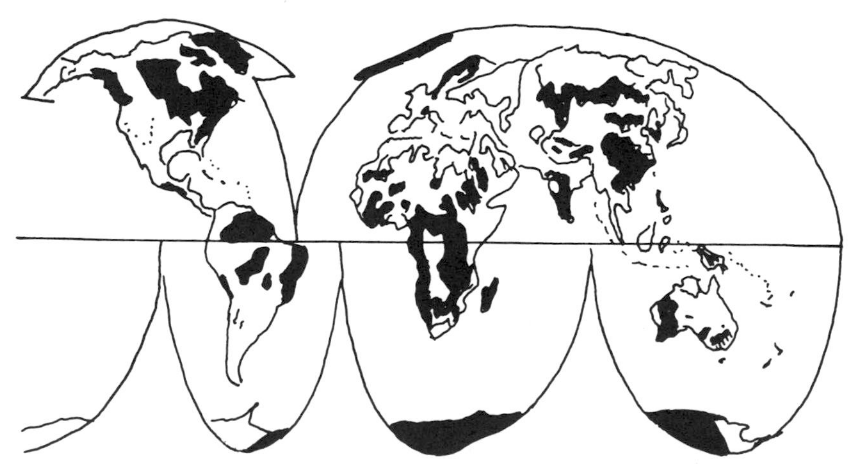

Fig. 3.4 Simplified geological map of the continents. Solid black denotes areas of exposed igneous and associated metamorphic rocks (adapted from Emery, 1966)

tions, primary deposits occur inland and only rarely are freshly eroded sediments containing valuable placer minerals discharged directly into the sea. Because of this, most heavier metal, and gemstone placers occur in fluvial environments and, generally, only the lighter heavy and more resistant of the valuable minerals, e.g. rutile, ilmenite, zircon etc., are found in economic concentrations on the beaches. Notable exceptions are the gold deposits formed along both present day and ancient beaches at Nome, Alaska, from source rocks and glacial debris located along the shoreline; and placers derived from the tin granites of south-east Asia where they extend out into the sea.

Linking the minerals in a placer unit to their respective source areas requires a full knowledge of the contributions made to the sediment flow by tributaries entering into the main stream and of the environments through which they have passed. While it is usually possible to identify specific features and to deduce the mode and approximate distance travelled, the source rocks may be widely distributed and the lithology might do little more than reflect the composition of a great many different rocks in a large drainage basin. For example, the valuable heavy minerals found along the coastline of Bangladesh between Chittagong and Teknaf have, as their immediate source rocks, the soft undifferentiated Tertiary sandstones of the Lushai and Chittagong Hill Tracts. These probably derived largely from the granites and gneisses of the Shillong Plateau and Mikir Hills in Assam, but vast quantities of sediments are also brought down from the Himalayas and studies of provenance have not yet been carried far enough for definite conclusions to be reached.

3.2.4 Connecting provenance with minerals characteristics

Variations in minerals composition and texture are usually related to the composition of the source rocks. For example, the percentage of thorium in monazite increases with increasing metamorphic grade of the host rocks (Overstreet, 1967). Hafnium may replace zirconium in zircon to the extent of several per cent depending upon the local supply of hafnium. The two are closely related, chemically and are invariably found together, although in different proportions particularly in the older geological deposits. The mineral hafnon ($HfSiO_4$) is found in Madagascar closely associated with zircon in Precambrian pegmatites and one occurrence of zircon from Krager in Norway, given the variety name alvite, contains over 4% hafnium.

The Pacific island arc magnetites have a high TiO_2 content (7–12%) that makes them unsuitable as prime feed for conventional blast furnaces, although they can be used in steel making by the more costly electric arc furnace methods. In Japan, steel makers blend some high titaniferous magnetite with their blast furnace feed but this constitutes the only market in the world and is currently restricted to 4 million tonnes per annum. Magnetite concentrations derived from igneous magmas usually contain less TiO_2 and may be acceptable for blast-furnace steel making; however, such materials are generally too low in density and magnetic susceptibility for coal-washing purposes. Contact deposits provide the required provenance for most washery grades of magnetite. In Australia the most important deposits have been formed in metasomatically replaced country rock, often limestone, by solutions originating from intrusives such as granodiorites.

In the case of ilmenite used for the production of pigment by the orthodox sulphate route, emphasis is placed upon a high TiO_2 content ($>50\%$) and an absence or very low content of deleterious trace elements like chromium. Direct chlorination processes and some of the newer upgrading processes are more tolerant of trace elements but a high TiO_2 content is still important. Important factors relating to the formation of high TiO_2 ilmenite are rate of cooling within magmas and the availability of oxygen. The relatively rapid cooling of lavas in volcanics or in minor near-surface intrusives, tends to produce ilmenite with inclusions of hematite and magnetite within the crystal lattice and hence a lower TiO_2 content, generally between 38 and 45%. The main source for high TiO_2 ilmenite appears to be major intrusives, either acid or basic, in which slow cooling is evident. Ilmenite grains lacking ex-solution features may derive from the coarser parts of thick flow sequences or from basic intrusives and Archaean granites.

High Ta:Nb ratios are typical of sub-alkaline granites. Pleonaste is always present in the sapphire gravels of the New England district and elsewhere in Australia (MacNevin, 1972). Alamandine garnets are most common in schists, gneisses and granulites and are rare in granitic and contact rocks.

Authigenic minerals

As distinct from detrital minerals, some of the lighter heavy minerals in placers may be precipitated authigenically from percolating solutions. In some cases small crystals already present in the sediments grow into larger crystals or aggregates of crystals; in others, new minerals form due to reactions between the original constituents of the sediments and fresh materials brought into the environment. Baker (1962) lists dolomite in Australian sediments as being almost wholly authigenic; occasional chlorites; sometimes tourmaline, anatase, fluorite, rutile, brookite, pyrite, zircon (overgrowths); rarely garnet, staurolite, zoisite, apatite and sphene. Generally however, the above minerals are more readily linked with specific groups of igneous rocks in associated pneumatolytic environments, and only occasionally with paragenetic processing.

3.3 PLACER FORMATION

Most placer minerals are sparsely developed as accessory types within their parent bodies and become of value only when released by weathering and subsequently concentrated in a favourable setting. They originate from a wide variety of primary source rocks ranging from simple quartz veins to complexes of magmatic and/or metamorphic origin and are also recycled through sedimentary rocks such as sandstones, conglomerates and tillites which, in places, provide important secondary sources. Economic concentrations occur wherever a catchment area of source rocks has yielded sufficient quantities of the valuable minerals and where physiography and climate have provided suitable conditions for deposition. The three main classifications for placers are Continental, Transitional and Marine.

3.3.1 Continental placers

The continental placer setting

Continental placers are formed within the constraints of ill-defined morphological systems in which the supply of valuable mineral particles commences with the surface weathering and erosion of the source rocks and ends when the particles have been so reduced in size during transport that they cannot be recovered economically by any known means. Common features of the environment are high unit values for the minerals, high waste:product ratios and a wide range of particle size, shape and distribution. Particle sizes for recoverable metallic and gemstone minerals range from a few microns up to coarse nuggets and crystals; waste materials are composed of mixtures of colloidal matter, clay, sand, gravel and boulders.

Selective deposition of the higher density, more valuable minerals, is confined very largely to the immediate and close environs of mineral-bearing rocks in areas of provenance. The lighter, less valuable 'amang' minerals rutile, zircon, ilmenite, monazite and xenotime travel further and may be concentrated in economic proportions in different environments. However, they also occur as minor constituents in continental placer accumulations and are normally separated out individually in the dressing sheds (see Chapter 7) in order to maximize recoveries of the high-value products.

The first stage in the concentration of placer minerals by natural processes takes place at the source and may, in appropriate circumstances, result in the formation of an eluvial placer.

Eluvial placers

Eluvial placers, sometimes referred to as residual placers, are weathered rock zones in which the parent materials have undergone chemical, mechanical and biological changes resulting in the loss of some of their non-valuable constituents by leaching and elutriation. Elements such as sodium, potassium, calcium and magnesium have high relative solubilities and are leached out preferentially to iron, manganese, aluminium and titanium. However, most minerals containing iron and manganese also break down reasonably quickly and the feldspars of granitic rocks are readily converted into soluble salts plus insoluble residues of clays, micas and other decomposition products. Percolating waters carry away the more soluble materials; fine particles are removed from the surface by sheet flow and rivulets or through the action of wind.

The most favourable conditions for formation are elevated ground surfaces where local depressions cover the mineralized zones. Favourable also are level surfaces that are covered by dense vegetation in regions of heavy sub-tropical rainfall. Rainwater penetrates more readily to the lower levels and massive surface run-off is reduced in rain forests and similar environments, where a profusion of tree and plant roots and other biological agents have shattered the near surface rocks. The weathering profile shows a progressive decrease in alteration with depth and a shape that is determined by the direction of the ground-water movement. Figure 3.5 illustrates a typical eluvial weathering pattern.

The surface of the soil zone is enriched in most eluvial placers partly because the chemical reactions are at an end and the soluble materials have been removed and partly because, through constant exposure to wind and rain, many of the less-dense particles of insoluble matter have also been carried away. Surface enrichment by deflation is exponential as described in Fig. 3.6.

Beneath the soil zone, the thickness of the weathering zone is determined by the level of the water table and the dimensions of the deep zone of water movement above the unweathered rock. The continued movement of ground water is necessary to sustain chemical actions and eluviation because otherwise a stagnation point is soon reached when no further reactions or removals can take place.

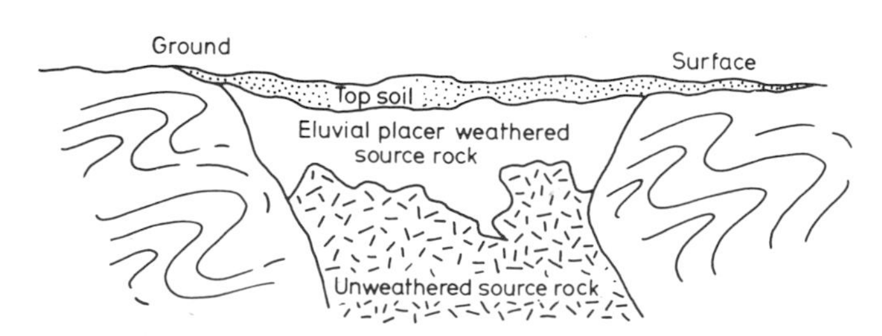

Fig. 3.5 Diagrammatic section across typical eluvial placer

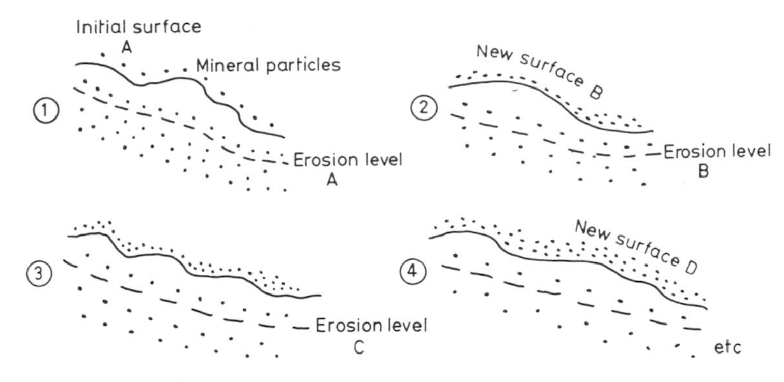

Fig. 3.6 Exponential pattern of surface enrichment

A zone of laterization is common in humid tropical regions dominated by ferric oxide and more or less hardened by exposure to the atmosphere. Some laterites, particularly those developed in the Tertiary period, have been found to contain particles of gold and, as already noted, opinions vary on whether the particles are relic in the laterized material from earlier erosion cycles or have been dissolved out of adjacent source rocks and re-precipitated along with the aluminium and iron oxides.

Eluvial deposits are generally limited in extent although there are notable exceptions such as in the weathering of diamondiferous kimberlite pipes *in situ*. The surface geometry approximates that of the primary body but becomes decreasingly regular with depth when additional constraints affect the movement of the percolating solutions. Changes in the mineralogy of the ore body within the weathering profile follow a similar pattern but although the profile is enriched the spatial distribution of the valuable minerals is virtually unchanged. Exceptions are found where the ground subsequently collapses due to the removal of material from under a resistant layer, and where particles of gold are dissolved and reprecipitated in secondary zones of enrichment at the water table.

Important examples of eluvial placers are found at Greenbushes in Western Australia (tantalite–tin); Round Mountain, Nevada, USA (gold); Pemali, Bangka Island, Indonesia (tin). Smernov (1976) refers to platinum eluvials in the Urals, cassiterite–columbite placers in the Transbaikal area of USSR and the baddeleyite placers of Brazil.

Colluvial placers

As distinct from the eluvial sub-environment where most movement takes place in the vertical plane, colluvial sub-environments are concerned with the mass movements of rock waste on slopes leading away from exposed source rocks. A number of transport mechanisms are involved: rainwater effects the movement of surface particles by impact when it splashes onto the soil and, by forming rivulets and sheet flow, washes the lighter materials away. Frost heave lifts sedimentary grains normal to the inclination of the ground and drops them vertically downslope when the frost melts. Masses of surface sediments may be washed down to form badly sorted cones of clastic materials at the bases of steep slopes during periods of high precipitation.

In addition to the circulation of water through pores and cracks in the weathering zone the water taken in by colluvial sediments moves downwards along seepage planes and fills the voids between particles to provide lubrication for the mass to move as a whole. The rate of movement is variable and sometimes abrupt depending upon the angle of the slope, the nature of the sediments and the intensity of the precipitation. The process is controlled by gravity and by the ability of the sediments to withstand the shearing stresses imposed upon them. In the general case, a reduction in mean particle size away from an outcrop is accompanied by an increase in clay content and, because of its increased binding effect, the thickness of the colluvium increases downslope. Depth is also a function of the removal of the colluvium at the downstream end. Movement is by sliding, mudflows and surface creep.

Sliding

Sliding takes place when mainly dry and inherently weak materials are able to move freely and rapidly down steep slopes. It may follow the removal of some downstream barrier holding the material in place or, where beds of material overlie less competent beds, these may fail when percolating waters lubricate slip planes and allow the whole bed to slide Sliding also occurs on precipitous slopes due to over loading by snow and ice, earthquakes, animals, sudden temperature changes and so on.

The very rapid movement of sliding debris is characterized by the generally equal distribution of velocity from top to bottom with, perhaps, a slightly higher velocity at the interface than at the top. It is a movement which is more conducive to mixing than to sorting and is typical of the screes and talus which accumulate at the foot of slopes.

Mudflows

As distinct from slides, mudflows occur when the internal friction of a mass of mud, earth and rock debris is reduced by the ingestion of excessive amounts of water. Mudflows are common on steep slopes in tropic regions where rock decay is rapid and the rainfall is torrential during monsoons and other wet periods. The movement may be as rapid as 10–15 $km\,h^{-1}$ and, in this state, 'solifluction' is the term used to describe the flow.

Surface creep

The motion on very gentle slopes is by creep and the division between eluvial and colluvial sedimentation is not always distinguished readily from surface features. In dry conditions on gentle slopes, surface creep is essentially slow and may be almost imperceptible except when observed over a long period of time. Gravitational forces are dominant at all times and animals traversing the slopes, trees swaying and winds blowing all exert some components of force which act downwards to assist the general movement.

Distribution of values

Lode formations on the sides of hills are usually more stable than the surrounding rocks which weather faster and to greater depths. Consequently, in addition to discrete mineral particles, colluvial sediments contain whole sections of rock fragments that break away and merge with the loosely packed spoil. Being heavy, they tend to work their way downwards towards bedrock but the sorting process is seldom well developed because of the lack of fluidity needed to allow the particles to move freely. Slope lengths and, hence, the alluvial trains are usually quite short although there are notable exceptions in some of the diamond areas of Africa and the drowned cassiterite deposits of the Sunda Shelf, in Indonesian waters.

Sedimentary patterns on steep slopes are more observable in respect of boundaries separating zones of provenance from those of transportation because the movements are more rapid; however, the flow paths are irregularly disposed. Ruhe (1975) describes nine basic slope geometrics resulting from three possible shapes: (straight, convex and concave) both laterally and in the vertical plane. The slope complexities affect both the distribution of energy governing individual rates and directions of movement, and also contribute to variations in the minerals distribution.

The intensity of energy on a straight even slope is the same across each horizontal plane and in ideal conditions, the path taken by the denser particles is generally as described in Fig. 3.7, the normally accepted two dimensional concept of colluvial placer formation. This supposes a slow downward settling of the larger and denser particles towards bedrock but, in fact, such simplistic differentiation is rarely seen. Normally the flow is

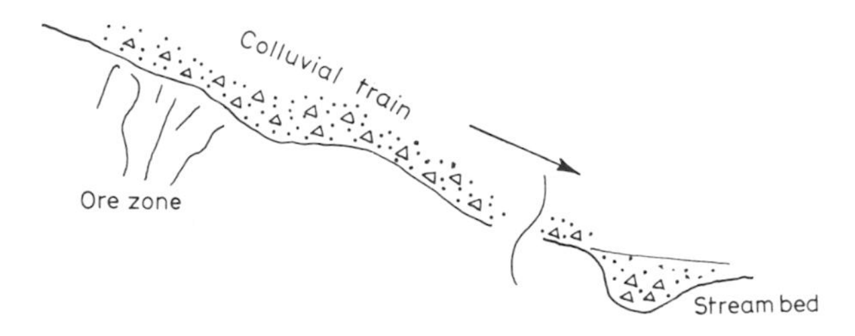

Fig. 3.7 Two-dimensional concept of colluvial transportation and sorting

three dimensional and sorting relationships are governed by slope lengths, the manner in which individual mass flows split or converge and by the differential fluidity that follows upon the uneven distribution of seepage and run off.

Orders of complexity involve both convex and concave cylindrical surfaces and the flow paths have components acting in the direction of the trough or away from the crest according to the curvature (Fig. 3.8). Sections across most colluvial placers conform locally to one or both of these configurations, however, many of the more extensive colluvial deposits occur on hill slopes that are doubly curved and on these the pattern of flow takes a variety of forms (Fig. 3.9).

Common to all slopes are changes in lithology due largely to chemical alteration and to contamination from barren materials weathered out of the bedrock. There are two distinctly different cases wherein:

(1) Colluvium from a single source exhibits a general falling off in grade downslope due to dilution and there need be no direct relationship between downslope values and the size and grade of the lode from which the values are shed.
(2) Colluvium from bedrock material that is mineralized, even sparsely, provides a tendency for increasing grades downslope. Rich colluvial placers are not necessarily related to any primary body of economic

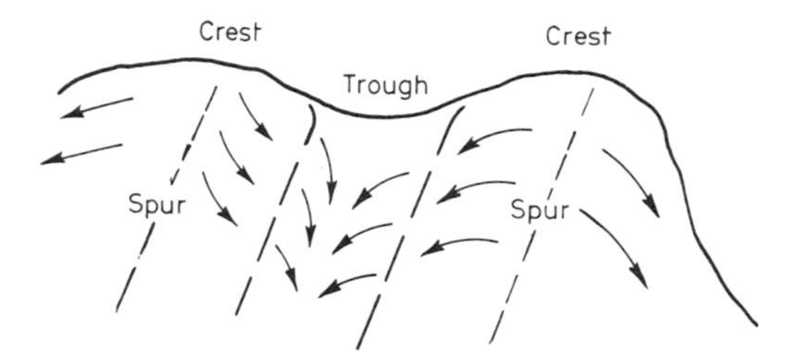

Fig. 3.8 Flow paths of colluvium towards troughs and away from crests and spurs

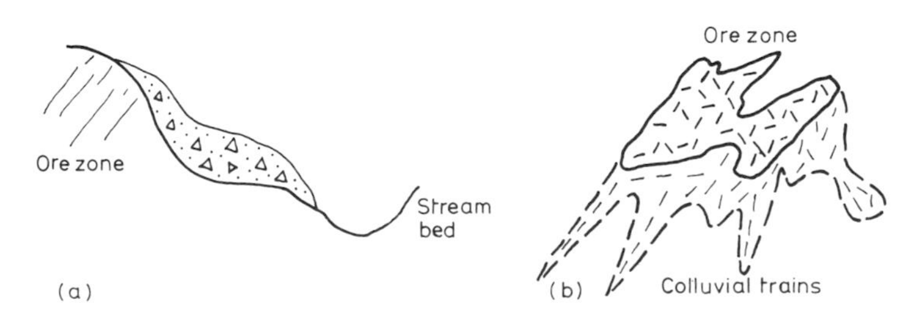

Fig. 3.9 Typical examples of colluvial trains on doubly curved slopes: (a) shows thickening where slope flattens, (b) shows tongues or trains of alluvium along steeper troughs

value and many of the cassiterite placers along the great Tin Belt of south-east Asia, as well as many extensive gold placers in India and in other parts of the world serve as prime examples.

Colluvial placers are seldom economical to mine except on a small scale (see Fig. 3.10); their main value is in providing an intermediate transfer mechanism between outcrops of source rocks, where the environment processes are mainly chemical, and the fluvial sub-environment in which the dominant forces are hydro-dynamic. The colluvial stage is sometimes eliminated when stream beds follow closely along the strike of the ore bodies or cut across them, but these are exceptional cases, and most important fluvial placers depend upon an extensive colluvial environment for their sediment supply.

Fluvial placers

The fluvial placer model

Fluvial placers are formed in networks of channels having steep sides and slopes in their upper reaches and lower parts that widen and flatten out with extensive beds of gravel and other sediments in which the active channels are contained. Some deposits are related to present drainage systems; in others the mineral particles have a long history of environmental change from when they were first set free from their parent rocks.

Fig. 3.10 Colluvial placer workings, Barjor, Madhya Pradesh, India

Ancient placers now take the form of deep leads, high-level terraces and other relict types perhaps completely away from any present-day fluvial system.

The fluvial placer model (Fig. 3.11) is in three sections: a source area in which the drainage channels rise; a transfer area in which sediment input closely matches the output and in which the major deposits are formed; a zone of widespread deposition which acts mainly as a storehouse of materials for subsequent processing in another environment.

The source area: Stream flow commences in the source area on slopes that are irregular in profile and variable in their resistance to erosion. Rills grow into channels wherever they are concentrated sufficiently in the one sector and, close to the source rocks, flow may only take place during periods of high precipitation or thaw. It becomes continuous wherever the subsurface water saturates the ground over a large area and is channelled beneath the surface to an outlet point to emerge as seepages and springs.

Fluvial channels in this sub-environment are characteristically narrow and deeply incised with steep gradients in which rapids and waterfalls are common. Stream beds are full of obstructions from boulders, rock bars and other impediments to the flow. Values are distributed sporadically throughout, with some of the larger nuggets and heavy minerals working their ways down into cracks in the bedrock or wherever some other feature

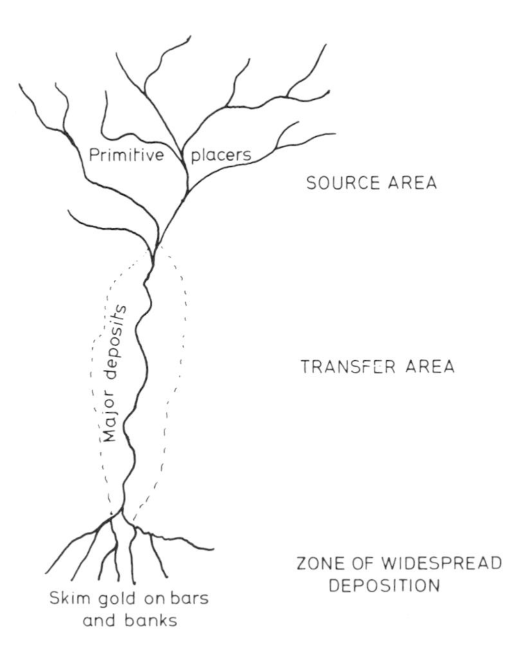

Fig. 3.11 The fluvial placer model

of the channel or its sedimentary fill provides a protective shield against the flow. Valuable particles remain in place until the larger sediments are reduced to more transportable proportions or until the stream cuts its way down through the fill to scour the bedrock surface and once more entrain the sediments at that level.

Transverse rock bars are common features (Fig. 3.12) and various forms of vortexing develop according to how they are oriented to the direction of flow. At angles between 45° and 90° Allen (1973) sees the development of rollers as closed loops and these may result in deep scouring and in the deposition of heavy particles on the downstream side of the bar. At smaller angles, the stream lines appear as open-ended helical spirals, thus providing for the mass transfer of both sediments and fluid along the vortex axes (Fig. 3.13).

Although relatively unimportant commercially, fluvial placers in the source areas represent the first stages of fluvial concentration and, where they occur abundantly in a large catchment area of source rocks, may be the forerunners of major concentrations downstream.

The transfer area: Many small streams contribute to the erosion of a large area of provenance and as they join together progressively at lower levels so are the channels enlarged and the total quantities of sediment in

Fig. 3.12 Cleaning around a rock bar cutting across a previous channel of the Sonakhan River, Raigarh Goldfield, Madhya Pradesh, India

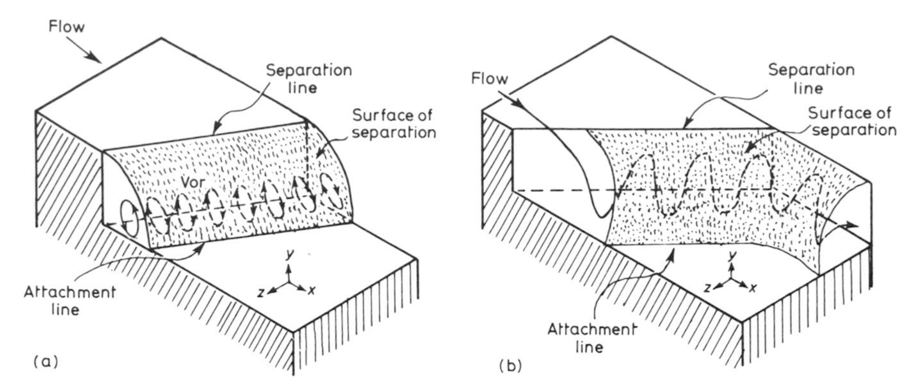

Fig. 3.13 Three-dimensional separated flows: (*a*) *roller,* (*b*) *vortex* (*from Allen, 1973*)

transport increased. The gradients flatten away from the source area and further downstream where the valleys widen and give additional freedom for lateral movement, incipient meanders cut into slopes to form side bars and shoals. Vertical erosion is less rapid than in the upper reaches and a measure of equilibrium is reached between sediment deposition and entrainment that is disturbed only during periods of flooding.

The major deposits are formed where the streams break out from the hills into large valleys or onto the plains and join other streams to form rivers. Sediment loads are deposited abruptly when the streams widen on flatter slopes and lose power, and channel systems become braided as do the larger mountain streams and streams that traverse glacial outwash plains. Depositional units take the form of alluvial fans, braid bars and flood plains; point bars are developed at inside bends and riffles may form downstream from any features affecting the direction of flow. Small islands appear in the wider sections as midstream bars. The characteristic patterns of development are illustrated in Fig. 3.14.

Zone of widespread deposition: This zone is typified by coastal plains lying between the mountains and the sea. The sediments are fine grained and deposition is by vertical accretion over wide areas that are inundated during periods of flooding.

At normal flow rates the streams transport only the smallest particles in suspension but when they rise and break their banks they carry with them large quantities of mud, silt and fine sand that are deposited when the waters recede.

With the exception of 'flood' gold that may sometimes be deposited in measurable concentrations along the banks of the streams or on skim bars, the valuable minerals are either too finely divided or too dispersed to be recovered effectively. Nevertheless, the zone of widespread deposition is

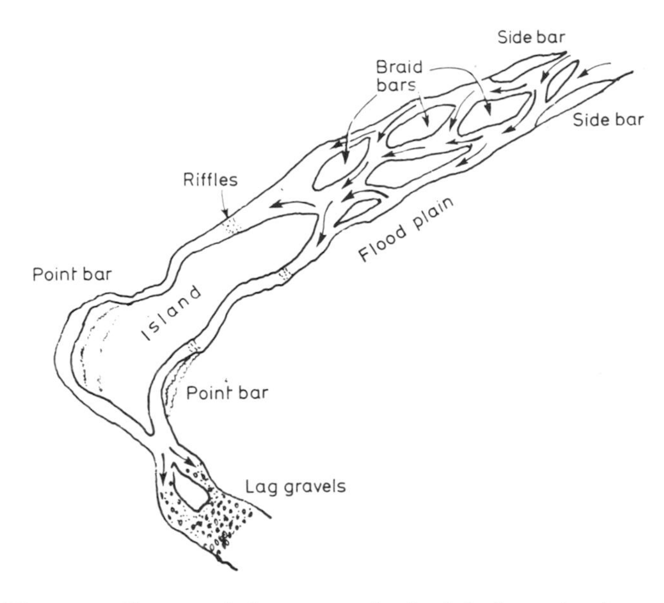

Fig. 3.14 Characteristic patterns in fluvial placer development

important as a vast reservoir of such valuable minerals as rutile, zircon, ilmenite and monazite. Although widely dispersed in this environment they will eventually be washed down to the sea and upgraded by wave action on the beaches. A trickle of such particles enters continuously into the marine environment via the active stream channels; enormous quantities are mobilized during periods of falling base levels.

Channel shapes and patterns
Fluvial channels are sinuous, straight or braided and each feature has a distinctive bearing on the distribution of heavy minerals in the stream channel.

Sinuous channels: Channel sinuosity is the normal response of a stream to variations in the resistance of bed and wall rock formations and is defined as:

$$S = \frac{l}{d} \qquad (3.1)$$

where l is the channel length and d the downstream distance described in Fig. 3.15. It provides a mechanism for absorbing excess stream energy by lengthening the flow paths, thus reducing gradients and increasing wall friction and other friction losses.

A common cause of sinuosity is the armouring of stream beds by layers of gravel. This protection against downcutting results in more energy being

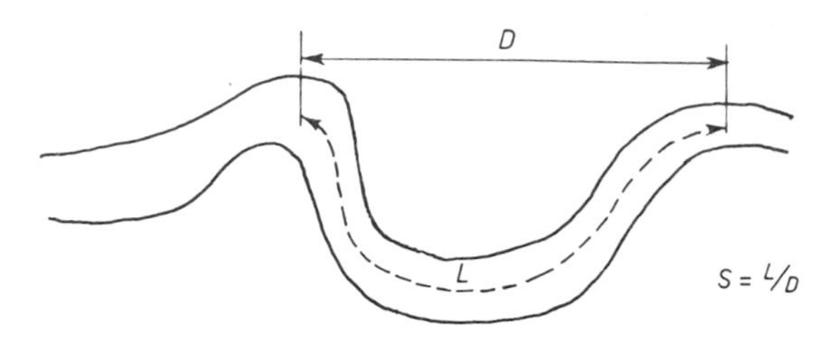

Fig. 3.15 Channel sinuosity

directed laterally against the walls, a slackening in stream velocity and the deposition of additional quantities of gravels and other coarse particles. A downcutting stream thus tends to become more sinuous with time as gravel accumulates along the channel bed.

Morphological features such as rock bars also cause diversions resembling sinuosity; however, extreme values for S are otherwise found only where streams traverse very fine sediments on flat slopes. In respect of heavy minerals transportation and sorting, placer accumulations are normally restricted in such streams to sections having sinuosities of less than 1.5.

Straight channels: It was noted by Leopold *et al.* (1964) that straight channels rarely approach a length of ten times the channel width and Reunech-Singh (1975) suggest that, even then, the thalwegs are mostly sinuous with deeper parts (pools) alternating with shallow parts (riffles). Figure 3.16 describes how a meandering thalweg can be developed by alternate bars in a wide straight channel and explains why, eventually, the banks opposite the bars are eroded and a straight section becomes sinuous. In fact, channels are rarely straight except where influenced by fault planes or other major lineaments. A classic example is found in the diversion of rivers along the San Andreas fault in California.

Braided channels: Braiding may develop in stream channels wherever the sediment load is in excess of what the stream can carry away. The excess load is dumped in mid-channel to form the characteristically complex patterns of small channels and mid channel bars from which a braided

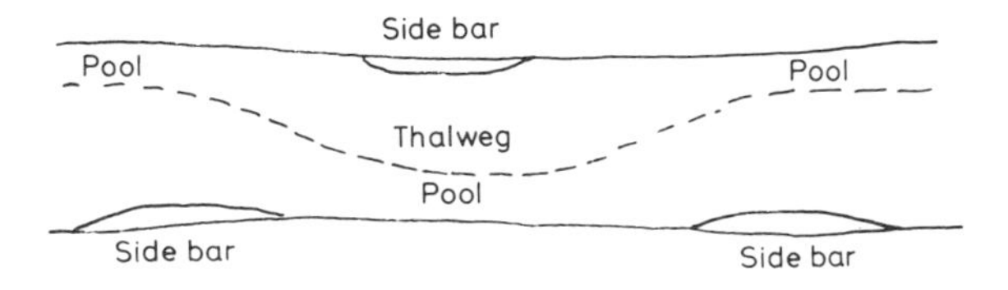

Fig. 3.16 Development of sinuosity in straight-channel section

stream gets its name. Favourable conditions for braiding occur when fast flowing streams widen abruptly on flatter gradients; where seasonal flow rates fluctuate widely and where the river banks are easily eroded. Falling base levels promote braiding in zones of widespread deposition particularly if the discharge is increased also by changing climatic conditions. Braided channels are found both in coarse alluvial fill and in the fine sediments of large deltas such as the Indus delta.

Deposition in streams

Deposition takes place in streams wherever the velocity of flow falls below its critical value for the size and distribution of the sediments in transport. Regionally, the larger and denser particles are deposited first; locally, deposition is less predictable because of local changes in the energy distribution in the stream. Hence, while observed stream sediment lithologies invariably show a general reduction in particle size with distance from the source, local particle size distributions may be quite erratic, particularly in the upper reaches of streams where the particle size differences are greatest.

As has already been demonstrated in Chapter 2 the power of a stream to move solid particles from a state of rest is directly associated with the velocity of flow and at every point there are some particles that cannot be transported further until they are either reduced in size or, for some reason, the power of the stream is increased sufficiently for movement to take place. Regionally there is a steady reduction in power, locally across any section of the stream there are wide variations due to eddying and turbulent velocity fluctuations resulting from differences in the geometry of the surfaces over which flow is taking place and from the swirling of waters between and around the larger particles. In such conditions, the local entrainment of particles is greater than can be sustained in the flow away from the disturbances and deposition takes place at another point to create a new set of conditions.

Reductions in stream velocity occur for a number of reasons of which the most important are:

(1) reduced gradients in the direction of flow
(2) loss of water through evaporation and seepage
(3) loss of kinetic energy due to sediment overloading
(4) fluctuations in run off due to changing rates of precipitation or freezing in the catchment area
(5) changes in stream direction
(6) changes in stream widths

Reduced gradients: Mountain streams commence their journeys to the sea down gradients as steep as 1:5 or more whereas on the coastal plains they flatten to as little as 1:5000. The consequent overall slackening in velocity superimposes changes on the main flow according to local changes in slope of the channel and its sedimentary fill. The distribution is most erratic where steep gradients are suddenly checked and sediments are dumped in their order of appearance without any opportunity for sorting.

Loss of water: For any given condition the velocity of a stream and hence its power varies with the rate of flow and its carrying power is reduced when water is subtracted by any means. Extreme examples of the mechanism are found in semi-arid desert conditions in which rivers flow only occasionally but sometimes with great force thus providing wide periodic differences in both flow rates and energy levels.

Loss of kinetic energy: Stream energy is lost continuously along the course of a stream for all of the reasons given in Chapter 2. Its availability is greatest in the headwaters of the stream and is of negligible proportions when it enters into the sea.

Fluctuations in run off: Rivers flow at minimum levels during periods of drought or when the headwaters become frozen and the water is stored up in a frozen state. Materials deposited during these times will be re-entrained as soon as the ice masses thaw or the rain falls steadily once more.

Changes in stream direction: Each change in the direction of a stream increases its length and frictional resistance thus decreasing its average gradient and carrying capacity.

Changes in stream widths: For the same rate of flow, an increase in the width of the stream increases its wetted perimeter. This results in increased energy losses due to friction and to a reduced carrying power because of the energy that is wasted.

Depositional land forms

Conditions of fluvial placer deposition are most diverse. The basic forms are alluvial fans, flood plains, braid and point bars, islands and lag gravels, terraces and there are many variations of each.

Alluvial fans (Fig. 3.17): These are deposited by streams flowing out from the mountains onto the plains. They occur in arid and semi-arid environments as well as in more humid conditions wherever the highlands provide a plentiful supply of sediments and the lowlands have ample space to receive them. They are favoured by streams that cut their way through narrow, steep gorges in a single flow and break up into a number of

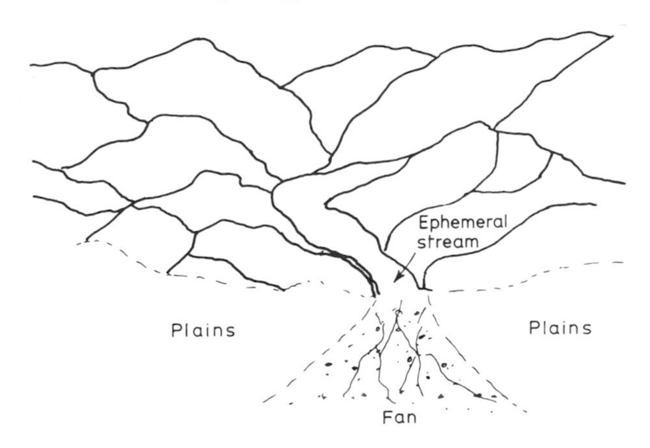

Fig. 3.17 Alluvial fan

tributaries on reaching the plains. The stream flow is usually either ephemeral or variable in its intensity. In the general case, the average size of the particles diminishes from the fan head to its toe and from bottom to top. Sediment reworking results in the development of channel braiding upstream and eventually in the formation of lag gravels and other fluvial facies downstream. Some fan deposits can be mined directly, however in most cases a considerable amount of re-working is needed before economic concentrations are formed.

The most outstanding examples of fan deposits are the Witwatersrand gold deposits which have so far supplied about 55% of all of the gold produced in the world. The coarser sediments appear to have been laid down by braiding channels filled with pyritic sand and small gravel and these extend for many kilometres in the finer materials. Schumm (1977) describes one channel having a maximum width of 100 m and depth 35 m and another up to 750 m wide and 85 m deep; both have been traced in mine workings for distances of more than 8 km. Major fan-derived deposits occur in a similar form in Ghana, Brazil, Canada, India, Gabon and Finland as well as in various parts of the USSR. According to Sestini (1978) their stratiform distribution suggests syngenetic origin and the weight of evidence points to palaeoplacer interpretation. More recent fan-derived deposits are mined by sluicing and dredging along sections of the Himalayan foothills in Uttar Pradesh, India and in the Choco region of Colombia where enormous quantities of gravels are mined for gold and platinum.

Flood plains (Fig. 3.18): These are relatively level areas bordering streams that overflow or break their banks during flooding. They originate from sediments within the stream channels and, thus, may contain fine particles

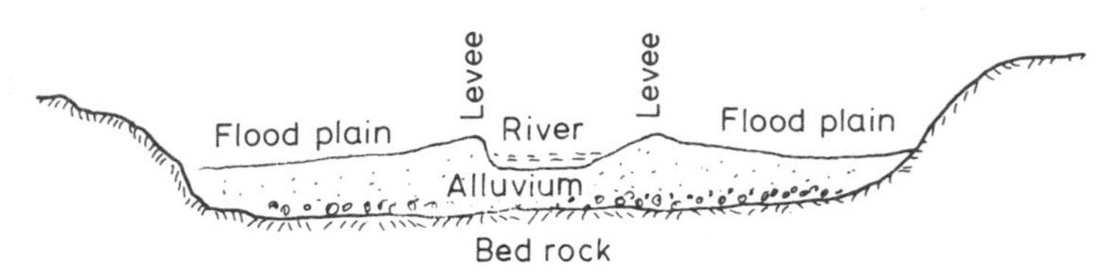

Fig. 3.18 Flood plain

of gold and other valuable placer minerals in sufficient quantities for them to be found by panning.

On flood plains such as those along the Punna Puzha river in the Nilambur Valley of Kerala, India, there is clear evidence of meanders migrating downstream being followed by upstream meanders that re-entrain the sediments deposited by previous flooding. The plains associated with streams in equilibrium in a valley can thus be considered as providing temporary storage for low-grade concentration of valuable minerals that will one day be upgraded to economic proportions. Flood plain concentrations, are rarely economic except on a small scale along the higher river banks where local miners may scrape the surface periodically after a few seasons of rain and wind have performed a natural upgrading of the surface sediments.

Purely aggradational flood plains result from rising base levels and are of less immediate interest to miners except in so far as they may have covered earlier channels containing economic concentrations. They assume importance only when, after long periods of aggradation, the base level is again lowered and the masses of sediments are re-entrained and re-worked.

Braid bars and islands: The bars and islands of braided streams are composed of mainly coarse sediment particles that are deposited in mid-stream as the streams progressively lose their power. The finer sediments, silt and clay, are carried on and settle out in more graded stream sections lower down or at sea.

Braid bars and islands are built up by lateral and vertical accretion and are predominantly lens shaped although of variable dimensions otherwise. They are subject to continual working and re-working in changing conditions and migrate downstream by eroding sediments from the upstream end and depositing them at the downstream end. Channel bars may also migrate laterally and new bars are created as others disappear. The pattern changes constantly and channels narrow, enlarge and alter course with every change in the volume of flow and the sediment load. Occasional braid bars become stable for a time when silt deposited during flooding is covered by vegetation, but eventually all of the material is moved downstream to where a steady reduction in size makes for better sorting. Figure 3.19 illustrates typical braid bars in plan and section.

Point bars: These form in successive stages from the local accretion of sediments on the inside of meanders in sinuous channels. They extend downstream from the point of maximum curvature of the meander bend. As they develop in size, point bars exhibit such other features as scroll-shaped ridges (scroll bars) separated by ox-bow lakes or marshy areas, natural levees, crevasses and splays. Each represents some adjustment of the stream bed to the forces imposed on it during flooding and, consequently, features that are built up during one wet season may be altered or destroyed in the next.

Point bars rely for their formation and structuring upon changes in stream direction and changes in the volumes and depths of flow. Some features result from lateral accretion during periods of stable flow, others are developed during flooding by both lateral and vertical accretion.The processes are complimentary: the accretionary material is eroded from the opposite banks and the changes in stream depth and direction set up the required mechanisms for lateral migration; because the bars extend further out into the stream, the flow is directed at a more acute angle against the bank and erosion increases. This continues until another set of factors: increased stream length, increased frictional losses and reduced gradients make it easier for the stream to develop in a different direction, which it invariably does. Figure 3.20 is a model of a coarse-grained point bar (after McGowan and Garner, 1970).

The distribution of values in river bars: Sedimentary bars of all types tend to become enriched at their upstream ends although the distribution of values is influenced also by the various accretionary processes responsible for their formation. Experience has shown that, in any bar, the highest

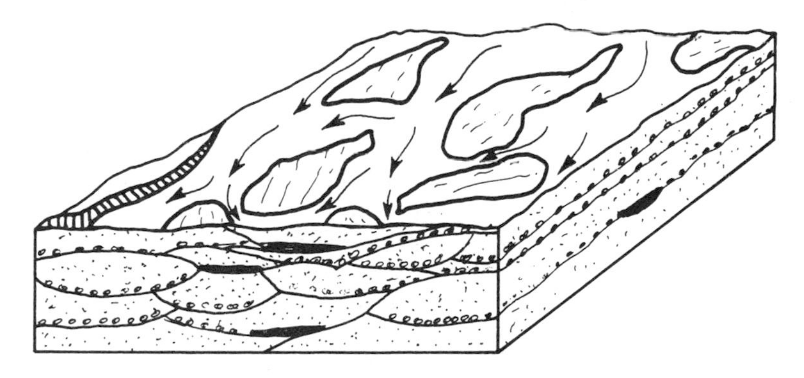

Fig. 3.19 Typical model of braided stream

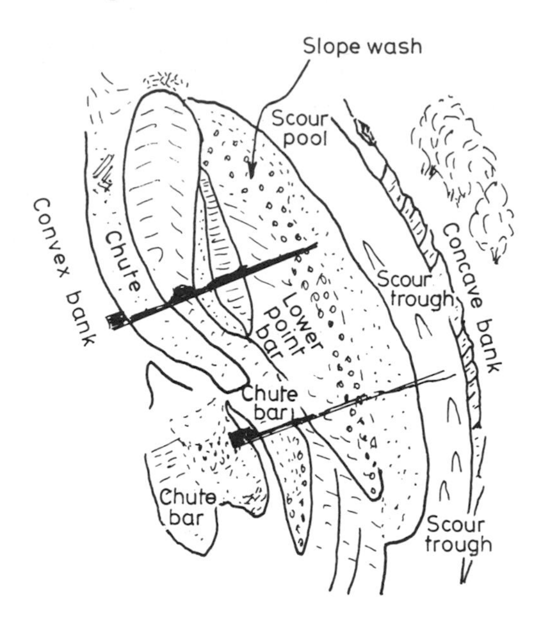

Fig. 3.20 Plan view of a coarse grained point bar (from McGowan and Garner, 1970)

values are associated with those sections where the energy of the stream has been greatest, hence splays are likely to have a higher heavy mineral content than beds of fine sand and flow across natural riffles developed in long profiles of streams provides ideal conditions for gravity concentration. Shallowness over the riffles leads to increased stream velocity and bed roughness, ideal conditions for saving coarse particles of gold. However, Wells (1969) injects a note of warning in respect of apparently rich concentrations of flood gold on accretionary or skim bars. He points out that this type of gold consists of minute particles so small that it might take 1000 to 5000 colors to be worth 1 cent, and that the underlying gravels may be barren.

Lag gravels

Lag gravels are the slowest-moving constituents of the bed load and are so called because they lag behind the bulk of the channel fill as it moves downstream. They may occur in any stream channel section but are most extensive and continuous in well-developed drainage systems in which the larger stones and boulders have been broken down to gravel size and where a general movement of much of the channel fill takes place during flooding without excessive scour. Such conditions are provided by streams that flow rapidly over relatively flat and uniformly sloping channel beds: deep swirling and eddying effects due to major irregularities in the bedrock geometry cause a re-mixing of the sediments and a dispersal of the valuable particles.

The mechanism of entrainment and differential settling is complex. At normal flow rates only the upper sediments are disturbed and the pebbles act as riffles to trap and hold back particles of the finer materials and heavy minerals. As flood waters build up and the stream velocity increases, the lower parts are increasingly dilated, thus allowing the larger and denser particles to move downwards. Dynamic forces are transmitted progressively further into the underlying sediments until eventually all of the channel fill may be disturbed. The grade of valuable heavy minerals tends to increase with depth although experience has shown that the distribution of values is not uniform over the whole bed and the lowest sections of a channel are not necessarily the richest. For example, Fig. 3.21 describes a narrow section of a tin placer at Tingha, New South Wales, Australia, where the richest values were found in hole E1 at a higher level than the lower-grade gravels in the adjacent holes, possibly due to the action of transverse currents and other components of the flow.

Regardless of the depth to which the gravels are disturbed, their transportation is a differential movement downstream with the upper layers moving faster than the lower ones. In such conditions, Schumm and Stevens (1973) found experimentally that the coarsest fragments tend to vibrate in place at high flow velocities without forward movement and that the voids become filled with finer particles amongst which are heavy minerals. A natural attrition takes place that eventually allows some

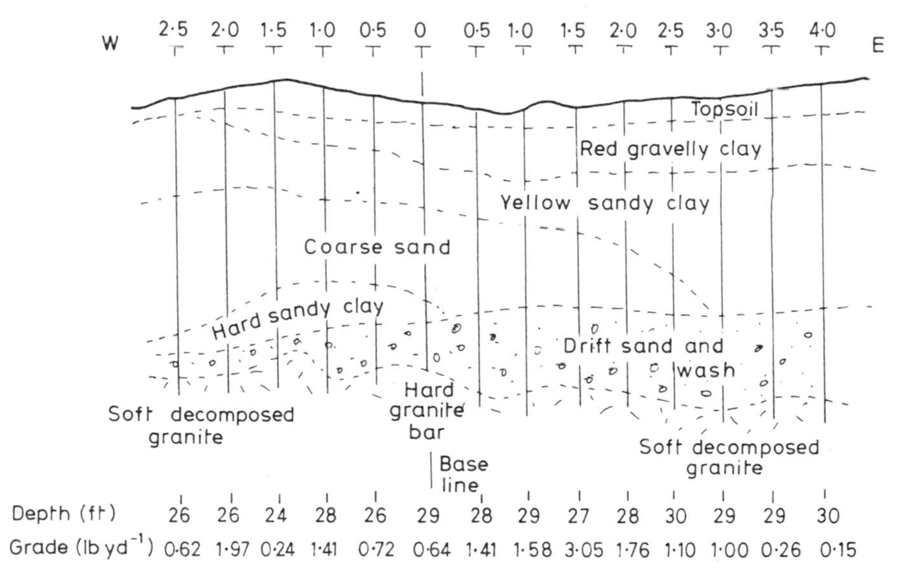

Fig. 3.21 Section across cassiterite placer, Tingha, NSW, Australia

movement and replacement of all of the lag, until the downcutting ceases and the incoming sediments commence to build up along the bed. The stream no longer has the power to disturb the particles at bedrock and no further enrichment takes place at that level.

The mechanism of enrichment was examined by Minter and Toens (1970) who found that a thick deposit of gravel with numerous thin layers has greater economic potential than a single thick layer. In flume experiments with particles of different sizes and densities, they examined the deposition of sand and magnetite particles in gravel beds and concluded that the heavy mineral grades were probably directly related to the number of micro layers in the bed. They observed that the deposition of sand is controlled by the surface layer of pebbles and that the average magnetite content of thick gravel beds was generally lower than that of the bed load supplying fresh heavy mineral particles. The tests showed a progressive increase in the magnetite concentration in gravel beds downstream and this was held to explain, in part, why the mid-fan parts of the Banket alluvial fan deposits of South Africa have a greater potential for economic concentrations of gold.

Valley fill placers

The general order of fluvial sediment emplacement in a valley commences with the large-scale production of waste rock during periods of diastrophism or climatic extremes and continues through various cycles of erosion, transportation and deposition in sequences that differ from valley to valley and from region to region. Differences in sedimentary behaviour stem from changing base levels and variations in stream power. A typical sequence may be as follows:

(1) A rising base level causes a river to slacken and deposit its transported load progressively upstream. Tributaries also choke up and, with time, the valley becomes filled with sediments that are relatively coarse grained and poorly sorted. Conditions are not conducive to the formation of placers but, nevertheless, the valley may become a storehouse of low-grade materials ready for sorting.
(2) Environmental changes lead to a falling base level. Increased gradients result in an increase in stream power and deep scouring occurs at the valley outlet.
(3) Tributaries are rejuvenated as the knickpoint of the incision moves up the valley.
(4) Surges of sediment brought down by the tributaries choke the main stream channel.
(5) Large-scale deposition occurs in the main stream channel which becomes braided.

(6) Renewed deep scouring takes place at the valley outlet and the sequence is repeated until equilibrium is reached at the reduced base level.

The depositional units are thus built up by lateral accretion, vertical accretion or by a combination of the two and, because each placer unit reflects a particular phase in the geological history of the valley, the spatial relationship between individual units and the distribution of values may be exceedingly complex. The cyclical inflow of sediments has not always been due to the same causes nor has it been derived proportionally from the same places or at the same rate. The same valley may have been filled and scoured many times. In some cases the supply of mineral-bearing sediments from the source areas may have been terminated abruptly by river capture and a typical example was the capture of the Delaware River, USA, by the Schuykill river which left, isolated, large ilmenite deposits to the east of its present course. Sections of a stream bed may similarly be isolated by avulsion and this is a common characteristic of placers in semi-arid climates such as those of Moolyella in the Pilbara district of Western Australia. Other deposits may have become buried by landslides, flood plains or lava flows (see deep leads p. 134); or eroded away and reconcentrated at lower levels. Regardless of their location, individual deposits at different levels in a valley almost invariably exhibit distinguishing features that can be in some way related to their geological history and in particular to the sedimentary response from climatic changes and extremes.

Valley-fill placers also reflect the resistance to change of valley walls and bedrock. Few large placers are built up directly from a single influx of sediments from source areas. In most cases they result from the working and reworking of vast quantities of sediments from previous erosion cycles, in valleys that have been able to expand with time in order to contain them.

It is becoming clear also that, while the major placer deposits are formed following large-scale diastrophisms that result in elevating great masses of land, the deposits are, themselves, laid down only in periods of relative calm. According to Schumm (1977) this has been demonstrated by Sigov *et al.* (1972) who concluded from studies of heavy mineral concentration in the Urals, that placers require major upheavals for their formation but are not formed while the movements are taking place because erosion and transportation at such times is too rapid. Formation occurs subsequently during periods of greater stability when the normal processes of weathering and erosion provide suitable environments for natural processing. This may also have been the general order elsewhere as evidenced by the world-wide formation of Tertiary placers following the formation of the Alps, Himalayas and Pyrenees.

River terraces

River terraces are remnants of alluvial valley fill left behind as benchlike landforms along the valley sides when streams incise their beds downward into the underlying rocks. They follow the course of the streams, approximating their gradients, and constitute all that remains of the valley floors when the streams were flowing at higher levels. Most terrace deposits result from the working and reworking of sediments that originally contained disseminated minerals of value and the richest sectors usually represent portions of old stream beds.

River terraces are distinguished from marine terraces by their gradients. This is because river terraces grade downwards along valley walls in the direction of the stream gradient when the terraces were formed, whereas marine terraces are at a common level according to the level of the sea at their time of formation.

River terrace sequences may be paired or unpaired (Fig. 3.22). Paired terraces represent periods in which the down cutting was temporarily stayed for one reason or another; unpaired terraces are representative of valleys in which the downcutting was effected by the meandering of the stream from side to side. Terraces formed entirely in alluvium are generally unsuited to the development of placer concentrations. A stream that cuts to bedrock at all levels is more likely to contain valuable deposits because all of the fill will have been re-sorted to reach the level of bedrock.

High-level terraces are not generally as rich as the lower-level terraces or present stream deposits which result from the working and reworking of much larger quantities of sediments. Exceptions occur where the low-level terraces take the form of flood plains raised above flood level by colluvium from the valley walls in which case they are unlikely to contain significant values; and where high-level terraces contain the remnants of deposits laid down during long periods of relative stability by streams in equilibrium with the inflow of new sediments and their working.

Deep leads

Deep leads, or high-level sub-basaltic placers, common in Victoria and eastern New South Wales, Australia, as well as in California, were formed

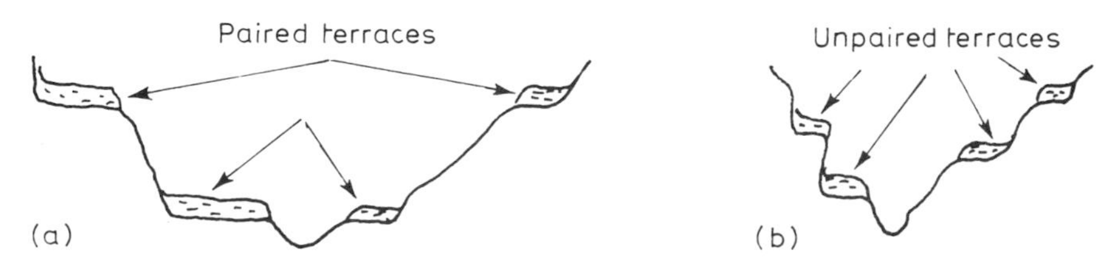

Fig. 3.22 Diagrammatic representation of terrace sequences: (a) paired, (b) unpaired

by the outpouring of Tertiary basalts over pre-Tertiary alluvial land forms. At Rocky River near Uralla in New South Wales, high-level gravels belonging to pre-Miocene drainage and capped by a residual cover of Tertiary (Miocene) basalt have been almost completely eroded away leaving remnants on the tops of hills as shown in Fig. 3.23. Pleistocene high-level gravels occur as terrace deposits.

Figure 3.24 (a), (b), (c) and (d) illustrate how such deposits occur. Figure 3.24(a) represents the pre-Tertiary surface with a fluvial placer occupying the lowest portion of the valley. An outpouring of tertiary basalts has mantled the area in Fig. 3.24(b) and stress fractures develop due to stretching along the higher portions. This initiates new erosion along the lines of weakness as shown in Fig. 3.24(c) and, finally the former channel deposits remain as high-level gravels protected by a capping of basalt high above present stream levels (Fig. 3.24(d)).

Deep leads may also be formed by overloaded rivers silting up their valleys in the absence of volcanic activity. Indeed, any type of placer deposit related to a river system may become a deep lead in appropriate circumstances. Common causes are a rise in base level and climatic changes resulting in increased precipitation in the catchment areas. As an example of this type of deposit, the Droogenveld deep lead in the Transvaal, reputed to be one of the richest diamondiferous leads ever found, extended over a distance of more than 3 km under, first a layer of rounded pebbles in a clayey matrix, then a thick cover of finer fluvial sediments and, finally, aeolean sand. Such deposits are preserved only as long as the base level remains high. Generally they are quickly eroded away in any fresh cycle of erosion following an epeirogenic uplift or fall in base level for any other reason.

Desert placers

Desert placer accumulations are formed around weathered outcrops mainly by gravitational forces and winds that winnow out the fines and blow them away. The finest particles may be transported globally to settle

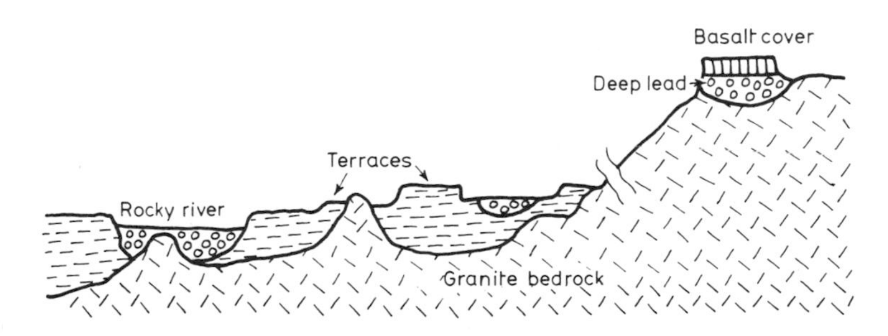

Fig. 3.23 Schematic cross-section of Rocky River, NSW, Australia, showing deep lead under basalt capping and present gold bearing streams

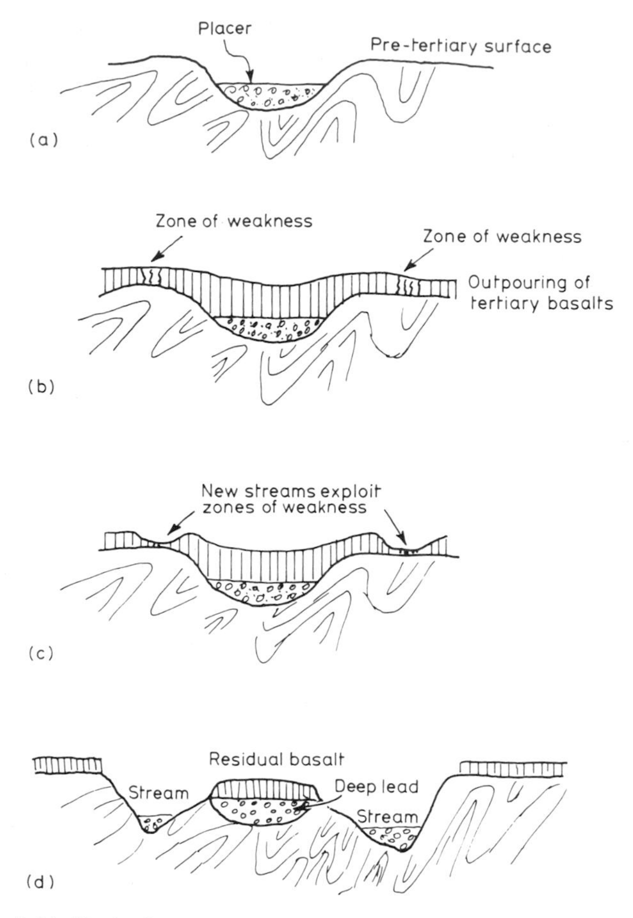

Fig. 3.24 Typical stages in the development of a tertiary deep lead

out as oozes and muds on the deep ocean floor, or as beds of loess on land. The larger sediments remain close to their source and the heavier fractions become further concentrated when occasional rains cause streams to flow in otherwise dry gutters and channels.

Morphological differences between individual desert placers appear to be due mainly to differences in aridity and the topography of their surroundings. In some cases differences have arisen because of differences in the environment when they were formed. Many show evidence of sustained fluvial action at some earlier period and age is obviously a vital factor.

Experience of desert placers in Australia has shown that similar principles can often be applied to their interpretation as to their counterpart types in higher rainfall areas. However, Tielman (1976) found more differences than similarities in his examination of the Basin and Range Province desert placers of the western USA. He found that 'wet' placers exhibit definite sedimentation patterns of sorting, rounding and particle distribution according to weight, rate of flow and channel gradient whereas gravels in arid deposits show little graded bedding and generally display a complete intermixing of fines, cobbles, rocks and an occasional boulder. The distinction applies equally to comparing colluvial placers with fluvial placers and perhaps some confusion has arisen from not distinguishing between the basically colluvial deposits of a very dry desert and sediments in channels many miles from the source lode. In Saudi Arabia, where intermittent flash flooding sets great masses of sediment into movement, a degree of sorting is seen to occur in channels leading away from auriferous gold reefs in most areas.

Glacial deposits

Glacial spoil gouged out from the valley floor and walls may be transported for considerable distances before being dumped in an unsorted heterogeneous mixture of detritus ranging from finely pulverized rock flour up to the largest of boulders. In provenance areas such glacial morraines may contain small patches of good-grade material but commercial-sized placers are found only where such deposits have been reworked subsequently by fluvial or marine action in glacial outwash plains and along beaches and nearshore sectors of a coastline. Singewald (1950) refers to mining in the morraines at Fairplay, Colorado, but points out that the most extensive workings were in the outwash aprons after fluviation.

In any glacial environment, the rate of movement of the ice depends upon the rate of snow precipitation and the slope of the valley floor. Prolonged periods of high snow fall and increased depths of snow cause the ice to flow faster and, because of the increased pressures, the glaciers extend further down their valleys. Flow rates diminish again during periods of lower precipitation and the glaciers recede, progressively dropping their loads of sediments as they retreat.

Glacial morraines are quite unstratified and the material is unsorted except when some of the finer particles are washed away by the melt waters. Dilution of mineralized sediments by barren material from the sides and floor is usually very high and, although specimens may be found containing valuable minerals, only rarely are concentrations formed. The velocity profiles of the ice flow are parabolic in both horizontal and vertical planes except where melt waters flow in channels along the bottom.

Summary
Table 3.3 summarizes the features of the main continental placer types and lists the valuable minerals most commonly found in each of the sub-environments.

Table 3.3 Deposition in the continental placer environment

Sub-environment	Valuable minerals	Deposit types	Characteristics
Eluvial	Au, Pt, Sn, WO_3, Ta, Nb, gemstones, all varieties	Weathered rocks *in situ* and downward vertical accretion	Dished appearance of outcrop weathering profile follows closely on the shape of the original ore body
Colluvial	As above	Colluvium	Deposits derived principally from surface creep and elutriation on slopes. Generally poorly sorted
Fluvial	Au, Pt, Sn, rarely WO_3, Ta, Nb, Diamonds, rubies, sapphires, chrysoberyl varieties	Channel deposits and valley fill deposits alluvial fans and flood plains	Present active stream channel deposits in lag gravels, point bars, braid bars. Lag gravels: downward vertical accretion. Bar deposits: vertical and lateral accretion and more transitory than lag gravels. Valley fill deposits provide abandoned channel segments such as terraces. Flood plains formed from fine-grained sediments deposited from suspended loads during flooding
Desert	Au, Pt, Sn, WO_3, Ta, Nb, gemstones all varieties	Colluvium and incipient fluvial deposits	Similar to colluvial placers and primitive fluvial forms
Glacial	Au (rarely) sand and gravel	Glacial tills and morraines	Unsorted sediments dumped progressively upstream as glaciers recede. Usually requires reworking in fluvial or transitional condition

3.3.2 Transitional placers

The transitional placer setting

Features of modern coastal land forms are closely related to the fact that oceans reached their present levels only about 5000 years ago at the end of the Holocene Marine Transgression. During the course of the transgression, which commenced about 20 000 years BP, sea level rose in a series of oscillations from a low of −130 to −160 m below present mean sea level and, in the process, inundated about 20% of the surface of the land. The newly submerged portions, now known as the continental shelves or platforms provide the setting for transitional placers found along the present low-lying coastal regions and shallow foreshore areas where a transition takes place from wholly marine to wholly aeolean conditions.

Marine forces acting along the margins of the land are influenced strongly by variations in solar and lunar attraction which affect the strength and height of tidal wave forms and accelerate the erosion and resculpturing of shore lines. Coastal physiography adjusts constantly to the forces acting and the most rapid changes take place when the ocean attacks the land successively at different levels. It is a daily occurrence because of the variability of waves, tides and winds. It is both seasonal and occasional when climatic changes and atmospheric disturbances cause storms at sea. On a geological time scale, tectonic uplifts and subsidences, isostatic adjustments and eustatic sea level changes during a succession of ice ages and interglacial periods of thaw have altered the levels of attack along shorelines for hundreds of millions of years. Early erosion surfaces have been subjected to a variety of changes affecting their final form.

Valuable minerals first become concentrated around the limits of wave action and the most important beach deposits occur as lenticular bodies at the base of frontal dunes exposed to conditions of high energy. Offshore bars result from the retention, in near shore areas, of material removed from beach deposits by wave action. Sand spits build out from headlands and become exposed to the sorting action of waves. Aeolean deposits are formed when mineral-bearing sands are blown up onto high ground by onshore winds. Transitional placers are the least stable of all placers although some fossil placers have survived in elevated positions from as long ago as Mesozoic or even Palaeozoic times. A generalized section across the transitional placer environment has already been illustrated in Chapter 1 (see Fig. 1.3).

Sediments of the environment

Transitional placer minerals of value are principally:

(1) Gold, platinum, cassiterite, rutile, ilmenite, monazite and magnetite.
(2) Gemstones, notably diamonds.

(3) Sand and gravel in suitable sizes for use as aggregates in the construction industry.
(4) Silica sand for glass making.

The metallic minerals are associated mainly with quartz sands although such other rock-forming minerals as garnets, pyroboles, etc., are also represented in varying quantities depending upon the nature of the rocks from which the sediments derived.

The initial heavy mineral content of the sediments is generally low, averaging perhaps not more than 0.01% of particles with densities greater than that of quartz. The quantities processed are very large and even where no important concentrations occur, deep concentrate shadows are common features on beaches between high and low tide level.

Particle sizes

Valuable heavy minerals along the beaches and in shallow water are mostly fine grained. Ranging in size from 250–75 μm, particles of these minerals become concentrated with somewhat larger but less-dense minerals of similar hydraulic equivalence. Whitworth (1958) has observed the average particle size of valuable heavy minerals along the coastlines of New South Wales and Queensland, Australia to be about one-half that of quartz particles. He noted further that, although there were considerable variations in particle size between the individual mineral suites, 'The grainsize in any one deposit is remarkably uniform owing to the long continued sorting action of the waves.' He attributed the regional size variations to the source rocks from which they were derived and to natural sorting in a previous cycle of deposition. Typical regional variations are described in Table 3.4.

Later work by the New South Wales Geological Survey has supported this view and there now seems little doubt that the above regional variations in particle size are related to relic patterns of sediment on the

Table 3.4 Regional variations in particle size of beach sand deposits northerly along the coastline of New South Wales, Australia

	Average grainsize – mean diameter (mm)		
Locality	Zircon	Rutile	Ilmenite
Dee Why	0.289	0.288	ND
Sawtell	0.143	0.141	0.153
Evans Head	0.192	0.223	0.221
Yamba	0.163	0.160	0.159
Byron Bay	0.205	0.218	0.213
Cudgen	0.177	0.190	0.182

inner shelf and not to either contemporary river supply or longshore movement over great distances. In fact, it is probable that the nature and distribution of most transitional placer sediments can be explained only in terms of the more recent Quaternary geology. Certainly, large reservoirs of sand accumulated during the Holocene Transgression appear to have been the main sources of supply for many of the present day beaches, regardless of their geographical location.

Sediment movement and distribution

Sediments entering into coastal marine areas are subjected to new and complex sets of high energy conditions resulting from the interaction of waves, tides, currents and wind. Particles fine enough to be carried in suspension are directed out into the ocean deeps. The more granular materials remain in close, if intermittent contact with the bed and are moved around locally in the direction of the waves and currents. Although the direct source of most beach sands is the shallow sea bed along which the sediments move, the continuity of the system depends upon the supply of fresh sediments to replace those that are carried away. Very favourable conditions apply in northern New South Wales and Queensland Australia, where fast-flowing rivers traverse vast areas of mineral-bearing source rocks. Enormous loads of sediment are brought down to the sea during periodic flooding and 1000 km of coastline has provided chains of beaches between rocky headlands and on offshore islands which provide natural entrapments for sediments moving northwards along the coastline.

Additional quantities of sediments are brought into the environment through the action of offshore winds; from the erosion of shoreline rocks and formations; and from the melting of glaciers along shorelines. Their impact depends to a large extent on whether they enter as widely dispersed particles along large sections of the coastline or from the local debauchment of streams. The former are assimulated gradually into the general pattern of littoral movement and have little short-term effects on coastal sedimentation. The impact of stream sediments entering into the environment may be much more immediate and complex within a short distance from the mouth, particularly where they form deltas.

Clearly, at any one time, there is a preferential movement of the finer particles towards deeper waters and distinct morphologies have been developed with increasing depths and distance from the shore. For example Davies (1979) has described three sedimentary zones in one section of the continental shelf of south-eastern Australia. In waters shallower than 40–60 m the sediments are mainly terrigenous. Carbonate compounds and clay minerals gradually increase seawards to depths of 120 m, and sediments in deeper waters near the shelf edge are composed largely of poorly sorted carbonates of various particle size.

This does not mean, of course, that the present distribution of shelf sediments should in all cases exhibit a similar gradation in the size and type of sediments away from the present shorelines; or that, at each previous period of stillstand, such conditions might have obtained along all shelves at those times. In fact, the Pleistocene sea level fluctuations were such that the coastlines of the world moved backwards and forwards across their shelf areas so that now, at any one point, the base sediments may comprise any of the gravels and other sediments of the coastal plains of those days. Some deposits have survived, covered by various thicknesses of later terrigenous and marine sediments; however, many have been destroyed and others, particularly the sand-sized particles, have been re-entrained and moved progressively towards the shoreline as sea level rose. Most marine geologists now agree that the onshore movement of these sediments has played a major role in the formation of even the most recent of the transitional shoreline placers.

Deposits of the transitional environment

Deposits formed in the transitional placer environment have characteristic properties that distinguish them from other placers. The geometry of the environment is usually less complex in the context of the Krumbein Model (Table 2.1) than that of the continental environment, the sediments are finer grained and more evenly shaped and the energy is applied more directly. In response, the strandline deposits are commonly lenticular in shape and deposits of the same age have a common base level corresponding to the level of the sea at the time of formation. Wind-blown deposits take the form of sand dunes that may be either structurally simple or complex according to the supply of sand, the availability of space for stacking, the extent of stabilizing flora and the strength and drying power of the winds.

Dynamic processes associated with the formation of placers in the transitional environment involve

(1) The action of waves, tides and currents along the foreshore (strandline deposits, spits and offshore bars).
(2) The action of predominantly onshore winds that blow sand up from the beaches onto higher ground (aeolean deposits).
(3) The combined action of fluvial, marine and aeolean agencies along some coastlines (deltaic deposits).

Strandline deposits

Ocean beaches are subjected to massive and rapid changes when the ocean attacks the land at various levels from day to day and season to season throughout their respective lives. During periods of build up large masses of sand are moved onto the beach. During periods of erosion the less-dense sand particles are selectively removed leaving concentrations of the heavier grains behind. Very large concentrations may be formed through a succession of these movements onto and away from the beach, or removed as quickly during severe erosion cycles to perhaps form again on the same beach or elsewhere. Variations in level have important consequences, both in the quantities of sediments affected and in the distances through which individual particles are moved. Body forces related to the gravitational effects of the sun and moon (see Chapter 2) have great significance by helping to determine the range of levels at which the ocean waves strike against the shore. Tidal currents play an important role in the distribution of sediments on platform areas and may have been even more important when the moon was closer to the earth.

Much depends upon the supply of sediments but other requirements are flatly sloping sea-floor gradients leading up to the shore and sufficient space onshore for the sand to accumulate. The heavy minerals become concentrated only in high-energy conditions; hence, massive and rapid changes take place during storms. It has become quite apparent that occasional storm action is essential for developing high-grade concentrations. For example, in the Bay of Bengal, where the monsoon season is accompanied by galeforce cyclonic winds and occasional storm surges cause great destruction in the low-lying coastal areas of the delta, some transient strandline accumulations approach 100% of total heavies.

Mechanism of concentration: When a spilling wave rolls in to the shore (see Chapter 2) it entrains sand in the surf zone and deposits it progressively as the wave loses energy in its run up the beach. Flowing back into the sea the water accelerates rapidly but, because much of the initial energy has been expended, only a portion of the sand is re-entrained and carried away. Each successive wave adds to the build up of sand along the foreshore, thus causing the berm to build out and steepen and become less stable. This has a twofold effect:

(1) Being at a higher level, the surface particles are more readily dried out by winds which can then blow them away, making it easier for the sea to bring in more sand.
(2) More sand is deposited closer to low-tide level with the result that shallow vertical faces are cut into the berm. This helps to promote scour so that, after a period of build up, increased quantities of sand are returned to the sea.

Sorting commences along the foreshore where the preferential removal of lighter, larger particles results in a steady increase in the heavy mineral content of the remaining fractions. Small dense particles are less affected by boundary shear or impact at the bed than are the larger low density particles which protrude higher into the flow (see Chapter 2). Once deposited, they are less-readily eroded than are the lighter, rock-forming minerals.

The process is accelerated when whole berms are cut back by wave action and rich concentrations of heavy minerals are built up during successive cycles of cut and fill. Tidal variations expose varying widths of the beach to wave action and, in an unequal measure of equilibrium between erosion and deposition some of the sand and heavy mineral particles are blown inland away from the action of the waves. Reservoirs of such materials build up in back beach areas for further processing by aeolean agencies or for subsequent return to the foreshore and to renewed wave action during storms. Typically for an active beach the sequence is as follows:

(1) At spring tides wave action extends to the upper foreshore with a net gain in heavy minerals and sand.
(2) At neap tides wave action is limited to a narrow strip of the foreshore. Concentrations developed further up the beach during the spring tidal period are not disturbed except by the action of the wind. In some cases, previous surface concentrations are covered by a protective layer of wind-blown sand.
(3) During periods of high onshore winds and storms the waves act further up the beach than before and, because of the increased energy expended in the process, much of the lighter material is removed leaving enriched layers of the heavier particles behind Fig. 3.25. The

Fig. 3.25 Magnetic concentrations on ocean beach at Nueva Gorgona, Panama

> level of the beach is lowered thus providing the appropriate conditions for subsequent build-up and for a repetition of the cycle.

On normally stable beaches the highest concentrations are found where the pounding of the waves has been most intensive and, except when the action is taking place, such concentrations lie buried under fresh supplies of sand brought in from the sea. The accretionary process continues until further storms reverse the trend and, if the overall pattern provides for a net movement of sediments towards the shore, very large deposits may be formed. Some concentrations along the New South Wales coast of Australia reached proportions of as much as 1 km to 2 km in length, 100 m wide and 3 m in maximum depth.

But not all cycles are accretionary and when cyclonic storms are associated with their centres close to the shoreline during periods of spring tide, an entire beach might be subjected to the most violent wave action, during the course of which more of the sediments are reactivated than at any other time. In terms of coastal erosion at such times, destruction is at its peak partly because of the great magnitude of the forces involved but also because of the rotational nature of the winds which tend to direct the waves successively at different angles to the beach as the storm centre moves one way and then another. With the changing angle of attack, whole beach areas that have taken decades and perhaps centuries to develop may be removed in a single storm.

It is to be observed also that since all coastlines are not the same, not all beaches re-act in the same way to similar sets of conditions. Thus, while build up in the foreshore area is normally a gradual process and, for the most part, storm wave concentrations result from berm erosion and the consequent selective removal of the lighter grains, this does not always happen. For example, in some cases the nearshore physiography is such that normal wave action results in building up a reservoir of heavy mineral particles in and around the breaker line with very little deposition on the beach. Such conditions arise when the near-shore bed gradient is slightly steeper than the critical slope for entrainment and transportation in traction (see Chapter 2) at normal wave heights, lengths and frequencies. During storms or high seas, high grade concentrates may then be deposited upon the beach within the space of a few hours to be as quickly removed when conditions change. This was illustrated graphically by Whitworth (1958) in reporting upon a 24 h observation of a small beach near the northern end of Tuggerah Lake early in September 1956.

> During a period of three hours the beach became well covered with heavy concentrate as a result of wave action during the afternoon. Low tide occurred during 3 p.m. and 4 p.m. as the concentrating action took place. The wind died away at nightfall and wave action dropped to a gentle lapping effect. By day break next morning (6 a.m.) after the high

tide which occurred about 10 p.m. no sign of the previous days concentration could be seen, the heavy sand having been sucked back into the surf zone during the high tide. Soon after daybreak, the wind freshened again and wave action increased resulting in the building up of a layer of concentrate within a period of three hours on a rising tide. After 11 a.m. the tide commenced to fall again and the concentrate was then left above reach of the retreating tide level.

Deposits on small and narrow beaches are the most transient. This is shown by the present-day movement of sediments along the offshore islands of the Indus Delta, Pakistan, where the author observed an actual transfer of heavy mineral concentrations from one part of an island to another and then on to an adjacent island between one high tide and the next. The end result was one beach completely denuded of sand, another on which a relatively barren frontal dune had been replaced by a thin layer of heavy minerals, and a third beach on which a steady build up of sand was taking place with little evidence of heavy minerals.

Wide beaches are more stable than narrow beaches because more space is available for storing sand and heavy minerals blown up from the main zones of activity. This sand tends to form series of low hind-shore dunes parallel to the shore. Wide beaches normally show a reduction in berm during the stormy months but regain their dimensions during the more clement seasons. Patterson (1965) describes beaches on which parallel dunes showing typical sub-aeolean bedding overlie lenses of wave-concentrated minerals. Individual fore-dunes may rise to more than 15 m in height and similar widths. A series of such dunes closely spaced may comprise a system extending for a kilometre or more inland. These are the beginnings of aeolean placer deposits.

Sand spits: The growth of sandspits out from headlands also provides favourable conditions for the formation of strandline deposits as at Punta Chame in the Bay of Panama where extensive deposits of magnetite sands have been mined. Typically in such cases, the individual strandlines are separated from one another by swamps and marsh land. Figure 3.26(a), (b), (c) and (d) illustrates the formation of such a sequence on the New South Wales coastline adjacent to an existing beach. The spit may eventually seal off the bay and the final result is a series of relict deposits protected by a sedimentary cover along the original shorelines and a series of strandlines deposits along the seawards portion of the spits representing successive stages of widening with longitudinal growth.

Offshore bars: Sand removed from the beaches during storms may be retained as bars between the breaker line and low-tide level. Plunging waves also develop sand bars on either side of the breaker line. None have yet been found of commercial interest.

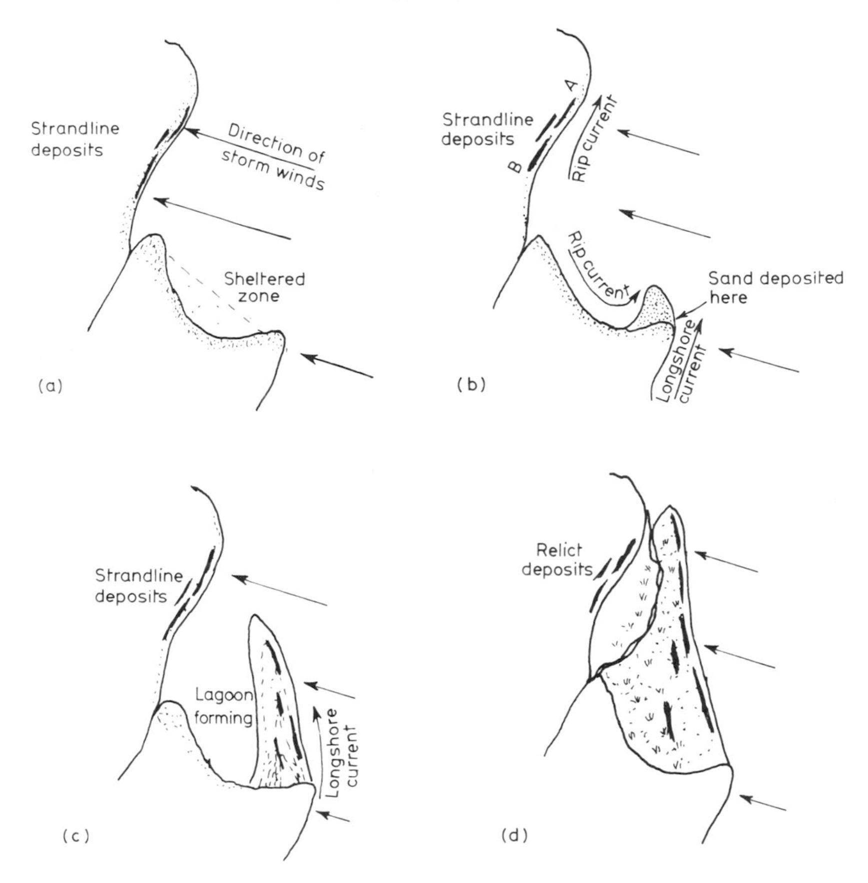

Fig. 3.26 Sand spit growth from headland. Heavy minerals deposits are formed on the beach sections exposed to the heaviest wave action. When the lagoon is finally silted up, a swampy back area is protected from the sea; original deposits (relict in (d)) are covered by sand that has migrated inland

Coastal aeolean deposits

When sand is thrown up onto a beach and allowed to dry out on the surface, the individual particles are blown about in the direction of the wind, either back into the ocean as part of the continual interchange of sediments between the land and the sea, or onto higher ground away from the action of the waves and tides. Sand dunes form according to process mechanisms that are described in Chapter 2 and upgrading occurs because the larger, less-dense particles are impelled forward at a faster rate than the smaller and denser particles which lag behind. Dunal systems are built up in stages. The sediments are mobilized when winds sweep across the ground at higher than critical velocities and large-scale movement takes place when the winds reach gale force. Particles come to rest during periods of calm and stability. In some cases (e.g. on narrow beaches or where the winds are predominantly offshore) a single frontal dune may parallel the coastline for several kilometers and be subject to continued cycles of build up and erosion without further transgression. Usually, however, there is transgression and as one set of dunes moves inland another forms to take its place. Individual dune characteristics vary according to; (a) the rate and supply of sand, (b) the strength, direction and drying power of the wind, (c) land surface geometry, (d) the growth rate of stabilizing flora.

Rate and supply of sand: This has a profound influence on the morphology of dunes. Where the supply of sand is limited and there is no impediment to movement on a flat surface, the pattern of deposition is a series of barchans (Fig. 3.27) which progress slowly across the surface of the ground in the general direction of the wind. Being isolated mounds of sand (Fig. 3.28) the wind resistance is greater in the central portion than along the flanks where the grains are more easily entrained by drag forces in the direction of the wind. Movement is assisted by gravitational forces and barchans are typically crescent shaped with the horns pointing down wind. In the central portion, the sand particles are propelled upwards along the windward surface of the dune and the drag forces are opposed by gravity.

In similar conditions but with a greater supply of sand, transverse ridges are developed with longitudinal axes that are normal to the direction of flow (Fig. 3.29). Systems tend to be evenly spaced and dune heights and spaces between dunes are functions of the strength and constancy of the wind. Aerial photographs frequently reveal them as major desert features although they may also form on a smaller scale in the hind-dune sections of wide beaches.

Strength, direction and drying power of the wind: Winds in temperate to semi-arid climates provide the most favourable conditions for large-scale

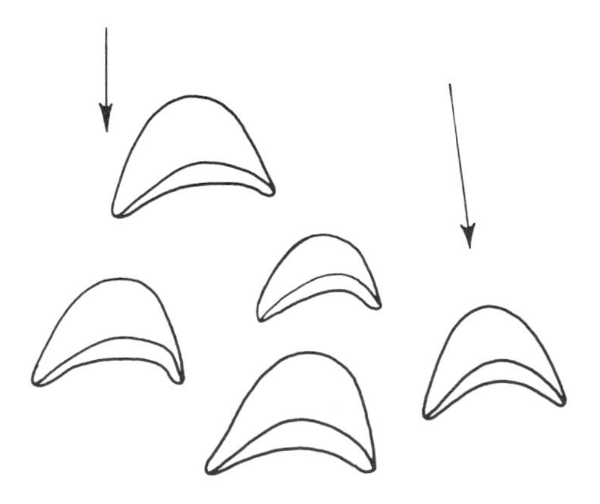

Fig. 3.27 Barchan dunes advancing in direction of wind

Fig. 3.28 Barcan dune near Walvis Bay, Namibia

Fig. 3.29 Transverse dune ridges, Saudi Arabia

aeolean dune development because they have great drying power for much of the year and tend, also, to blow unidirectionally for long periods of time. On the other hand strong winds in the tropics are usually associated with heavy rains that saturate the ground and resist wind erosion through packing and surface tension. Strong winds in sub arctic and arctic conditions have reduced drying power and the ground retains moisture for most of the time.

Land surface geometry: All dunal systems develop their own geometry and individual migrations across their surfaces reflect the local conditions. For example, parabolic or U-shaped dunes are formed when the wind is funnelled into a closed corridor and can escape only by sweeping upwards to conflict with the surface winds (Fig. 3.30). The resulting cross winds, eddies and turbulent mixing of air streams cause a lateral displacement of the sand face at the head of the corridor and ridges are formed that become narrower and sharper as sand is winnowed away from the sides. With time, smaller migrations are initiated on either side of the parabolic basin at the head of the corridor and individual transgressions become oriented obliquely to the direction of the prevailing wind. The process continues until a relief passage provides an easy escape route for the winds.

Seif dunes on the other hand are elongated masses of sand, having their main axes oriented in the direction of the prevailing wind. Their lateral configurations are irregular because of the effects of cross winds of varying strength and direction and geometric complexities. Seif dunes may develop from other dunal forms or in the shadow of an obstruction and systems may assume major proportions. On Stradbroke and Moreton Islands, Queensland, Australia massive seif dune migrations have been developed with individual dune lengths of up to 600 m and with depths of up to 15 m within total systems as much as 5 km in length. Lambert (1975, personal communication) has described them as having broad, strong profiles with, at times, laminated patterns due to the influence of flank

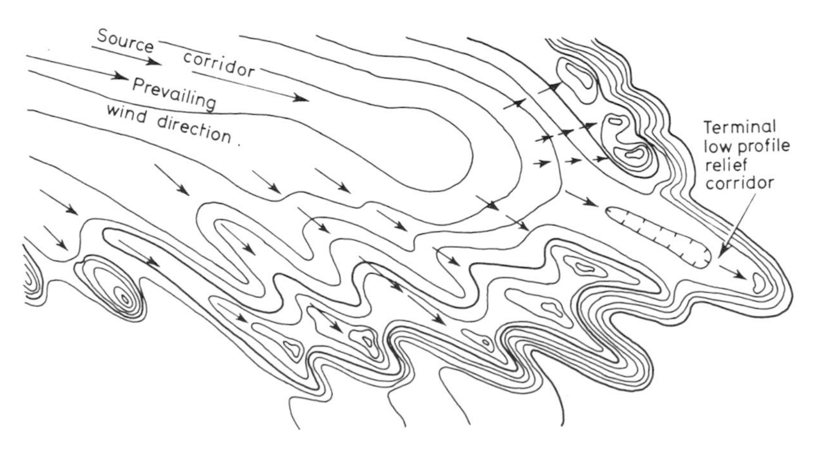

Fig. 3.30 Sequential lateral displacement of dunal formation as successive live transgressions migrate via the source corridor

winds. In his opinion, the systems were formed during periods of higher wind velocities than are encountered today, although in the same direction.

Growth rate of stabilizing flora: Large dunal systems are built up in stages from elements of many different types of stationary and migratory dunes. Distinctive characteristics result from the varied manner in which the dunes pile up on one another and from the complexities of such modifications to the developing pattern as blow outs, slumping, cut and fill. The process is an intermittent one when periods of high winds and violent storms are followed by intervals of calm and stability and major systems may be developed wherever the climate favours the rapid growth of stabilizing flora. Typical grasses on the windward slopes are marram (*Amophila arenaria*) and sand spinifex (*Spinifex hirsutus*). Conifers such as *Casuarina equistifolia* flourish on the leeward slopes together with such stunted shrubs as Planchon stringybark (*Eucalyptus planchoniana*) and various heath-type species.

Individual migrations tend to become fixed in such conditions and subsequent migrations build up one on top of the other. The plant life at each stabilized horizon dies and deposits become separated from one another by beds of peat and other rotted vegetable matter. Usually this means, also, that the underlying sand becomes affected by some degree of organic cementation and in some instances sections of the substrata become heavily indurated. Such effects have important process implications and measures for dealing with sands so affected are described in Chapter 7. They also contribute to the retention of water in dunes. Figure 3.31 shows a typical series of perched water tables lying on top of peat horizons above the true water table.

Very arid coastlines, such as that of the Skeleton Coast of Namibia, have no stabilizing growth between successive migrations to impede their free movement. Consequently, the sand is in a more constant state of movement as the individual dunes migrate inland. Between Swakasmund and Walvis Bay, series of transverse dunes crowd together behind the shoreline and then separate out into barchan forms which move forward more rapidly

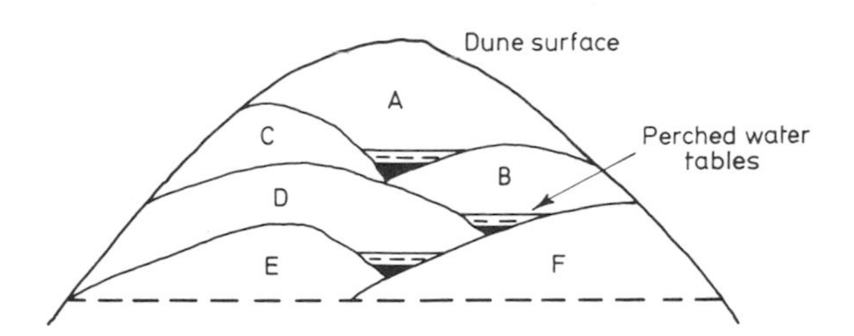

Fig. 3.31 Perched water tables on peat beds in composite dune

than the main mass. They disappear several kilometres inland when first the light and then the heavier particles are swept away across the barren desert floor (Fig. 3.32). In a more temperate climate most of this sand would have been retained along the coastline as in the Richards Bay–St. Lucia area on the opposite coastline of Africa. Here the dune systems are generally less than 0.5 km wide but heights of 100 m and more are common. On Stradbroke and Moreton Islands, Queensland, where individual migrations may become fixed in their locations in a matter of months only, the sequence now rises to as high as 275 m above present sea level.

Mechanism of concentration: Sand has been accumulating along some coastlines throughout the Pleistocene and successive climatic changes and tectonism have resulted in some dunal structures that are highly complex and ore zones in which the distribution of valuable minerals is equally complex. Such dunes may be stationary or migratory. Stationary dunes form by accretion whenever there is sufficient time between individual migrations for the normal, more gentle winds to deposit sand evenly over a mound, or where some barrier to the flow cannot immediately be overcome. Migratory dunes, i.e. those which move forward bodily, do so by virtue of the wind picking up sand on the windward side and depositing it over the crest and down the slip face on the leeward side. This process involves the subtraction of sand from one side and the deposition of an equal amount on the other to keep the whole mass moving

Fig. 3.32 Remnant of heavy mineral on desert floor inland from the coastline, between Swakasmund and Walvis Bay, Namibia

ahead. Each type has its own characteristic structure: stationary dunes are built up in more or less concentric layers following the curvature of the dune, while migratory dunes are in layers that parallel the slip face. Figures 3.33(a) and (b) are schematic representations of structure in the two main types.

Sand grains move forward largely by saltation and surface creep and are sorted according to their physical properties and the geometry of their environs and other process elements. For example, in the formation of some stationary dunes a gentle winnowing of the particles leaves concentrations at the base of the windward aspects. In most cases, however, the migration of windblown particles is effected through the transfer of sand from the windward slope to the leeward of a dune. In the area of erosion, the lighter silica particles in saltation leap higher and travel further with each movement, thus allowing a progressive build up of the valuable heavies in the slower-moving portion. Smirnov (1976) suggests that this concentration is confined to the tail of individual dunes and to small hollows and depressions in deserts. This is an oversimplification when applied to coastal aeolean deposits in which some very large concentrations have been built up as integral parts of whole dunes.

Such concentrations are associated with migratory dunal systems in which, although the movement is slower, the heavier particles continue on and pass over the crest to come to rest on the leeward slope. Successive periods of erosion, transportation and deposition result in a series of individual concentrations in each of the active dunes. Additional sorting takes place during each of the formative periods when the mineralized sections are remobilized and swept forward. Figure 3.34 describes the erosion of a migratory or transgressive dune from which, in the zone of erosion, the lighter particles are removed preferentially from the surface leaving the heavy mineral seams relatively intact. The seams protrude a centimetre or so above the surface and the ground assumes a 'dished' configuration. The raised portions are then more quickly eroded by the wind in their exposed positions; the heavies are again blown forward and the process is repeated.

However, sorting is not confined only to the preferential deposition of the valuable (heavy) and non-valuable (light) fractions, taken as a whole it occurs also in respect of individual heavy mineral particles having size, shape and density differences. For example, on Stradbroke Island,

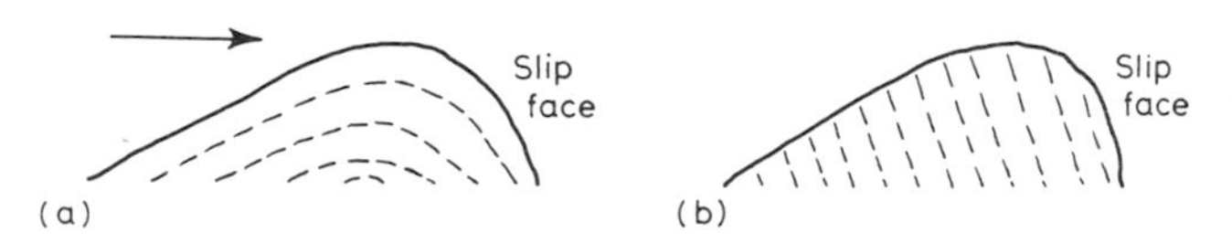

Fig. 3.33 Structures of (a) stationary (b) migratory dune

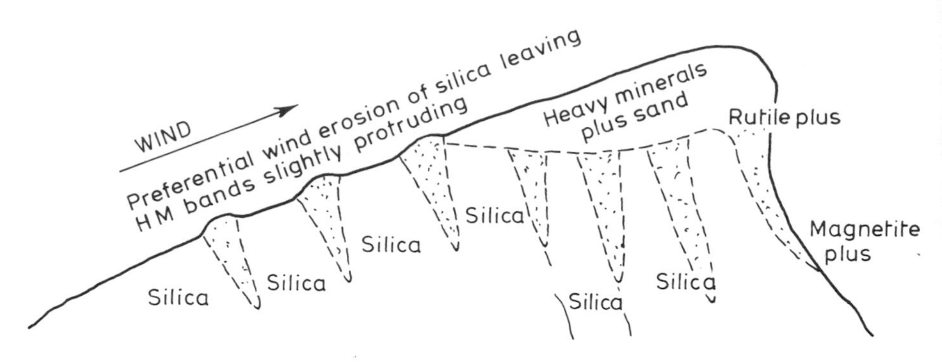

Fig. 3.34 Truncation and redeposition in transgressive dune, the deposition of alternate layers of sand and heavy minerals due to varying wind velocities

Queensland, zircon particles travel shorter distances in saltation than particles of rutile which, in turn, are deposited behind the coarser magnetics and rock forming minerals which are deposited further down the leeward and flank slopes. Hence, in any dune configuration on Stradbroke Island, and in the direction of travel, the tendency is for a preferential zoning of zircon along the crests, then rutile and finally the magnetic and rock-forming minerals down the slopes. Figure 3.35 illustrates a typical substrata section showing the relative distribution of the various mineral types.

The pattern is repeated for each migration so that, in any vertical section, both grades and minerals distribution vary within wide limits. On Stradbroke Island the rutile content of the valuable heavy mineral concentrations is between 13 % and 20 % in individual samples. The zircon content is similarly variable and the pay content of the heavy mineral assemblages differs from place to place by as much as 35 %. Figure 3.36 is a plot of grade variation with depth, as the average of a number of boreholes in high level dunes at Cape Morgan, South Africa.

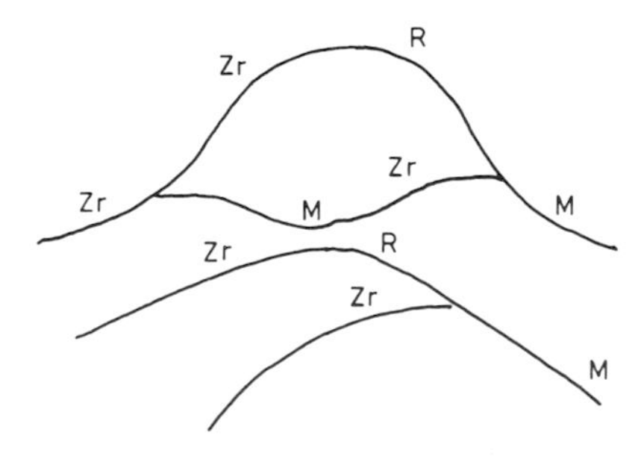

Fig. 3.35 Distribution of dominant concentrations of zircon rutile and magnetics in typical sub-strata section of high level dunes on Stradbroke Island, Queensland

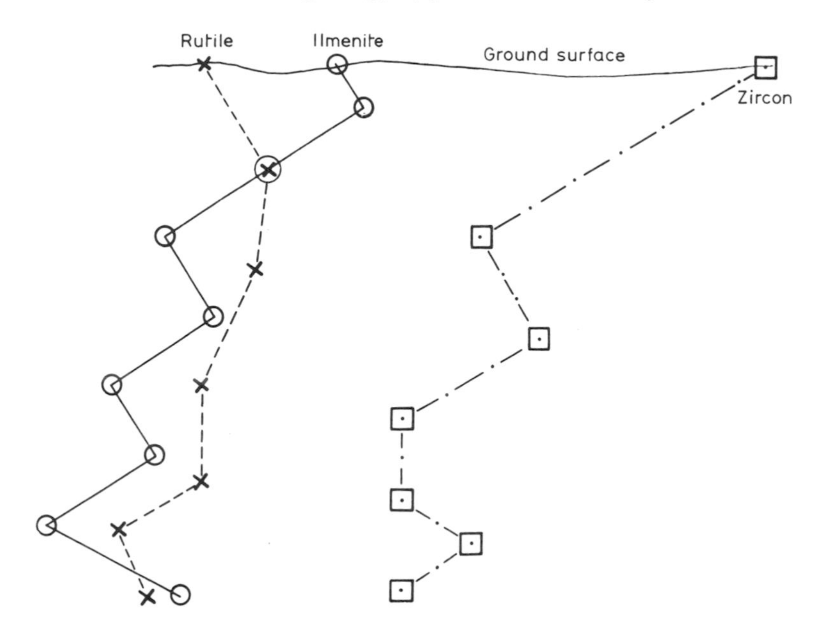

Fig. 3.36 Variations of minerals distribution with depth, Cape Morgan, South Africa

Fossil dunes: These dunes represent the eroded remnants of shoreline and aeolean placers formed along coastlines before either tectonic uplifts or falling sea levels left them elevated above the particular environment within which they had their beginnings. In Western Australia, fossil strandlines are found over a distance of more than 1000 km along the coastal plains from Geraldton down to the south-western tip of the continent. Early strandline deposits have been identified at successive levels above present sea level of 120, 110, 104, 92, 87 and 85 in the sectors between Perth and Geraldton and south of Perth at levels of 76, 40, 35, 28, 4.8–6.0, 3.0, and 1.0–0 m respectively. Some incision of fossil strandlines takes place where newly developed streams cut through them but significantly large sections may be preserved, particularly in large areas of low relief.

Deltaic placers

Deltas are built up from sediment-laden streams that branch out and change course frequently on flat sea-floor gradients. Old channels silt up when new channels form and, with time, islands of sedimentary fill become part of the extending mainland and lose their separate identities. In the Indus Valley of Pakistan, port facilities used by shipping only a few hundred years ago now lie as much as 60 km inland. The present main mouth of the

Indus River is several kilometres distant from the position it occupied only a few years ago, and branches from this channel constantly change as new islands are formed and the delta builds out.

Mechanism of concentration: The sediment load of a river is governed by its transporting ability and when fast-flowing streams reach the sea, the abrupt change in velocity causes the sediments to settle around the mouth. The coarser particles are deposited first and are either dispersed by wave and tidal action to make way for renewed deposition, or remain in place so that a delta is formed. The finer particles carry further out to sea and settle in deeper waters. The ultimate form of deposition is determined by many factors including the energy of the coastline, the presence or otherwise of submarine canyons leading out into the depths and the quantity of sediment to be moved. Although deltas tend to form wherever tidal amplitudes are small they also form when the sheer volume of sediments disgorged by such rivers as the Indus and Ganges is so great that the amounts dumped are far in excess of what can be transported away.

Characteristics of the deltaic sub-environment are gently sloping sea floor gradients; networks of channels and currents, some of which reverse with the tides; generally low tidal ranges and calm seas, except where occasionally whipped into action by violent storms.

Observed patterns of mineralization along the seawards aspects of delta islands are related to a combination of sea-bed gradients and currents. For example, differences in chenier type sand dunes along the sea fronts of one string of ten islands in the Indus delta are marked. The islands have been numbered in order 1–10. Island 1 with a steeply sloping approach has the most extensive frontal dunes but they are barren. Islands 2, 5, and 9 with gentle approach slopes have less sand, but are the most heavily mineralized. Evidence from earlier islands that have become terrestrial suggests that some mineral deposits have been preserved but have been subsequently covered by silt; others have been blown about and dispersed by the wind. Delta islands along their foreshores are in a constant state of flux in respect of marine dynamics and the annual changes in size and shape are considerable.

3.3.3 Marine placers

Marine placers are mainly drowned deposits that were formed originally under transitional or continental conditions during periods of emergence. The morphology of any shelf area may thus be considered in the general terms of underwater extensions of onshore deposits and relicts of beaches, valleys and hills with terraces and strandlines corresponding to periods in which the sea level remained static over long enough periods for placers to form. Most continental shelves were exposed to sub-aerial weathering for only brief periods – less than 25 000 years for the latest glacial stage and similar time spans for the earlier ones. This has been the main limiting factor determining the extent to which some source rocks were weathered.

The lithology of drowned placers is usually recognizable as to the source areas thus making it possible to identify separately, the sediments from beaches, rivers, slopes, and glaciers even though they may have been reworked to some degree during submergence.

Drowned deposits of transitional origin

The search for economically viable deposits of the heavy minerals rutile, zircon, monazite and ilmenite offshore from where large onshore deposits are known, has so far met with little success. The testing of submerged beaches along the New South Wales coastline has yielded only low values, less than 1 % of total heavy minerals, and significantly lower proportions of the valuable minerals rutile and zircon (usually much less than 50 % of the heavies), even though some of the worlds major deposits were formed onshore in the same localities. However, two distinct placer types were identified offshore in the Tweed Heads district of New South Wales namely; (a) superficial blanket type deposits 2–4 m thick, usually containing less than 1 % heavy minerals, and (b) long, narrow seams covered by 5 m or more of barren sediments.

As described by Jones and Davies (1979) these latter deposits represent former strandlines but although one recorded sample gave a value of 50 % heavy minerals over a 2 m interval, more than 80 % of the heavies comprised particles of amphibole, magnetite and ilmenite. These minerals were probably derived from nearby Tertiary volcanics, and apparently are unrelated to the deposits on the adjacent beaches.

One exploration company claimed to have outlined a resource containing 375 million tonnes plus a further indicated 500 million tonnes of mineral sands off the New South Wales coastline. However, most of the work was done in water deeper than 30 m, the holes were widely spaced and the sampling techniques were suspect, being largely suitable only for reconnaissance (see Chapter 4). In the central area, high proportions of rutile plus zircon ranging up to 50 % of the heavies were encouraging, but the average valuable heavy mineral content was only about 0.80 %. Elsewhere the average grade of around 0.20–0.22 % heavy minerals was well below the levels needed for economic viability in the environment at this time.

No rational explanations have yet emerged to explain shelf morphology fully and, although linear features believed to be related to Quaternary low sea level stillstands have been identified, as yet there is no simple picture to present the general case. Schofield (1976) argues that the so-called drowned beaches and sand spits off the east coast of Otago, New Zealand, are actually in the process of formation under present hydraulic conditions. He suggests that the continental shelf should not be thought of simply in terms of a little modified, drowned landscape but rather as a seascape in equilibrium, or approaching equilibrium, with present sea level. On the other hand, while there is good reason to believe that most of the heavy minerals remain in the active surf and beach zones during submergence and that the bulk of the sand follows the shoreline as it moves inland, there is also ample evidence elsewhere of relict strandlines offshore that now lie well below the active sedimentary levels. In some cases the deposits may have been trapped behind a steeply sloping backshore or in an embayment surrounded by low cliffs; others may have accumulated on the seabed from the movement of ocean currents alone, although this does not appear very likely. Whatever the causes, many of the answers have yet to be found. Company operations have shown only minor enrichments in the shallow near-shore sediments. As Jones and Davies point out, during much of the Pleistocene, sea level fluctuated at levels that place the sediments at depths beyond the reach of present-day mining techniques and consequently there is little incentive for more detailed studies.

Drowned deposits of continental origin

Minerals distribution patterns that are readily identified along foreshore areas, and to a lesser extent along the landward side of the breaker line, are much less understood in deeper waters where accessibility is a major problem. Geological evidence of the extension of metallogenic belts offshore is found in many shelf areas such as off the coastline of eastern Canada and north eastern USA (Emery and Uchapi, 1965) but again the generally thick sedimentary cover and deep water puts many of these extensions beyond the reach of present-day exploration and mining techniques (Chapters 4 and 5).

The marine placers of south-east Asia

Economically viable concentrations of cassiterite and other heavy minerals occur along portions of the Sunda Shelf, Malacca Straits and shelves of the Andaman Sea along the coastlines of Thailand and Burma. They result from the weathering of tin granites in a belt of Mesozoic folding stretching northwards from the tin islands of Bangka and Belitung, Indonesia, through Malaysia, Thailand and Burma to the Himalayas. They are currently the most important shelf areas in the world in respect of valuable cassiterite placers. Underwater extensions and relics of beaches, valleys and hills with terraces and strandlines correspond to periods during which the sea level remained static for prolonged periods.

Sujitno (1977) has recognized three placer types in the Indonesian sector:

(1) Primitive placers, formed entirely under terrestrial conditions
(2) Intermediate placers in which the progression–retrogression of shorelines due to eustatic changes in sea level has resulted in several phases of upstream deposition
(3) Modern placers, in which the depositional environments were alternately terrestrial and marine.

Seismic records have revealed two distinct classes of marine sediments in this area, old and young, and three types of bedrock: granite, sedimentary and younger sediments from the peneplanation of Sundaland (Tertiary sediments). The offshore tin deposits are found in:

(1) Quaternary valleys which may contain either the primary source rocks at various stages of weathering or fluvial concentrations in old streambeds traversing the valleys. Sea level changes have resulted in modifications to the original sedimentary pattern and in the acquisition of additional features from subsequent cycles.
(2) Terrace deposits, both continental and marine, that have been partially reworked in surf and swash zones.
(3) Eluvial and colluvial deposits, locally known as 'Kulit' and 'Kaksa', which are representative of the early peneplained surface of Sunda land and may be expected to retain some of the original features.

The distribution of values in drowned valleys

Depending upon the extent and number of the sea level changes that have taken place, the stratigraphy in offshore mineralized valleys may follow the same general vertical sequence as described by Sujitno, i.e. young alluvial, young marine sediments, old alluvials, old marine sediments, basement.

Placers are found at basement and in some of the higher alluvial beds but are normally absent from the marine sediments laid down during periods of stillstand.

The lateral distribution of heavy minerals in such deposits normally follows the direction of the dominant current axes. Occasional small local concentrations may be aligned in almost any direction; however, regionally they follow the direction of the main flows.

Sujitno (1977) published a simplified model describing the placer deposits and processes leading to their formation (Fig. 3.37) and one describing the best and least-favourable locations according to levels or erosion and proximity to source rocks (Fig. 3.38). These models probably present the general case for drowned deposits of the heavier and more valuable continental placer types in any shelf areas.

Offshore sand and gravel

The third, and currently the most important offshore placers from an

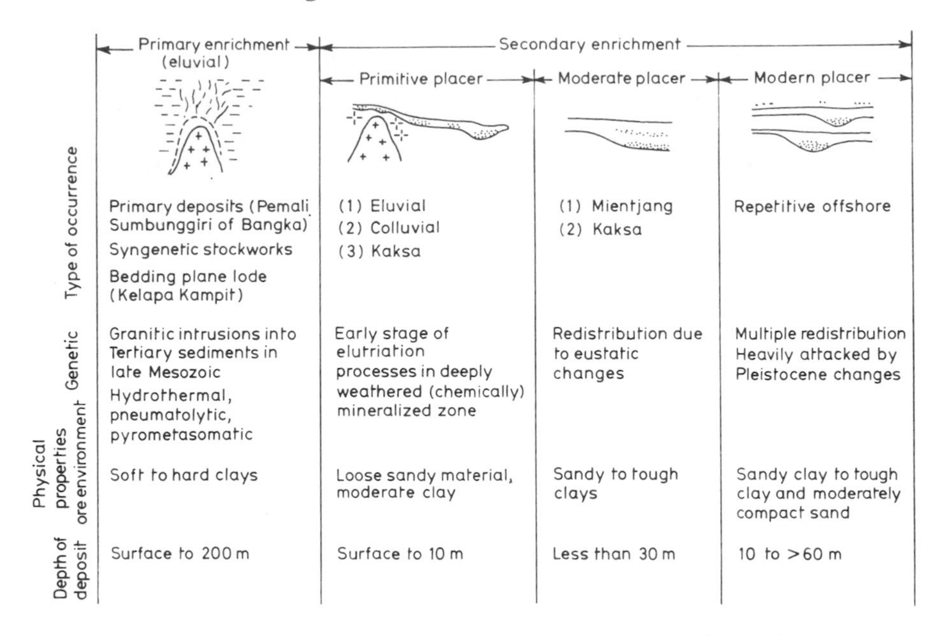

Fig. 3.37 Simplified model of tin depositional processes (after Sujitno, 1971)

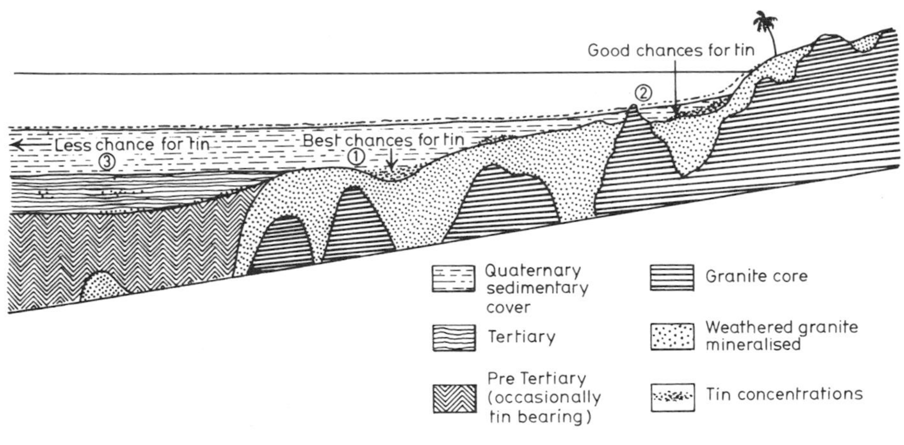

Fig. 3.38 Simplified model for offshore tin placers erosion levels and proximity to source rocks

economic viewpoint, are the deposits of sand and gravel found along continental shelves throughout much of the world. Most of the gravel beds have originated from former metamorphic terraines weathered under subaerial conditions. The total supplies are virtually inexhaustible but nearly all shelves are overlain by a deep cover of very fine marine sediments and probably within the foreseeable future only those deposits that crop out on the ocean floor or have only a very thin silt cover will be workable.

REFERENCES

Allen, J. R. L. (1973) *Physical Processes of Sedimentation*, Earth Science Series 1. Unwin Uni Books, London.

Baker, George (1962) *Detrital Heavy Minerals in Natural Accumulates*, Aust. Inst. Min. Met, Melbourne.

Chayes, F. (1964) A petrographic distinction between cainozoic volcanics in and around the oceans, *Journ. Geophys. Res.* **69,** 8.

Davies, P. J. (1979) Marine geology of the continental shelf off south east Australia, *BMR Bull.* **195,** Bur. Min. Resources, Canberra.

Emery, K. O. (1966) Geological methods for locating mineral deposits on the ocean floor, Trans. 2nd. Mar. Tech. Soc. Conf. in *Exploiting the Ocean*, 24–43.

Emery, K. O. and Uchapi, E. (1965) Structures of Georges Bank, *Marine Geol.* **3,** 349–58.

Evans, David Le Count (1981) Laterization as a possible contributer to gold placers, *E&M.J.*, 86–91.

Hess, H. H. and Poldervaart, A. (1967) *The Poltervaat Treatise on Rocks of Basaltic Composition*, Vol. 1. Interscience Publishers, New York.

Jones, H. A. and Davies, P. J. (1979) Preliminary studies of offshore placer deposits eastern Australia, in *Marine Geology No. 30*. Elsevier, Amsterdam. pp. 243–268.

Landsweert, D. R. (1869) Metallic deposits–gold nuggets theory as to their formation, *Mining and Scientific Press, San Francisco*, **17,** 82 and 274.

Leopold, L. J., Wolman, M. G. and Miller, J. P. (1964) *Fluvial Process in Geomorphology*, Freeman, London.

Liddy, John C. (1972) Mineral deposits of the south western Pacific, *Mining Magazine*, March.

MacNevin, A. (1972) Sapphires in the New England district, N.S.W., *Records of Geol. Surv. of NSW*, **14,** Part 1.

McGowan, J. H. and Garner, L. E. (1970) Physiographic features and stratification types of coarse grained point bars. Modern and ancient examples, *Sedimentology*, **14,** 77–111.

Minter, W. E. L. and Toens, P. D. (1970) Experimental simulation of gold deposition in gravel beds, *Trans. Geol. Soc, South Africa*, 89–98.

Noakes, L. C. (1976) *Review of Provenances for Mineral Sands and Tin in Southeast Asia*, CCOP (XIII)/38 (ESCAP), Bangkok.

Overstreet, W. C. (1967) The Geological Occurrences of Monazite, *Geol. Surv. Prof. Paper 530*, US Govt Printing Office, Washington DC.

Patterson, O. (1965) Exploration for rutile and zircon deposits, in *Exploration and Mining Geology*, Vol. 2. *Eighth Comm. Min & Met. Congress Australia & New Zealand*, 336–42.

Rao, K. B. (1975) A tantalite pegmatite from Bihar, India, *Journ. Geol. Soc. India*, **16,** 3, 310–16.

Reineck, H. E. and Singh, I. B. (1975) *Depositional Sedimentary Environments*, Springer-Verlag, New York.

Ruhe, Robert V. (1975) *Geomorphology*, Houghton Mifflin, Boston.

Schofield, J. C. (1976) Sediment transport on the continental shelf, east of Otago – a re-interpretation of so called relict features, *N.Z. Journ. Geol. & Geophysics*, **19**(4) 513–26.

Schulling, R. D. (1967) Tin belts on the continents around the atlantic ocean, *Ec. Geol.* **62,** 540–50.

Schumm, S. A. (1977) *The Fluvial System*, John Wiley, New York.

Schumm, S. A. and Stevens, M. A. (1973) Abrasion in place: a mechanism for rounding and size reduction of fine sediments, *Geology*, VI 37–40.

Sestini, G. (1978) Sedimentation of a 'paleoplacer' the gold bearing Tarkwaian of Ghana, in *Encyclopaedia of Earth Sciences Serv.*, (eds Rhodes W. Fairbridge and Joanne Bourgeois). pp. 275–305.

Sigov, A. P., Lomayer A. V., Sigov A. V., Storozhenko L. Y., Khrypor, V., and Shub, I. Z. (1972) Placers of the Urals, their formation, distribution and elements of geomorphic prediction, *Soc. Geogr.*, **13,** 375–87.

Singewald, Q. D. (1950) Gold placers and their geologic environment in Northwestern Park County Colorado, *US Geol. Surv. Bull.* **955**-D.

Smirnov, V. I. (1976) *Geology of Mineral Deposits*, MIR Publishers, Moscow.

Sujitno, S. (1977) *Challenging Elements in the Offshore Tin Resource Potential of Indonesia*, CCOP/TAG(x)30, Bangkok.

Taylor, R. G. (1976) *Tin Deposits – Geology and Exploration*, Aust. Min. Found. Ltd, Adelaide.

Wells, John H. (1969) Placer examination, principles and practice, *Tech. Bull No 4*; U.S. Dept. Int. Bur. Land Management, Washington DC.

Whitworth, H. F. (1958) The zircon rutile deposits on the beaches of the east coast of Australia with special reference to their mode of occurrence and the origin of the minerals, *Dept. of Mines, N.S.W, Techn Rept. No. 4*, 7–60.

4

Exploration and Prospecting

4.1 INTRODUCTION

The search for placer minerals was once considered a simple matter of panning upstream towards the source and loaming on the sides of hills to find the main accumulations. Sampling was carried out using hand methods of boring and pitting and the samples were dressed by hand or in crude sluice boxes. Up until the mid-eighteenth century there was no estimation of ore reserves as such and a prospector having located a deposit, immediately settled in and commenced mining. If the returns were low, the prospect was abandoned and prospecting was resumed elsewhere; if the returns were satisfactory the deposit was mined in its richest sections and then the prospector-miner moved on.

The adoption of more scientific methods of stream sediment analysis, shortly after the turn of the century, led to a gradual evolution in techniques. The introduction of new and costly mining plant demonstrated the need for more systematic exploration and more reliable estimates of ore reserves to eliminate unnecessary risk. At first, prospectors were expected to be their own geologists, mineralogists, chemists, miners and valuers but the inadequacy of any one individual's ability to perform all of the functions well was soon made apparent by the many failures of dredging companies floated on optimism rather than on facts.

As a result of these failures, specialists began to appear in various branches of placer technology and exploration became separated from prospecting in the context that *exploration* became the investigation of large areas in order to understand the surface and near-surface geology; and *prospecting*, the testing of individual deposits to prove their worth. Exploration, as such, now embraces studies of soils and rocks, geomorphology, sedimentology, geophysics, geochemistry, paleoclimates and the geological history of the environments; prospecting involves the physical processes of boring, pitting, sampling, sample dressing and engineering.

This new scientific approach does not mean, of course, that geologists are concerned only with exploration and engineers only with prospecting although it is often wrongly assumed that this should be so. In fact, the two areas of responsibility are clearly interdependent and, although some aspects are undoubtedly of more interest to one than to the other, neither function can be handled effectively alone. Prospecting tools are used by explorers to confirm or deny early hypotheses and experience in their use is the basis for selecting methods and equipment for the final prospecting stage. Common to both is that success or failure depends upon the ability to measure accurately and to take samples that truly represent what is being sampled. Nothing that follows, however skilfully done, is of any use if this ability is lacking.

Practice varies according to the environment and some exploration programmes in remote areas still offer scope for the older methods of prospecting. However, most deposits with surface expressions have already been found and more than just 'rule of thumb' methods are needed for modern exploration programmes, particularly offshore where exploration involves regional aspects of provenance and a multi-disciplinary study of the Quaternary geology of the coastal and shelf areas to determine the possible existence of favourable morphogenic features inherited from the past. The initial work is done on a regional scale through the cooperation of governments and international agencies, oil and minerals exploration groups and universities, and from a sharing of all knowledge as it is acquired. The scope is international because costs and physical resource requirements are far beyond the capacity of any one agency. Its findings assist local explorers by indicating favourable target areas within the field of search.

Minerals distribution patterns are readily identified along foreshore areas, and to a lesser extent along the seaward side of the breaker line: they are much less understood in deeper waters where accessibility is a major problem. For example, although geological evidence of the extension of metallogenic belts offshore is found in many shelf areas, such as off the coastline of eastern Canada and north-eastern USA and offshore Burma, Thailand and Malaysia, the generally thick marine sedimentary cover puts these extensions beyond the reach of present-day exploration and recovery techniques. Such deposits, for which there is no clear lithology, and placers that have survived a long geological history of environmental changes are the most difficult to find and prospect. Formed in one environment and perhaps modified in others, these deposits now lie hidden beneath various thicknesses of sediments, volcanics and vegetation onshore and by oceanic waters and marine sediments offshore. Discoveries of such deposits are made by professional geologists and engineers; prospecting in the old sense of the word is more a weekend hobby for amateurs than the way of life it was for so many of the old timers.

4.2 EXPLORATION

Modern explorers have at their fingertips an ever increasing array of technological aids and deposits are now being found in locations that were not considered prospective even as short a time as a few years ago. It is timely, therefore, to take an inventory of current procedures and to realize that, more and more, geology is becoming the essential central feature in all placer exploration. Important contributions are made in such other fields as geophysics and oceanography but it is primarily upon geomorphological and sedimentary grounds that a target is defined, the operations planned and decisions made upon the services to be used and the fitting of their respective contributions into the overall pattern of exploration. It must be realized, however, that while the appropriate skills have either been developed or can be acquired, the overall need for increased investment in exploration will be met only if potential investors are assured that geologists will spend their money wisely and that all of those factors determining the respective priorities of individual prospects will be subjected to the closest scrutiny before final decisions are made. This assurance can be given only by geologists who are as much aware of the economics of the search as of the geotechnics involved.

Exploration is a search for the unknown and, for placers, the areas of study extend from the slopes in high mountain ranges to the sea floor along the margins of the land. In some cases exploration is aimed at finding new deposits; in others it involves locating fresh channels for testing in already known drainage systems. Past failures do not mean, necessarily, that a particular area lacks promise for the future because the methods may have been unsuitable or the geology poorly defined. For example, in describing a field geology oriented approach to programming a modern mineral exploration survey Goosens (1980) gave the example of a geochemical programme in Ecuador where the stream sediments were analysed only for Cu, Pb, Ni, Co and a few for As and Ag, thereby missing a potentially major tin deposit found later by a prospector using only a pan to guide him in his search.

4.2.1 Ore reserves and resources

Explorers are concerned with both ore reserves and resources and, hence, there must be a clear understanding of what is meant by the terms and of their interchangeability.

(1) A deposit constitutes a reserve of ore if it can be exploited at a profit under existing economic circumstances and available technology. It is finite in both quantity and grade and the means for recovering and marketing the valuable constituents are necessarily proven.
(2) Resources, on the other hand, are independent of time and present economics and consist of both known and as yet undiscovered deposits irrespective of grade and the technology to deal with them.

Government and company attitudes naturally differ greatly in respect of their individual resource policies and economic standards. Governments consider both the present and future availability of certain strategic minerals and will either seek to stockpile the requisite amounts or aim at having the ability to produce adequate quantities in times of emergency regardless of cost. A valuable national resource may contain only trace quantities of the minerals being sought if they cannot be obtained from any other source.

Company policy is aimed, essentially, at making a fair profit from the results of its exploration programmes and decision-making is influenced primarily by market considerations in both the short and long term. A currently non-exploitable resource may be explored further if shortages are likely in the foreseeable future; an expected overabundance may point to the desirability of utilizing exploration funds in other ways.

The terms ore reserves and resources are interchangeable because the boundaries distinguishing one from the other shift according to prices and policy decisions that are, themselves, variables of changing economic, political, and technological trends.

Total resources are divided on a global scale into a number of categories. Taking as an example the tin resources of the world, the US Geological Survey Professional Paper 820 has made a classification in terms of reserves, conditional reserves and undiscovered resources. Table 4.1 is an

Table 4.1 Tin reserves and resources, the Asian scene (000 tonnes)

Country	Reserves		Conditional Reserves		Undiscovered Resources	
	Measured plus indicated	Inferred	Para marginal	Marginal	Hypo-thetical	Specu-lative
Burma	250	250	—	—	250	250
China (PR)	500	1000	1000	1000	1000	1000
Indonesia	500	1860	540	540	—	—
Malaysia	600	230	—	1000	1500	1000
Thailand	217	1000	1860	—	1500	1000
Other	17.5	90	—	—	—	100
	2084.5	4430	3400	2540	4250	3350

extrapolation of data from Table 135 of that paper and covers the Asian scene.

The classification 'undiscovered resources' considers that entirely new deposits will be found, particularly in the remoter regions of the world, and the above predictions which, as their heading suggests are highly speculative, are based upon hypothetical models and empirical relationships now coming into prominence in a number of fields, related and otherwise. Rudenno (1978) has applied one of the laws, Ziptf's Law, to predict world wide the size and grade of primary tin deposits yet to be found. He foreshadows the discovery of an additional fourteen deposits, each containing in excess of 40 000 tonnes of tin and suggests from the distribution of already known deposits, that two of them will be found in Indonesia, perhaps in the offshore regions. If this is so it can be expected that at least some of the primary deposits will have related placers, some of which have still to be discovered.

4.2.2 The placer exploration model

This model is designed to show how exploration leads to prospecting through the various branches of geology. The two functions are distinguished primarily on the basis that exploration involves mainly geological expertise while engineers are more concerned with physical aspects of boring, sampling and evaluation. The three main divisions of the model described in Fig. 4.1, are provenance, geomorphology and history; the main associated technologies are sedimentology, geochemistry and geophysics as tabulated in the accompanying notes.

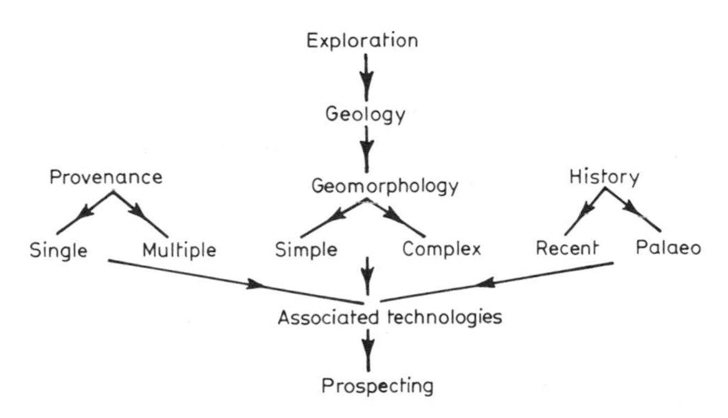

Fig. 4.1 The placer exploration model

Exploration model notes

Function	*Division*	*Notes*
Exploration geology	Provenance	The zone of provenance may be a single large lode formation or a complex of de-

Function	*Division*	*Notes*
		posits spread over a wide area. The quantity of valuable mineral released from the source rocks is the most important factor, not necessarily the grade of the host materials. Large placers result only from large provenances.
	Geomorphology	Economic concentrations may occur wherever the physiography and climate have provided suitable conditions for deposition. A high relief favours rapidly flowing streams, good sediment transport and sorting. In such conditions, erosion is rapid, fresh surfaces are exposed to weathering more often and even occasional flooding may form channels within which segregation can take place. A low relief favours deep weathering in some environments but not in desert or semi-arid conditions.
	History	Recent stream channels in arid conditions are shallow, with sediments that are poorly sorted. In such environments much of the mineral is still locked up in fragments of quartz. Early miners probably exploited many of the colluvial and recent fluvial deposits but in many cases lacked the technology for mining channels deeper than 4–5 m. Exploration should be concentrated in areas of provenance having histories of deep weathering in more humid climates and subsequent high relief. In these circumstances, significant placers may have been formed and covered by later, barren sediments.
Associated technologies	Sedimentology	How sediments behave once they are formed depends upon how much energy is available in the form of running water and wind. Where this is limited, deposition tends to be concentrated near the source rocks and placers are unlikely to be formed except in the colluvium and in fans. Sorting is progressively well de-

Function	*Division*	*Notes*
		veloped away from the source only if sufficient running water is available with sufficient energy to transport the sediments in mass and to assist in breaking them down into smaller sizings. Sedimentation is thus linked with provenance, geomorphology and history.
	Geochemistry	Mainly particulate for gold, platinum and tin and aimed at detecting particles in stream sediments, fans, terraces etc. However, some tracer elements and minerals are important (see pp. 185–87)
	Geophysics	To define paleochannels and buried placers. Gravity methods may be tried first for large flat areas followed by seismic, resistivity etc. as seen to be applicable.

In spite of the many advances made, there are still some areas where technology has so far failed to provide many of the tools needed, at costs that can be supported by the ongoing programmes. In this regard, recent technological achievements suggest that economics rather than the physical problems involved may be the main limiting factors, particularly offshore where specific problems relate to sampling in the deeper shelf areas and difficulties in penetrating acoustically opaque sedimentary horizons. As in the past, the solutions to these and many other problems may come eventually from large experimental programmes in other fields of sufficient importance to justify the very high costs.

4.2.3 Programme logistics

Commencing the Programme

The obvious starting point for exploration is the collection of historical and background data. Primary sources of information are government agencies for aerial photographs, satellite imagery, geological reports and surveys, and records of previous operations. Reference is made also to libraries in universities and research centres. In some cases miners and local residents supply important information from their own knowledge and experience and there may be some analogous models to study from other fields. Both onshore and offshore the overall impression is considered favourable when there is direct evidence of:

(1) A substantial provenance in an appropriate geological setting (Chapter 3).

(2) A prolonged weathering regime in a mainly temperate to tropical climate (Chapter 2).
(3) A dynamic transport system with favourable conditions for deposition (Chapter 2).
(4) A favourable sequence of land and sea movements (Chapters 1 and 4).
(5) Meteorological and oceanographic conditions favouring the use of currently available technologies for prospecting and mining offshore (Chapter 4).

Consideration is then given to individual sectors in order to select the most prospective areas for future study. Certain minimum standards are set and areas that appear unlikely to fulfil them are either rejected forthwith or held for future examination. Priorities are set for the remainder in terms of possible size and value and, with the scope of the exercise known, the regional programme can be planned. It is realized, of course, that initial judgements are tentative, at best, and consequently such programmes normally have sufficient inbuilt flexibility to allow changes of direction when required and for the initial target quantities to be scaled up or down according to what is learned.

The first task is to reconstruct the geological history of the environment at each stage of its development in order to predict the existence and possible locations of economic deposits. Detailed studies investigate the geomorphic history of the drainage systems with particular reference to tectonic implications, climatic changes and changes in base level. Close attention is paid, offshore, to the chronology and absolute dating of sedimentary units and geomorphic surfaces and to the Quaternary history of the shelf and coastal regions.

The visual presentation of data

A basic requirement is a map, however primitive, of the area to be explored. National maps relating specifically to detrital heavy minerals, if available, are probably on a scale of 1:1 000 000 or even 1:5 000 000. Such broad coverage usefully documents regional trends and sediment distribution but it cannot cope with the details required for analysing future minerals potential in individual areas. For large areas of 5 000 km^2 or more the key index maps may be either 1:250 000 or 1:100 000 depending upon the availability of information and the main purpose of the investigation. 1:250 000 maps correspond to blocks of 1° latitude and $1\frac{1}{2}$° longitude; 1:100 000 maps correspond to blocks of $\frac{1}{2}$° latitude and $\frac{1}{2}$° longitude.

Maps are the most efficient means for comparing two-dimensional data from different sources and they offer presentations that can as easily be understood in the boardroom as in the office. Maps as large as 1:250 000 can be used in areas of provenance to distinguish rock units according to age and rock type, and to demonstrate the main topographical and drainage systems. 1 : 100 000 maps provide more detail in areas of geological interest and are more effective as a framework for interpretation. Both scales are generally suitable for correlation with available remote sensing imagery and with maps produced by other organizations.

Compilation of onshore data

River systems are plotted on regional maps in as much detail as practicable to show their relationships to the general form of the landscape and its drainage. Important factors influencing the transport of detrital minerals are indicated by form lines or hachures and access is indicated by showing major roads, railways and port facilities in bold relief. Appropriate symbols are used for rock, sandy or muddy shorelines, raised shorelines and major areas of alluvial flats. (See appendices at end of chapter.)

Land geology is generalized to emphasize geological sub-divisions thought to constitute different provenances. In this regard, recent volcanics which shed only titaniferous magnetite may be represented as one unit whereas, igneous bodies that may constitute important sources for a number of detrital heavy minerals are differentiated where possible. Attempts are made also to delineate sediments, such as sandstones, which are the immediate source rocks for detrital heavy minerals found on beaches. Where the provenances for detrital minerals are only partly understood it is desirable to show as many geological sub-divisions as practicable on the scale of the map.

Scales ranging down to 1:10 000 are used for compiling data from field plotting and for the geological interpretation of aerial photographs. A geomorphological map will recognize terraces, bars and other alluvial landforms in individual placer environments. Buried channels may sometimes be identified from changes in vegetation along the course of underground streams or from thermal scanning. Appropriate symbols are used to delineate such features.

Map compilation can be made on black and white or dyeline copies of the base on which geological boundaries and appropriate letter symbols form the basis for hand colouring. The colours selected should preferably follow the international colour scheme adopted for regional mapping. Goosens (1980) suggests that the most suitable paper is polyester paper for tracing the main lineaments and other structural features from both black and white, and colour photographs.

Compilation of offshore data

Offshore data of importance to placer exploration programmes include bathymetry, the character of bottom sediments, occurrences of offshore bars and reefs, submerged strandlines and drainage channels, thickness of sediments above basement rocks (or seismic basement), indications of net sediment transport and notation of tidal data. Bathymetry is shown as isobaths if possible and isobaths of 20, 200, 1000 and 2000 m are used in relation to map scale to provide a basic profile. Spot soundings are used where isobaths cannot be drawn.

It is critical for practical reasons to differentiate wherever possible between basement rocks that are covered by less than 20 m of sediments and those that are covered by sediments of greater depth. This is because where the sedimentary cover exceeds 20 m there is a probability that the basement rocks were not exposed to sub-aerial weathering during the Pleistocene and, hence, have not shed detrital minerals during the brief periods of shelf exposure. Colour conventions suggest pale brown for basement rocks, pale yellow to denote areas of less than 20 m of sedimentary cover over basement rocks and pale green for areas where the sedimentary cover exceeds this amount.

Data on currents has a significant bearing on the accumulation of detrital minerals in the direction of net sediment transport. The effects can be deduced in a number of ways from studies of current directions, velocities and frequencies; and from studies of wave refraction and reflection patterns relating to features of the shoreline.

Tidal information, dominant wind directions, storm intensities and frequencies are important factors affecting both exploration and subsequent operations. Basic information on such features can be shown on small inset maps drawn to emphasize tidal ranges around the coastline and principal wind directions seasonally.

Using detrital heavy-mineral maps

Deposits of detrital heavy minerals studied at various stages of exploration require further definition in terms of heavy mineral composition, grade and quantity. Orders of quantity, in terms of cubic metres of material, can be symbolized on a map but as Noakes (1972) points out, grade and mineral composition of detrital heavy mineral deposits commonly include a number of valuable mineral types and become much too complicated for clear symbolization on regional maps. He suggests providing this information by numbering each detrital heavy-mineral deposit and relating each number to a table on which the available data are shown. This table can be shown as an inset on the map or be reproduced on the back of the map.

Detrital heavy-mineral maps are comparatively new on a national level and certain compilations suggested by Noakes have broken new ground during the past decade. His proposals for concepts and presentation, included as an appendix to this chapter, although to a certain extent experimental, nevertheless, constitute one of the more important advances in formalizing the techniques of placer exploration in modern times.

Noakes has addressed his efforts primarily to the compilation of national maps of sufficiently large scale to be of maximum use in the exploration and development of detrital heavy minerals within the national sphere. The logical extension is for the plotting of more detailed information on a smaller scale. Good interpretation demands maps, photographs and features that can be related to one another by overlays; survey control points that are plotted strategically in respect of possible sample grids and, for marine studies, shore station sites that are located suitably along the shorelines for accurate positioning. Precise survey control is essential so that index sheets and overlays match, thus allowing the geology, drainage and topographic features to be related. Scales as large as 1:10 may finally be adopted for the presentation of subsequent boring and sampling results in assay sections.

4.2.4 Exploration tools

Remote sensing and satellite imagery

Techniques of remote sensing developed over the past decade or so include side-scanning radar, side-scanning sonar, thermal scanning, infra-red photography and multi-spectral imagery. Used in conjunction with existing geological maps they are useful complementary tools and may sometimes reveal data that are not apparent from surface observations alone. Goosens (1980) suggests that the best approach for large areas is to first draft a topographic map with the contour levels and drainage patterns plotted on a scale of 1:250 000 to 1:500 000 and then to superimpose the satellite data. According to Carter (1979) the most sophisticated of current Landsat images are with Computer Compatible Tapes (CCT). These tapes permit the measurement of the brightness value of the sediments and the formation of ratio images in areas of provenance to enhance the display of hydrothermally altered rocks.

Some of the most important achievements of oceanographic mapping have resulted from the ability to take measurements of the topography of the ocean floor using side-scanning sonar and echo-sounding devices originally used for tracking down submarines in wartime. These systems use reflected sound impulses to produce images of submarine topographical features and methods have been improved to resolve small objects projecting upwards by as little as 3 cm above the bed and to cover the sea floor in sweeps of 30° in width and 10° in depth from the centre.

Remote sensing techniques are also used to monitor sedimentation and turbidity levels offshore in order to follow the effects of currents transporting sediments, and to trace offshore extensions of onshore geological structures. Services provided by the Earth Resources Technology Satellite (ERTS), Synchronous Earth Observation Satellite (SEOS), Nimbus and other sponsored systems are rapidly improving and projects can now obtain data on placers in shallow waters using photographs taken from spacecraft. On land, thermal scanning techniques, providing shades of light for heat and dark for cold, are used to define near-surface groundwater channels; however, they are very sensitive to climatic conditions at the time of measurement.

Aerial photography

Aerial photography supplies fill-in details not provided by currently available satellite imagery or side-scanning radar. Experience has shown its value in recognizing the various geomorphological units associated with continental and transitional placer environments. Colour photography is generally more effective than black and white in delineating most surface features and for recognizing changes in vegetation relating to different lithologies. However, photographic costs vary enormously and the possible benefits of colour presentation must always be weighed against the additional costs involved.

In considering the economics of placer photogrammetry, there is scope also for investigating the respective merits of various lens types in a range of conditions. Macdonald (1981) cites the following quotations for a base area of 600 km × 80 km of aerial colour photography in 1975:

$3\frac{1}{2}$ in (8.09 cm) superwide colour – 480 exposures – US$8500
6 in (15 cm) normal wide colour – 1220 exposures – US$22 340

Black and white coverage was available at the time from government sources at US$1.10 for each exposure and company response was to defer the decision on how to proceed with colour photography pending a preliminary interpretation of the black and white prints alone.

A wide spread of aerial photographs and satellite imagery is already available from various sources and many operators are reluctant to incur the costs of separate programmes. Regional surveys require the preparation of base maps, stereo plotting and contouring for control purposes and few exploration groups have the necessary expertise. Specialist organizations having access to very sophisticated techniques of photogrammatic interpretation and plotting, are normally required for both the original work and for updating it as fresh information comes to hand. Consequently, it is important for both scientific and economic reasons to resolve any differences regarding the instruments to be used and the scope of the exercise before incurring the high costs of the specialist services.

Underwater photography

According to Shepard (1973), the earliest records of undersea photography relate to studies by Boutan in Algeria in 1883 using a protected camera in shallow water. Modern underwater cameras have pressure protection for almost any depth and automatic film advancement for taking a great number of photographs during the one submergence. They are equipped with electronic flash units and control their height over the bottom using a sonic sensing device known as a 'pinger'.

In a technical cooperation project, 'Mineral potential in the Malacca Strait', between the Federal Republic of Germany and CCOP (ESCAP) countries in 1976, underwater television scanning was carried out using a TC125 Camera by Hydro Products, San Diego. A FOK-1B camera of IBAK, was combined with the television system to take black and white or coloured underwater pictures of interesting features.

Geophysical systems

Geophysical systems play increasingly important roles in placer exploration, particularly offshore and, when directed towards specific objectives, provide fundamental data that cannot be achieved as well in any other way. Because of this, manufacturers are now taking a greater interest in problems specific to placer environments and instrument capabilities are improving in terms of resolution, penetration and definition, at the same time becoming increasingly miniaturized electronically (Macdonald 1981).

It must always be remembered that geophysical programmes are essentially exercises in mathematics related to geology. In practice, the task of obtaining accurate measurements is not too difficult, but the standard of interpretation depends upon the accuracy of the relationship. The geologist has a responsibility to define the target and describe its likely geology; the geophysicist must clearly understand what information is required from his work so that he can conduct his survey accordingly. Success depends upon the adequacy of the data and this involves recognition of the precision with which measurements are to be made and the use of instruments capable of working to the required detection limits.

Whatever method is chosen, reliable measurements must be recorded within the time, cost and purpose of the exercise or all expenditure is jeopardized. Consequently, in any new areas, orientation surveys will be carried out both for instrument calibration and to test the applicability of the method for the particular terrain. Usually, two or three days should be allowed for equipment testing and for the establishment of operational procedures.

With placers there is generally a lack of clear lithology; methods that work for one drainage system may not find suitable responses in another. The principal ground geophysical systems are seismic, gravity, electrical, magnetic and radiometric. Aerial magnetometry and spectrometry are currently the main airborne systems. Regional programmes are normally combined with photogeology, remote sensing, and occasionally with geochemistry. Onshore or offshore, their task is generally the same: i.e. determine the depth to bedrock; indicate, within the known geology of the area, what the bedrock may be composed of; measure the depth of weathering of bedrock; measure the total sediment thickness and the thickness of the individual layers; identify such features as buried river channels, fossil strandlines, early erosional surfaces, basalt cappings etc.

Specific applications of geophysical methods onshore relate to the tracing of ancient stream courses for which there are no surface indications. Their formative period probably followed the general pattern outlined at the commencement of this chapter but, in time, stream courses may be changed radically by avulsion, river capture, tectonism etc. (see Chapter 3) and present drainage patterns may bear little or no relationship to the earlier channels. In such cases the results from boring and pitting may be quite inconclusive and some other methods are needed to fill the gaps.

Daly (1965) saw two main possibilities in exploring for deep leads by geophysical means: the detrital and weathered material in the old channel filling might be less dense than the unweathered bedrock; or the elastic properties of the unconsolidated fill might differ considerably from those of the unweathered bedrock. He pointed out that the first possibility provides a basis for applying gravity methods, the second for applying seismic methods but warned that:

(1) In wide leads the payable gravels may be concentrated in a meandering channel on the weathered surface which has no counterpart on the unweathered surface in which case geophysical methods would not indicate the presence of the channel.
(2) Owing to non-uniform weathering of the bedrock geophysical surveys may indicate a channel in the unweathered surface which does not exist on the weathered surface.

The principal offshore systems incorporate:

(1) Shallow seismic surveys to establish the distribution and thickness of sediments overlying the acoustic basement and thus to indicate broad areas where economic deposits might have accumulated in Quaternary times.
(2) Studies to determine the character and minerals potential of areas of bedrock that might have produced economic concentrations during Quaternary erosion cycles.
(3) Studies to determine the age, characteristics and structure of possible host rocks, and environments in which the several types of primary and secondary deposits occurred on shore as a guide to offshore prospecting.
(4) Studies to establish the magnitude, age and sequences of Quaternary strand-line movements and their influence on detrital heavy mineral concentrations.

Various levels of success, both onshore and offshore, depend largely on the difficulties encountered. The starting point is a map that seeks to define the horizontal and vertical extent of the features identified. An order of priority can then be drawn up for more detailed investigations using one or more of the following geophysical systems.

Seismic
Elastic waves travel at different velocities through different materials thus making possible the measurement of depths to key horizons having rather different velocities to those of bedrock or adjacent strata. The waves are initiated at measured intervals along a traverse and their arrival times are determined at fixed points along the same traverse. Velocities are generally greater in igneous and crystaline rocks than in sediments. Parasnis (1979) gives typical values in Table 4.2 in which (a) compressional velocities are measured in the direction of wave propagation, and (b) shear velocities are measured normal to the direction of wave propagation.

Onshore: In favourable conditions the drop-hammer method may be applied to shallow continental placers in order to delineate earlier channels and measure the depths of gravel and other fill. The method fails when key strata have similar responses as, for example, when velocities in a decomposed granite bedrock are similar to those in an overlying clayey gravel wash. Some difficulty is found in measuring channels that are deeply and narrowly incised into the bedrock but drop-hammer work is relatively quick and inexpensive and the method should usually be tried ahead of any other method for exploring continental placers.

Modern seismic instruments have signal enhancement properties provided by an automatic gain control in the amplifier. They are also equipped with filters to exclude background noises from such sources as people walking nearby. Individual wave energies are generally quite small but when signals are built up from a repetition of impulses at each striking point, and registered as total signals on a cathode ray display, they

Table 4.2 Elastic wave velocities in various media ($m\,s^{-1}$)

Medium	Compressional velocities	Shear velocities
Air	330	—
Sand	300–800	100–500
Water	1450	—
Glacial morraine	1500–2700	900–1300
Limestones and dolomites	3500–6500	1800–3800
Granites and other deep-seated rocks	4600–7000	2500–4000

magnify differences that might otherwise be not readily observed. The instruments are robust and easily carried by field crews.

A geophone spacing of 15 m is usually sufficient to estimate the depths to bedrock when the seismic wave velocity in the weathered and detrital layers are known. These velocities may be determined from weathering spreads shot with a geophone spacing of about 3 m.

Offshore: Seismic methods in shallow offshore areas of the marine placer environment utilize measurements of seismic reflection and refraction to differentiate between individual sedimentary horizons and to establish the thickness of sediments overlying the acoustic basement. Shockwaves generated by an explosion or series of explosions propagate at different velocities in the various media through which they travel. The waves received at the sensors are recorded differentially according to whether they have been refracted through denser layers or reflected from the interfaces between layers.

Waves are refracted back to the sensor only from layers that have increasing velocity signatures with depth; refraction profiles do not extend to layers in which the velocities are lower than in an overlying bed or in layers that are acoustically opaque. Ryall (1977) describes a reconnaissance geophysical survey offshore Kedah, west coast of Peninsular Malaysia, where approximately 15% of the records gave very little information for this reason. He attributed the masking effect to muds containing bubbles of natural gases, predominantly methane, produced by the decay of organic matter and, at the same time, referred to similar examples in other parts of the world.

Telemetric sonobuoy systems provide a means for recording wide-angle reflection and refraction data and for calculating the velocity of sound in unconsolidated marine sediments as well as in hard rock. They are used to indicate the strength of bedrock and depth of weathering for dredgability studies and have also been found useful in the search for relict strandline deposits and mineralized bars offshore. Modern systems are reasonably cheap to operate and have been found suitable for determining geological features that could formerly be determined only by extensive boring, much of it 'wildcat'.

A number of methods are used for producing appropriate wave forms in offshore investigations. Van Overeem (1960) describes a sonic transducer used to locate bedrock off the coasts of Belitung and Singkep Islands, Indonesia. The Malacca Straits survey, 1976, used an airgun system consisting of a triggering unit (quartz clock and ignition device), pre-amplifier, filter and final amplifier, and an EDO Western recorder. Electrical discarge units (sparkers) find wide applications for marine seismic profiling. Sparkers generally allow deeper penetration in the sub-bottom although with less vertical resolution.

Gravity

Gravity methods rely upon Newton's Law of Gravitation which states that the mutual attraction between masses is directly proportional to the product of the masses and inversely proportional to their distances apart, i.e. they measure gravitational differences between individual points of measurement and a base point. Corrections are applied to account for differences in latitude, elevation, topography and instrumentation. Anomalous differences are generally quite small and interpretation is very sensitive to density values. Table 4.3 gives the range of densities of some typical materials. As a general guide however, values for use in any application should be measured on site and not merely assumed.

Gravity methods are very sensitive to topographic highs and lows and some readings may require corrections of an order many times those of the anticipated anomalies. Their main application in placer exploration is to delineate drainage patterns over wide areas of moderate relief for follow-up seismic investigations. Peterson *et al.* (1968) claimed an accuracy within $\pm 10\%$ in determining bedrock configurations beneath exposed Tertiary gravels in Sierra Nevada using shallow seismic refraction in combination with gravity. Antung and Hanna (1975) investigated coastal magnetite sands at Meleman, Lamajang, East Java and found correlation between the amplitudes of computed and observed anomalies and their best estimate of the sand density ($2.75\,g\,cm^{-3}$) agreed generally with density measurements in the dunal area.

Daly (1965) describing a geophysical survey at Ardlethan, New South Wales, Australia, found that a maximum gravity anomaly of about 0.6 milligal over the deep lead was large enough for accurate measurement and that the gravity and seismic results were in close agreement. However, the resolving power of the gravity method was lower than that of the seismic method so that in one case where gravity indicated one channel, the seismic method indicated two. His conclusion was that the best procedure appears to be to use gravity for reconnaissance and seismic for follow up where gravity anomalies indicate the presence of deep leads.

Electrical

Electrical methods used in placer exploration are, (a) self potential, (b) earth resistivity, and (c) induced polarization.

Table 4.3 *Densities of materials* ($g\,cm^{-3}$)

Material	Density
Sand dry	1.40–1.65
Sand wet	1.95–2.05
Limestone	2.60–2.70
Granite	2.50–2.70

Self potential: Self potential methods measure the natural potential difference between ground stations associated with ground electrical currents. They are of most use in revealing the presence of large graphite or sulphide bodies and are less useful for sediments where ground waters contain salts or where local geological or topographical features interfere.

Earth resistivity: This method measures the distribution below the surface of potential (either DC or low frequency AC) introduced into the ground by means of two electrodes and is commonly used in water surveys or prospecting for conductive beds of material. A resistivity survey was used successfully at Tingha, New South Wales, for delineating channels cut into a granite bedrock in a deep lead cassiterite system under a cover of basalt. However, in this particular case the density of sampling was apparently too low because subsequent underground mining showed that while the main features had been identified, some important details had been missed.

Induced polarization: Induced polarization methods rely upon the fact that if the flow of a ground current is interrupted, the voltage does not drop to zero immediately but decays, after an initial drop, to some fraction of its commencing value, in a manner that correlates with the content of conducting minerals in the ground. According to Parasnis (1979) the method sometimes affords better resolution than resistivity methods but its diagnostic capabilities, nevertheless, are much in question unless the geology is well known.

The experimental application of magnetic-induced polarization has given encouraging results on both high- and low-grade beach sands and has given better definition than some other geophysical methods such as electrical-induced polarization (EIP). The Bureau of Mineral Resources, Canberra, has conducted a research programme on the application of the method to relict strandlines in offshore areas and, according to Noakes (1977), the anomalies appeared to be mainly produced by the ilmenite content of the heavy minerals. Electrical resistivities for some rocks and sediments are given in Table 4.4.

Table 4.4 Electric resistivities

Rocks and sediments	Electric resistivities (ohms)
Sandstone	35–4000
Limestones	120–400
Moist sand	100–10 000
Clays	1–120
Glacial morraine	8–4000
Granite	5000–1 million

All electrical interpretations assume flat surfaces and regular layers. Much work is needed initially to determine the rock signatures but when they are known, various applications of the method can be used for determining the thickness of overburden and gravels and depths to bedrock. Although interpretation calls for a high degree of expertise, costs are generally low, the procedures are simple and they do not require trained observers.

Magnetic

The concept of a general geomagnetic field with a definite orientation at each point on the surface of the earth was first introduced by William Gilbert, physician to Queen Elizabeth I, in the year 1600. Since then, instrumentation has been gradually developed for measuring horizontal and vertical forces and declination, and eventually to locate geological features such as faults and intrusions as well as deposits containing magnetite.

Magnetic methods rely upon differences in the local geomagnetic field due to differences in the magnetic properties of some minerals. In placer exploration, sediments containing significant proportions of magnetic minerals may provide measurable anomalies. For example airborne magnetometry has been used successfully in the USSR to locate kimberlite plugs through double (plus and minus) magnetic anomalies of 500–700 gammas where abundant magnetite or ilmenite is present. However, both surface and airborne magnetometer surveys have generally been less rewarding than hoped for, possibly because of mass effects from particle dispersion. Based upon iron = 100, common placer minerals in order of relative magnetic susceptibilities are shown in Table 4.5.

Table 4.5 *Common placer minerals listed in order of relative magnetic susceptibilities*

Mineral	Chemical composition	Relative density	Relative magnetic susceptibility
Magnetite	Fe_3O_4	5.2	40
Ilmenite	$FeO. FeTiO_2$	4.5–5.0	11.67
Garnet	Various iron, calcium silicates	3.5–4.0	6.68
Haematite	Fe_2O_3	4.9–5.3	4.64
Monazite	$(Ce\text{-}La\text{-}PO_4)$	4.6–5.4	4.11
Rutile	TiO_2	4.2–4.25	0.94
Zircon	$ZrSiO_4$	4.2–4.9	0.47
Quartz	SiO_2	2.65	0.40
Cassiterite	SnO_2	6.8–7.1	0.13

Electro-magnetic
When an electro-magnetic field is impressed upon a ground surface the resulting currents give rise to secondary electro-magnetic fields that distort the primary field and so reveal the presence of conductor minerals. In aerial surveys, the method succeeds best where ground surfaces are flat or generally undulating. Its main application in placer exploration is for studying the geology of districts containing old shorelines and in areas of provenance. For raised beach investigations flight elevations are preferably around 60 m for the 'bird' (sensor) at say, 25 m. Samples may be taken at 15 m intervals with precisions down to 2 gammas.

Radiometric
Radiometric methods are applicable to near-surface placers containing some minerals that are radioactive. Ground radiometric surveys are generally more effective than airborne surveys and the technique is valuable for placer diamond prospecting because of the common association of diamonds with uranium and thorium minerals. It also has value in tracing beach sands containing monazite or radioactive zircon and, although monazite normally has a higher level of radio-activity sometimes, even when both are present, zircon gives the highest response. For example, Meleik (1978) found a stronger correlation between radioactivity and zircon than between radioactivity and monazite in the Damietta beach sands of Egypt. He considered this to be due partly to the presence in the sand of more than three times as much zircon as monazite but partly, also, to differences in concentration because of differences in size and shape. The monazite particles being more rounded were more evenly distributed than the zircon particles which were more elongated.

The above investigation in Egypt was carried out using a scintillometer with four sodium iodide crystals measuring 8 cm in diameter and 5 cm in thickness optically coupled to one photomultiplier tube yielding a sensitivity of approximately 200 Hz/μr/h. The net gamma-ray flux from the ground surface was measured at an altitude of 70 m along 260 flight lines in two directions 250 m apart perpendicular to the shoreline, with a tie line parallel to the shore.

Airborne surveys record uranium, thorium and potassium anomalies and may also reveal hidden geological contacts and major fault structures. For example, in northern and north-west Australia, many silicified fault zones yield counts of 3–4 times background. Near-surface concentrations of monazite or other radioactive minerals sometimes give measurable responses that can be further investigated on the ground. Total count scintillometers with 5 cm crystals are suitable for such investigations.

Seaborne spectrometer surveys have similar objectives and Street (1977) describes a towed seabed gamma-ray spectrometer that continuously monitors four energy regions in the gamma-ray spectrum characteristic of:

total gamma ray radioactivity
potassium (K–40)
uranium (Bi–214)
thorium (Ti–208)

The detecting probe is enclosed in a 30-m long weighted flexible PVC tube known as an 'eel' that protects it from damage and at the same time allows it to ride over obstacles on the sea floor. According to Street it can clear obstacles 7 m high when towed at 4 knots with 7 m of hose trailing along the sea floor.

Survey control and positioning

Onshore

Strict survey control is essential and all measurements should be referred to a common datum so that individual features on different maps can be compared on the same scale by overlays. This is particularly important when exploring for deposits related to levels of stillstand during the Quaternary or tracing the drainage channel patterns in buried fluvial placer systems. In a recent example of a small fluvial placer in Colombia, precise levelling distinguished earlier channel remnants at various stages in the development of the valley and identified a major splitting of the main channel at bedrock. The present stream pattern gave no hint of what had happened and grid boring had suggested that one split was a branch channel from a tributory before precise levelling resolved the matter.

Offshore

Exploration and the subsequent exploitation of drowned placers along the continental shelf requires an ability to relocate sampling positions and to co-ordinate the results of such systems as seismic reflection profiling, side-scan sonar, marine seismic refraction and underwater photography. In the majority of cases the maximum allowable tolerance is ± 5 m and generally the work will be shared between more than one vessel.

Satellite navigation systems are not sufficiently accurate for pinpointing drill hole positions. Electronic distance measurement systems such as the Autotape, Electrotape and Tellurometer are restricted to operations within 35–40 km from the shore. Lasers have been used for extreme accuracy in line-of-sight surveys and short-range observations. The Raydist System used in the Malacca Straits survey had a range of 50–70 nautical miles and a theoretical positioning accuracy of 2–3 m but was found at that time to be affected seriously by atmospheric effects, especially thunderstorms.

Location markers

Limits for sampling deposits that lie within line of sight, say 18 km from the shore, are defined using buoys anchored to the seabed by cables. On fields less than 8 km square the buoy pennants should rise at least 3 m above the waterline to compensate for the effects of the earth's curvature and, once accurately located from the shore, may be used as markers for all future gridding. In marking out a mining field, three or more buoys are placed at opposite extremities of the field and just outside of it to avoid the need for relocation as mining proceeds.

The accuracy of positioning from moored buoys depends upon the depth of the water and the strength of the waves and currents; underwater sonar devices are used where a higher degree of accuracy is justified by the additional cost, or where the prospective area is within a shipping lane. Such acoustic methods, developed during the Second World War for anti-submarine warfare and torpedo control, were effective even at that time to a distance of up to 30 km.

The principles have been extended to peaceful uses, a major advantage of the system being the ability of sound waves to travel through water with very little attenuation. According to Barton (1979) acoustic positioning systems have an absolute accuracy of the order of 1% of the water depth and, more importantly, a repeatable accuracy of about 0.2% of the water-depth measurement. Sound may be concentrated at a point using wide-angle reflectors and a vessel having receivers in its hull can calculate its position very closely by correlating the intensity of sound from a number of fixed emitting stations.

Underwater radio transmitters are sometimes substituted for buoys to avoid navigational hazards; however, these can be used only for positioning close to to the marker because radio signals fade quickly in water and acoustic positioning is preferred.

Accurate positioning is a costly exercise and Barton refers to a tremendous growth rate in the application of microprocessors to replace mini-computers or to undertake roles that were uneconomic with mini-computers. In his view systems using microprocessors are currently less flexible than those relying upon mini-computers but cost reductions in the future are likely to outweigh any disadvantage arising from the lack of flexibility.

Economic considerations may sometimes dictate the use of cheaper and less-efficient systems until a discovery is made. Thereafter, reliable positioning is essential for correlating individual measurements and for preparing working plans and sections. An ideal positioning system is expected to satisfy the following basic requirements; accuracy and speed of reckoning, continuous availability, simplicity of operation, freedom from interferance, multi-user operation.

Few systems are available that might, within the financial limitations of a typical marine placer exploration programme approach these standards; nevertheless there are minimum requirements that must be met. It is preferable for all systems to be modular with individual modules small enough for ready transportation and be capable of installation in vessels with restricted space. Towed systems should be small enough to be man-handled in and out of the water to avoid using mechanical lifting gear which may cause mechanical damage. There should be ready access to spares and to experienced electronics technicians, preferably associated with the supplier, when breakdowns occur. Generators, batteries etc. should be standard types for the region and use standard voltages. The navigation system should operate at a frequency that will not interfere with ordinary ships radar or radio. Short-wave navigation systems utilize operating frequencies that are subject to overcrowding and interference and skilled operators are needed for both operation and interpretation.

Geochemistry

The main branch of geochemistry applied to placer exploration is 'particulate' geochemistry and simple methods such as panning stream sediments have been used for exploration for hundreds of years and are still of great value. For example, in exploring for diamonds, kimberlite-derived ilmenite and pyrope garnet are common tracer particles; both minerals are very resistant to chemical and mechanical breakdown and may be well represented in sediments at great distances from the kimberlitic source rocks. Closer to the target area, the presence of quantities of large-sized particles of these minerals may also indicate good concentrating conditions for the diamonds if they exist (Lord, 1976). In a typical reconnaissance programme in Tanganyika, 13 kg samples were taken on a 1 mile square grid closed to 0.2 mile and then to 0.1 mile. Samples were screened at +5 mm, −5 +1 mm and −1 mm sizings for mineralogical examination.

Niobium has been used as a pathfinder element for gemstone placers and their associated pegmatite source rocks in the State of Kerala, India, where two major fans of gem gravels appear to emanate from the hills (Ghats) along the Tamil Nadu border, east of Trivandrum. As a pathfinder element, niobium is more likely to be significant if the carrier mineral is pyrochlore or columbite–tantalite than if it is rutile, perovskite or any titanium mineral because niobium is widely dispersed in metamorphic and igneous rocks in small amounts carried by titanium minerals.

Generally, however, where the pegmatite provenances are known and the search extends away from them, other elements such as tungsten, cobalt and nickel may be of more interest than niobium. Niobium minerals such as columbite have little chemical mobility and dispersion and wider spectrums of granitic elements Be, Li, Zr, Y, Th, U etc. are more likely to delineate anomalous areas.

While geochemistry in placer exploration is largely restricted to particulate geochemistry the explorer often has the task of locating primary deposits in areas of provenance and he should understand the rudiments of geochemical soil sampling in order to take an intelligent interest in programme planning. Pan sampling, whereby heavy minerals are recovered in a prospecting dish while loaming up a hillside or sampling sediments in streams, is usually adequate for outcropping reefs but such particulate methods may fail in more complex situations, as for example where the outcrops are masked by weathering or transported sediments. In such cases, preliminary orientation sampling will influence the choice of method on the basis of analytical costs, detection limits, differential mobility and required precision.

Soil sampling involves the chemistry of soil horizons differing widely in their physico-chemical properties and thickness. The chemistry of the soil horizons varies greatly with depth and samples are best taken at the surface or at bedrock. They are generally taken according to lithology, sometimes at pre-determined levels but never in the B horizon in tropical climates such as in Australia because of the lack of distribution equilibria. Figure 4.2 is a typical section describing the three horizons A, B and C.

Pathfinder elements: Pathfinder elements for placer minerals are listed in Table 4.6.

Pits and trenches are dug in anomalous areas to look for discontinuities in the section, transported overburden may be a problem and if the geochemistry changes abruptly it is probably for this reason. Tin-bearing placers for which the parent rocks contain fluorine may be located by ground-water surveys, a method that has also led to some success in finding sedimentary uranium deposits in various parts of the world.

Geochemical analyses: The choice of an analytical method is usually determined according to required levels of precision. There are three

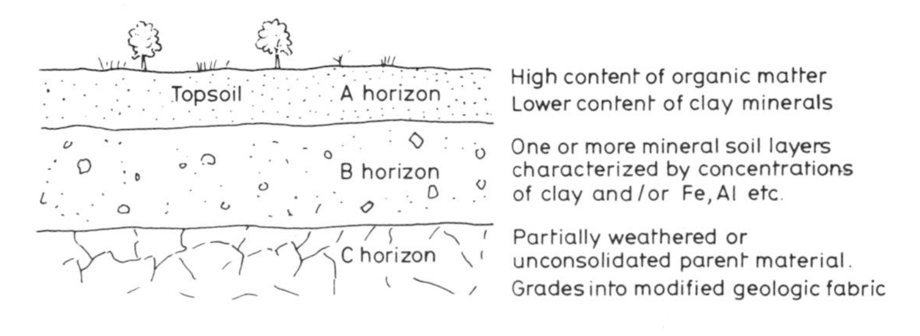

Fig. 4.2 Geochemical soil horizons (A, B and C)

Table 4.6 Pathfinder elements for placer minerals

Target mineral	Pathfinder elements and minerals	
Gold	Arsenic, copper, uranium	
Tin	Boron, molybdenum, fluorine	W and Sn colloidal in fine sediments. On tracing back there is a need at some stage for particulate geochemistry
Tungsten	Boron, molybdenum	
Platinum	Chromium, copper, nickel, cobalt, palladium.	
Diamonds	Pyrope garnet, ilmenite, diopside, phlogopite, uranium	
Gemstones (Pegmatite)	Niobium, and granitic elements	

principal methods:

Atomic absorption spectrometry: precision ± 2–10%.
Optical emission: precision $\pm 50\%$.
X-ray fluorescence: precision ± 2–10%.

Of these, atomic absorption finds the most general application. However, all methods are improving and it is usually advisable to experiment with various techniques initially to be assured that the selected method is best for the purpose.

4.3 PROSPECTING

4.3.1 Introduction

Selected areas are prospected to determine deposit dimensions in terms of volumes, grades and geometry and to identify any morphological or lithological features that might affect proposed methods of mining and treatment. Such features include individual zones of high and low mineral content, the presence or absence of a water table, the nature of the bedrock, diagenetic effects and the contamination of particles through textural or other associations, and coatings.

In conjunction with the field work, financial studies examine the economic practicability of what is being done in respect of what might be achieved and each additional cash commitment must be seen to be justified before going ahead. Explorers are, and should be, optimists by nature but the exploration data are normally sketchy and prospectors are constrained to deal with facts. They must be prepared to call a halt to further expenditure the moment it becomes apparent that a prospect is subeconomic. A decision to continue can be given only when there is a clear indication of possible viability. In this respect, the most exasperating prospects are those that always seem to offer better results from the next batch of samples but never quite attain the levels hoped for. Generally, in such cases, the indecisions arise from inadequate geology or through the use of inappropriate methods and equipment for sampling.

Although there is some overlap, the three main stages in placer prospecting exercises are; (a) reconnaissance, (b) scout testing, (c) close testing.

4.3.2 Reconnaissance

The reconnaissance phase of a prospecting programme is directed primarily towards solving problems of geological and geophysical interpretation and towards testing hypotheses formulated from the results of regional exploration. It does not attempt to define the limits or levels of payable values but, rather, to build up an appropriate model that can be used for such purposes.

The tools of reconnaissance are simple. Prospecting dishes, crudely built sluices and winnowing devices suffice for testing most shallow surface placers. Mechanical and hand auger drills are used for largely qualitative sampling down to bedrock in the deeper deposits and to investigate the lithology of the beds. Pits for sampling are normally sunk only in shallow ground but may also be used for testing deep leads where cheap labour is readily available.

In order to achieve reconnaissance objectives as quickly and cheaply as possible, samples are taken first from what appear to be the most favourable sectors, based upon geological deductions. If the results are favourable further samples are taken progressively according to the results of previous samples until sufficient additional data have been accumulated to justify entering into the next phase of the project. If the results are unfavourable, the project might be abandoned forthwith. Important factors having a bearing upon decision-making at the reconnaissance stage of testing are as follows.

Availability of water for mining and treatment

Indications from exploration data are firmed up by additional work in the form of hydrology studies aimed at locating sufficient supplies of water for all purposes. Preliminary mineral-processing tests are carried out to determine the amounts needed for slurrying the wash and for gravity concentration. Losses are inevitably high because of seepage, evaporation and retention in slimes and, although some water may be reclaimable, the overall requirements are seldom less than 150 l m^{-3} of ground treated and may rise to more than 1500 l m^{-3} for some very clayey materials.

Climatic conditions

Extremes of climate add significantly to the difficulties and costs of mining in some areas through freezing or flooding and because of the physical difficulties of operating a surface mine in either sub-zero or intensely high ambient temperatures. Evaporation, a prime cause of water loss in desert regions, is aggravated by high temperatures, low humidities and high winds. Cyclonic storms disrupt beach-mining operations along some coastlines such as along the coastline of Bangladesh where island and beach-mining operations are restricted by the weather to non-monsoonal periods. It is important to note, moreover, that in such extremes of climate marine processes that are accretionary in one season may be erosive in the next and, hence, estimates of ore reserves might also change significantly with the seasons. Flooding in some areas may also restrict operations and add to the difficulties and costs of programme logistics.

In the absence of adequate safeguards any of the above possible effects of climatic extremes may endanger an operation to the point of failure. Even commencing an operation in an unfavourable season might in some cases strain the financial resources of an exploration group because of equipment and production losses that were not budgeted for.

Access roads and infrastructure

One aspect of infrastructure costing often underestimated in early evaluation exercises is that for the construction and maintenance of roads and other means of access. Rule-of-thumb estimates for access roads around housing settlements, plant buildings and for outside services may be reasonably based upon experience elsewhere, however, it is much less simple to predict the costs of building and maintaining access roads and facilities for mining purposes and for the movement of mechanical drilling equipment in swampy ground and other adverse surface conditions. Such roads are of a temporary nature only and a balance must be struck between spending either too much or too little on them.

Masking of deposits

Deposits lying buried beneath considerable thicknesses of masking sediments or volcanics may require extensive geological and geophysical studies in order to obtain a preliminary picture of the relevant channel configurations. However, while a reasonable measure of success is achieved for such methods when dealing with relatively uncomplicated successions, in extreme cases the geophysical methods fail and the channels can be located only by probing with a drill or by conducting underground mining operations.

Transitional placers relict from previous environments and now located inland from the coast may be completely masked by later windblown materials and the first task in reconnaissance is to locate the ancient beaches and wave platforms on which the deposits were laid down. Fossil beaches and platforms relate, in such conditions, to earlier periods of stillstand and, once found, precise levelling is needed to test their continuance.

Spot sampling

Particulate geochemistry includes panning and, during reconnaissance, it is important that the prospector has some means for making spot estimates of gold and other precious metal samples in the field. Spot analyses are made most easily by panning off samples and counting the number and size of the 'colours' observed through a magnifying glass as compared with sets of standard-sized particles prepared in the laboratory. For example one such set used by a UN project team in South America was based on the premises shown in Table 4.7.

Measured quantities of sample are taken, for example a normal-sized prospecting pan may contain about 1/200th of a cubic metre of loose gravel. If when panned down to a rough heavy-mineral concentrate, the sample contained one medium, two fine and ten very fine particles its grade would be $[(1 \times 6) + (2 \times 1.56) + (10 \times 0.31)] \times 200 = 2.444\ \mathrm{g\ m^{-3}}$ approximately.

Another useful spot-sampling technique has been developed for making approximate determinations of the Ta_2O_5 and Nb_2O_5 contents of tantalite–columbite concentrates based upon relative density measurements. The relative density of a sample of the concentrates is determined by water displacement and related to the curves described in Fig 4.3 to read off the approximate composition. For example clean concentrates having a relative density of 5.5 will average closely 13.7% Ta_2O_5 and 66.2% Nb_2O_5.

4.3.3 Scout testing

The decision to undertake a scout-testing exercise is based upon evidence from reconnaissance and, since considerable expense is involved, the programme will be adopted only if there is a reasonable chance of a successful outcome. To be successful, the results from scout testing must confirm the initial predictions of possibly economically viable quantities

Table 4.7 Particulate gold estimation data

	Description				
	Coarse	Medium	Fine	Very fine	Flour
Mesh (USS)	+10	−10+20	−20+40	−40+60	−60
Aperture (mm)	+2.00	0.85/2.00	0.425/0.85	0.25/0.425	−0.25
Av. weight (mg)	—	6.00	1.56	0.31	0.05
Av. no/g	—	170	640	3200	20 000

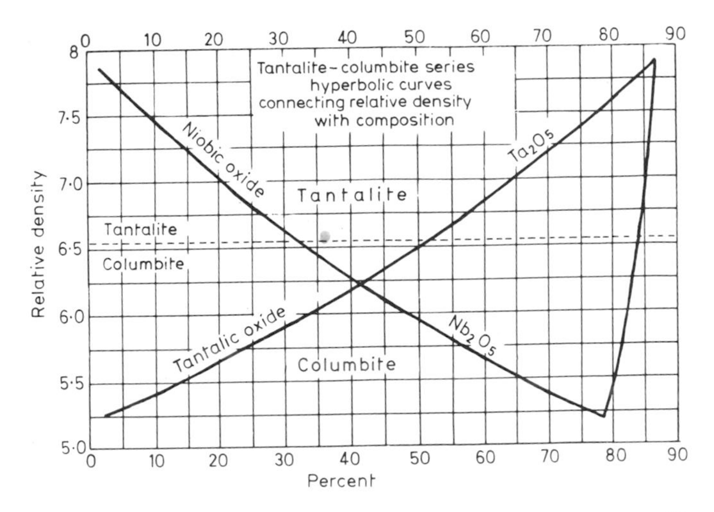

Fig. 4.3 Tantalite–columbite series – density versus composition

of ore and lay the foundations for obtaining the final proofs. In this respect there is still a further stage of close gridding to be completed before proceeding to final evaluation.

Estimated costs for the exercise come under careful scrutiny. Minimum logistic requirements are a project headquarters, camp facilities, equipment for drilling, sampling and sample dressing, vehicles for mobility and whatever back up facilities are needed for what has to be done. The establishment costs are inevitably high and planning needs to be detailed in order to avoid waste.

Except where certain geostatistical methods are to be employed (see Chapter 8) the normal procedure is to lay out an appropriately designed grid over the whole of the area to be tested and to sample first at wide, but geometrically spaced, intervals in and around the margins of the deposit.These are the scout samples and if they yield satisfactory results the grid will later be closed progressively until all of the required data are obtained. Sampling points are located at the intersections of co-ordinates dividing the prospective area into rectangles for large deposits or at intervals along lines spaced regularly across and normal to the longitudinal axes of narrow placers. The sample grids are called 'rectangular' grids and 'line' grids respectively.

Rectangular grids

Rectangular grids are used for setting out the sample positions geometrically over very large deposits or deposits that are relatively uniform in grade. The grid dimensions are selected to allow a considerable reduction in spacing using whole numbers to facilitate measurements and calculations. Typically, scout holes are spaced initially at 400 m intervals to allow supplementary holes to be drilled at 200, 100, 50 and 25 m intervals. By contrast, a 450 m scout bore spacing would reduce in an unwieldy progression through intervals of 225, 112.5, 56.25 and 28.125 m, thus adding to the tedium of survey calculations.

Line grids

Long, narrow deposits are best tested using line grids for laying out the sampling positions. Most stream placers and strandline accumulations fall into this category. Lines are marked out at regular intervals along baselines that are angled where necessary to follow major changes in direction. The scout hole and line intervals are similarly designed for ready reduction as for rectangular grid spacings, the initial spacing being determined largely according to the geology of the deposit.

Placer deposits that cannot be tested immediately by either rectangular or line scout gridding include those that have only been explored geophysically and require confirmatory drilling and sampling to check the supposed drainage pattern; and those for which systematic sampling to a grid pattern is impracticable for topographical reasons or because sections of the deposit have already been exploited. Such deposits may need to be sampled more randomly at the scout boring stage to gain further knowledge of the existing conditions at bedrock before determining a final test pattern.

Essentially, the scout-testing phase of an exploration programme is designed for the dual purposes of determining suitable standards and conditions for all future sampling, at the same time minimizing as cheaply as possible the risks of entering prematurely into a high-cost feasibility study. Hence, sample intervals are made large in the first instance but not so large that significant concentrations might be missed, or that positive recommendations cannot be made at the conclusion of the scouting exercise. There is, at the same time, adequate provision to terminate the project abruptly if it becomes clear at any time that the investment criteria of the undertaking are unlikely to be met. The adoption of standard methods and techniques from the outset will allow all of the scout sample results to rank equally with the results from close gridding. Some initial experimentation may be needed to establish the required level of confidence in what is being done, but if then a project measures up to the overall project requirements it should be possible to state with reasonable confidence; (a) the approximate dimensions and grade of the deposit, (b) the most suitable spacing of holes for close gridding, (c) the extent to which the methods and equipment already used might be modified to improve the quality of sampling or reduce its costs in subsequent studies, and (d) the extent to which mineral dressing and other problems might require special attention in the form of bench-scale or pilot-scale investigations (see Chapters 7 and 8).

4.3.4 Close testing

The scout-testing phase ends and a formal valuation featuring close testing commences with the deposit defined roughly in three dimensions. Cost is still an important consideration but it is false economy to take short cuts at this stage just because it is cheaper to do so, or to accept grid spacings that are excessively wide when basic uncertainties have still to be resolved. On the other hand it is wrong to take more samples than are needed because statistical methods of assessment are now available from which to determine the adequacy of sampling progressively as testing proceeds. This has not always been the case: Palmer (1942) advocated closing the grid spacings used for sampling some deposits until quantity and grade estimates made from the results of alternate boreholes agreed with estimates made from the results of the remaining bores. This was a conservative approach favoured at the time but it obviously required at least twice as many holes as would be needed today.

Grid closing

A grid is closed sufficiently when all of the constituent parts of each section of a deposit have equal chances of being represented by samples that are taken without bias.

Rectangular grids

The first step in the reduction of sample spacings in a rectangular grid is to make the points of intersection of the diagonals of each rectangle the point of the first reduction and to fill in the additional sampling points as illustrated in Fig. 4.4. Depending upon the uniformity of the results, some sectors may require more rigorous sampling at closer intervals than others.

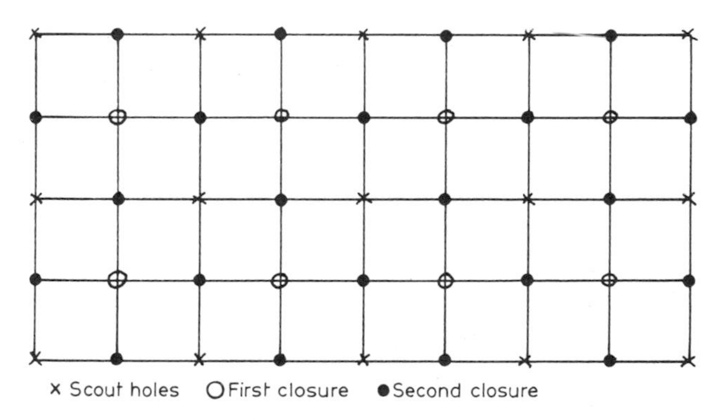

Fig. 4.4 Closing pattern for rectangular grid

Line grids

Line intervals commencing at say 400 m may be reduced eventually to 50 m or less and holes along the lines from, say, 20 m down to 5 m. Ultimate spacings are influenced by the narrowness of the pay streaks and, as a very general rule, the spacings will be reduced until the pattern of distribution of the values is seen to follow in logical progression from each line to its neighbours. Again, in closing in the grid, the density of sampling may be closer in one sector than in another (see Fig. 4.5).

4.3.5 Pitting

Pits are sunk either mechanically or by hand. The choice of method is usually governed by cost, particularly where cheap labour is readily available, otherwise the availability of suitable equipment, ground conditions and depth are the main considerations.

Hand pitting

Hand pitting is best suited to dry shallow ground in remote areas although it may be used in a variety of circumstances. In wet or running ground the excavation is supported by close-set lagged timbering or by sectionalized steel cylinders known as caissons. The size and shape of pits vary but range generally from about 1.2 m to 1.5 m diameter for circular pits. Rectangular pits, 1.2 m × 0.8 m have sufficient space for dewatering by bailing or pumping. It is most important that the pit walls are kept vertical for the full sampling depth.

Sectionalized caissons are fabricated in identical matching sections with lugs that allow them to be bolted together. During sinking, new sections are added from the top and when the bottom edge is undermined the cylindrical body moves downwards under its own weight or is driven down by heavy blows. Plate 3–4 mm thick is suitably sized for most cylinders. Common dimensions are 0.5–1.0 m long with diameters of 1.00–1.25 m for the cylindrical sections. A small winch and tripod is used for handling the individual sections. Figures 4.6 and 4.7 illustrate two

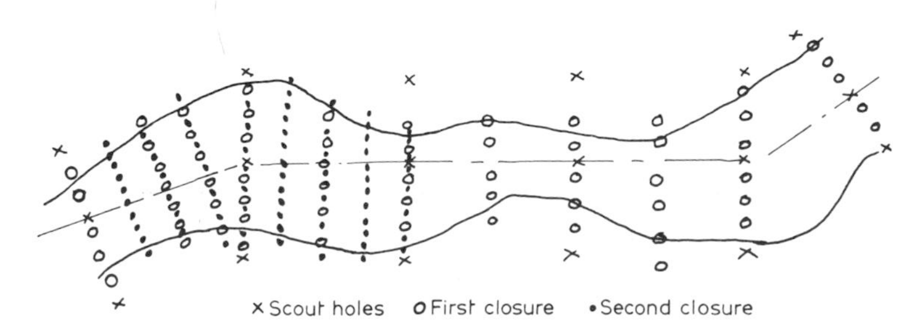

Fig. 4.5 Closing pattern for line grid

Fig. 4.6 Commencing sinking along the edge of an island in Punna Puzha, Kerala State, India

Fig. 4.7 Excavating gravels from inside of the caisson during sinking. Photograph shows pump suction line used for dewatering

stages of caisson pitting in a river bar in the Nilambur Valley, Kerala, India.

Telescopic caissons are fabricated so that each succeeding section is slightly smaller in diameter than the preceding one. Individual sections are dropped inside the others as sinking proceeds. The hole reduces in diameter with depth and the method is effective to maximum depths of about 15 m. Flush jointed caissons can be used in favourable conditions for sinking holes to greater depths but the bottom sections are more difficult to extract from free running ground. Some contamination results from surges of material from around the caisson walls and because of swell but this material can usually be extracted separately and discarded without affecting the accuracy of sampling. When sampling is complete the individual segments are recovered from the bottom upwards thus providing safe working conditions.

Timbered pits and caissons yield reasonably true sample cross sections except where the ground contains many large stones that jut out from the walls. Some contamination is inevitable in such cases.

Mechanical pitting and trenching

Back-acting hydraulic excavators and older-type back hoes are used extensively for sampling ground that holds up well enough for regular shaped trenches to be cut by such means. Nine to ten metre digging depths are achieved by some of the larger machines with bucket widths of up to 1 m. Vertical channels can be cut in the walls for sampling, but generally it is more satisfactory to trim the walls of the excavation and process all of the material extracted. Hydraulic excavators provide very fast digging and Wells (1969) describes the sinking of thirteen shafts ranging in depth from 4–12 ft (1.2–3.7 m) in 6 h using a tractor-mounted unit with a 38 in (approx. 1 m) bucket.

Hydraulic excavators also provide a means for digging pits and Applin (1976) reports that the manual excavation of deep samples in diamondiferous areas of Sierra Leone has now been largely superceded by the use of hydraulic excavators either track or wheel mounted, equipped with long, jointed booms supporting hydraulically operated clamshell grabs, 1 m in diameter. As an example of the type, the Poclain excavator using a long mount as an extension to the boring or earthmoving clamshell can dig to maximum depths of around 16 m. Figure 4.8 illustrates the action of an hydraulic excavator in pit-digging service.

Bulldozers are useful reconnaissance tools for cutting shallow trenches to depths of 3–4 m but the trenches are usually limited in length to pushing distances of 50–60 m and depths of 3–4 m. In most cases, back hoes will do the same job cheaper and to greater depths.

The main application for bulldozers in placer exploration is for roadmaking, site preparation and opening up large pits for inspection. In

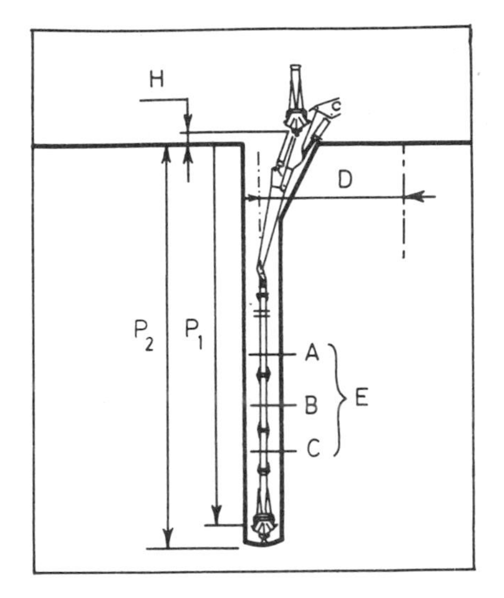

Working ranges with two-piece 2.65 m (8'8") + 2.90 m (9'6") boom and articulating cylinder.
Length of extensions: A : 2.50 m (8'3")
B : 1.75 m (5'9")
C : 1.25 m (4'1")

P_1 : digging depth without entrance hole.
P_2 : digging depth with entrance hole.
With 5.40 m (17'9") boom, 3 m (9'10") stick and 3B assembly, a maximum depth of 13.20 m (47'4") can be attained.

	2,20 m (7'3") Stick		3,00 m (9'10") Stick	
E	2B+C	2A	3B	A+B+C
D	3,90 m (12'10")	3,90 m (12'10")	4,65 m (15'3")	4,65 m (15'3")
H	0,65 m (2'2")	0,40 m (1'4")	0,80 m (2'8")	0,55 m (1'10")
P1	11,95 m (39'2")		12,50 m (41')	
P2	11,95 m (39'2")	12,20 m (40')	13,10 m (43')	13,35 m (43'10")

Fig. 4.8 Poclain 160 excavator in pit digging service

many diamond exploration programmes, they are used for stripping and site clearing while back-acting type hydraulic excavators are used for sampling.

4.3.6 Boring

Placer boring methods are classified broadly under five main headings: hand boring; percussion and hammer drilling; auger and rotary drilling; vibro drilling; pit-digging.

Hand boring

Hand drills are used largely for transitional and shallow marine conditions and occasionally for shallow fluvial deposits. The Banka type has been used extensively in countries where labour is cheap and where climatic or topographical conditions preclude the normal use of mechanical methods. With various refinements, such as hydraulic rotation and jetting, the drill has been marketed world wide under such trade names as 'Empire' and 'Conrad Banka' (Fig. 4.9). The Banka drill was invented by a Dutch mining engineer, J. E. Akkeringa, who worked on the Bangka Island deposits in Indonesia during the last century.

Banka drills

Essentially the method entails using a steel cutting tool screwed to the end of a string of steel rods inside lengths of 10–15 cm diameter casing. A run of casing has a hardened cutting shoe attached to its bottom end and a

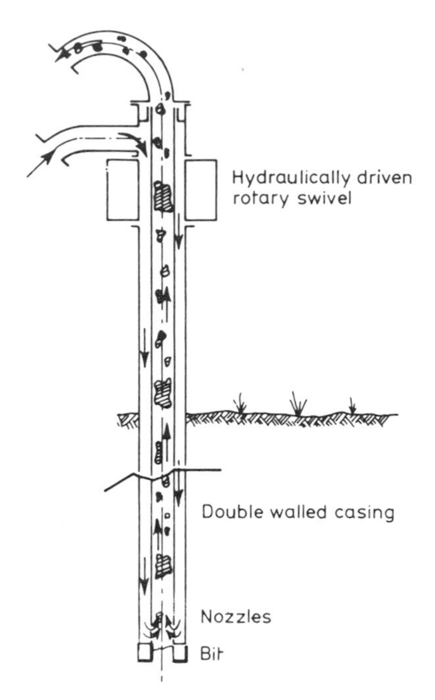

Fig. 4.9 Conrad Banka drill arrangement

platform clamped near the top on which the drillers stand. Penetration is effected by rotation under the combined weight of the casing and crew or, in difficult ground by driving, using a heavy block attached to the drill stem. A drive head is screwed onto the top of the casing and the block is lifted and dropped to give a pile-driving effect.

When the casing is driven into the ground it cuts out a cylinder of earth which is broken up by the cutting tool and brought to the surface. A hand sludge pump, consisting of a steel cylinder fitted with a clack or ball valve, is attached either to a rope or to another, lighter, string of rods and pumped up and down to take on its load. When full, it is withdrawn and the sample is discharged into a measuring vessel and the volume checked. A hydrostatic head of water is maintained inside the casing to inhibit surging. The method is described fully by Harrison (1954).

Banka drills are also used extensively in shallow offshore areas but not in the deeper platform sectors where mechanical drills are mounted on platforms designed to cope with heavy swell. Four inch (100 mm) Banka drills operating from pontoons have been used for many years in Indonesian waters and in recent years sample recovery has been effected by water jetting in dual tube units. Special bit design has helped to overcome such problems as underflushing and blockages between the inner and outer casings.

Hand augers

Hand augering and sludging provides the simplest and cheapest method for sampling most mineral sands placers. Hand augers are available in a number of designs but the most satisfactory is a hardened-steel shell type, 50–100 mm diameter. The shell section is up to 40 cm long and is coupled to aluminium extension rods for lightness and portability. Including the handle section, augers are suitably dimensioned for most purposes if made in the same length as the extension rods, say 1.5 m. Each extension rod has a male and female end and additional lengths are added as the hole is deepened.

Applying slight body pressure and taking care to keep the tool vertical, the auger is rotated by hand some five or six turns. It is then withdrawn from the hole and the sample cuttings are shaken out onto a 2 m × 2 m canvas sampling sheet. The process is repeated for the full sample interval after which the accumulated sand is mixed, coned and quartered down to a suitable size for transportation to the laboratory.

Hand augers are satisfactory only in damp unconsolidated material; dry sand is not retained in the shell and the sides of the hole cave. When this occurs, water may be trickled down the rod into the hole to dampen the sand so that drilling can proceed, but the accuracy of sampling is reduced.

Depending upon the quality and texture of the sand, one driller can usually bore to a depth of 10 m in less than 30 min but it is usual for two to three men to work as a team for both boring and sampling on holes to 20 m and four men thereafter. Progress becomes slower with increasing depths and a 20 m hole may take between $1\frac{1}{2}$ and 2 h to complete. In the high level dunes at St Lucia in South Africa, contract parties of local Zulu tribesmen were able to complete a 100 m hole in 8–10 h, but this was exceptional.

Sand deposits overlain by beds of hard soil or clay are penetrated initially using a clay auger of such dimensions that the sand auger follows without jamming. Clay augers are fabricated more robustly than sand augers because of the higher stresses involved and are more open in construction to facilitate emptying. Neither sand nor clay augers are suitable for drilling through gravel beds although occasional small stones can be picked up with the finer sediments. Coarse sand always presents difficulties in hand augering because of the lack of cohesion between grains.

Hand sludgers

The limit of accurate shell auger sampling is reached at the water table. Below this level sampling is made possible only through the use of casing. The casing is made slightly smaller in diameter than the auger hole to provide clearance for it to be run freely. As with the Banka drill, the casing is forced down into the waterlogged sediments by steady pressure and rotation. A casing clamp is fastened to the upper end of the casing as a platform for the driller to stand on and so that leverage can be applied when turning the casing. The drillers weight provides the steady pressure for rotation.

The sludge pump consists of a barrel, approx. 40 mm OD, having a removable nose piece at the bottom end fitted with a clack or ball valve. It is attached through a female thread at the upper end to extension rods or to a rope socket. In operation, the sludge pump is never allowed to approach closer than 50–70 mm from the mouth of the casing shoe and a high-water level is maintained within the casing to provide a positive static head against surging.

A specific problem near Chisimaio in Somalia was overcome by reversing the normal order of boring and sludging. In the stabilized dunes the top 3 m of sand was too dry for augering and, in some cases layers of very dry sand also occurred at greater depths. To cope with the upper layer, two 1.5 m lengths of 100 mm diameter aluminium casing and a 75 mm diameter sludge pump were used to sludge down to a depth of 3 m. At this depth the sand contained sufficient moisture for it to allow augering to proceed. Only in one or two instances was the surface depth of dry sand found to exceed 3 m and the greatest depth cased in any one hole was 6 m.

To cope with the problem of dry sand at depth, small quantities of water were trickled carefully down the extension rods to wet the sand. About 1 pint of water was used for each run of approximately 5 cm and this allowed boring to proceed slowly until naturally damp sand was again encountered. Figure 4.10 illustrates an augering operation following sludging to about 2 m in very dry surface sand, Somalia.

Sludging is usually restricted to depths of about 5–10 m for purely physical reasons although greater depths have been obtained. In hard materials a star drill is fitted to the end of an extension rod to break through into the underlying softer beds. Progress is slower than for augering but in Australian conditions two men can sludge to a depth of about 6 m in 2 h. Sample splitting may be carried out as for the auger samples after first decanting off the excess water. Figure 4.11 is an illustration of a standard set of hand augering and sludging equipment (Macdonald, 1973).

The advantages of hand boring are: ready portability in any type of country; low capital cost; low operating costs; accurate sampling in suitable ground conditions and simplicity. The disadvantages become apparent as costs rise with increasing depths and sampling accuracy falls away in stoney or wet ground. Very sticky clays are almost impossible to bore either by hand augering or sludging because of high viscous shear.

Mechanical boring

Mechanical drill rigs have been developed or adapted for taking representative samples in most conditions. The main types are: percussion or churn

Fig. 4.10 Hand sludging in coastal dunes near Chisimaio, Somalia

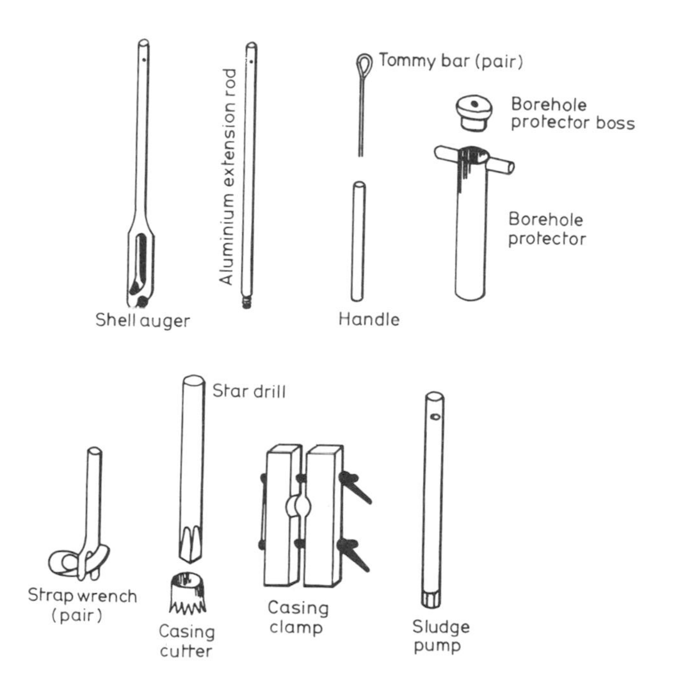

Fig. 4.11 Hand-boring and sludging equipment for beach sands

drills, auger drills, reverse flow drills, vibro drills, hammer drills and pit-digging drills.

Percussion drills

The percussion (churn) drill is a mechanical adaptation of the Banka type hand drill and is used extensively in continental placer exploration. Various rigs have been developed to drive casing through clay, rock and sediments. Depths are restricted generally to 30–50 m, however, they are capable of much greater penetration and one stratigraphic hole near Metung, Victoria, Australia was sunk through sediments to a total depth of 1782 ft (543 m).

Casing, normally up to 20 cm diameter, is driven vertically downwards by successive heavy blows from a drop hammer. The hardened cutting shoe has a slightly larger diameter than the casing to provide clearance between the outside of the casing and the earth. The casing is joined in lengths that are screwed to butt up against one another in a socket so that the impact is transmitted directly from casing to casing and not through the threads.

The casing is driven approximately 15 cm for each run and the plug of material forced into the casing is drilled out using chopping bits and bailers. Care must again be taken to ensure that all drilling and sludging takes place within the casing to minimize contamination. An exception is where boulders or large stones are encountered and these must be broken up ahead of the casing before it can be driven further.

Vacuum-type bailers are best for most purposes because they provide rapidly accelerating flow at the inlet to lift the particles of heavy minerals from the bottom surface. This results in higher overall yields of the valuable mineral types and less carry down of values. The driving-chopping-bailing procedure is continued until the desired depth is reached. All of the material collected for each sample interval is collected in a calibrated vessel, decanted and the volume measured.

Penetration rates vary with the type of ground. Ten metres per shift of 8 h is possible in good ground but progress falls away sharply with increasing depths and difficulties in handling the casing. This operation is probably the most time consuming of all and, even in the most favourable circumstances, 15 m of casing may take as long as 2 h to withdraw.

Churn drills are manufactured in many sizes. Small Hillman and Cyclone drills known as 'Airplane' drills (Fig. 4.12) can be broken into pieces small enough to be carried by relatively light aircraft, hence the name. Larger drills have evolved from the early Keystone types. Diamondiferous gravels are sampled along the Birim River in Ghana using 22 in (55.88 cm) steel casings and 20 in (50.80 cm) diameter bailers. Again, however, much of the time is taken up with emplacing and withdrawing casing. The method is sometimes practicable when other

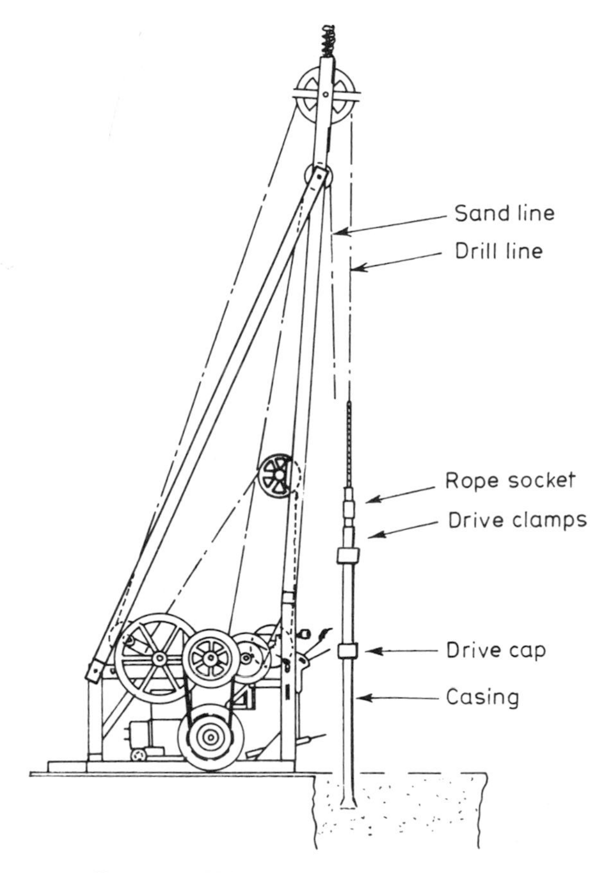

Fig. 4.12 Hillman airplane placer drill

methods fail but it has many disadvantages. One of these is the need to apply corrections to volume measurements. Figure 4.13 illustrates the use of an 'airplane' drill for sampling gold alluvials at Rio Uraudo, Colombia. The sample is taken by emptying the bailer into a sample trough as shown (Fig. 4.14).

Volume corrections: each time the casing is driven, the depth of the hole is measured from the surface of the ground to the top of the solids. The volume of solids in the casing is then calculated and compared with the theoretical volume for the depth driven. Sample recoveries are normally lower than theoretical volumes for the size of the casing shoe and various pipe factors have been adopted to compensate. The best known is the Radford factor described for various casing-shoe diameters in Table 4.8. Radford factors are applied as positive corrections to the recovered volume on the assumption that the lost portion is of similar grade to that

Fig. 4.13 Sampling gold placer in Rio Oraudo Valley, Colombia using airplane drill

Fig. 4.14 Emptying the bailer into the sampling trough

Table 4.8 *The Radford Factor*

Diameter of cutting edge of shoe	Factor	Diameter of cutting edge of shoe	Factor
4.00 in (10.16 cm)	0.077	6.00 in (15.24 cm)	0.173
4.25 in (10.80 cm)	0.086	6.25 in (15.88 cm)	0.187
4.50 in (11.43 cm)	0.097	6.50 in (16.51 cm)	0.203
4.75 in (12.07 cm)	0.108	6.75 in (17.15 cm)	0.218
5.00 in (12.70 cm)	0.120	7.00 in (17.78 cm)	0.235
5.25 in (13.34 cm)	0.132	7.25 in (18.42 cm)	0.252
5.50 in (13.97 cm)	0.145	7.50 in (19.05 cm)	0.270
5.75 in (14.61 cm)	0.158		

recovered. The normal rise is 1.4–1.6 ft (42.7–49 cm) per 1.0 ft (30.48 cm) of drive for a 7.5 in (19.05 cm) cutting shoe.

A different form of correction may be applied by measuring the volume recovered after each drive and adjusting the sample values directly according to the ratio between the actual and theoretical volumes. However, all such arbitrary corrections are subject to error, particularly if they are applied in new locations for which there is no previous experience in mining to guide the explorer. Colp (1976) states the case: 'Usually, the adjusted value given to a particular placer deposit depends on the engineer's powers of deduction and experienced judgement, rather than on the rigid application of a particular formula or formulas.' Wells strikes a similar warning note that 'placer drilling requires specially trained personnel and that "correcting" the drill results is a specialty that is not within the experience range of most mining engineers'. Breeding (1976) states that he has seen and used positive corrections as high as 240% – this must have taken great courage as well as requiring a great deal of judgement.

Other disadvantages in the system are:

(1) Accurate sampling calls for the recovery of all material in the path of the cutting shoe. Inevitably, some is pushed aside and where large gravel and stones are encountered, the chopping bit must work ahead of the casing, thus driving broken material and, perhaps values, into the walls.
(2) The method leads to considerable sample deformation and to a marked destruction of texture. Accurate logging is difficult.
(3) The cuttings are slurried for recovery by bailing or jetting and some degree of sorting in the column is inevitable. Higher density particles have a tendency to be carried down to either 'salt' the lower portions or be lost. The development of improved types of vacuum bailers has only partly solved this problem.

Auger drills
Auger drills may be of two general types; (a) solid augers and (b) hollow augers.

Solid augers: A solid auger comprises a tungsten-tipped cutting head at the end of a stem, so formed that the drill cuttings travel to the surface along flytes arranged in a helical path around it. Boring is by rotation with pressure, extra flyte sections being added as the hole is deepened. The drive is commonly through the power take off of a truck or tractor. If the former, an oversize clutch is introduced so that both borer and hydraulic lifting gear can be operated from the power take off, thus facilitating sampling by allowing the spoil-laden auger to be raised to the surface without rotation. Auger diameters may be as large as 60 cm for shallow depths but 20-cm augers are satisfactory for reconnaissance generally. Figure 4.15 illustrates the auger drilling flyte arrangement.

The most important applications are in dry or semi-dry clays and augers fitted with hydraulic lifting gear can bore up to 40 m of clay alluvium in 8 h. Drilling is faster in beach sands, perhaps as much as 100 to 125 m per 8-h shift but the method is less accurate than many others in this service.

Sampling is either 'continuous' or 'dead stick'. In continuous sampling the cuttings travel to the surface along the flytes and are collected continuously during drilling. However, it is impossible to judge accurately, the interval from which each sample is obtained and there may be considerable contamination from the walls. Probably the worst feature is the uneven representation of sample material from beds of different textures. For example, in very decomposed granite bedrock the drill may penetrate for several metres while only recovering the loose waterlogged pay dirt lying at some interval above it.

'Dead stick' augering is more accurate. In this system the auger is run for a depth of 1.5 m or some other pre-determined interval and is then

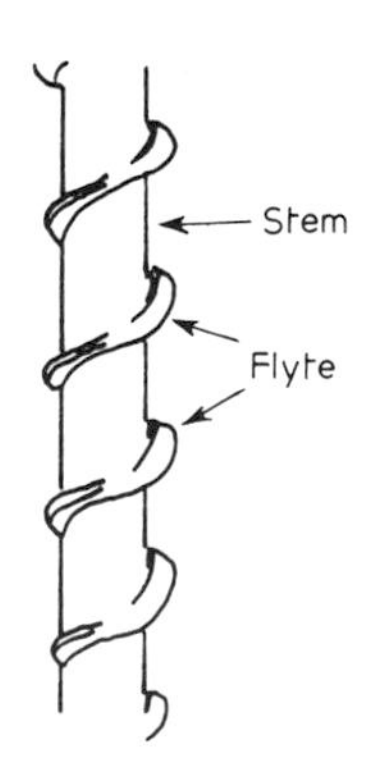

Fig. 4.15 Solid auger flyte arrangement

pulled without rotation, to recover the sample. The sample is retained as a plug on the leading auger flytes and is taken only from the groove, extreme care being exercised to clean off any possible contamination. The procedure is repeated with minimum rotation to the total depth of the hole so as not to disturb the sample unduly.

Solid augers are useful tools in reconnaissance because of their speed of operation and simplicity but the method usually holds too many risks of error to be used in final valuation studies. Solid augers are quite unsuited to even moderately stoney ground or where the material to be sampled rests on a hard bottom. In free sandy material the heavier particles tend to settle out preferentially in the flytes.

Hollow augers: In hollow auger drilling (Fig. 4.16), a cutting auger rotates around the axis of the sampling device and a little above it to ream out the hole for the barrel to pass. A core may then be taken in a relatively undisturbed state. Extension augers of the same diameter are fitted progressively after each sampling to form continuous flyte configurations for removal of the waste cuttings from around the core.

Samples are taken by forcing a non-rotating sample tube with a retaining ring down into the soil and withdrawing it when it is full. If solid rods are used, the entire string is withdrawn for sample recovery. However, drilling rates which may be as low as 2 m h^{-1} in clay, are greatly improved with the use of hollow stems and wireline sampling and may be as high as 15–20 m h^{-1} in sand and 6–10 m h^{-1} in clay.

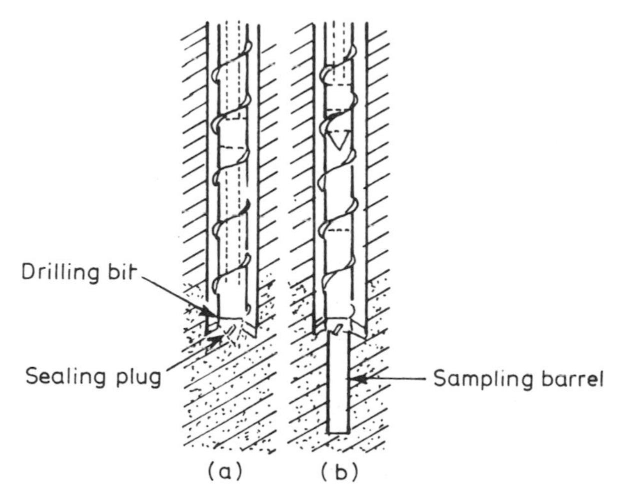

Fig. 4.16 Hollow auger equipment for undisturbed sampling: (a) shows sealing plug in position during preliminary drilling, (b) shows the sealing plug replaced with sampling barrel which is hammered into the sediments ahead of the drilling bit

Hollow auger sampling can also be carried out by bailing or jetting to recover the sample material. In this case the sample is recovered in a broken state but penetration rates are much higher. Sampling is reasonably accurate in stone-free ground provided the values are evenly distributed but it is not generally a preferred method because core diameters are generally limited to less than 7 cm.

The Sandrill: A power-driven auger known as the 'Sandrill', designed and manufactured in Australia for sampling high-level dunes, penetrates to depths of 100 m and more in increments of 1.5 m for the full depth of the hole. Drilling rates of up to 30 m h^{-1} are common in good conditions. The Sandrill is constructed with an outer casing having removable hardened cutters within which counter-rotating flytes draw the sample up into the sample barrel. Extensions of aluminium tubing are fitted with an inner drive tube on ball bearings. Figure 4.17 shows a Sandrill being used for testing dune formations in the Thar Parkar Desert, Pakistan.

The machine consists of an engine-powered hydraulic pump unit with a 10 HP motor direct-coupled to the pump. The complete unit is fitted with stretcher-type handles and, with inbuilt reservoir, filters and controls, weighs approximately 73 kg. Two men can work the rig but for maximum footage four to five men work as a team. Two drillers carry out the physical

Fig. 4.17 Testing dune formation in Thar Parkar Desert, Pakistan using a sandrill

processes of boring and sampling, two the work of sample splitting and logging while the fifth man empties the auger, cleans it, and has it ready for replacement when the next one comes up. Each sampling cycle occupies between 2 min and 10 min depending upon the depth from which the sample is being taken.

Reverse flow drills

An adaptation of hollow auger drilling, this method involves rotating both outer and inner casings while water or air is forced down the annulus between the two tubes. The fluid returns to the surface through the inner tube, drawing the sample up with it (see Fig. 4.9, the Conrad Banka drill). A packed water swivel attached to the top end of both pipes allows them to rotate freely and independently without loss of fluid. Return water passes to a settling tank for sample recovery; compressed air to a cyclone separator.

In early models, the inner casing was kept at a height of about 50 mm above the casing-shoe inlet. Air or water pressure was then adjusted to avoid losing sample material and to prevent surging. Drilling mud was sometimes used to facilitate the recovery of the cuttings and sampling rates were of the order of 10 m per 8 h shift with drilling depths to 40 m below the water table.

The method lost favour because of problems in taking reliable samples but, since then, manufacturers have overcome many of the features which had bought the method into disrepute. Remedial measures taken include routing the compressed air or water through the cutting bit and projecting it upwards into the inner barrel to draw the sample with it. By dropping the inner casing onto the floor of the bit, extraneous material is prevented from passing into the system while the sample is being pumped to the surface.

Vibro drills

In vibro drilling (Fig. 4.18) a vibrator with a directional effect transmits impulses to a drilling device which may be a screw auger, shell auger, sludge barrel or chisel. Screw augers are used in friable clayey soils and shell augers in loose sandy soil. Sludge barrels with foot valves are driven by percussion into saturated sandy soils, fine-grained gravels and muds. Chisel bits are used for breaking up hard layers and boulders after which the fragments may be extracted using a sludge barrel or shell auger. For most beach-sand work, the impulses are transmitted to a string of casing having a cutting shoe and an inner tube through which the spoil is jetted to the surface.

Because of the oscillations there is a marked reduction in shearing stresses between the earth and the casing and the drill easily penetrates the soil or waterlogged sediments under its own weight. Rates of

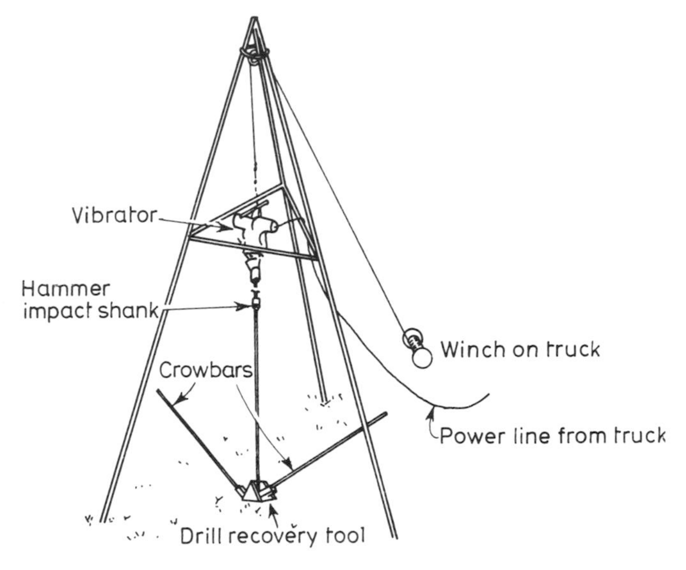

Fig. 4.18 The vibro drill

penetration vary with increasing depths and practical depth limitations are usually around 20 m. Although rates may be as high as 5–6 m min $^{-1}$, drilling costs rise steeply with increasing depth because of the time spent in handling the casing sections and sampling. The method encounters many difficulties when drilling through sticky clays and alternating beds of sand and clay.

Depending upon whether disturbed or relatively undisturbed samples are to be recovered, either of two main types may be used. Hill (1976) describes a low-frequency vibrating drill (Amdrill) for taking samples in the form of slurries and a high-frequency type (Vibro-corer) for coring in selected conditions. The choice of method is governed by whether the need is for close geological interpretation or only to estimate depths and grades. Costs for handling undisturbed core samples are very much higher than for disturbed samples.

Hammer drills

Hammer drills using pile-driving mechanisms to drive the casing are generally effective for testing gravels and other materials that cannot be sampled using vibro drills. The Becker '180' hammer drill has drilled at the rate 30–100 m per 10 h shift in glacial drift and till containing many large boulders and around 70 m per shift offshore (Colp, 1976).

In the Becker system, twin tubes are driven without rotation using a Link Belt diesel hammer delivering around 90 blows min $^{-1}$ at 7600 ft lb per blow. The drill pipe has a patented cast-steel shoe measuring 6.625 in

(16.83 cm) OD by 4.00 in (10.16 cm) ID through which air and/or water under pressure is introduced into the inner pipe. Materials entering into the shoe are lifted to the surface more or less continuously during sinking and recovered as cyclone underflows into settling vats. Figure 4.19 describes the recovery mechanism.

Depth limitations offshore depend largely upon the ocean conditions but reasonable swell can be tolerated. A Becker drill, operating from a barge off Phuket Island, Thailand, sampled tin gravels in water depths to 35 m and sediment thicknesses of as much as 40 m. A similar rig on the US Bureau of Mines research vessel *Virginia City* off the American coastline is reported to have sampled in water depths of 30–45 m and sediment thicknesses of 15–30 m.

Although the hammer drill has a high rate of penetration in a variety of conditions both capital and operating costs are very high and the rigs can be justified only for major undertakings. Penetration rates are high in favourable ground but must be reduced substantially to allow good volume recoveries in loose, free-running ground. Other disadvantages include the possibility that rocks may block the shoe unnoticed and interfere with

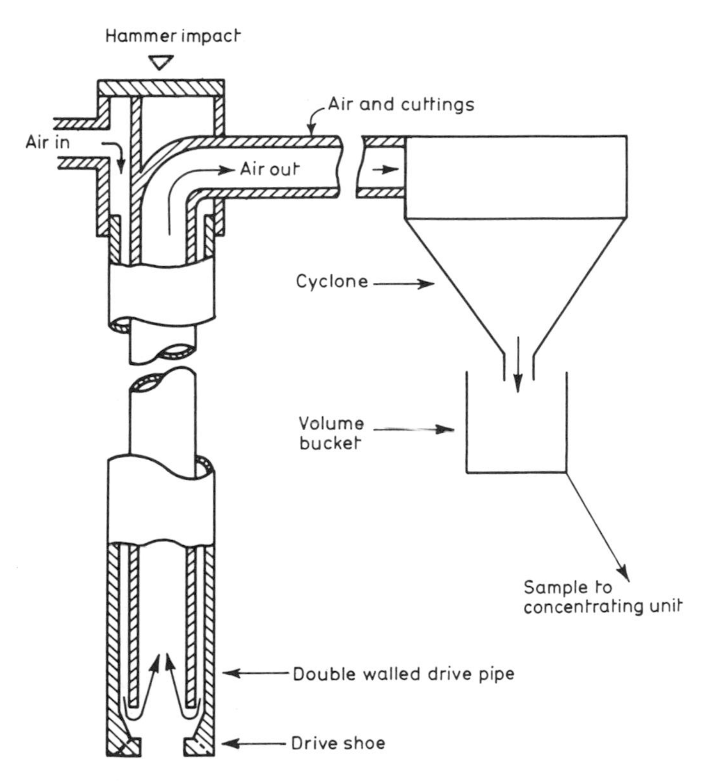

Fig. 4.19 The Becker hammer drill – recovery mechanism

core recovery, and the somewhat excessive length of the discharge system that may result in the contamination of succeeding samples by mixing with previous ones. Most problems can be largely overcome by careful attention to procedures laid down by the manufacturers.

Pit-digging drills

Pit-digging drills are amongst the most satisfactory types for sampling placers because of their ability to take large-sized samples. There are a number of designs of which two of the most popular makes are the 'Calweld' and the 'Conrad'.

The Calweld drill: The Calweld rig (Fig. 4.20) utilizes a ring and pinion drive gear. The ring gear (approximately 1.4 m diameter for normal-sized buckets) rotates a drive kelly attached to the bucket. The kelly is telescopic and extends to about 20 m before additional drill stems need to be added. An hydraulic dumping arm allows the bucket to be hoisted to either side of the drilling rig. The spoil is discharged onto the ground or into a bin or truck, 3–4 m away from the hole. The weight of the kelly and bucket (approximately 1.5 tonnes for a 90-cm bucket) supplies the necessary downward thrust.

'Earth' buckets are used in gravel-free soils. The cutters are located in the base of the bucket which is hinged and operated by a quick-opening

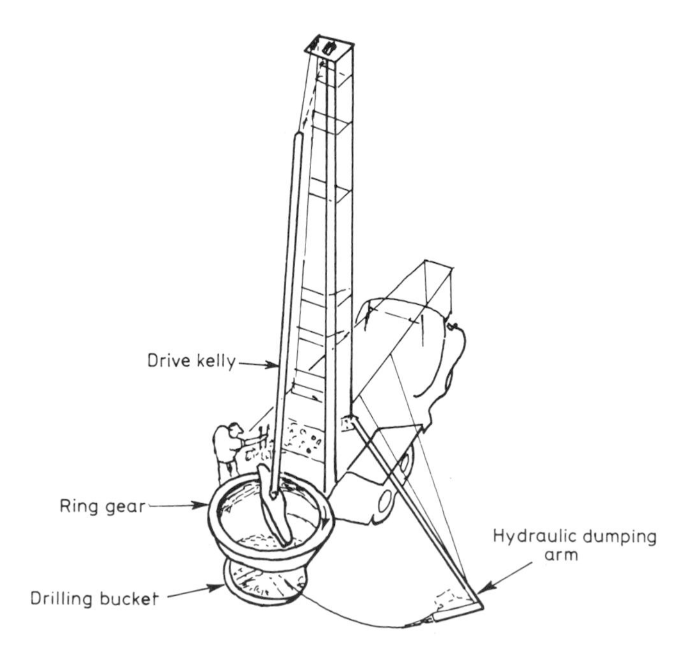

Fig. 4.20 Calweld bucket drill

mechanism for rapid discharge. Ripper teeth are used on the cutter face and reaming blades at the sides provide clearance. Massive hard bands of induration are drilled using a chopping bucket fitted with five to six hardened teeth.

Rock buckets are similar to earth buckets but have one large opening in the base to accommodate stones up to 30 cm in diameter. Clam-shell buckets or 'orange peel' grab types may also be used for drilling and sampling in stoney ground. A complete rock core can be taken using a calyx-type bucket in some sedimentary formations.

Bucket diameters range from 45 cm to 2.5 m and in favourable dry conditions drilling rates of 60–70 m per shift of 8 h are common. In wet ground necessitating casing, the rates are reduced to half or even less.

The advantages of bucket drills are as follows:

(1) Provided that access is satisfactory large samples can be taken in almost any type of onshore placer ground. The bore holes are sufficiently large that, if boulders are encountered, they may be blasted with explosives and removed in pieces.
(2) Bore-hole logging can be carried out visually within the hole and thus can be accurate to within a few centimetres.
(3) The volume of the hole can be measured physically and accepted as the true volume without need for correction.
(4) Suitable quantities of material can be recovered quickly for pilot-scale testing.

The Conrad drill: This bucket-type drill is also used for taking large samples in both wet and dry conditions. It is a development of the Banka hand drill and can be mounted on a four-wheel-drive truck or tractor for use in open, well-drained country, or on platforms or pontoons in marshy conditions.

Boring is effected by rotating the casing under its own weight and the spoil is excavated using a clamshell or pivotting bucket. Stones up to 25 cm in diameter present no difficulties in holes 50–60 cm in diameter. Holes may be drilled to 25 m in average conditions but the penetration rates fall away with increasing depth because of casing handling problems. In a series of tests conducted by the Dorset Tin Dredging Company at South Mount Cameron, Tasmania (Macdonald, 1966), the penetration rates for one borehole were logged as shown by Table 4.9.

Sample volumes approximate measured casing volumes if care is taken to maintain a positive hydrostatic head within the casing. In tests at Dorset the variation was less than 1%, well within the limits of acceptability for this type of operation.

Another of the pit-digging drills is the KLAM. A unit described by Colp (1976) drives 1.5 m sections of casing 91 cm diameter made from 9.5 mm

Table 4.9 *Example of Conrad drilling*

Depth (feet)	Material	Average penetration rates
0–3	Surface sand	$4\,ft\,h^{-1}$
3–4	Cement – 'billy'	
4–7	Sand	
7–13	Sandy clay	
13–29.5	Silt and sand	
29.5–30	Light gravels wash	
30–50	Silt and sand	$2\,ft\,h^{-1}$
50–61	Tight silty material	$1\,ft\,h^{-1}$
61–63	Clay	
63–64	Medium sand	
64–66.5	Shingle	
66.5–67	Soft bedrock	

Note: 1 ft = 30.48 cm

steel plate vertically into the ground. A 'Klam' bucket dumps the sample material onto a specially designed, pivotting, pan-type bucket for transfer to the truck. The bucket has a 91 cm × 91 cm bottom; it is 20 cm high at the back and the sides taper down to about 15 cm at the front. However, although once again the method is suitable for taking large samples, the penetration rate is very slow (less than $1\,m\,h^{-1}$) in most conditions.

4.3.7 Sampling

The ultimate purpose of sampling is to obtain a comprehensive set of data with measurements that are reproduceable regardless of the computer and with observations that supply such additional information as may be required for mine planning and plant design. Sampling is expected, also, to provide data towards understanding the geomorphic processes that have resulted in the present state of being of the deposits, processes that may have been influenced by such environmental features as palaeoclimates, changes in base level and tectonism. A knowledge of how the various processes have acted in the past is a vital element in predicting fresh discoveries in the future.

Accurate logging is essential and, for placers, the recorded information should describe the physical characteristics of the payable horizons, overburden and bedrock, stratigraphic relationships between key beds, the position of the water table and observations concerning possible grades. Reasons must be given for any sampling difficulties encountered and any divergence from normal procedures is noted because valuation as an analytical process assumes standard conditions of data collection unless otherwise informed.

Access for sampling

Access for sampling is by bores, shafts, pits or other excavations and careful consideration is given to selecting suitable methods and equipment for individual circumstances. The evidence may come from previous workings but, generally, some studies are needed to determine the parameters involved. This may be in the form of trial excavations or some inexpensive form of boring such as augering.

Large-diameter shafts, pits and trenches are suitable for most relatively dry placers containing large stones and gravel and may be preferred to boring in shallow surface alluvials, particularly where labour is cheap. Bulk samples are the only practical means of valuing diamond deposits or deposits in which the distribution of values is excessively erratic. In deep, or wet ground the presence of quantities of boulders up to 30 cm in diameter will automatically limit the choice to pit-digging drills.

Topographical features such as swamps, dense jungle, streams and ravines also effect sampling and influence the selection of equipment. Heavy drill rigs lack mobility in swampy country and, where water covers the ground, boring will usually be done from a platform that is either floating or supported on legs. Uneven 'flag' rock bottoms, pinnacles of limestone or trap rock and potholes inevitably lead to errors in valuation if the borehole spacing is not close and, even then, sampling is adversely affected because of the difficulties experienced in recovering the bottom values.

A section of this chapter has been devoted to describing the various types of drilling equipment commonly used for prospecting; placer sampling is now discussed along with typical analytical procedures under the following main headings:

(1) Sample dressing.
(2) Analytical methods.
(3) Sampling errors and the reliability of sampling.

Sample dressing

Samples are dressed (i.e. processed) in order to reduce their size to manageable proportions for subsequent assay. Sample dressing procedures also provide physical data for the design of mining and treatment plant including information on the particle-size distribution of both valuable and non-valuable rocks and minerals, ore washability characteristics, slimes content and recoveries.

Placer samples fall generally into two groups distinguished by the size range of their particles and, hence, of the environments within which the deposits occur.

Continental placer samples

Continental placer samples contain such minerals as gold, platinum, cassiterite, wolfram and gemstones in a heterogeneous, usually claybound, wash. Compared with the volumes of associated non-valuable particles the quantities of valuable minerals are very small. The valuable particle sizes range from very fine (less than 50 μm) sizings to coarse nuggets and crystals in wash composed of colloidal matter, fine clays, sand, gravel and boulders.

Depending upon their volumes, samples are dressed by hand or in mechanical devices designed specifically for the type of material to be treated. The relatively small measured samples recovered from auger drilling or from percussion drilling may be slurried in small mechanical puddlers but are usually dressed manually using small sluices or pans. Fine gold particles are collected by amalgamation or by cupellation with lead foil. Concentrates of most other minerals in this group are recovered by gravity along with other heavy particles in the form of rough concentrates containing up to 30 or 50% of valuable particles. Parks (1957) describes a sampling cutter for sampling sludges from drilling but such devices are seldom used for sampling placer sludges.

Pit samples generally have volumes of 0.5 m^3 or more and are best treated by mechanical means in a pilot treatment plant. The material is first puddled to free all the values from their clayey matrices. It is then screened, and finally subjected to some form of gravity concentration. The flowsheet in Fig. 4.21 illustrates a sampling plant used for testing cassiterite placers in the New England district of New South Wales. A Calweld drill was used to recover the samples and from five to eight samples (averaging 0.75 m^3 in volume) were processed during each shift of 8 h.

The sample was sluiced from the measuring hopper through a 30 mm aperture grizzly into a pump hopper. The oversize was collected for examination, measurement, and further scrubbing, where necessary. The undersize was pumped through a 3 in (7.62 cm) gravel pump to a 4 ft (1.22 m) long, 1.5 ft (45.72 cm) diameter trommel fitted with a 6 mm aperture screen, lifters, and high-pressure water jets.

The oversize from the trommel was scrubbed in a vortex mixer made from an old flotation cell and again screened at 6 mm to reject the oversize for measurement and examination. The -6 mm material combined with the -6 mm material from the trommel was deslimed in a 30 in (76.2 cm) diameter settling cone. The slime passed by gravity to a 10 ft $\times$ 1.5 ft (3.05 m $\times$ 45.72 cm) streaming box fitted with corduroy strakes to hold any fine tin passing over with it. The underflow from the cone was fed to a two-hutch Ruoss type jig, having a total screen area of 900 cm^2. Haematite pebbles were used for the jig-bed ragging.

Jig tailings were scavenged in a 10 ft $\times$ 1.5 ft (3.05 m $\times$ 45.72 cm) sluice

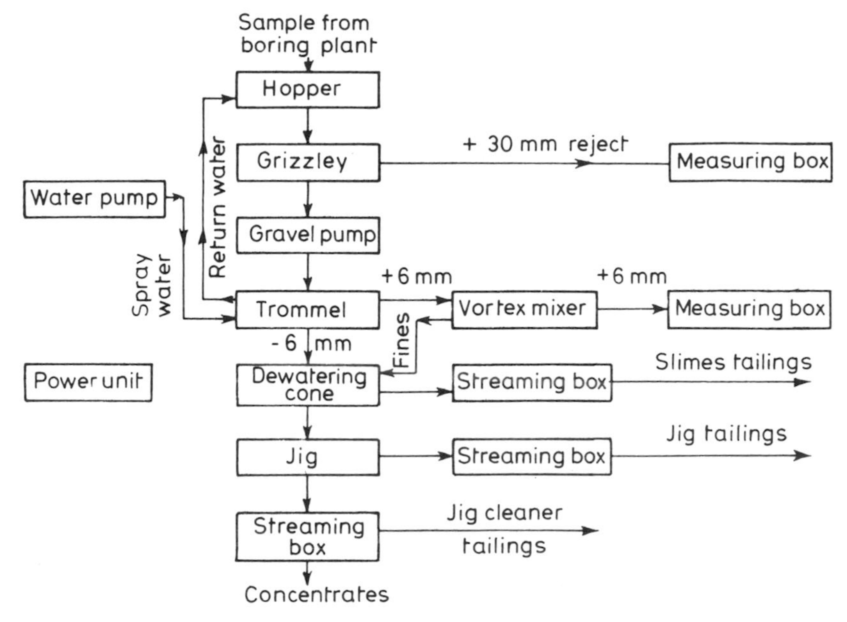

Fig. 4.21 Flowsheet of sample dressing plant

box fitted with riffles. The finished tailing combined with the slime and ran to waste. The jig-hutch products were upgraded in two streaming boxes arranged in series, the concentrates being panned to a final concentrate for chemical assay.

This design is suitable for any continental placer samples, with the proviso that, in testing auriferous placers, the streaming boxes are replaced by sluice boxes using corduroy strakes or sacking to catch the fine gold. Undercurrent sluices might also be incorporated for recovery of very-fine gold. Coarse gravel is kept away from the undercurrent boxes so as not to disturb the fine gold as it settles out of the slurry.

The cost of this plant, including a reconditioned 25 kVA alternator, was \$4600 in 1968. If all new materials and equipment had been used, the cost would have been between \$7000 and \$8000. It was operated by three men who shared all the duties of sample dressing, recording and packaging.

In operating such a plant, the concentration criteria require to be such that for each particular type of material an insignificant amount of recoverable mineral is lost. Dressed cassiterite concentrates may range from 10% to 50% Tin, but rarely higher. Gold is concentrated only to the extent that the concentrate contains a high percentage of the associated heavy minerals regardless of the grade.

Pit samples taken from accumulations in some present day stream beds are more readily fluidized and, for these, much of the preliminary mechanical puddling operation can be eliminated. Figures 4.22 and 4.23 illustrate methods of sample dressing developed for sampling the river bars in Punna Puzha in the Nilambur Valley of Kerala State India and at Barjor in Madhya Pradesh.

Transitional placer samples

Transitional placer samples have a short size range and contain valuable minerals which are generally much less dense than the continental type minerals. Because of this, fine and even particle size-range samples are split in the field down to a suitable size for transport to the laboratory. The original sample is taken from a small-diameter borehole and two, or at most three, splits are usually sufficient.

Field samples are reduced by coning and quartering or by repeated passes through a riffler of the Jones type. A sample weighing not more than 2 kg remains. Using a Jones splitter and large trays for drying samples over open fires, one field sampler can split and bag 30–40 samples per day. These samples are either reduced again in the laboratory to weights of between 50 and 200 g for heavy-liquid separation, or are concentrated on a laboratory shaking table. Heavy-liquid separation gives a higher yield of valuable heavy minerals than does tabling although tabling results are normally closer to what might be achieved at plant level. Either method is

Fig. 4.22 Sampling river gravels in Nilambur Valley, Kerala, India

Fig. 4.23 Sampling clayey gravels at Barjor, Raigarh Goldfield Madhya Pradesh, India

acceptable on its own but serious errors may occur if both methods are used in the same valuation.

Analytical methods and procedures

Samples received at the laboratory are checked in and the details are entered into a sample register to record:

(1) Sample number, location, date of sampling and date received.
(2) Sample details, e.g. line, hole and interval.
(3) Remarks on condition of sample as received or any other relevant matters.

Analytical methods and procedures from this point onward vary according to the sample type and minerals to be analysed. Generally, however, the first step is to prepare a heavy-mineral concentrate using some form of gravity concentration and then to apply the appropriate physical or chemical procedures indicated by the nature of the exercise. For example, Fig. 4.24 (after Macdonald, 1973) describes a typical sampling flow sheet for heavy-mineral sands containing rutile, zircon, ilmenite, magnetite and monazite.

Gold samples that have been reduced to a heavy-mineral concentrate may in some cases be further upgraded using a super panner type

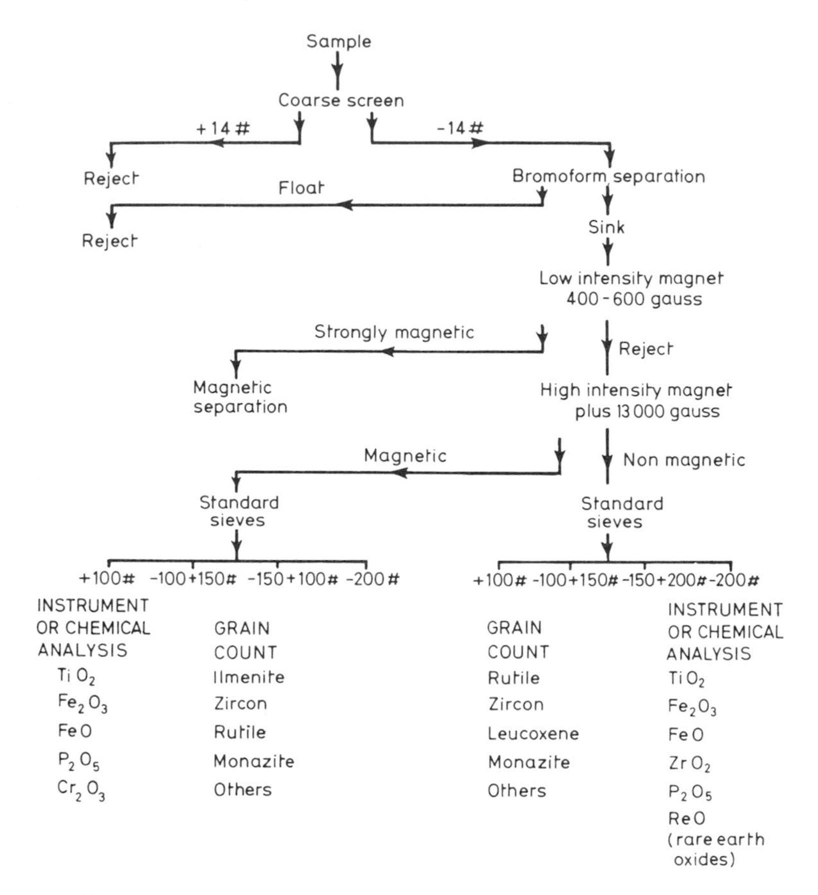

Fig. 4.24 Typical flow sheet for heavy mineral determination

concentrator or amalgamated directly with mercury to recover the gold. The procedure is to place a small globule of mercury into a miniature prospecting pan together with the concentrate and sufficient water to allow the mixture to be agitated freely. Agitation by hand is continued until all of the gold has been picked up by the mercury. The amalgam and any residual mercury beads are recovered by decantation taking care, always, to carry the process out over another receptical in case of spillage. The gold is then recovered by dissolving out the mercury in dilute nitric acid using gentle heat if necessary to speed the reaction. The gold residue is washed and annealed at red heat to drive off the last traces of water and mercury. Any risk of losses through splattering during the annealing process can be avoided by adding a drop of acetone to the gold residue before heating.

Samples containing the valuable light heavy minerals rutile, zircon, ilmenite etc. or any other discrete particles which can be identified using a microscope may be analysed by 'grain counting', by instrument or chemical methods of analyses, or by a combination of methods.

Grain counting

Grain, or more accurately particle counting, finds important applications in minerals laboratory practice wherever the samples contain discrete particles of minerals that can be identified readily using a binocular microscope to view the particles displayed on a microscope slide.

The portion of sample from which the slide is to be prepared is mixed carefully on a sheet of clean paper and flattened out to present an even surface. A small drop of a saturated solution of sucrose in water, or similar mounting medium is smeared over the surface of the slide. The slide is pressed firmly down onto the surface of the sample and then tapped lightly to dislodge any particles that are not in actual contact with the film on the glass. Alternatively, the sample material is spread over the surface of the slide by taking a small portion on to a spatula and distributing it evenly from a height of about 3 cm onto the slide. It is best for counting purposes if the individual particles are not in contact with one another.

If it is necessary to preserve the slide this can be done by placing eight to ten drops of a mounting medium such as benzyl benzoate onto the slide, using another slide as a cover strip.

A magnification is selected for counting such that the number of particles in the field is in the range 25–100 particles. The particles in 10–15 fields are counted on each slide to give a count per slide of 250–1000 particles. The fields are selected at random to cover the whole area of the slide using cross hairs or sections to assist counting.

A variation of the normal grain-counting procedure, called 'point counting', samples the mixture at a number of points. It utilizes a cross-lined graticule fitted in the focal plane of the eyepiece and the intersection of these lines becomes the focal point for counting. The specimen moves beneath a stationary point along a series of linear traverses and the accuracy of the analysis is a function of the number of points sampled. An electrically triggered specimen carriage attached to the stage of the microscope can be adjusted to move the carriage sequentially at intervals of 0.16 mm or similar both north–south and east–west to cover the slide.

Corrections are applied for relative density in relating the number of valuable particles to those of the non-valuable types. The method used is tabulated as an example in Table 4.10.

Instrument analysis

Modern techniques for the routine analyses of mineral samples include; (a) atomic absorption spectrometry, (b) arc emission spectroscopy, (c) X-ray fluorescence, (d) polarography, (e) neutron activation, and (f) colorimetry.

Each of the methods has certain advantages and one may be preferred to

Table 4.10 *Specimen particle count tabulation*

Sieve fraction (μm)	Weight pct (a)	Mineral variety	Grain count (b)	Relative density (c)	(b) × (c) (d)	% of fraction $\frac{100 \times (d)}{\sum(d)}$	% of sample $\frac{(a) \times (d)}{\sum(d)}$
+246	32	Zircon	232	4.5	1044	22.4	7.2
		Rutile	210	4.2	882	18.9	6.0
		Ilmenite / Magnetite	517	4.9	2533	54.3	17.4
		Monazite	40	5.0	200	4.3	1.4
		Others	2	3.0	6	0.1	0.03
				$\sum\{(b) \times (c)\} =$	4665	100.0	32.03
−246 +104	58	Zircon	207	4.5	932	19.4	11.3
		Rutile	180	4.2	756	15.7	9.1
		Ilmenite / Magnetite	572	4.9	2803	58.3	33.8
		Monazite	59	5.0	295	6.1	3.5
		Others	7	3.0	21	0.4	0.23
				$\sum\{(b) \times (c)\} =$	4807	99.9	57.93
−104	12	Zircon	139	4.5	626	12.5	1.5
		Rutile	151	4.2	634	12.6	1.5
		Ilmenite / Magnetite	702	4.9	3440	68.5	8.2
		Monazite	53	5.0	265	5.3	0.6
		Others	19	3.0	57	1.1	0.13
				$\sum\{(b) \times (c)\} =$	5022	100.00	11.93

Therefore approximate composition of heavy minerals

Zircon	7.2 + 11.3 + 1.5	= 20 %
Rutile	6.0 + 9.1 + 1.5	= 16.6 %
Ilmenite–magnetite	17.4 + 33.8 + 8.2	= 59.4 %
Monazite	1.4 + 3.5 + 0.6	= 5.5 %
Others	0.03 + 0.23 + 0 + 0.13	= 0.39 %
Total		101.89 %

another on the basis of rate of throughput, sensitivity or specificity in any particular application. All suffer the disadvantage of not being able to distinguish between minerals except where the distinction can be inferred from the elemental composition of the mineral. For example, zircon in a sample is almost certainly the only zirconium-bearing material present and can thus be assessed directly. Rutile is commonly associated with other titanium-bearing minerals such as ilmenite, anatase and sphene and its content in a sample cannot always be calculated directly from the titanium analysis. Grain counting is commonly used as an adjunct to instrument analyses to estimate the proportions of each of the minerals of a particular elemental group. The apportionment of the instrument result is then made accordingly.

Atomic absorption spectrometry: This method is used widely for analysing placer minerals because of its high throughput and moderate sensitivity. Some interferance due to other elements that are taken into the solutions prepared for analysis can be corrected by making suitable adjustments to the standards or by suppression of the interfering ions through the addition of other reagents.

Limits of sensitivity	0.5 ppm
Throughput	200 samples h^{-1}.

Arc emission spectroscopy: One advantage of this method is the simplicity of sample preparation compared with most other methods of instrument analysis. It involves only grinding, compounding with graphite and transferrence of the mixture to an electrode. The resulting spectrum, registered on a photographic plate, reveals all of the other emissive elements present. Quantitative data are obtained by measuring the degree of opacity of the lines of the spectrum. An automatic densitometer is used for more precise analysis.

Limits of sensitivity	about 1 ppm
Throughput	4–6 specimens h^{-1}

X-ray fluorescence spectrometry: This technique is highly specific and is readily automated for semi-continuous throughput. However, careful sample preparation is needed to obtain uniform particle size and repetitive counting must be carried out for extended periods to analyse concentrations containing less than 500 ppm.

An XRF procedure referred to as XRF scanning is used for the qualitative and approximately quantitative definition of all of the heavy minerals in a given sample. It is used to confirm microscopic identifications of some minerals that may be hard to distinguish by optical means. Typical examples include differently coloured particles of cassiterite that commonly resemble rutile and other opaque minerals such as xenotime.

Analysis by X-ray fluorescence depends upon the irradiation of the sample with X-rays. The X-rays cause the atoms contained in the sample to fluoresce with radiations of characteristic wave lengths which are quite specific to the individual elements present. The intensity of the fluorescence at any particular wave length is proportional to the amount of the element present. The composition of the sample is then determined by measuring the intensities over the range of individual wave lengths.

Limits of sensitivity	10 ppm
Throughput	4–20 specimens h^{-1}

Polarography: Prior solution of the sample is necessary as for the atomic absorption method but subject to this being done, the throughputs are quite high. Various modifications to the original DC instrument include AC polarographs, square-wave differential with or without cathode ray oscilloscope and pulse instruments. The method can be very specific and it has good sensitivity.

Limit of sensitivity	0.2 ppm
Throughput	100 specimens h^{-1}

Neutron activation: This method is comparatively new to minerals processing laboratories. It is the most sensitive and specific method available but is unlikely to come into common use, at least in the immediate future, because of the high cost of the neutron sources and because specialist skills are required for operation. Special safety precautions have to be taken to avoid radiation hazards.

Limit of sensitivity	0.001 ppm
Throughput	100 specimen h^{-1}

Colorimetry: This method is specific for tin assays using a variety of reagents such as dithiol, gallein and haematoxylin. It suffers from a rather complicated sample preparation procedure and by the need to use mixtures of reagents, some of which are expensive or unstable.

Limit of sensitivity	about 5 ppm
Throughput	100–105 specimens h^{-1}

Chemical determinations

Because they are more time consuming, chemical methods of analysis have been largely superceded in placer laboratories by the instrument methods described. The chemical section of such laboratories is now given over largely to sample preparation for instrument analysis and to such determinations as those of ferrous iron and total insolubles in ilmenite concentrates and total insolubles in monazite concentrates. Typical methods for these determinations are as follows.

Determination of ferrous iron: The sample is dissolved in sulphuric acid under conditions that preclude any oxidation of the ferrous iron already present. The solution is then titrated with standard potassium permanganate. The apparatus illustrated in Fig. 4.25 consists of:

1 × 500 ml Erlenmeyer flask
1 × 250 ml beaker
1 water-cooled reflux condenser (Davey, Liebig or similar), delivery tube, bunsen burner, tripod, asbestos gauze, retort stand, clamps and rubber stoppers suitably drilled.

Solutions used are:

Sodium bicarbonate 10%:	dissolve 100 g in sufficient water AR quality or equivalent to produce 1 litre of solution
Sulphuric acid 1:1:	500 ml AR quality or equivalent. RD 1.84

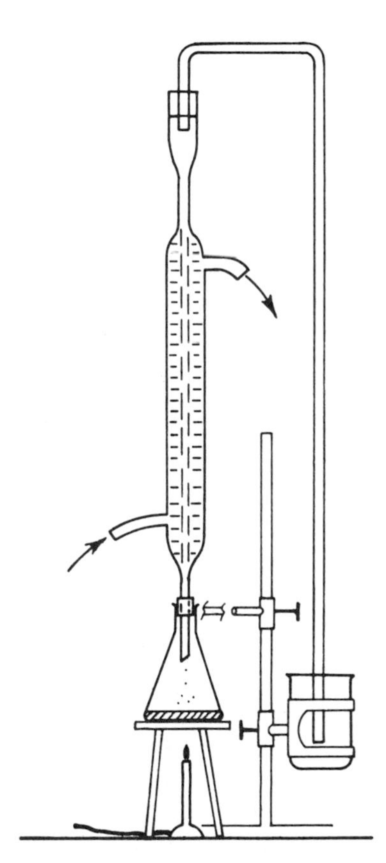

Fig. 4.25 Apparatus for determining ferrous iron

Potassium permanganate:	standard N/10 solution. One millilitre is equivalent to 0.007 185 g FeO.

The method is to weigh out accurately 0.5 g of ground sample and transfer to the 500 ml flask. Add 50 ml 1:1 sulphuric acid and fit condenser and delivery tube as in Fig. 4.25.

Transfer 100 ml of sodium bicarbonate solution to the 250 ml beaker and support the beaker so that the end of the delivery tube is dipping into the solution but is clear of the bottom of the beaker. Apply gentle heat and swirl the contents of the flask to ensure that the sample is thoroughly wet. (Failure to do this may result in some of the sample adhering to the sides of the flask and being incompletely dissolved.)

The sample is then allowed to reflux gently for 1.5–2 h. The solution is cooled and the flask is removed from the condenser. The sample is diluted to approximately 200–300 ml with previously boiled and cooled distilled water and titrated immediately to a permanent pink colour using potassium permanganate.

$$\% \text{ FeO} = \frac{\text{ml N/10 KMnO}_4 \times 0.007\,185 \times 100}{\text{sample weight}}$$

Total insolubles in ilmenite: any material not soluble after fusion with pyrosulphate and treatment with 1:1 hydrochloric acid is regarded as insoluble.

The method is to fuse 10 g of unground ilmenite with 50 g of potassium pyrosulphate in a 500 ml Erlenmeyer flask and allow to cool. Add 200 ml 1:1 hydrochloric acid and heat to dissolve all soluble matter. Decant the clear liquid.

If any free ilmenite still remains it is fused with a further 10 g of pyrosulphate, dissolved in 1:1 hydrochloric acid and washed six times by decantation with hot water.

After drying, the insolubles are transferred to a watch glass and weighed. The percentage of insolubles is calculated by difference in weight and, if appropriate, a heavy-liquid separation can be carried out on the insolubles to determine the silica content.

Total insolubles in monazite: Monazite is dissolved by fuming with perchloric acid. The insoluble material is recovered by filtration. Minerals normally associated with the magnetic fraction in which monazite occurs are to all intents and purposes insoluble under these conditions.

The method is to weigh out 1 g of monazite concentrates and brush into a 150 ml squat beaker, add 15 ml of perchloric acid (70–72%), cover with a watchglass and simmer gently on a hotplate for 1.5 h.

Cool the solution and add 40 ml of 1:4 nitric acid and 3 ml of hydrogen peroxide (100 volumes). Filter under vacuum through a tarred Gooch crucible (porosity 4), dry and weigh to constant weight. Porosity 4 is very fine and some analysts prefer to use porosity 3 although it may tend to give lower results.

Since

$$\% \text{ insolubles} = \frac{\text{crucible weight gain} \times 100}{\text{sample weight}}$$

it is not essential to use exactly 1 g and it may be preferable to use the actual sample split whether it be 0.9835 or 1.1153 or any other such quantity.

However, extreme care must be taken in handling the perchloric acid; it is explosive. The addition of 10 ml of concentrated nitric acid makes it safer to use but if this is done the watchglass cover should be omitted in the first step.

Sampling errors and the reliability of sampling

Sampling errors may result in the rejection of valuable properties or in the waste of large sums of money spent in developing uneconomic deposits. They can usually be avoided, or at least minimized, if the reasons are understood. Some errors are common to all placer types, others are specific to particular placers in their individual environments. Some are errors of omission, some are errors of commission while some are errors of judgment.

Errors common to all placer environments

The reliability of sampling increases with the number of samples taken and, hence, with reductions in the volumes of influence assigned to each sample. A case can also be put for increased reliability with increased sample volume. There are, however, economic considerations and while in both cases the ultimate sample has a volume to volume of influence ratio of 1:1 this, of course is quite impracticable and at some point a compromise must be made between sample cost and reliability.

Incorrect sample dimensions: For each deposit there is an optimum density of sampling that is based upon the geology of the deposit and the selection of methods that will hold sampling errors within pre-determined limits. Consider, for example, two adjacent samples each representing the same volume of influence. One is a pit sample having a cross-sectional area of 1 m^2 and a sample interval of 1 m; the other is a borehole sample of diameter 10 cm and the same sample interval of 1 m.

Pit sample volume	$= 1 \times 1 \times 1 = 1\ m^3 = 10^6\ cm^3$
Bore sample volume	$= (11/14) \times 10^2 \times 100 = 7857\ cm^3$
Therefore ratio of volumes	$= 127{:}1$

It is contended, nevertheless, by some mathematicians that the prime consideration in sampling is the number of samples taken rather than the size of individual samples and there are grounds for this view when sampling populations of similar-sized particles. For example it is common practice to evaluate mineral sands deposits using small 50-mm diameter holes. Such deposits are closely sized in a fine-particle size range and statistical methods have been applied with a great deal of success.

It is quite a different matter to extend the same principles and reasoning to long size-range sediments in the continental placer environment where the valuable particles are such high-value minerals as gold, platinum and diamonds. It is particularly difficult to do so when the main values are contained in narrow paystreaks, gutters and potholes and where interpretation is made more difficult by the presence of large stones and boulders. Such impediments to drilling may be wrongly assumed as being part of the bedrock whereas, in fact, they lie upon the bedrock and, in doing so, may conceal the presence of concentrations of valuable minerals around their bases. The result is an underestimation of both average grades and volumes.

Geostatistical methods referred to in Chapter 8 take the 'nugget' effect into account when evaluating such deposits but even so they are subject to considerable error if the size of the samples taken fails to reflect adequately the geology of the deposit and the geometry of its surroundings. It is well established, for example, that in sampling some diamond placers, sample volumes as large as several hundred cubic metres may be required in order to recover and represent fully the many different grades and values of stones.

Consider, also, the effect on grade estimations of one particle of gold valued at 1 cent, i.e. 1 mg Au with gold valued at \$10/g. If such a particle is added to or lost from the above pit sample the grade of the sample is raised or lowered by only $\$0.01/m^3$. Added to or subtracted from the above borehole sample the grade differential becomes $\pm \$1.27/m^3$. Clearly, if the misplaced particle was valued at \$1 instead of \$0.01 or was, instead, a small diamond valued at \$1, the resulting sampling errors would be scaled up or down by a factor of 100. Figure 4.26 illustrates the effects of one particle valued at 1 cent on the calculated values of samples of various volumes from $0.1\ m^3$ up to $1.0\ m^3$ to show the extreme vulnerability of small samples to errors. Figure 4.27 shows how samplers in the Ghanaian diamond fields dress samples in the field.

In most cases, a reliable placer valuation will rely upon sample sizes and spacings that have been proven by adequate test work to be appropriate to the particular circumstances. For example Wells (1976) refers to a 15-year period (1945–60) in which the Natonas Company while operating with five dredgers having a combined throughput of $207 \times 10^6\ yd^3$ ($158 \times 10^6\ m^3$) obtained an overall recovery of 104% of the prospect values. All sampling

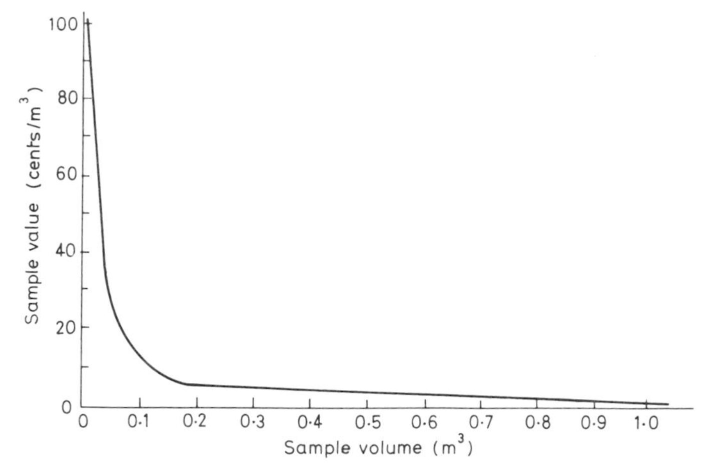

Fig. 4.26 Plot of sample volume versus sample value

Fig. 4.27 Dressing diamond bearing concentrates by hand, Birim River Flats, Ghana

was done using standard placer-type churn drills providing 7.5-in (190.5 mm) holes. The use of short, 1–1.5 ft (304.8–457.2 mm), sample intervals combined with the careful measurement of sample volumes obtained from each increment contributed to the reliability of the evaluation and made possible a profitable mining operation.

Relating these figures to the graph drawn as Fig. 4.26, it is probable that the total sample volumes for each hole were of the order of 0.29 m^3 per metre of depth. This places the samples on the flatter curve section and obviously for the number of samples taken, i.e. the density of sampling, the sample size was adequate. Perhaps had the samples been larger, fewer samples need have been taken. However, for the recorded density of sampling any decrease in sample volumes would have increased the rate of failure markedly and, for sample volumes less than 0.1 m^3 a faulty valuation could almost have been assured.

Wells qualified his remarks by referring to the low price of gold and the wide experience and skill of the placer drillers and managers of those times. He suggests that today this is not the case in placer gold prospecting ventures and, hence, any gold valuation based upon drilling should be viewed critically until the procedures used and all related factors have been weighed objectively and found to be adequate.

Faulty analytical procedures: The first consideration in sampling is to select suitable equipment and develop techniques for taking the samples that are in keeping with both the environment and the specific purposes of the investigation. This requirement applies equally to analysis as to the physical taking of the samples in that the analytical methods used should not run contrary to the methods that will finally be used for recovery on a commercial scale. For example, some of the valuable constituents may be locked up in particles of non-valuable materials and not be recoverable by gravity methods of treatment. For this reason alone, and there are others, fire assays and methods of instrument analysis used as standard procedures for analysing hard-rock ores, give totally misleading results on most placer ores.

Salting: A deposit is said to have been salted when boreholes are deliberately sited in more-favourable locations, when quantities of the valuable minerals are introduced into samples after they have been taken, or when false measurements are substituted for true ones. It must always be assumed that salting will be attempted and precautionary measures include systems of control samples in which blanks are given normal code numbers and submitted along with other samples, and statistical checks on symmetry. Any samples not conforming to the general pattern are retaken and re-analysed under a different code number known only to the person checking.

Carelessness: Carelessness includes such faults as undue haste in preparing and dressing samples; accidental contamination; using approximate measurements instead of true ones to facilitate calculations and confusing sample labels. If sampling methods are not standardized and rigorously policed some, if not all, of these faults will inevitably occur.

Inaccurate measurements: Inaccuracies in marking out, levelling or measuring distances results in errors in calculating areas and volumes of influence and in geological interpretation.

Drilling onto a hard bottom: Values are lost when drilling onto a hard bedrock unless it can be cleaned up by hand. This can be done only if pit-digging drills are used to make holes large enough for human access. Auger drilling onto hard basement will always result in a loss of sample and, in general, an underestimation of values.

Use of pipe and other factors: Where possible, the need for arbitrary corrections should be avoided by using systems that do not require corrections. Where corrections are necessary, the factors should be determined by careful studies in controlled conditions.

Inaccurate logging: Failure to recognize and delineate payable horizons and basement rocks may result in excessive dilution or, conversely, in a loss of mineral during both sampling and subsequent mining.

Unsuitable boring equipment: Representative samples are unlikely to be taken if unsuitable gear is used. The type of hand auger suitable for mineral sands sampling (Fig. 4. 28) is quite unsuitable for sampling in other placer environments.

Uneven distribution of values: Where the geometry of the bedrock is such that rich accumulations settle in potholes and other natural traps, the sampling pattern must be very close to avoid major errors. Such deposits tend to be undervalued unless very large bulk samples are taken.

Not allowing for dilution and batters: No placer is mined entirely within its payable limits. Batters must be allowed for along the sides of the excavation and a significant level of dilution results from mining along the boundaries between the gravels and barren overburden and from excavating into the bedrock. The amount of overburden to be allowed for depends upon the stratigraphic contours. A minimum of 25 cm above and below the gravels is normally included in the material excavated for treatment except where the bedrock is too hard to be ripped.

Ignorance: Probably more errors result from ignorance than from all of the other causes put together. All personnel should be fully trained in their

Fig. 4.28 Collecting hand auger sample from fossil strandline deposits, Phore, Pakistan

respective duties and should not only understand how to perform the tasks assigned to them, but also why they should be done in a particular way.

Additional errors that apply specifically to the individual placer environments are as follows:

Errors in continental placer sampling
These are due mainly to:

(1) Splitting samples containing particles of widely different sizes. Such samples should be dressed as a whole.
(2) Cutting irregularly shaped pits and channels. All such samples are liable to error if the pit walls are not regular and vertical or if the channels are not regular in depth and section. The best method of sampling is to treat all of the material removed from a geometrically true section having vertical walls.

Errors in transitional placer sampling
These relate most importantly to grain counting and unnoticed changes in lithology.

Grain counting: This finds an important application in minerals laboratory practice for spot testing and for use in conjunction with instrument methods of analysis. It is not favoured on its own for final analytical determinations because certain factors tend to make the procedure unreliable. Important amongst these are:

(1) The sample population viewed is extremely small compared with the bulk of the sample.
(2) The particles counted under the microscope vary in size, shape and density. Possible errors can be minimized by dividing each sample into a number of sieve fractions and counting each of the fractions separately.
(3) Direct measurements, if they are taken, are in two dimensions only.
(4) Errors may result from faulty identification of individual minerals.

Failure to recognize changes in lithology: Errors arise from the failure to recognize and evaluate changes in the lithology of individual strata during drilling, failure to recognize grain coatings or textural and other associations affecting marketability. Valuations may fail because operators expect similar responses to those of placers in other districts and have neglected to carry out pilot-scale investigations before designing the plant.

Errors in marine placer sampling
Specific sampling errors may be due to:

(1) Loss of heavy minerals during decantation: this is a problem peculiar to sample dressing on Becker drilling rigs operating offshore in bad weather conditions.
(2) Irregular sample recovery due to the methods used: there is often a tendency for heavy minerals to be under-represented in the overlying sediments and for heavy minerals in the lower sediments to be over represented because of carry down and sorting in the column.

REFERENCES

Antung, M. and Hanna, W. F. (1975) Geophysical investigations of coastal magnetite sands at Meleman, Lumajang, East Java, *Geol. Surv. of Indonesia, Seri GeoFisika No. 5.*

Applin, K. E. S. (1976) Exploration for alluvial diamond deposits, in *A Short Course on Placers, Exploration and Mining*, University of Nevada, USA.

Barton, P. H. (1979) Dynamic positioning systems, *ERT*, July 34–7.

Breeding, W. H. (1976) The evaluation of placer deposits, in *A Short Course on Placers, Exploration and Mining*, University of Nevada, USA.

Carter, W. D. (1979) Significant results from the use of remote sensing in the exploration for Precambrian mineral deposits, *UN Seminar on Precambrian Mineral Deposits*, Moscow.

Colp, Douglas B. (1976) A placer gold evaluation technique comparison between a hammer and klam drill, in *A Short Course on Placers, Exploration and Mining*, University of Nevada, USA.

Daly, J. (1965) Geophysical prospecting for deep leads, in *Exploration and Mining Geology*, (ed. D. J. Lawrence), AIMM, Melbourne.

Goosens, Pierre J. (1980) Programming modern mineral exploration survey, a field geology – oriented approach, *Memoires of the Institute of Geology*, University of Louvain, Belgium.

Harrison, H. L. H. (1954) *Valuation of Alluvial Deposits*, 2nd edn, Mining Publications, London.

Hill, J. C. C. (1976) Economic evaluation of a marine alluvial deposit using vibratory coring, in *A Short Course on Placers, Exploration and Mining*, University of Nevada, USA.

Lord, John F. (1976) Placer mining evaluation, methods and applications, in *Placer Exploration and Mining Short Course*, Mackay School of Mines, Nevada.

Macdonald, Eoin H. (1966) The testing and evaluation of Australian placers, *Proc. Aust. Inst. Min & Met*, 218, Melbourne.

Macdonald, Eoin H. (1973) *Manual of Beach Mining Practice. Exploration and Valuation*, 2nd edn, Dept. of Foreign Affairs, Aust. Govt. Printing Office, Canberra.

Macdonald, Eoin H. (1981) Offshore minerals other than hydrocarbons in southeast Asia, in *Southeast Asian Seas, Frontiers for Development*, (eds Chia Lin, Sien and Colin McAndrews) McGraw Hill, Singapore, 51–74.

Macdonald, Eoin H. (1983) Placer exploration and mining, in *AGID Guidebook on Mineral Resources Development*, (ed. Woakes, M.), in press, Bangkok.

Meleik, M. L., *et al.* (1978) Aerial Ground Radiometry in Relation to the Sedimentation of Radioactive Minerals of the Damietta Beach Sands, Egypt, *Economic Geology*, Vol.73, Economic Geology Publishing Co., Lancaster P.A., U.S.A., pp. 1738–1748.

Noakes, L. C. (1972) *Proposed Legend for Mapping Detrital Mineral Deposits*, CCOP (ESCAP), Bangkok.

Noakes, L. C. (1977) Provenance for mineral sands and tin, *Tech. Bull. No. 11*, CCOP (ESCAP) Bangkok.

Parasnis, D. S. (1979) *Principles of Applied Geophysics*, 3rd edn, Chapman and Hall, London.

Palmer, A. G. (1942) The estimation of gold and tin alluvials in Malaysia, *Proc. Inst. Min & Met Aust*, 128, Melbourne, 201–20.

Parks, Roland D. (1957) *Examination and Valuation of Mineral Property*, Addison Wesley Publishers, Massachusetts.

Ryall, Patrick J. C. (1977) A reconnaissance geophysical survey offshore from Kedah, west coast of Peninsular Malaysia, *CCOP Newsletter*, 4 Dec. Bangkok. 69–73.

Rudenno, V. (1978) *The Probability of Economic Success in Exploring for Primary Tin Deposits*, School of Mining, University of New South Wales, Australia.

Shepard, Francis P. (1973) *Submarine Geology*, Harper and Row, New York.

Street, W. R. (1977) *Seabed Gamma Ray Spectrometer*, Dept. of Energy, Thames House South, Millbank, London.

Van Overeem, A. J. A. (1960) Sonic underwater surveys to locate bedrock off the coasts of Billton and Singkep, Indonesia, *Geologie En Mignboun*, 39c Jaagang, 464–71.
Peterson, D. W., Yeend, W. D., Oliver, H. W. and Mattick, R. E. (1968) Tertiary gold bearing channel gravel in northern Nevada County, California, *US Geol. Surv. Circ.* 566, 22 pages.
Wells, John H. (1969) Placer examination, principles and practice, *Tech. Bull No. 4*, US Dept. Interior. Bur. Land Management, Washington, DC.
Wells, John H. (1976) Some notes on placer sampling, in *A Short Course on Placers, Exploration and Mining*, Mackay School of Mines, Nevada.

APPENDIX 4.1 PROPOSED LEGEND FOR MAPPING PLACER MINERAL DEPOSITS AND ENVIRONMENTS (AFTER NOAKES 1972)

Shorelines and alluvial areas

Narrow sandy beach

Wide sandy beach

Beaches and rocky headlands

Rocky shoreline

Rocky shoreline with few beaches

Raised beach

Muddy shoreline

Shorelines and alluvial areas (*contd*)

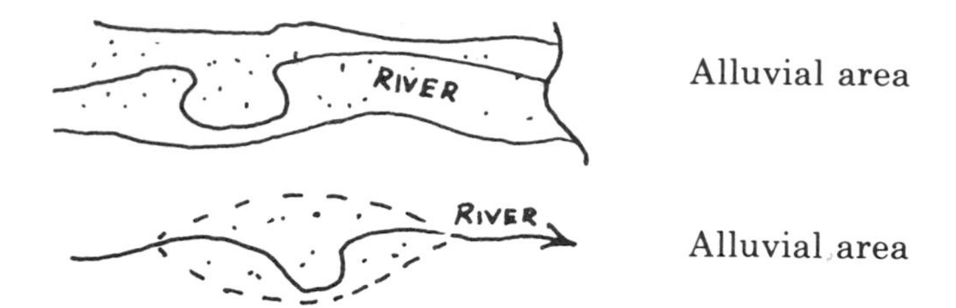

Bathymetry (in blue)

0.20	Sounding in metres
——— (thin and thick lines)	Isobaths (20, 200, 1000, 2000 m or at closer intervals if practicable – isobaths in blue with the 200 m isobaths thickened for emphasis)

Black and white conventions

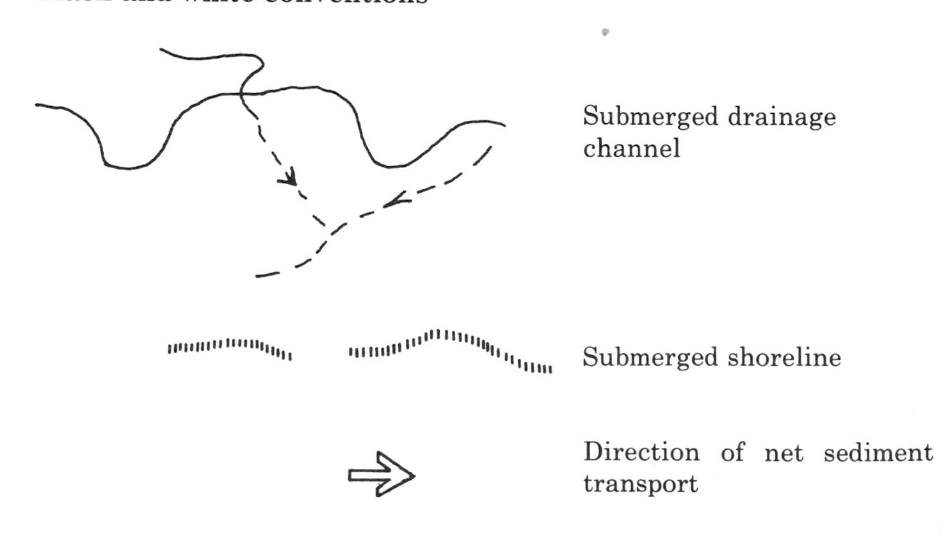

Character of bottom sediments

Spot samples		or	
G	gravel		gravel
S	sand		sand
M	mud		mud
R	rock		metamorphic rocks
	rock reef*		granitic rocks
	coral reef*		other igneous rock
	sand bar*		bedrock undifferentiates

*Standard symbols from hydrographic charts.

Colour convention

Thickness of sediment above basement

Pale brown – basement rocks exposed on the sea floor.
Pale yellow – cover of up to 20 m of sediment over basement rocks.
Pale green – cover of more than 20 m of sediment over basement rocks.

APPENDIX 4.2 DELINEATION OF DETRITAL HEAVY-MINERAL DEPOSITS

Prospect or unworked deposit

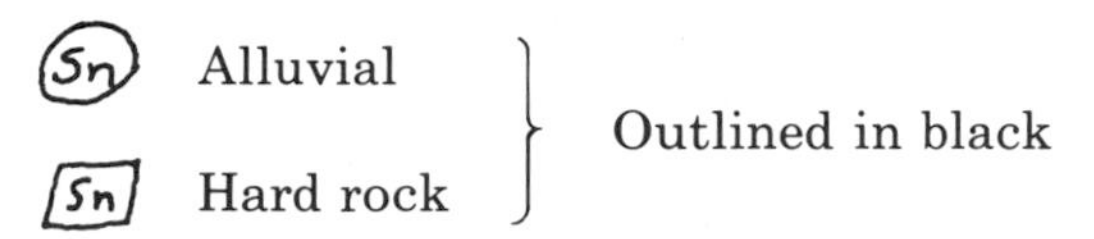

Mines

Alluvial
Hard rock

Outlined in: red – major deposit
blue – intermediate deposit
yellow – minor deposit
black – uncertain?

(See text for categories)

Previously mined

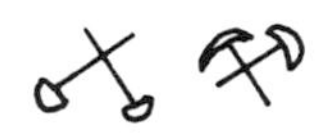

Currently mined

Reference number to refer to accompanying table.

Symbols

Au	Gold	G	Garnet
Sn	Tin	Ta	Tantalite
Fe	Magnetite, Hematite	W	Tungsten
Tf	Titaniferous magnetite	Cr	Chromite
Im	Ilmenite	Di	Diamonds
Zr	Zircon	Gs	Gemstones
Ru	Rutile	Pt	Platinum group
Mz	Monazite		

5

Centrifugal Slurry Pumps and Pumping

5.1 INTRODUCTION

From the first recorded use of a pumping device for conveying solids–fluid mixtures in pipelines (see Chapter 1) placer technology has come to rely more on pumps for solids transportation than upon any other form of materials handling and slurry pumps are now the central features in most placer operations. Typical examples are found in wet mining methods that require slurries to be pumped away from the face (suction dredging) or shortly thereafter (bucket line dredging and hydraulic sluicing). Concentrating plant applications include controlled density pumping; hydrocyclone sizing, dewatering and desliming; concentrate stacking and tailings disposal. In all of these operations total dynamic heads rarely exceed 60 m of water and, for them, the centrifugal slurry pump system is favoured against any other system of pumping because, at low to medium heads, it offers the best combination of solids transportation characteristics, ease of operation and maintenance, and low unit pumping costs.

5.1.1 Availability of data

Manufacturers have access to a great deal of information on their own pumps and those of other manufacturers that is not readily available to the operators and hence they should always be consulted on any new pumping installation at the stage of final design. On the other hand, the operator needs to carry out many preliminary studies, as knowledge of the deposits and its economics is being developed. In order to do this and to examine and evaluate critically what is later proposed, he must have reasonably comprehensive information on such matters as power to throughput ratios, pump characteristics, wear and other relevant factors such as delivery dates and the cost and availability of spares. First cost as a factor is seldom very relevant in the long term and the cheapest proposal may subsequently prove to be the most expensive.

Centrifugal slurry pumps are used widely in the chemical and allied industries and a great deal of experimentation has provided manufacturers with empirical data from which, given adequate job descriptions, responsible and competitive alternatives can usually be proposed for any normal undertaking. For unusual problems, test rigs are available from which much of the required information can be obtained although there may be some practical difficulties in simulating actual work situations on a pilot-plant scale. The alternative of conducting scientific investigations in a commercial plant is seldom possible for operational reasons.

It is unfortunate, therefore, that not all slurry manufacturers publish sufficient data for valid comparisons to be made between their pumps and those of others. This is an unacceptable situation to the designer who will normally consider only those pumps for which there are adequate performance data from which to base his calculations. In this chapter, the underlying principles governing slurry pumping as applied to placer technology are discussed using data supplied by Warman International of Sydney, Australia. Charts and conversion factors supplied as appendices to the chapter can be used to translate metric units into other units if required.

5.1.2 Reliability of data

The application of engineering principles governing centrifugal water pumps to slurry pumps and pumping installations was initially thought to require only simple corrections based upon differences in slurry densities. However, anomalous results soon demonstrated the inadequacy of such measures and it is now realized that many more factors are involved than can be evaluated exactly at the present state of knowledge. In this text both empirical and arbitrary methods are applied freely wherever the problems cannot be approached from theory alone. Slurry pump design in complex pumping situations, such as are found in suction dredging, is highly dependent upon the experience of the manufacturer in related undertakings.

Many opportunities for error are inherant in the basic data. To begin with, all pump performance curves are developed from clear-water performances only. The tests are carried out in water research and other laboratories having appropriate test-rig facilities and accuracies are subject to the tolerances allowed for the specific test conditions.

Scope for additional error arises when similar tolerances are applied to estimating the effects of solids concentration on pump performances. The most sensitive factors are those relating to slurry friction-head losses, total dynamic heads, pump speeds and power consumption on assumptions of proposed system details that may, themselves, be subject to some uncertainties.

However, the modern pump manufacturer is usually able to refer to a range of slurry pumping case histories and test results to reduce the margin of error and there is a reasonable confidence that actual pump performances will not fall outside of the allowable limits except where the duty data have been wrongly supplied. On the other hand, plant engineers seldom have access to such information and generally their selections should be confirmed by the manufacturer before being put to the test.

5.2 CENTRIFUGAL SLURRY PUMPS

Centrifugal slurry pumps differ in design and construction largely according to the maximum size of the particles to be pumped. Gravel (dredge) pumps feature large clearances for the passage of gravel and cobble-sized fragments. Sand pumps are restricted to handling finer grained sediments containing only occasional small pebbles. Although not strictly classified as centrifugal slurry pumps, jet lift pumps are included because they incorporate multi-stage centrifugal water pumps and find use in some placer applications.

Pumps are designed for either horizontal or vertical mounting. Horizontally mounted pumps offer the widest range of applications; they cost less and are easier to maintain in general service. The vertical types are preferred wherever difficult suction conditions can be avoided by using a submersible pump or where the headroom is limited in confined spaces. Modern gravel-pump operations utilize vertical slurry pumps almost exclusively for elevating slurries to the treatment plant; they also feature in most milling designs.

The main limitations to the use of centrifugal slurry pumps are found in pumping slurries through very long pipelines and against high static heads. Individually increased pressures are obtained only at the expense of reduced flow rates and, multi-staged, each pump can contribute only a small increment of pressure if total flow rates are to be maintained at a constant level. A multiplicity of centrifugal pumps is needed for very long pipelines and, for such applications, they are disadvantaged against positive displacement pumps which can individually develop very high pressures at constant rates of flow. A considerable amount of literature is available to describe the pumping of slurries through long pipelines and a comprehensive list of references is given in Wasp *et al.* (1977). Placer technology, for the most part, is concerned only with the centrifugal types of slurry pumps.

5.2.1 Dredge and gravel pumps

Gravel pumps are designed for handling long particle size range sediments and are the most versatile of all slurry pumps. They have the ability to pump mixtures of sand, mud and gravel-containing particles that are too large for most other pumps; as dredger pumps they must cope with large instantaneous variations in slurry concentration, particle size and suction head and adjust, when dredging in marshy ground or offshore, to the effects, also, of variable amounts of trapped or dissolved gases. The gases are mainly methane and nitrogen and Scheffauer (1954) refers to extreme conditions in which hopper dredgers are equipped with gas removal systems to take the gas from the suction line before it reaches the pump.

5.2.2 Sand slurry pumps

Sand slurry pumps are used for pumping relatively clean, free-flowing sands in concentrating plant service and in some beach-mining operations. They differ in construction from gravel pumps in that, normally, the pump casings are split and impellers and liners are supplied in moulded rubber except where sharp objects such as shells, wood fragments and similar trash may be present, when special steel alloys are used to resist the wear. Sand pumps are usually lighter in construction than gravel pumps and have narrower passages.

5.2.3 Jet lift pumps

Jet lift pumps are used mainly for elevating solids–fluid mixtures in conditions unsuited to conventional gravel pumps and one service in which they clearly outperform any other type of pump is in the selective recovery of gravels that lie buried under considerable thicknesses of silt and mud offshore.

Jet lift pumps also find application as diver-operated units for the small-scale dredging operations referred to in Chapter 6 and for cleaning up at bedrock in conjunction with larger dredging projects.

The same principle is employed in the use of high-pressure water injection, using a venturi device at the suction pipe inlet to boost pump performances in some deep dredging operations. The benefits increase rapidly below digging depths of around 6 m, however, for such depths, the designer can also adopt a submerged slurry pumping system and will probably have little hesitation in selecting such an alternative for new installations. The main applications for jet-assisted pumps are found in adapting the older type dredgers for deep dredging. An obvious disadvantage is in the dilution of the slurry flowing. For example, one manufacturer of jet boosters recommends the injection of 40% of the total water as high-pressure jet water thereby allowing only 60% of the total water to be used for drawing solids into the suction-pipe inlet.

5.3 PUMP EFFICIENCIES

5.3.1 Centrifugal slurry pumps

Centrifugal slurry pumps are designed and rated from tests with water from which the results are presented in the form of general head-quantity curves and plots relating to pump efficiencies, thrust, horsepower requirements and cavitation. These data can be used to predict the behaviour of mixtures of solids in water using information obtained in test rigs having the means for measuring power consumption, torque, pressures and wear. Variously designed impellers and casings are tested and performance curves are adjusted to allow for the effects of increased slurry densities, viscosities and slippage between the particles and the fluid. The effects are greatest during periods of acceleration and deceleration on entering and leaving the pump. Modern design is aimed at reducing energy wastages in the form of:

(1) Hydraulic losses resulting from eddies, turbulence, impact and friction along the solid boundaries
(2) Leakage losses resulting from providing additional clearances for the passage of solids
(3) Mechanical losses including bearing, stuffing box and disc friction losses which are functions of design

The three measures of efficiency are, thus:

(1) The hydraulic efficiency which is determined by comparing the total head developed by the impeller with that delivered by the pump.
(2) The volumetric efficiency which relates the amount of fluid set into motion by the impeller with that delivered by the pump.
(3) The mechanical efficiency which compares the amount of energy provided by the prime mover with that employed usefully in pumping.

Hydraulic efficiency

The hydraulic efficiency e_h is expressed as:

$$e_h = \frac{H}{h_i} \tag{5.1}$$

where H is the total head developed by the pump and h_i is the total head developed by the impeller.

Volumetric efficiency

The volumetric efficiency e_v is given by the expression:

$$e_v = \frac{Q}{(Q + Q_L)} \tag{5.2}$$

where Q is the rate of discharge and Q_L is the amount of leakage through the clearance spaces, i.e., the amount circulating within the pump.

Mechanical efficiency

The mechanical efficiency e_{mech} is a function of the friction between the moving and stationary parts of the pump. It is to be noted that the presence of solid particles in the flow increases the frictional losses above those for water alone.

Gross efficiency

The gross efficiency e is, then:

$$e = e_h \times e_v \times e_{mech} \tag{5.3}$$

This is the measure of efficiency adopted for use in determining the power required to drive a centrifugal slurry pump. The equation for power is

$$P = \frac{Q \times H \times S_m}{101.97 \times e} \tag{5.4}$$

where P is the shaft power in kilowatts, Q is the flow rate in litres per second and S_m is the relative density (specific gravity) of the slurry.

From an operational standpoint, pump efficiency is also a function of the ease with which individual pumps can be assembled, disassembled, repaired and maintained. Simplicity is the keynote of modern design and manufacturers devote much time and effort to making their pumps, particularly the larger ones, easy to deal with in all circumstances. For example, Warman International has patented a gravel-pump design that features a clamp-ring arrangement for the pump head to hold the wearing parts together and locate them with the adapter plate. The entire head is assembled with only a few heavy bolts to tighten and the discharge outlet can be positioned circumferentially at any desired angle. Suction and delivery-pipe fittings are designed to allow pumps to be maintained without having to disconnect either delivery or discharge pipes away from the pump.

5.3.2 Jet lift pumps

Jet lift pumps operate on the principle that when a high-velocity jet of water is injected into the intake of a venturi chamber, the velocity gradient between the incoming stream and the surrounding fluid gives rise to shear stresses and accelerative effects that lower the pressure in the inlet section of the venturi below that of the surrounding fluid. The shape of the venturi (Fig. 5.1) provides for a rapid acceleration of the low-velocity component followed by a gradual deceleration to allow an efficient transformation of kinetic energy into pressure energy. The rate of increase in momentum depends upon the carrying capacity of the jet and upon the rate of change of pressure (pressure gradient) between the low- and high-pressure sections. Sand and gravel drawn through the suction inlet into the venturi are elevated to the surface through the delivery line. Since there are no moving parts in contact with the slurry the only limitations to the size of the solids passing through the system are those imposed by the throat diameter of the venturi. An 8 in (20.32 cm) jet lift pump will normally transport particles up to 6 in (15.24 cm) in diameter without blockage. However, such pumps are also fitted with clean-out hatches to clear blockages caused by irregular-shaped particles or trash entrained in the feed.

Jet lift pumps are the least efficient of all centrifugal type pumps and, generally, the total jet system efficiency from prime mover, through the water pump and jet lift is around 20% or less, whereas a comparable submerged pump system has an overall efficiency several times greater. However, jet lift pumps can usually pump at higher solids concentrations than other dredger pumps and some hydraulics engineers claim that in terms of solids pumped per unit of power, jet-pump efficiencies actually compare favourably with the efficiencies of other pumps in many applications for which they are not currently considered. High pressure water jets can also be used to improve the suction conditions in centrifugal slurry pumps operating against high suction lifts as described in Chapter 6.

From basic principles, jet-pump efficiency:

$$e_{\text{jet}} = \frac{W_w}{W_s} \times \frac{H_d}{(H_w - H_d)} \times 100 \tag{5.6}$$

where W_s is the jet water flow in kg s^{-1}; W_w is the induced slurry flow in kg s^{-1}, H_d is the net jet pump head in metres of slurry and H_w is the net jet pump head in metres of water.

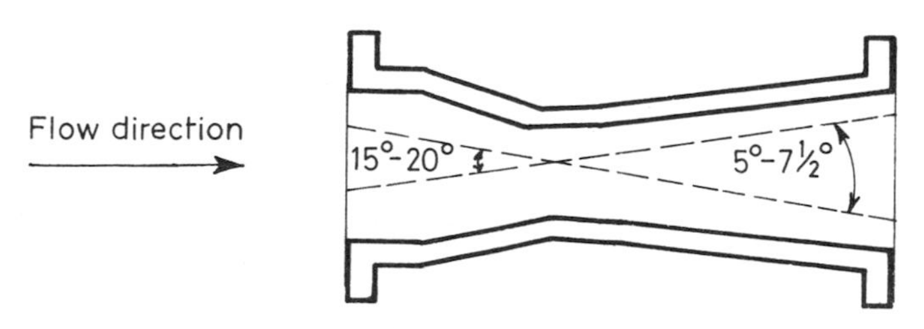

Fig. 5.1 Typical venturi shape

5.4 DESIGN CONSIDERATIONS

Features of design that allow centrifugal slurry pumps to handle efficiently slurries of highly abrasive, high-density pulps are concerned primarily with the construction of suitable pump heads and liners, impellers, shaft and bearing seals, feed and priming arrangements and materials of construction.

5.4.1 Pump heads and liners

Pump heads and liners are subjected to large impact and abrasion forces due to the presence of solids in the flow and operators in the placer mining industry have benefited greatly from the experience gained in developing anti-abrasive materials (see materials of construction) and in the reduction of turbulence within the pumps. This has been achieved by designing for large radii and by constructing volutes with gradually increasing cross-sections in the direction of flow in order to provide a more efficient conversion of kinetic energy into pressure energy. For example, the area of the throat section in hopper dredger pumps is normally about 65% of the discharge-pipe diameter. The pump cross-sections are approximately rectangular with large corner radii.

Dredge and gravel pumps are said to be of uncased design when the shell is cast in segments, usually three, which must be replaced or repaired when worn through. No liners are used and, hence, design features that provide for ease and speed of assembling and disassembling for repairs and maintenance assume considerable importance. Basic to this requirement and for additional protection against excessive wear is for all of the mating faces of the linear segments to be precision ground to ensure gap free, face to face fitting. Apart from avoiding abrasion at the joints the parts should fit together instantly without the need for additional fitting.

Modern split-casing, sand-slurry pumps have separate liners on the intake and shaft sides of the pumps. Warman pumps also feature a volute liner with wearing parts that are separate and capable of independent replacement or repair. The parts are all contained within an outer split casing, the two halves of which are drawn together using a limited number of bolts, two of which are positioning bolts to ensure accurate axial alignment. Virtually any desirable combination of rubber or rubber-like synthetic materials and metal liners can be made available for any operational requirements. Modern sand-pump design allows for vertical mounting if required and for the discharge branch to occupy any one of a number of positions.

5.4.2 Impellers

Impeller design considers the provision of minimum clearances for the maximum size of the solids to be passed, at the same time minimizing hydraulic and volumetric losses due to leakage, eddying, and back flow. It is aimed, also, at providing slurry flow characteristics that can be maintained reasonably over a long impeller life. Wear is considerable in most solids pumping applications and the designer must make due allowances for the resulting lowering of efficiency in order to avoid too frequent replacements and excessive replacement costs.

Pump impellers for dredging duties must be able to handle all types of debris, gravel, stone etc. and pass, freely, the largest particles entering into the pump. In achieving maximum bore dimensions, some early manufacturers worked to rule of thumb standards that required a vane spacing such that an uninterrupted view could be obtained through the blades. In modern configurations impellers are multivaned according to the pump dimensions and duty and a compromise is effected between the most efficient blade configuration and the required minimum allowable bore. Such impellers are usually of closed design and feature side sealing vanes on the shaft side of the impeller to relieve the pressure on the gland and to help exclude solids from the gland area. Side-sealing vanes are placed on the suction side of the impeller to eliminate leakage back to the intake through the running clearance of the impeller.

Sand-slurry pump impellers have similar side-sealing vanes and most manufacturers supply interchangeable rubber-covered or hard-metal impellers of either open or closed design.

Johnson (1975) offers basic guidelines for dredger-pump selection and design as follows:

(1) The efficiency increases with the impeller speed up to the point of cavitation when it falls away. It will be noted in this regard that drag-suction dredger pumps having a high suction lift to pump against will reach cavitation point at much lower speeds than will dredger pumps operating against low suction lifts.
(2) Impeller blades should have a good smooth curvature and, for a four-vane impeller should have not less than a 90° lap.
(3) Efficiencies decrease with increasing impeller widths, but clearances must be maintained in order to pass the largest particles.
(4) Efficiencies decrease with wear in impellers and volutes.
(5) Efficiencies increase with increased flow rates up to a maximum after which they decrease.
(6) The entrance angle design at the eye of the pump has a marked influence on efficiency and cavitational tendencies.

Impeller speeds

Slow impeller speeds are significant factors in reducing the wear on surfaces in contact with the slurry, however, each application has specific features that govern the speed at which the pump must run. For example, suction-cutter dredgers operate mainly under conditions of low suction and high delivery heads and high impeller speeds are needed to develop the required high discharge pressures. Drag-suction dredgers, on the other hand, pump against high suction lifts and low-delivery heads and the best characteristics for these pumps are obtained with impellers and volutes that are designed for low peripheral speeds. In such pumps the impeller speeds are generally within the range 150–300 rpm with peripheral velocities up to 7.6 m s^{-1}.

Specific speeds

The specific speed index N_s is used by manufacturers to gauge the relationship between the rpm of an impeller N, the discharge rate Q and the total head H and is expressed by the formula:

$$N_s = \frac{NQ^{0.5}}{H^{0.75}} \tag{5.7}$$

Impeller speeds and diameters are related and one cannot be selected without arbitrarily selecting a value for the other. Pump manufacturers use some form of the formula:

$$DN = KH^{0.5} \tag{5.8}$$

in conjunction with equation (5.7) to arrive at a suitable compromise. D is the impeller diameter and K is a dimensionless value determined by the manufacturer from experience. For example, the Corps of Engineers, US Army aim for their dredger pumps at a value of K between 1575 and 1900 and between 150 and 300 rpm for N (Scheffauer, 1954). For practical design, the highest speed is used within the above range commensurate with the head to be developed, having regard to the fact that high specific speeds indicate a sensitivity to cavitation on high suction lifts.

Note: Cavitation occurs in a pump when the impeller speed is excessive compared with the velocity and pressure of the incoming slurry.

5.4.3 Shaft and bearing seals

Centrifugal slurry pumps are equipped with variously designed shaft sealing systems to keep solids away from the stuffing box and bearing assemblies and to prevent excessive shaft-sleeve and gland-packing wear. The two main types of seal are gland seals and centrifugal seals.

Gland seals

Gland-packed pumps with water seals can be used in any pumping installation provided that an adequate supply of fresh water is available for sealing. Gland water sealing is essential for pressure-fed installations, such as are used in stage pumping and where suction lifts are required, as in dredging.

Water-sealed gland arrangements are used in dredger pumps wherever the positive intake head exceeds 25 % of the total dynamic head and for all suction dredging duties. Variously sized pumps have gland water requirements from as little as $1.5\,l\,s^{-1}$ for small pumps up to $3.8\,l\,s^{-1}$ and more for the larger pumps. Common to all gland-sealed slurry pumps is the requirement of sealing water to be delivered at a pressure of at least 35 kPa in excess of the delivery pressure on the pump being sealed.

Sealing arrangements generally follow conventional stuffing box design with some variations in lantern ring and neck ring combinations depending upon the suction duty. The sealing water must be available at full pressure before starting a slurry pump and the supply should be continued until the slurry pump delivery pressure has dropped to zero on shutdown. Flow indicators installed in the sealing water lines ahead of the stuffing box are recommended for all important installations to allow operators to visually check that water is flowing at the required rate at all times.

Centrifugal seals

Centrifugally sealed slurry pumps operate without the need for packing or sealing water by using an expeller mounted on the shaft in front of the stuffing box to return any leakage back into the main flow. The typical sectional arrangement provided in Warman Series A type SC centrifugally sealed pump heads is illustrated in Fig. 5.2.

Centrifugal seals are suitable for most gravity fed slurry pumping installations. With the pump at rest, leakage is prevented by grease-lubricated packing with a neck ring, lantern ring assembly as shown. In operation the expeller, in conjunction with expeller vanes on the back face of the impeller, ensures a completely leak-proof shaft seal and requires no gland water. Practical limitations of centrifugal shaft seals are determined by the impeller speeds which must be sufficiently high to

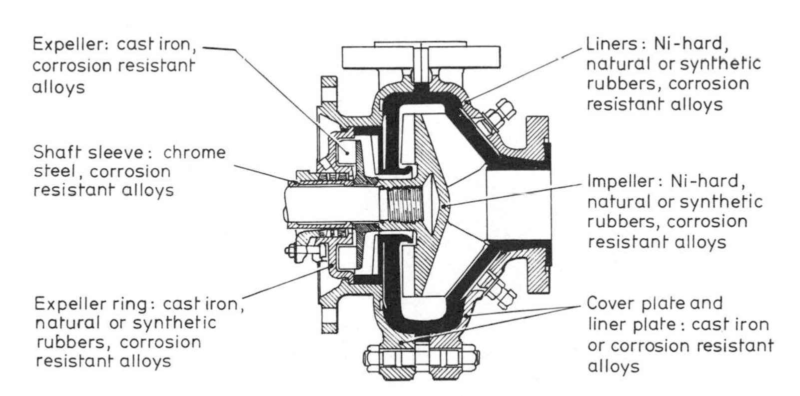

Fig. 5.2 Typical sectional arrangement of centrifugally-sealed pump head

generate the pressures needed in each case to return the leakage back into the main flow. Below this critical speed, the centrifugal seal will fail.

Bearing seals

Bearing seals are normally designed as double piston ring type labyrinth inner grease seals. The internal labyrinth for dirt exclusion gives maximum protection to the bearings.

5.4.4 Feed arrangements

All centrifugal pumps are highly sensitive to their feed arrangements and, as a result, most pumping troubles originate on the suction side of the pump. Pumps may be gravity fed or suction fed.

Gravity feed

Figure 5.3 illustrates the standard arrangement for feeding a centrifugal slurry pump from a feed bin installed above the centre line of the pump inlet. The downward slope of the intake pipe from the feed box to the pump is essential for handling materials that settle quickly and minimum slopes of 30° are recommended. Gibault joints installed in the intake and delivery pipes, as shown, ensure speedy dismantling of the pump for inspection or servicing without disturbing the other arrangements. The bottom of the feed box slopes towards the pump intake to avoid settling; the intake pipe is made as short as practicable to reduce friction losses.

It is always beneficial to streamline the flow from the feed bin into the suction pipe although the benefits may not be readily observable under normal pumping conditions. However, where pumps block frequently due to surging of feed into the bin the problem can often be corrected by

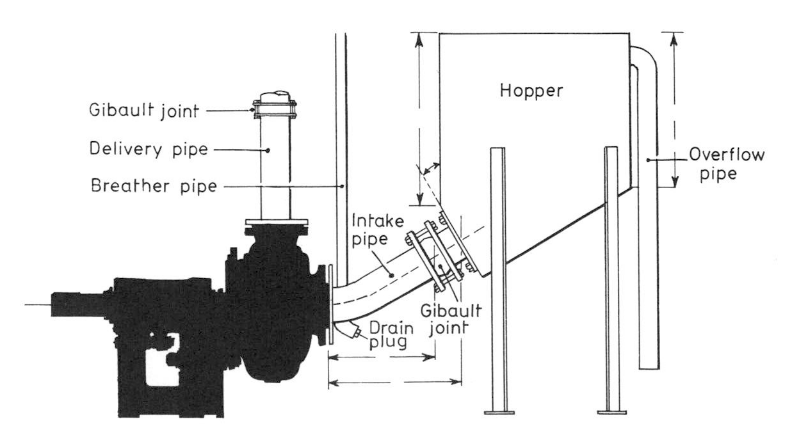

Fig. 5.3 Gravity-fed pump installation

improving the entry conditions. This is illustrated by the differences in discharge rates through straight edged and streamlined orifices under a constant head of water. Figure 5.4 illustrates the limiting effects of the vena contracta formed just below a sharp-edged orifice compared with flow through a streamlined orifice under the same head.

Suction feed

For any given slurry and rate of flow, pump characteristics vary according to the total suction and delivery heads and the most careful attention must be paid to achieving good hydraulic conditions on the suction side of the pumps because it is here that more troubles originate. Pumps are designed for either positive or negative suction heads and interruptions to the flow may occur for a number of reasons of which, apart from excessive impeller and liner wear, the most important are:

(1) Bad entry conditions resulting in vortexing or to the restriction of free flow.
(2) Leakages of air into the system from faulty joints, seals or other defects that allow air to enter into the pump volute.
(3) Excessive velocities in the suction piping or excessive suction lifts causing cavitation.

Differences in flow rates are sometimes less marked under pump suction conditions than are differences measured in terms of suction head losses. This was shown by Macdonald (1965) in studying the effects of variously shaped nozzle profiles on mass flow rates and developed pressures in both the straight ended and streamlined model types described in Fig. 5.5. The results presented in Table 5.1 demonstrate clearly that any degree of streamlining is beneficial to the inlet flow characteristics and that, in the models tested, the circular arc profile provided the least resistance to flow. Figure 5.6 compares the results of the straight-ended nozzle with those of nozzles having straight taper, quarter ellipse and circular arc inlets. The pressures compared in Table 5.1 were measured at a distance of six pipe diameters from the nozzle inlet where the flow pattern was shown to be fully developed.

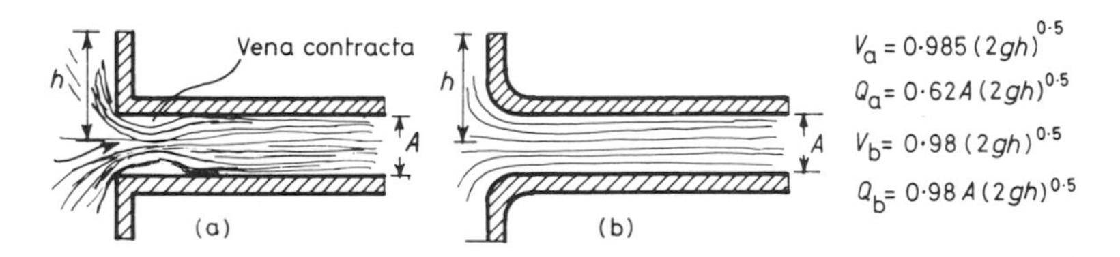

Fig. 5.4 Formation of vena contracta and restricted flow due to bad inlet conditions (a); streamlined flow without restrictions in good inlet conditions (b)

Figure 5.7 (a) and (b) represent typical suction-feed arrangements for pumping slurries from sumps installed below the pump intake centreline. Features of the two arrangements make various provisions for

(1) Priming the pump at start up.
(2) Bypassing the flow back to head feed to allow the pump to run continuously without having to re-prime during intermittent flow conditions.
(3) A water jet in the system to agitate and fluidize the solids settling out at the bottom of the sump.
(4) A coarse screen installed in the bin to catch any tramp metal or other objects that might fall into the sump and cause blockages in the pump.

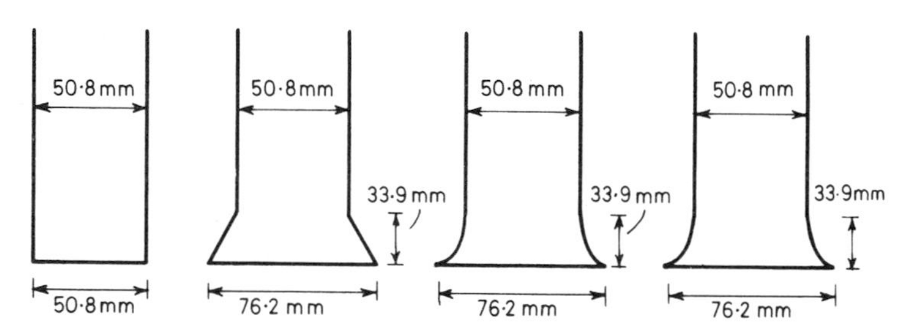

Fig. 5.5 Dimensions of test nozzles: (a) straight ended, (b) straight taper, (c) quarter ellipse, (d) circular arc

Table 5.1 Mass flow rates and pressures at suction-pipe inlets of various shapes

Nozzle shape	Percentage solids by weight (C_W)	Mass flow rate (lb s^{-1})	Negative pressure energy in water gauge ($\Delta p/\gamma$)	Kinetic energy ($\frac{1}{2}\rho V^2$)
Straight edged	1.0	15.6	35.3	125
	10.8	16.8	37.3	131
	36.8	18.3	38.8	135
Straight taper	Nil	15.7	27.4	128
	10.0	16.4	28.0	130
	36.3	18.8	31.2	143
Quarter elipse	1.0	15.7	26.8	127
	9.6	16.4	29.0	131
	38.1	18.7	29.7	140
Circular arc	0.8	15.8	26.3	130
	17.5	17.1	28.0	137
	37.9	19.0	29.0	140

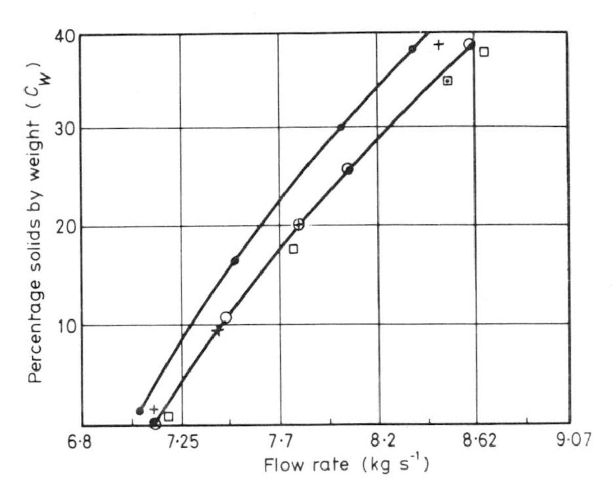

Fig. 5.6 Plot of mass flow rate versus C_W *for nozzle shapes described in Fig. 5.5*

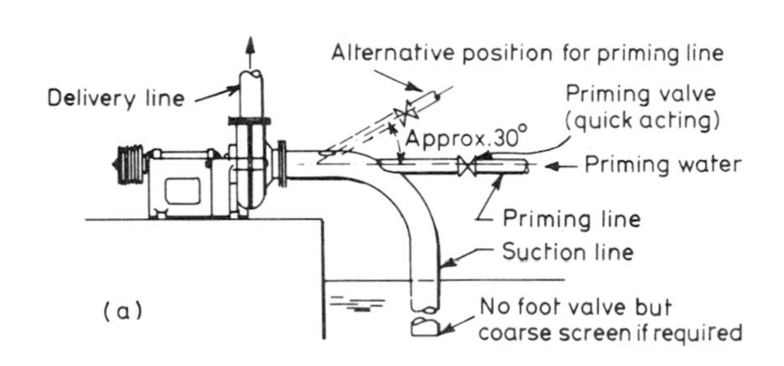

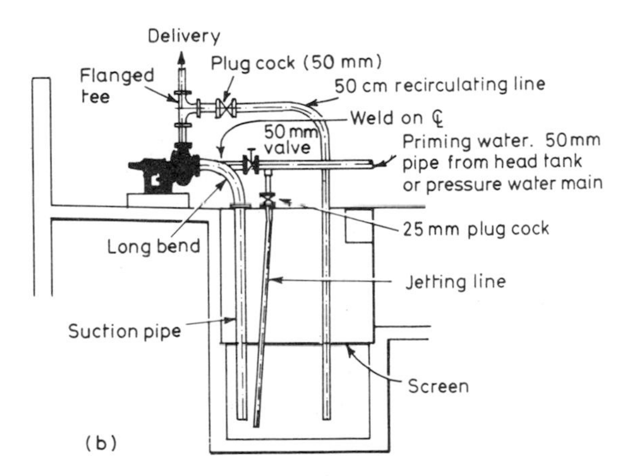

Fig. 5.7 (a) Typical suction feed arrangement for pumping slurries from a sump located below C_L *pump inlet, (b) same arrangement as for (a) with additional provision for bi-passing the flow and for agitating settled solids in the sump*

(5) Reduction in the length of the suction line in the horizontal plane to a practical minimum in order to avoid settling in the pipe and unsteady conditions of flow into the pump.

Suction lift

The velocities required at the suction-pipe inlet to lift solid particles into the stream entering the pipe are functions both of the physical properties of the particles entrained and of the efficiency of the cutter head or other entraining mechanism. There are obvious advantages in keeping such velocities to a practical minimum because, as Pekor (1973) points out, higher velocities are bought and paid for by higher entrance losses, thus, using up a significant percentage of the available energy on the suction side of the pump.

Additional losses affecting the pumps lifting power relate to the density of the slurry, skin friction in the suction pipeline and the vertical height between the suction pipe inlet and the centreline of the pump suction inlet. As a result, the practical limit for suction lifts in dredging service may not be more than 60% of what it would be for clear water. Taking as an example the theoretical maximum suction lift for water at 15°C at sea level, i.e. 760 mm of mercury or 10.36 m of water: the maximum suction lift that can be recommended for a slurry of relative density 1.6 in such conditions is:

$$\frac{10.36}{1.6} \times 0.6 = 3.9\,\mathrm{m}$$

Regardless of design however, there is a limiting depth for dredging using a pump mounted in the hull of a dredger and pumps that operate against higher suction lifts are mounted on the ladder close to the suction inlet or are jet assisted as already described. Even within the available limits imposed by atmospheric pressure there is still an advantage in mounting the pump close to the pond bottom. According to Sheehy (1976) when dredging to a depth of 10 ft (3.05 m) as compared to 50 ft (15.24 m), the percentage of solids at the 10 foot depth will be approximately twice that pumped from the 50 foot depth, this being related to the effective available negative pressure at the eye of the pump.

Suction-pipe diameter

From the foregoing it is apparent that, for good design, the suction pipe diameter should be large enough to pass the required quantity and size of material at a flow velocity that will easily lift all of the particles into the pump without excessive slip; but not so small as to use up an excessive proportion of the available suction head. According to Erickson (1956) common suction to delivery ratios in slurry pumps range from 1.26:1 to 1.56:1 and these ratios are not selected arbitrarily but only after careful studies to provide minimum suction head values commensurate with efficient pumping performances.

However, practical considerations in selecting suction-piping sizes relate also to the availability of standard pipes and fittings and in the British system, sand slurry pumps are normally designated 4×3, 6×4, 8×6 etc, the digits referring to the diameters of the suction and delivery branches in inches. In most gravel pumps a suction pipe of the same diameter as the delivery line is used to avoid the possible intake of particles that are larger than the minimum size that the pump and delivery piping can pass. A higher suction-pipe velocity and increased friction loss in the suction pipe is accepted to prevent such particles from entering into the system and causing damage.

Critical submergence

Care is needed when pumping from a shallow sump or in dredging at shallow depths to avoid vortexing and, hence, sucking air into the pump. Such vortices are frequently observed to form and then disappear to re-appear again, perhaps, around the nozzle of a suction inlet when careless or inexperienced dredger operators are taking surface-stripping cuts, or when changing boundary conditions around the nozzle cause temporary fluctuations in the critical depth of submergence. Vortexing results in unsteady pipe flow, fluctuating loads and sometimes loss of prime.

5.4.5 Priming

Centrifugal slurry pumps are not self priming but rely upon a water supply from an elevated tank or auxilliary water pump for start up on suction-lift duties. Foot valves are seldom used because they add to the suction head and are subject to blockages because of the restricted openings. The interruption to smooth flow conditions also imposes an extra head loss and reduces the amount of energy available for doing useful work.

Direct priming

Priming water is injected into the suction line and is directed towards the eye of the impeller as illustrated in Fig. 5.7 (a), (b). Recommended flow rates are of the order of 25% of the nominal capacity of the pump being primed under a head of at least 6 m (20 ft). Priming by means of water introduced through the delivery line is not recommended because of the possible formation of air locks in the pump volute. The procedure in priming is to:

(1) Open the priming value.
(2) Start the pump.
(3) Close the priming valve when the pump has picked up its load.

Vacuum priming

Vacuum priming is sometimes used when a separate and adequate supply of fresh water is not available. It requires a valve, as shown in Fig. 5.8 to give a positive shut off in the delivery line when priming. The pump is primed by exhausting air from the suction line and pump casing using a hand-operated primer. The operation consists of:

(1) Close delivery valve and open primer valve.
(2) Work hand primer until only water discharges from primer.
(3) Close primer valve and start pump.
(4) Open delivery valve and start pumping.

5.4.6 Materials of construction

Pump impellers, liners and piping have to withstand varying degrees of chemical corrosion, abrasion and impact. In pumps the main points of wear coincide with zones of low pressure where the slurry is locally accelerated along the leading edges of vanes and this may result in excessive local erosion of both impellers and liners. Critical wear points of pipelines are just downstream from flanges and at fittings and bends and the most general wear takes place along the bottom of the pipe because of the greater amount of contact between the solids and the pipe walls in the invert of the pipe. Corrosion is a problem in intermittently operating pipelines and in pipelines that do not run full, particularly in beach-mining operations where salt water is used for slurrying.

Present bowl designs have evolved from pumps in which the casings were cast in one piece. These pumps were hard to assemble and maintain and the bowl was discarded when worn, even locally, thus wasting a great deal of metal. Cast iron was used at first, then cast steel, some were made in two sections and eventually liners were introduced to take up the wear. Special alloys now give a longer life under severe conditions of impact and

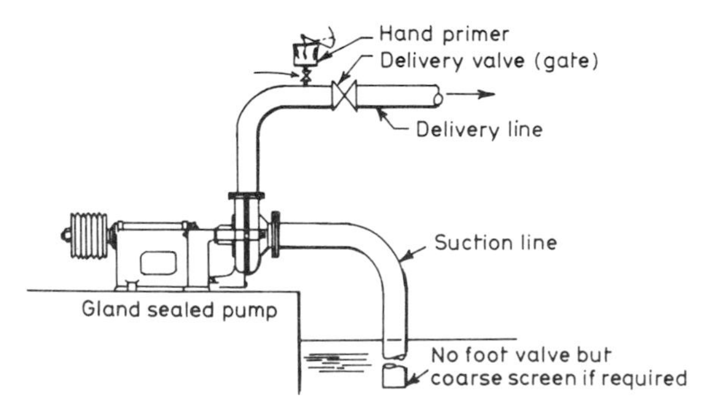

Fig. 5.8 Typical arrangement for vacuum priming

abrasion. Typical of these are Nihard type II, chrome alloy iron, manganese steel and cast steel.

5.5 CAVITATION IN PUMPS

Purely mechanical faults caused by wear or faulty installation can usually be eliminated or remedied without too much trouble. Cavitation, when it occurs, is a direct result of inadequate or faulty plant design such as in using a low-lift pump on a high-lift service or running a pump at an excessive speed to try and achieve increased throughputs for which the pump is not specifically designed. In such cases, cavitation is invariably followed by a reduction in output, a loss of developed head and excessive erosion of impellers and liners. The effects are less extensive and less abrupt in slurry pumps than in water pumps in similar adverse conditions because of the wider impellers and greater clearances. Nevertheless they still lead to higher maintenance costs, lower performances and sometimes to total failure.

Vaporization occurs in any fluid medium in motion wherever the static pressure falls to the level of the vapour pressure and, in any centrifugal pump, variations in pressure occur along the path of the fluid as a result of the dynamics of the motion. The lowest pressures are at the inlet to the impeller where the fluid is locally accelerated along the leading edges of the vanes and vapour-filled cavities are formed wherever the pressure energy is no longer sufficient to maintain cohesion between particles, or contact between the fluid and the solid boundaries along which it travels.

The onset of cavitation is usually in regions where high velocities lead to absolute pressures that approach or equal the vapour pressure of the fluid as it exists at that time. A further increase in velocity does not reduce the pressure but results in 'boiling' and pump characteristics fall away sharply. Small bubbles of vapour form near the points of lowest pressure and move on with the flow to collapse subsequently in zones of higher pressure.

5.5.1 Cavitational effects

Cavitational effects in slurry pumps as in other hydraulic installations are due in part to the violence with which cavities in the fluid collapse and partly to the wastage of space occupied by the cavities before they collapse. When collapse occurs along a solid boundary, cavities with lives of 0.04–0.05 s exert forces that have been measured as high as 7.8 tonnes cm^{-2} and theoretically may rise to almost 12 tonnes cm^{-2} (Peterka, 1953). Tiny pieces of material are torn out of the surfaces of impellers and linings causing pitting and eventual failure. There is a marked drop in efficiency when the useful cross section of the flow is reduced due to separation from the walls and from the formation of vapour in the fluid which increases its volume and reduces its mass. Cavitation is usually but not always accompanied by audible crepitation.

The Thoma cavitation index

The Thoma cavitation index σ has proved useful in the selection of pumps for specific purposes. Consider the suction dredger arrangement illustrated in Fig. 5.9. Application of the energy equation between a point a on the liquid surface and the point of minimum pressure at the pump eye e, neglecting all losses between the two points, yields the relationship:

$$\frac{P_e}{\gamma}+\frac{V_e^2}{2g}+H_s=\frac{P_{atm}}{\gamma}+0+0 \tag{5.9}$$

where γ is the specific weight of the fluid, P_{atm} is the atmospheric pressure and P_e is the absolute pressure.

Cavitation will occur at *e* if the pressure at *e* is equal to or less than the vapour pressure P_{vap}.

If $P_e = P_{vap}$, σ, the Thoma cavitation index, is expressed as:

$$\sigma=\frac{V_e^2}{2gh}P_{atm}-P_{vap}-\frac{\gamma H_s}{\gamma H} \tag{5.10}$$

For cavitation-free performance the suction lift must be so restricted that the value of σ is greater than σ_{cr} the value at which cavitation commences. Sharp bends, sudden changes in cross section and badly designed inlet nozzles may all lead to cavitation and care must be taken to

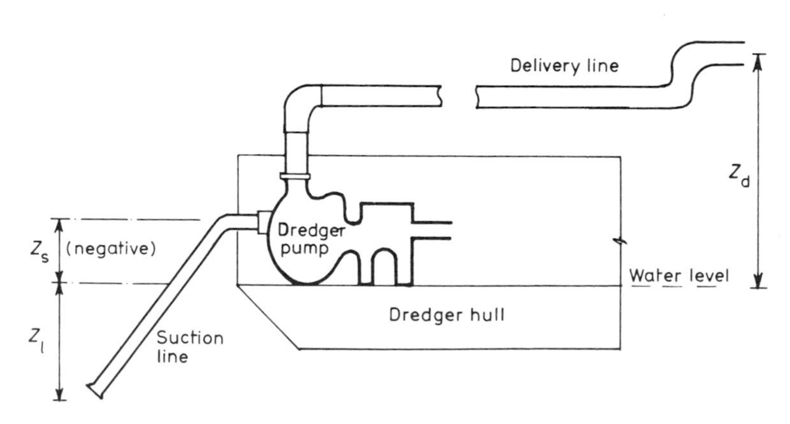

Fig. 5.9 Suction dredger – diagrammatic arrangement

guide the flow of liquid so that at no point is the pressure insufficient to maintain fluid contact with the walls.

Entrained air and gases

Wherever dredgers mine into bank material some air is entrained and becomes part of the flow. Similarly, as already noted, gases such as nitrogen and methane that are present in some materials below water level are also incorporated in the slurry. The effect of this entrainment is to raise the value of σ_{cr} so that cavitation occurs at higher absolute pressures than if no extraneous gases were present. At the same time, the presence of entrained air increases the compressibility of water (0.1% air by volume increases the compressibility tenfold) and the collapsing forces are smaller and do less mechanical harm.

Net positive suction head

The pump manfacturer, in recommending a pump for a particular high-suction duty is concerned with the net positive suction head (NPSH) required at the pump intake to avoid cavitation. Based upon Thoma's principle, the available NPSH in a suction slurry pump is, in effect, the difference between the static inlet head and the vapour pressure of the fluid. In practice, it is calculated from the gauge reading at the free-liquid surface, the total losses in the suction pipe including inlet losses, the static head from the free surface to the pump centreline and the density of the slurry in the suction pipe. In order to avoid cavitation the inlet pressure to the pump must be maintained above a certain critical value and suction pump performance curves should always show the NPSH required for safe operation.

The operating range of a dredger pump for a given slurry concentration thus lies between the speed at which the larger particles begin to sink back against the flow in the suction pipe (i.e. the lower critical velocity) and the speed at which cavitation is about to occur (i.e. the upper critical velocity). Within these limits, the range of slurry concentration may extend from nil concentration to more than 60% of solids by weight for short periods with a normal range between $C_w = 15\%$ and $C_w = 35\%$. Well-designed, modern dredgers can hope to achieve average slurry densities of around $C_w = 30\%$; the older dredgers were seldom able to average better than $C_w = 20\%$ in general service.

Whatever the conditions, continuous cavitation must be avoided and, hence, a combination of suction system and pump characteristics is chosen such that the available net positive suction head ($NPSH_a$) exceeds the required net positive suction head ($NPSH_r$) by a reasonable margin of safety (usually at least 1 m) for the proposed duty point or for the whole of the proposed range of duty points. To achieve this, the $NPSH_a/Q$ characteristic curve of the suction system should be designed initially so as to provide the highest practical value or values of $NPSH_a$ for the required duty. The $NPSH_r/Q$ characteristic curve or curves of the pump at any given speed (or over any given range of speeds) represents the actual values of $NPSH_r$ published in pump performance curves. The value of $NPSH_r$ for any point on a given pump performance curve is fixed and cannot be altered. The only way in which the $NPSH_r$ may be changed for any given duty point is to substitute a different type of impeller or a different type or size of pump.

NPSH required ($NPSH_r$): in an example set out in the Warman Slurry Pumping Manual, Fig. 5.10 illustrates portion of a typical pump performance curve, showing H_w/Q characteristics (points A–C); the NPSH required by the pump is represented by the family of $NPSH_r$ value curves, each intersecting the H_w/Q curve. These $NPSH_r$ values (metres) are plotted against Q to develop the $NPSH_r/Q$ curve for any pump speed. The $NPSH_a$ values have been calculated as described later (see calculations pp. 280–93) and are also plotted against Q.

From $Q = 0$ to Q_{NPSH} (corresponding to the intersection of the $NPSH_a$ and $NPSH_r$ curves) i.e. for performance from point A to point B, $NPSH_a$ exceeds $NPSH_r$ indicating cavitation-free performance until flow rate Q_{NPSH} is reached.

Beyond Q_{NPSH} the $NPSH_r$ is greater than $NPSH_a$ by an ever increasing amount and cavitation would occur as emphasized by the shaded area.

Note: Where pumping is carried out at any significant elevation above sea level, the $NPSH_a$ should be carefully evaluated for all duties including gravity-fed installations to ensure that its value is not less than the $NPSH_r$. Figure 5.11 is a plot of barometric pressure versus altitude. (See also tabulation in appendix at end of book.)

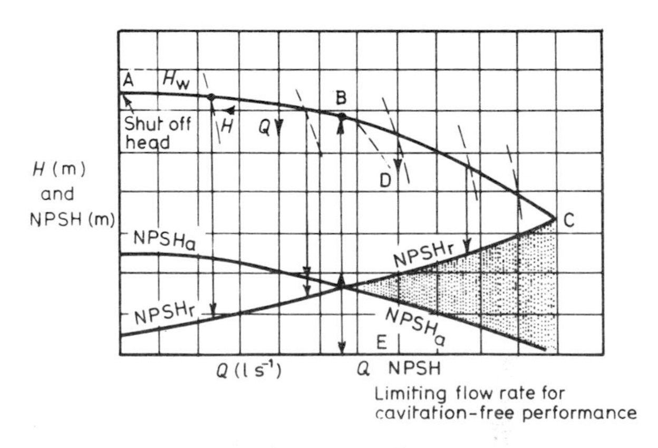

Fig. 5.10 Typical pump performance curve

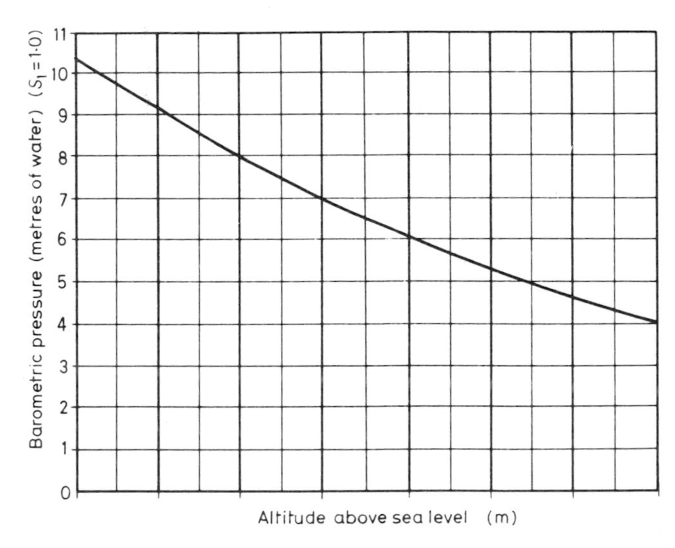

Fig. 5.11 Plot of barometric pressure versus altitude

5.6 HYDRAULIC TRANSPORTATION OF SOLIDS IN PIPELINES

The nature of the flow of slurries is governed by the physical characteristics of the solids contained in the slurry and in Chapter 2 it has been shown that, for open channel flow:

(1) More energy and higher velocities are needed for transporting large particles that move by rolling and sliding than for finer particles moving in suspension or by saltation.
(2) The larger particles in saltation have smaller trajectories than smaller saltating solids.
(3) Shape factors influence the behaviour of individual particles in a fluid according to departures from sphericity.
(4) The behaviour of individual particles in swarms of particles in a fluid is influenced by the group behaviour of the swarms.

Solids particle movement in pipelines follows the same general pattern, with some differences due to the confined nature of the flow and to the more general movement of the particles in saltation or suspension.

5.6.1 Types of slurry flow

Slurry flow in pipelines is classified as homogeneous flow, heterogeneous flow or combined homogeneous and heterogeneous flow largely according to the size and size range of the solids in transport. A slurry is homogeneous when all of the solids are so fine that they remain in virtual suspension with the fluid at rest. It is heterogeneous when the solid particles are so large as to concentrate in the bottom part of the pipe during transport. In conditions of combined homogeneous and heterogeneous flow, some of the larger particles are carried along in suspension with the finer particles at low velocities.

Homogeneous flow

Homogeneous flow takes place with slurries in which the particles are of Stokesian proportions or in sufficiently high concentrations for individual particle settling velocities to become less important than the interaction between the particles resisting their settling. Homogeneity thus assumes a uniform distribution of solids in the fluid across any section of the flow and a sufficient degree of turbulence at low flow rates to keep all of the particles in suspension.

Heterogeneous flow

With increased particle size or density, more energy is needed for transportation and heterogeneous flow is characterized by the manner in which the individual particles move at different flow rates:

(1) At low flow rates the larger particles form a sliding, rolling bed and the head loss is made up of the sliding friction of particles along the lower portion of the pipe; friction at the interface of the sliding bed and the upper faster moving stream; and friction between the upper, more homogeneous flow fraction and the pipe.
(2) At higher velocities, the turbulence increases and lift forces become greater; more particles are lifted from the bed load and movement is increasingly by saltation. There is less friction along the lower portion of the pipe and increased friction between the fluid and the surface of the pipe.
(3) With fully established heterogeneous flow, all of the particles move either in suspension or saltation and fluid friction at the pipe walls is the main source of energy loss.

The head loss characteristics during different phases of heterogeneous flow have been demonstrated in experiments carried out at the British Hydrodynamics Research Association (BHRA) Laboratories in Cranfield, Bedfordshire (Anon, 1969). These experiments showed that while the loss of head due to friction between the particles and the walls is reduced by increasing saltation, the head loss due to fluid-pipe friction increases with increased flow velocities. As a result of the two opposing head losses a characteristic head loss–velocity curve is obtained as illustrated in Fig. 5.12. This particular curve is plotted for one specific slurry concentra-

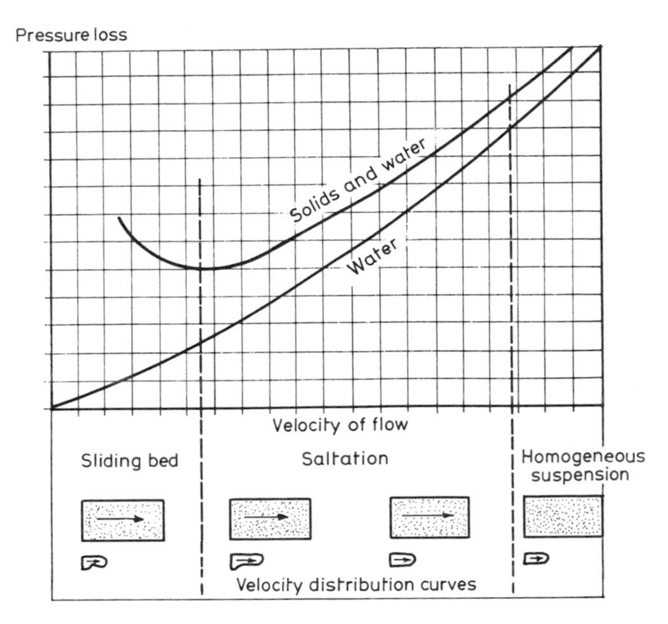

Fig. 5.12 Characteristic head loss–velocity curves for water, and solids and water

tion and density; it would have to be re-plotted and another curve obtained if the concentration of solids in the fluid was increased or decreased.

Combined homogeneous and heterogeneous flow

Certain minimum flow velocities are required to prevent settling and blockages in heterogeneous flow. Such velocities are reduced when the slurry contains a large proportion of solids smaller than about 0.125 mm. The slurry then assumes the form of a high-density fluid with non-Newtonian characteristics (see Chapter 2) and some larger particles are carried along at lower velocities in either a suspended or an increasingly saltating state. Obviously over the range of combined homogeneous and heterogeneous flow there are gradations in which conditions tend to favour one state of flow or the other and the extent to which transportation by saltation is safe, i.e. without causing blockages, depends very much upon the design and configuration of the pipelines and upon the pressure at the pump delivery.

Pseudo-homogeneous flow

Slurry flow characteristics are complex functions of the size-range distribution and densities of the solid particles in transport. Heterogeneous mixtures and homogeneous mixtures when combined behave essentially as homogeneous mixtures above a certain velocity and this pseudo-homogeneous state has important applications in such aspects of placer technology as 'slimes' disposal systems in which a predominance of particles smaller than 62.5 μm (the upper limit of the Stokesian range) carry along with them particles as large as 125 μm. Referring to current experience of long-distance pipelines and to the values for optimum solids concentration by weight and volume as described in Table 5.2, Cabrera (1979) suggests that relatively thick slurries may be pumped satisfactorily through long-distance pipelines with mixtures of solids that are inversely proportional in size to their relative densities e.g. coal 4 mm, magnetite 150 μm.

Although Table 5.2 refers only to pumping through long-distance pipelines it has important implications for tailings disposal systems in placer operations where water shortages demand that as much water as possible is retained in the plant for re-use.

The effects of solids concentration

Ross (1961) described a scheme for the joint pumping of residual slimes from various gold mines in the Welkom area of Orange Free State Province of South Africa and made the following observations:

(1) *Density of the slimes*: if this dropped below 1.35, particle settlement commenced, however, incipient blockages were cleared almost immediately by raising the density to about 1.58.
(2) *Solids concentration*: at concentrations of 45–70% of solids by weight settling velocities become less important and some larger particles are also held in suspension at low velocities.
(3) *Flow velocities*: the best average flow velocity was found to be 3.96 ft s^{-1} (1.21 m s^{-1}) at a slimes density of 1.46. For slurries intermediate between true slimes and heterogeneous mixtures, the mean pipe velocity of the mixture increases with the percentage increase of solids concentration in the lower layers.
(4) *Head losses*: head losses for slurries in the homogeneous region are

Table 5.2 Typical solids concentration for hydraulic conveying

Material	Specific gravity	Percentage solids		Particle top size (μm)	Mesh top size (BSS)
		Weight	Volume		
Gilsonite	1.05	40–45	39–44	4700	4
Coal	1.40	45–55	37–47	2300	8
Limestone	2.70	60–65	36–41	310	42
Copper	4.30	60–65	26–30	230	65
Magnetite	4.90	60–65	23–27	150	100

Source: Cabrera (1979), from table in *Mining Magazine*, April 1972).

usually obtained in terms of water gauge by multiplying the head loss for clear water at the same velocity, by the relative density of the slurry.

5.6.2 Flow in suction pipes

Experimental work by Bonnington and Denny (1959), Hattersley (1960) and others has shown that swirl outside the entrance to a vertical suction pipe continues into the pipe, first as a free vortex (tangential velocity V_t proportional to $1/r$) and further along the path as a solid vortex (V_t proportional to r). Bends in the system resulting in secondary flow cause further swirl and high-velocity regions on the outside radii (Fig. 5.13). For these reasons wear at slurry suction inlets and at bends is invariably higher than along straight lengths of the pipe.

Swirl has an effect also, on pump performances. Small random deviations from no whirl entry may not be significant but, particularly with pumps of high specific speed, strongly whirling flow adversely affects pump characteristics. Pre-rotation of flow in the same direction as the pump rotation results in reduced output and lower power usage; pre-rotation of flow in the opposite direction to the pump rotation results in increased output with a corresponding rise in power consumption. Entry swirl may persist for 50 diameters or more, although axial velocities quickly become uniform.

Macdonald (1965) found that pressure measurements were reproduceable within $\pm 2\%$ at approximately 5 diameters in from the suction inlet regardless of changing boundary conditions and could be directly related to the mass flow rate of the slurry.

Swirl

Swirl around the nozzle of a dredge pump is due largely to assymetrical boundary conditions and is subject to the dampening effects of suspended

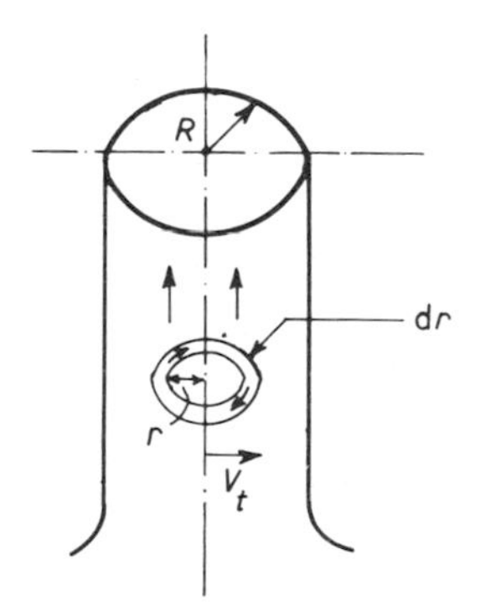

Fig. 5.13 Swirl in a suction pipe

solids and by passage through trash screens of various dimensions. It fluctuates widely except when induced by revolving mechanical cutters and may reverse its direction of rotation from time to time. As already noted, rapidly changing boundary conditions may, in extreme cases, result in the formation of 'tramp vortices' that form as cores from the dredge pond surface to the lip of the inlet nozzle. Eddies develop in suction pipes wherever bad entry conditions, excessive velocities or sharp changes in direction cause swirl. Vortices require energy to form and sustain them and they result in deviations from design conditions of flow.

Similar observations apply to slurry pump intakes in the mill and experience has shown that blockages are fewer and wear on pump impellers and linings is reduced when inlet flow is channelled smoothly into the pumps. Good hydraulics at the suction-pipe inlet depends on the promotion of streamline flow and this in turn is dependent upon the boundary configuration. A suction pipe with a straight-ended entry has a low pressure zone just inside the intake and the formation of a *vena contracta* follows separation from the walls thus reducing the effective inside diameter of the pipe. Experimentally it has been found (Macdonald, 1966) that the best characteristics for inlet flow are obtained with nozzle shapes between the limits $r-0.4D$ and $r-0.55D$ where r is the radius of curvature of curved inlet and D is the suction-pipe diameter. An important observation in this work was that, in the inlet zone, the wall pressures increased with the density of the slurry whereas in the zone of stability, some five diameters along the pipe, the wall pressures decreased with increasing slurry densities within the range tested. This supports the previous explanation for increased wear by abrasion at the entrance to the pipe in some installations.

5.6.3 Flow in delivery pipes

The flow characteristics of slurries in delivery lines are influenced by whether the direction of flow is horizontal, vertically up or down, or at some intermediate inclination.

Horizontal pipelines

In a horizontal pipeline the particles tend to settle through gravity in a plane perpendicular to the pressure gradient. Re-mixing is achieved through turbulence and two critical flow velocities are important. The first is the critical velocity below which solid particles settle from the liquid and partially or totally block the pipe. The second is the minimum velocity of homogeneous flow for the particular slurry. Between these two limits is the region of heterogeneity in which the solids move at a velocity below that of the liquid but do not settle out. Smith (1955) experimented with sands at several size gradings and at concentrations up to 27% by volume in 5 cm and 7.6 cm diameter pipes. He measured the pressure drops at velocities from 0.914 to 2.438 m s^{-1} and found that, at high velocities, the correlation between friction factor and NR was the same for a slurry as for a liquid (the density being taken as that of the mixture ρ_m and the viscosity μ, that of the liquid). As the flow velocity decreased the pressure gradient showed a progressive departure from that for water and finally reached a value many times greater than for water alone. This phenomena has already been illustrated by the U-shape of the curve in Fig. 5.12.

The tendency for coarse solids to concentrate in the lower halves of horizontal pipelines slurry flow in such pipelines results in increased rates of wear along the bottom part of the pipeline. Increased velocities distribute the solids more evenly across the section of the pipe and reduce wear, however, they also lead to higher power usages and to steeply falling pressure gradients away from the pump.

Vertical pipelines

When a mixture of solids and water passes upwards or downwards in a vertical pipeline, the gravitational forces act in the same plane as the pressure gradient, adding to the one and subtracting from the other. Because they are moved by Bernouli forces towards the central, high-velocity region of the flow, the larger solids tend to behave more predictably than in horizontal pipelines. The effect is to minimize coarse solids contact with the walls of the pipe thus reducing skin friction and the pressure drop due to skin friction. In some vertical flow lines the value of V_{cr} may be reduced by one or more metres per second.

On the other hand, slippage between solids and fluid in vertical upward flow results in a slight increase in the volumetric concentration of the solids above the delivered concentration. The water travels slightly faster than the solids and friction losses include a small element for skin friction at the particle boundaries. In downward vertical slurry flow the solids move faster, slower or at the same velocity as the fluid depending upon the velocity of the fluid and the particle sedimentation rates.

Sloping pipelines

Solids–fluids flow in sloping pipelines tends to follow the pattern of behaviour of similar flow in either vertical or horizontal installations depending upon the angle of inclination. Worster and Denny (1955) have shown that the excess flow is equal to the sum of the excess pressure drops in the corresponding horizontal and vertical pipes joining the same mid-points. This is illustrated schematically in Fig. 5.14.

5.6.4 Calculation of total head

Centrifugal pumps may be designed for either a gravity feed (positive suction head) or a negative suction duty such as in dredging or densifying.

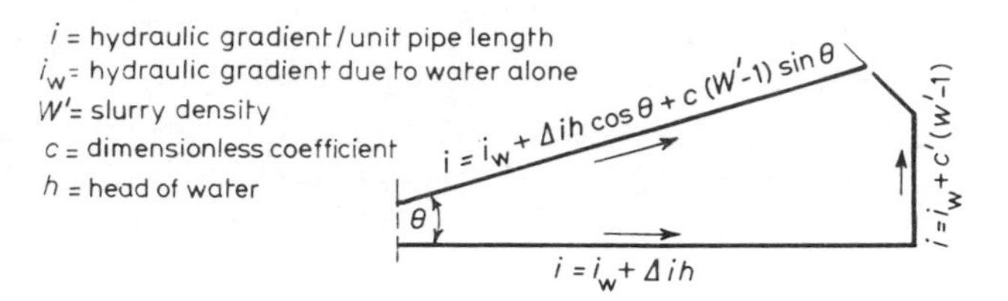

Fig. 5.14 Hydraulic gradient in sloping pipelines

Two or more pumps may be placed in series to obtain higher heads but a single pump is normally sufficient for most placer applications. In all cases, however, the pumps are designed to deliver the slurry against a total head which is a combination of static, friction and velocity heads.

Velocity head

Calculated from the term $V^2/2g$, the velocity head is the height through which a body would have to fall freely in order to attain the velocity of flow at the point of measurement. If the velocity head in the delivery pipe is greater than in the suction pipe, the velocity head is the head added to the delivery head to accelerate the flow from V_s to V_d.

Static head

This is the vertical distance from the nozzle inlet to the centreline of discharge multiplied by the density of the mixture being pumped, minus the vertical distance from the nozzle inlet to the surface of the dredge pond, multiplied by the density of the fluid in the pond.

Friction head

Friction losses in pipelines, bends and fittings result from skin friction due to contact with the pipe walls, frictional drag between fluid and solid particles moving at different velocities and intermolecular fluid friction due to eddies and turbulence. The friction head increases with the velocity and viscosity of the fluid and surface roughness. It is allowed for in design by the introduction of a friction factor f which is a function of the Reynolds Number (NR) and relative roughness ε in the expression:

$$f = \varphi(NR \cdot \varepsilon) \tag{5.11}$$

where ε is defined as the ratio of the height κ of the projections causing roughness and the diameter D of the pipe so that:

$$\varepsilon = \frac{\kappa}{D} \tag{5.12}$$

and

$$\mathrm{NR} = \frac{VD\rho}{\mu}$$

At velocities greater than the critical settling velocities for heterogeneous solids–water mixtures, the friction factor f is given in Darcy's relationship

$$H_f = \frac{flV^2}{2gD} \tag{5.13}$$

based upon the average pressure drop P along a length l of a pipeline of diameter D and may be written for slurries:

$$\frac{P}{l} = f\rho_m \frac{V_{cr}^2}{2gD} \tag{5.14}$$

At high velocities, as has already been noted, the pressure gradient is little more than for water alone because of the laminar sublayer of water along the walls in turbulent flow. As the turbulence diminishes, a critical velocity is reached and the friction factor increases rapidly with diminishing velocity (see Fig. 5.12). It is dependent, in this regime, on the concentration and density of the solids rather than on the velocity of flow. For all classes of flow, the head loss thus varies, (a) directly with the length of the pipe, (b) closely with the square of the velocity, (c) almost inversely with the diameter, (d) according to the roughness of the surfaces, (e) independently of the pressure intensity, and (f) according to the density and apparent viscosity of the slurry.

Although tables prepared from the empirical formula of Williams and Hazen are in common use, Darcy's formula is generally the most reliable method for estimating H_f for water. The Williams and Hazen formula:

$$f = 2.083\left(\frac{100}{C}\right)^{1.85} \frac{Q^{1.85}}{d^{4.3655}} \tag{5.15}$$

estimates the friction losses in pipelines using a range of C values for different pipe materials. Q is the quantity of water flowing in the pipeline, d is the diameter of the pipeline, f is the friction factor and common values for C are shown by Table 5.3.

These values are used in respect of variously sized pipelines to prepare

Table 5.3 *Williams and Hazen C values*

Pipe material	C value (new)	C value (used)
Asbestos-cement	150	140
Fibre and plastics	150	140
Seamless steel	140	100
Cast iron	130	100
Wood stave	130	110
Rubber-lined piping or hose	130–100	100
Concrete	120	100

plots for ready reckoning. However, not all actual internal diameters of commercially available pipes are represented, individual judgement is often required for selecting the appropriate C value and no allowances are made for the abrasive and polishing effects of slurries of granular materials in water. Values for relative pipe-wall roughness (κ/D) vary according to the continuity of the flow and, for example, values of κ/D for commercial steel pipes are in fact the same as for asbestos cement and polythene plastics when pumping abrasive slurries continuously. On the other hand internal rusting and scale pipelines that carry flow only intermittently can increase the κ/D values quite significantly on start up. Figure 5.15 has been prepared by Warman Equipment Limited for estimating H_f and for the subsequent construction of the system resistance curves for clear water and category A slurries. They rate the accuracy of the plots at around $\pm 5\%$ for good conditions but warn that experience and skill in the interpretation of duty data and in the use of many of the empirical and arbitrary factors are necessary for proficient hydraulics engineering.

Entrance and other losses

It is important to note that each change of direction or the installing of additional fittings in a pipeline is followed by an additional head-loss due to friction, impact and changes in momentum. Typical effects are grouped in Table 5.4 in terms of the equivalent lengths in metres of straight pipe sections for a range of fittings and configurations. Using this table the frictional head loss for 200 m of 200 mm-bore piping having three long-radius bends and one fully open diaphragm valve would be:

$$200 + (3 \times 4.27) + (1 \times 19.81) = 232.62 \text{ m}$$

of straight 200 mm-bore piping.

Other typical head loss items are:

(1) Loss of head at entrance.
(2) Loss of head at exit.
(3) Loss of head due to sudden contraction.
(4) Loss of head due to sudden enlargement.

These losses are based upon the velocity of the slurry at the points where they occur and have been grouped by Warman International in Table 5.5 to demonstrate the approximate proportions of the velocity head that apply in certain conditions. V indicates the upstream velocity and V_1 the downstream velocity in groups 1–4. The conversion head required for equipment requiring a pressure head when handling a slurry mixture is dealt with in Group 5.

The types of valves and fittings suitable for slurry pipelines are limited.

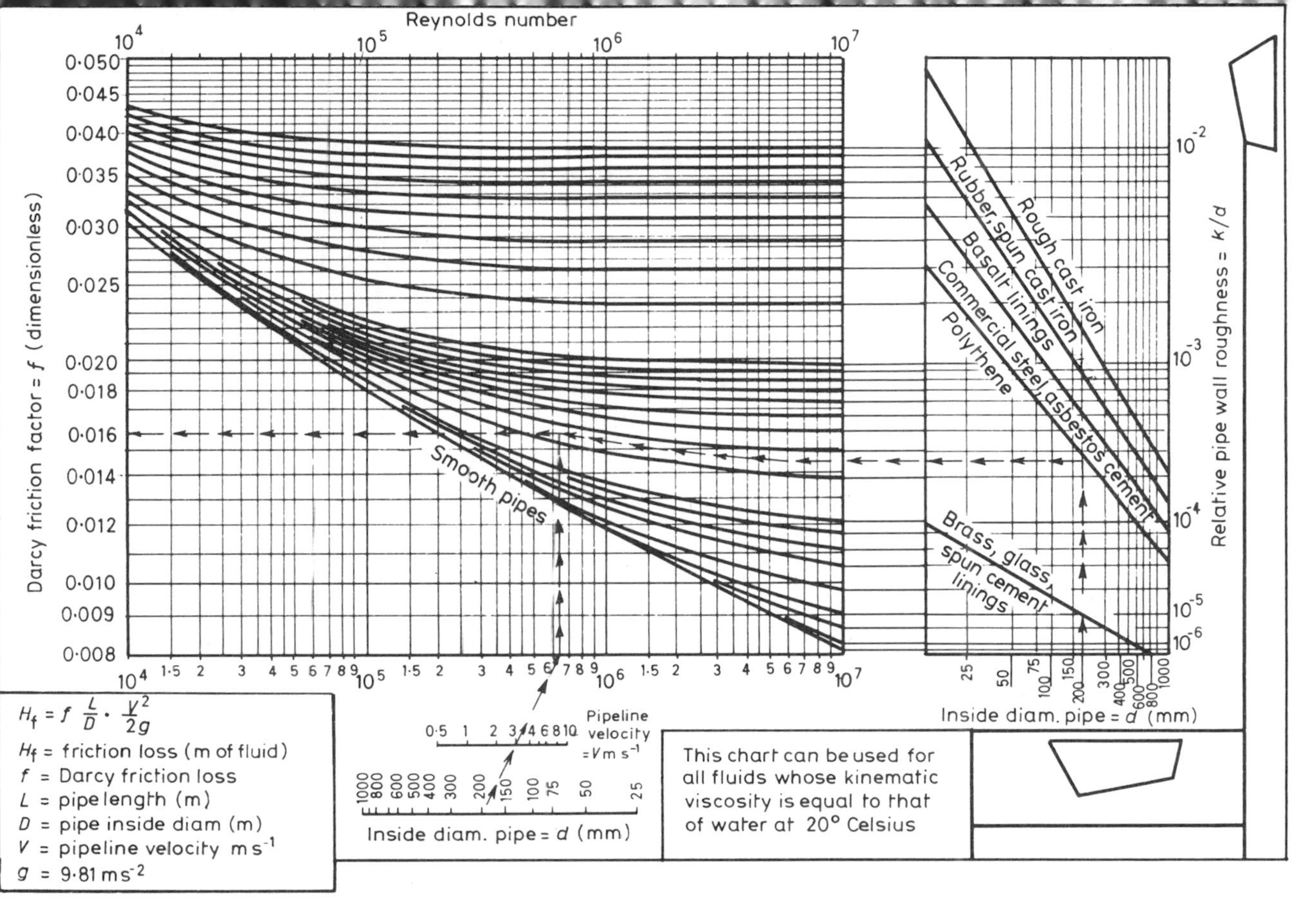

Fig. 5.15 Darcy friction factor f based upon Durand and Condolios Formula (1952)

Table 5.4 Friction losses in valves and fittings expressed in terms of an equivalent length of straight piping of the same diameter

Pipe size NB	Long radius bend	Short radius bend	Elbow	Tee	Rubber hose	Diaphragm valve full open	Full bore valve round way	Plug-lub valve rect. way	Tech. Taylor valve* ball type	Pipe entrance	Pipe Exit
	Radius More than 3 × NB	Radius is 2 × NB			Minimum Radius 10 × NB					$K_s = 0.5$	$K_d = 1.0$
										Sharp edged	
mm	Equiv. length in m of straight pipe giving equivalent resistance to flow										
25	0.52	0.70	0.82	1.77	0.30	2.56	—	0.37	—	0.6	1.2
32	0.73	0.91	1.13	2.36	0.40	3.29	—	0.49	—	0.8	1.5
40	0.85	1.10	1.31	2.74	0.49	3.44	1.19	0.58	—	0.9	1.9
50	1.07	1.40	1.68	3.35	0.55	3.66	1.43	0.73	—	1.3	2.6
65	1.28	1.65	1.98	4.27	0.70	4.60	1.52	0.85	—	1.8	3.5
80	1.55	2.07	2.47	5.18	0.85	4.88	1.92	1.04	0.20	2.2	4.4
90	1.83	2.44	2.90	5.79	1.01	—	—	1.22	—	2.6	5.2
100	2.13	2.77	3.35	6.71	1.16	7.62	2.19	1.40	0.23	3.1	6.2
115	2.41	3.05	3.66	7.32	1.28	—	—	1.58	—	3.6	7.1
125	2.71	3.66	4.27	8.23	1.43	13.11	3.05	1.77	0.30	4.1	8.2
150	3.35	4.27	4.88	10.06	1.55	18.29	3.11	2.13	0.37	5.0	10.1
200	4.27	5.49	6.40	13.11	2.41	19.81	7.92	2.74	0.62	7.2	14.3
250	5.18	6.71	7.52	17.07	2.99	21.34	10.67	3.47	0.61	9.4	18.9
300	6.10	7.92	9.75	20.12	3.35	28.96	15.85	4.08	0.76	11.9	23.5
350	7.01	9.45	10.97	23.16	4.27	28.96	—	4.88	0.91	14.3	28.3
400	8.23	10.67	12.80	26.52	4.88	—	—	5.49	1.04	16.5	30.5
450	9.14	12.19	14.02	30.48	5.49	—	—	6.22	1.16	18.9	37.8
500	10.36	13.11	15.85	33.53	6.10	—	—	7.32	1.25	21.9	43.6

* Tech-Taylor valve is a ball type changeover device used only on the delivery side of the pump

Table 5.5 Other typical head loss items in slurry pipeline flow

Group	Item	Head loss
1.	Loss of head at entrance: From pump hopper to pump or from storage tank to pump	
	(a) Flush connections $K_s = 0.5$	$0.5 \frac{V_1^2}{2g}$
	(b) Projecting connection and dredge suction pipes $K_d = 1.0$	$1.0 \frac{V_1^2}{2g}$
	(c) Rounded connection $K_s = 0.5$	$0.5 \frac{V_1^2}{2g}$
2.	Loss of head at discharge: $K_d = 1.0$ From discharge line to storage tank other equipment or dam	$1.0 \frac{V^2}{2g}$
3.	Loss of head due to sudden contraction: K_c is a factor depending on ratio d_1/d_2	$K_c \frac{V_1^2}{2g}$
	where d_1 is the large diameter and d_2 the small diameter as illustrated below	
	Ratio d_1/d_2: 12, 14, 16, 18, 20, 25, 30, 40, 50 Factor K_c: 0.08, 0.17, 0.26, 0.34, 0.37, 0.41, 0.43, 0.45, 0.46	
4.	Loss of head due to sudden enlargement:	$\frac{(V - V_1)^2}{2g}$
5.	Cyclone and other equipment operating under pressure P: (ρ = density)	$\frac{P}{\delta\rho}$

Items in general use
Valves: plug lubricated valves, full-bore rubber-lined valves, pinch valves and diaphragm valves except on the suction side of the pump. Special valves used in changeover devices on the delivery side of the pump. Note, however, that the use of valves for regulating slurry flows should be avoided where possible. *Fittings*: long-radius bends, reinforced delivery hoses, tees only if essential to the layout and make-up pieces for changeover devices.

Items to be avoided
Valves: gate valves and globe valves; diaphragm valves and pinch valves on the suction side of the pump, foot valves and check valves. *Fittings*: elbows and sharp bends; complicated and long horizontal pipe work on the suction side of the pump.

5.6.5 Limit deposit velocity

For every combination of pipe and solids geometry and slurry concentration there is a certain critical velocity below which the solids will settle out and flow will cease. This velocity is called the limit deposit velocity and, probably the best-known formula for determining limit deposit velocities in heterogeneous flow is that of Durand and Condolios (1952). It is based largely upon average conditions and is a useful guide in many small applications. However, the theoretical work for the derivation of this expression was done using graded materials and, in planning large installations, it is advisable to develop the operational and design parameters in a test rig designed to simulate conditions in the proposed undertaking.

Using slurries with closely sized particles ranging from 20 μm (0.02 mm) up to 4 in (100 mm) and horizontal pipes from 38.1 mm (1.5 in) up to 0.584 m (23 in) diameter, Durand and Condolios developed their empirical expression:

$$V_{cr} = F_L \left[2gD \left(\frac{\rho' - \rho}{\rho} \right) \right]^{0 \cdot 5} \tag{5.15}$$

where V_{cr} is the critical or limit deposit velocity; F_L is a dimensionless constant dependent upon particle size and solids concentration; g is the acceleration due to gravity, 32.2 ft s $^{-2}$; D is the diameter of the pipe in feet; ρ' and ρ are the densities of the solids and water, respectively. The formula is based upon the well-known expression for a body falling freely in air:

$$V^2 = 2gh \tag{5.16}$$

by substituting the inside diameter of the pipe for h and applying a correction for the increased viscous drag and buoyancy in water.

In their calculations, Durand and Condolios used the volumetric concentration of solids and fluid taken as

$$C_V = \frac{\text{Volume of solid material}}{\text{Volume of mixture}}.$$

Figure 5.16 gives values of F_L for various volumetric concentrations from 2% of solids by volume up to 15% of solids by volume to show that the effects of concentration of solids by volume C on the friction factor F_L is a maximum in the size range 0.5 mm to 2.54 mm.

For particle diameters of more than 0.1 in (2.54 mm) particle size effects are negligible because of the high velocities needed to sustain the solids in suspension. Note that the particle diameter is the 'effective' diameter, that is the diameter of a sphere that falls with the same terminal velocity as the particle.

Durand's formula (5.15) can be written for slurries of sand of RD = 2.65 for the metric system:

$$V_{cr} = F_L \cdot 5.97 D^{0.5} \tag{5.17}$$

by substituting for $g = 9.81$ m s $^{-2}$ and with D in metres.

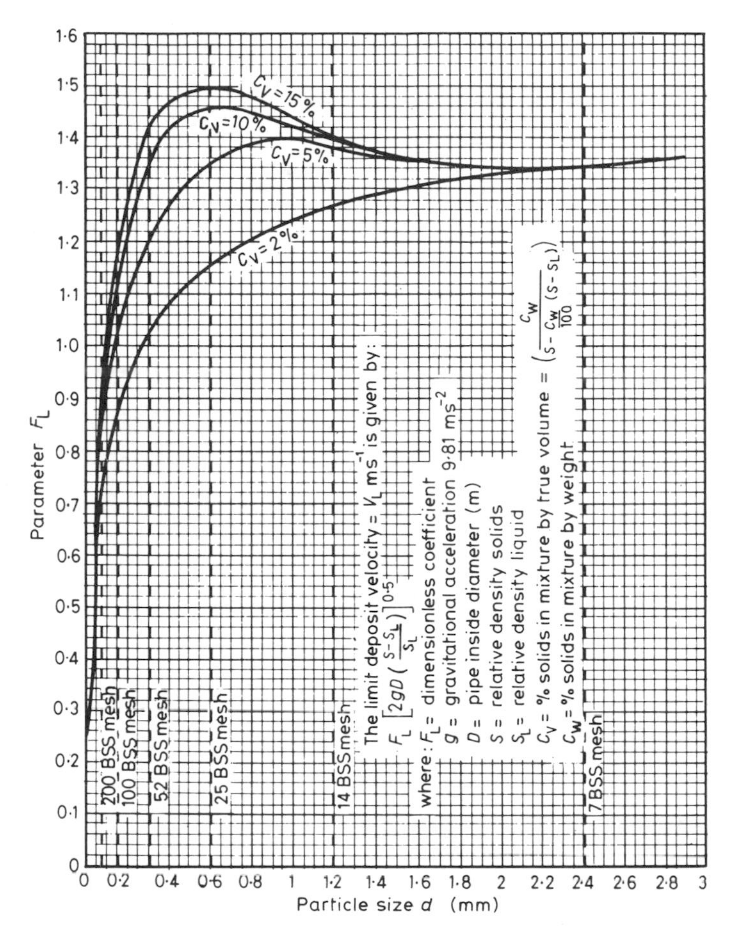

Fig. 5.16 Limit deposit velocity parameter F_L as a function of particle size for closely graded solid particles in water

The limit deposit velocity V_{cr} obtained from this formula is the lowest safe pipeline velocity above that at which deposition occurs when pumping mixtures of closely sized particles. It is a higher value than when F_L is obtained as a function of average particle size for solids of widely graded sizing because, as has already been noted, the presence of fine particles in a solids–fluid mixture adds to its carrying power. It is customary in preliminary design calculations to use the Durand formulae for graded particles and to add at least 0.3 m s^{-1} onto the calculated value as an added safety factor. In final design, pump manufacturers, such as Warman International, use lower figures based upon experience and test work, but still add a factor of safety to cover contingencies.

The method for selecting the most suitable pipe diameter for a given rate of flow Q is explained in the following example for $Q = 104 \, \mathrm{l \, s^{-1}}$.

Pipe diameter d (mm)	150	200	250
D (m)	0.150	0.200	0.250
$V_{cr} = 5.97\, D^{0.5}$ (m s^{-1})	2.31	2.67	2.99
Flow rate (l s^{-1})	41	84	147
Comments	unsuitable	suitable	unsuitable

Explanation:

$$Q = AV = \frac{d^2\, V}{1273} \tag{5.18}$$

and, at the required flow rate of $Q = 104$ l s^{-1}

$V = 5.88$ m s^{-1} in the 150 mm diameter pipeline – excessively high
$V = 3.31$ m s^{-1} in the 200 mm diameter pipeline – safe for $V_{cr} = 2.67$
$V = 2.12$ m s^{-1} in the 250 mm diameter pipeline – excessively low

Note: H_f for slurry may be taken as numerically equal to H_f for water in the initial stages of calculation.

5.6.6 Pipeline materials and construction

Before the advent of specialized materials for flow lines, operators, from their experience, rotated the piping periodically to distribute the wear and turning schedules were based either upon a time limit, say every 3 months according to the quantity of solids pumped or according to measurement. Typically, therefore, either an ultrasonic thickness gauge was used to determine the most economic wear pattern or the pipe was turned three times, in succession 180°, 90° and 180°, at regular fixed intervals of time or quantities passed.

The modern approach to minimizing pump and pipeline wear in addition to turning is to use very hard alloys such as Ni-Hard and ceramics that resist the abrasiveness of the softer slurry particles or elastic, deformable, comparatively soft materials such as rubber, polyethylene and polyurethane which have a high resilience so that particles tend to bounce off rather than to cut. Table 5.6 (Anon, 1981) describes the relative abrasion resistance in a typical laboratory wet-sand test. Based upon carbon steel 100, the wear rates show clearly the reason for the present dominance of polyether elastomers.

Urethanes may be based upon polyethers or polyesters but those based on polyesters are the least stable because, when they hydrolyse, an acidic by-product is produced which catalyses further hydrolysis. Polyether urethanes also react with water but at a very much slower rate that is insignificant in terms of a normal working life.

Certain characteristics of polyether urethanes make them particularly valuable in many aspects of placer technology:

(1) They can be cast around a steel core or against a steel shell, and will also bond to aluminium, plastic and fibreglass. This is subject to the

Table 5.6 Relative abrasion resistance: typical laboratory sand test (wet)

Materials	Percentage of weight loss
Polyether elastomer (Elastoglass)	7
Ni-hard	18
Polyethylene plastic	21
Trelleborg rubber	46
Linatex rubber	55
304 Stainless steel	78
Carbon steel	100
Hard neoprene rubber	800

requirement that the pipe can be rotated at a temperature of about 80°C to produce a satisfactory cast lining.

(2) Elastoglass pipe systems combine excellent abrasion and chemical resistance with high strength, light weight, and the flexibility to form long-radius bends.

(3) They bond strongly to steel, generally exceeding the strength of the urethane material itself. One important application is in the moulding of pump impellers. In ideal conditions, urethane impellers may outlast either Ni-hard or soft-rubber impellers.

(4) They provide high abrasive resistance qualities for cone separator inserts and other wearing parts (see Chapter 7).

(5) Polyether urethanes can be machined, drilled and shaped as required.

Note: Very long lives are obtained for urethane-lined pipes by turning them frequently as described.

5.6.7 Flow measurement and control

Without measurement there can be no adequate control. In some dry-mining installations solids concentration may be held steady at a predetermined value by mechanically controlling the inflow of solids and fluids into the system. In suction dredging and gravity concentrating plant, however, this is not practicable and, for proper control of pulp densities and pumping, there must be a means for the instantaneous measurement of both flow rate and concentration. This involves the primary characteristics of size, time, force and mass from which may be determined the required dimensions of velocity and density.

Early gravimetric and volumetric measurements required the flow to be diverted to a collecting vessel during a measured period of time to determine its mass. Weightometers continually reacted to changing densities in a flexible section of flow line through a suitable system of levers. This method was independent of the rate of flow provided that no impact forces were present and that the pipe ran full. Combinations of pitot and static tubes were used to measure flow velocities directly, the quantity rate being determined by integrating the instantaneous velocities over a full traverse. This method was reasonably satisfactory for homogeneous suspensions but for heterogeneous mixtures the measurement of spot velocities was complicated by the following difficulties:

(1) If the aspect of the probe was larger than the grain the probe disturbed the flow.
(2) Velocity head instruments are sensitive to the presence of solids in the flow regardless of their density. They are also affected by the density which, in heterogeneous mixtures, varies instantaneously at every point of measurement and varies from point to point across any section of the flow.

Many other forms of metering have also been devised and a 'Whistle Flow Meter' of a type developed in the Laboratoire Dauphinois D'Hydraulique (Neyrpic, Grenoble) France was used in experimental work already referred to (Macdonald, 1966). This was a nozzle of eccentric dimensions named because of its resemblance to a whistle (Fig. 5.17). It provided an upstream contraction and a convergence of such shape that flow was streamlined along the boundaries and the fluid stream was discharged without separation.

Few of these devices are still used in commercial undertakings and most installations are now controlled by magnetic flow meters and radiation detector density gauges.

Magnetic flow meters

Measurement of the velocity of flow of pulps by magnetic flow meters depends upon Faraday's Law of Magnetic Induction which states that the voltage E induced in a conductor of length D moving through a magnetic field H is proportional to the velocity V of the conductor. C is a dimensional constant. The pulp is the conductor and the magnetic flux is created by a transmitter, the EMF being sensed by two electrodes.

$$E = CHDV \tag{5.19}$$

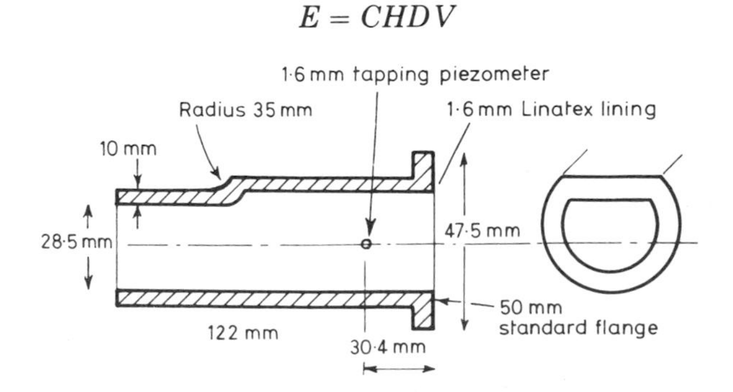

Fig. 5.17 Section through whistle flow meter

The transmitter line is clamped around a section of the flow line and is not in contact with the slurry to be metered. Consequently, there are no restrictions in the flow, no loss of head and no pressure taps to plug, or obstructions to bridge. On the other hand problems such as magnetic particle build up must be recognized and eliminated in individual calculations.

Radiation detector density gauges

This type of density gauge measures and records continuously the density or percent solids concentration of a slurry. It consists of a radioactive source, a radiation detector amplifier, a power supply indicator and a suitably calibrated recorder and is built around a short section of the pipeline. The radioactive isotope source and the detector are shielded to avoid radiation hazards.

The instrument is calibrated to record any change in mass per unit volume of the flowing pulp by measuring the amount of radiation being absorbed as it passes through the pulp. The radio isotope source is shuttered to direct gamma rays through the pipe and its contents. The residual radiation is measured by the detector and converted with an electrical signal which is then fed into the analysing and recording instrument. Since the amount of absorption is a function of the slurry density the difference between emitted and recorded radiation can be analysed in terms of the desired units of concentration.

Combined flow and density measurement

Measuring the quantity of solid material passing through the pipeline per unit time requires the recorded density and flow data to be fed to a single analysing and integrating instrument which calculates the dry tonnes of solids in the flow. Early installations were subject to error through the frequency and magnitude of surging in the flow lines and because of fluctuations in the magnetic and gamma ray absorption properties of the material being pumped. However, these difficulties have now been largely overcome and the advantages of not being in contact with the slurries being measured have brought them into universal use. Modern instrumentation features automatic zero and gain control of the flow signal and push-button calibration through a microcomputer which processes the mass and flow signals continuously to provide the mass flow information. In less automated plants, the operators can be trained to make plant and process adjustments according to the instantaneous readings. The adjustments are made as soon as the recordings are seen to deviate from design levels.

Avoiding and clearing line blockages

One of the most troublesome aspects of handling slurries of granular solids and slimes in pipelines is the ease with which some of the lines tend to block, and the amount of operational time that is lost in clearing them. Common causes are power failures, the migration of solids to low points on shutdown, sudden surges of high-density particles, worn impellers, carelessness, and leaks in the pipeline. Whatever the cause, a major blockage may take several hours to clear. It is not likely to happen in a fully automated plant, but it might occur if the electronics fail. Blockages occur all too frequently in process plants with little or no instrumentation and where the design has been inadequate.

The blocked section can usually be identified by tapping the pipe with a hammer. It may then be replaced by a new section and cleared at leisure, or cleared on the spot using a hose to wash the material out. A high-pressure water pump is sometimes used to force the plug out in the form of a sausage. This method can be applied to blockages in a long line of pipes by breaking the line at several points and treating each section individually.

In slimes circuits, a nearly choked pipeline may be more readily cleared using a gritty slurry rather than clean water. Some engineers contend that this produces an abrasive or scouring action on the deposited particles as well as providing greater mass and momentum. See also, the effects of solids concentration in Chapter 7.

5.7 PRACTICAL PUMPING CALCULATIONS

Calculations are based upon:

(1) Total solids to be pumped within a given time, their particle size range and densities.
(2) Total suction and delivery heads.
(3) The geometry of suction and delivery lines.
(4) Flow velocities and pipe friction factors.
(5) Pump characteristics,
(6) Required pulp density.

5.7.1 Effects of different slurry concentrations

Determining how much water is needed in the system is of paramount importance. As already pointed out (p. 000) early suction dredgers, by the crude nature of their construction and techniques, were constrained to pump at average slurry concentrations of around $C_W = 20\%$. With the added control available to modern dredging systems, operators can now expect to average around $C_W = 30\%$ of solids by weight. Consider what this means for, say, a dredging rate of 200 TPH of dry solids, first at $C_W = 20\%$ and secondly at $C_W = 30\%$.

(1) At $C_W = 20\%$, the amount of water in the slurry $= 200\,\dfrac{(100-20)}{20}$

$$= 800\ \text{TPH}$$
$$= 222\ \text{l s}^{-1}$$

(2) At $C_W = 30\%$, the amount of water in the slurry $= 200\,\frac{(100-30)}{30}$

$$= 467 \text{ TPH}$$
$$= 129\,\text{l s}^{-1}$$

The power consumed at the pump spindle is calculated from the equation:

$$P = \frac{QHS_m}{1.2e} \tag{5.20}$$

where P is the power (kW), Q is the quantity flowing (l s^{-1}), H is the head (m), S_m is the relative density of the fluid or mixture flowing (water = 1.0) and e is the efficiency (say 50%).

Considering only the water contained in the mixtures and assuming a total dynamic head H of say 30 m.

$$222\ \text{s}^{-1} \text{ requires } P = \frac{222 \times 30 \times 1}{1.2 \times 50} = 111\,\text{kW at the pump spindle}$$

$$129\ \text{s}^{-1} \text{ requires } P = \frac{129 \times 30 \times 1}{1.2 \times 50} = 64.5\,\text{kW at the pump spindle}$$

Therefore additional power required for pumping at the lower density of $C_W = 20\% = 46.5$ kW.

The cost differential is not for power alone because, when appreciable quantities of water are present in the system over and above the amount needed for safe transportation, there will also be increased capital costs for larger pumps and pipelines and, perhaps also, for the structures on which they are mounted. Furthermore, it is always more difficult to control a plant in which excessive quantities of water are to be rejected and the usual consequence is either the acceptance of additional losses of valuable minerals in the overflows from bins and cyclones or the provision of costly additional circuitry in order to recover the losses.

Primary head feed slurries pumped to cones or pinched sluice concentrators are controlled within a desired range of 57–65% of solids by weight according to the particular item of plant and the characteristics of the solids (see Chapter 7). Sizing and desliming cyclones also require the feed rate and slurry density to be controlled and a 20–40% solids concentration is common for most separations.

5.7.2 Design considerations

The design of a slurry pumping system for a suction-dredging operation entails consideration of a number of important principles covering a range of somewhat diverse duties. The dredger pump is required to pump against a fluctuating negative suction head and values for the available and required net positive suction heads are basic features in the design. Pumps delivering to gravity treatment units do so at controlled water–solids ratios and draw their feed materials from densifying bins. Fine tailings for jet stacking are densified and pumped at high concentrations to a discharge nozzle for discharge at high velocity. Concentrates composed of relatively fine high-density particles may be pumped ashore through a floating pipeline and calculations must take account of optimum slurry densities for cyclone dewatering and of dewatering cyclone characteristics.

Preliminary design calculations usually assume standard conditions for temperatures and pressures, however, in final studies, the manufacturer will require to know more about the climatic and other environmental conditions in which the units are to operate. Barometric pressures are of particular importance, as already noted, and Fig. 5.18, which is a plot of absolute vapour pressure against temperature for pure water, illustrates graphically the steep rise in the absolute vapour pressure associated with high water temperatures. Values are based generally on water at 20°C but in some attritioning processes where live steam is used to promote chemical actions, slurries may have to be pumped at high temperatures.

The following calculations relate to the design of a dredging and slurry pumping system for a 200 TPH suction cutter dredger and floating treatment plant for which an outline of the basic details is given in this Chapter and in Chapter 7. The prototype was proposed for treating fine tailings, predominantly quartz sands from previous tin dredging operations, and, for the purpose of the exercise, the average relative density of the solids is taken to be 2.65.

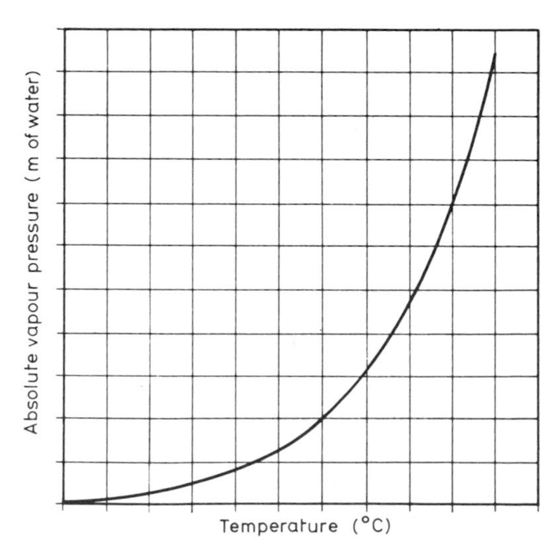

Fig. 5.18 Plot of absolute vapour pressure versus temperature

5.8 THE DREDGER PUMP

5.8.1 Duty

To dredge and pump 200 TPH of free-running sands using a pump selected from the Warman Series 'A' dredge and gravel pumps. The dredging arrangement is illustrated in Fig. 5.9. The duty details are shown in Table 5.7

5.8.2 Slurry dimensions

(1) Weight of solids in slurry = 200 TPH

(2) Weight of volume of water equal to volume of solids $\frac{200}{2.65} = 75.47$ tonnes

(3) Weight of water in slurry of $C_W = 30\%$ $200\frac{(100-30)}{30} = 466.67$ tonnes

(4) Total weight of slurry, add (1) and (3) $200 + 466.67 = 666.67$ tonnes

(5) Total weight of equal volume of water, add (2) and (3) $75.47 + 466.67 = 542.14$ tonnes

(6) Rate of slurry flow (Q), multiply (5) by 1000 and divide by 3600 $\frac{542.14 \times 1000}{3600} = 151\,\mathrm{l\,s^{-1}}$

(7) Relative density of slurry S_m, divide (4) by (5) $\frac{666.67}{542.14} = 1.23$

Table 5.7

Material: slurry of sand and water	Pumping system	
	Suction	Delivery
Relative density $S = 2.65$ $S_l = 1.00$ $d_{50} = 200\,\mu$m (widely graded) $C_W = 30\%$ (max.) Water temperature = 25°C (max.) Barometric pressure head = 10.34 m (water) Vapour pressure head = 0.32 m (water)	$Z_s = -0.5$ m $A = 3.0$ m (max.) $L_s = 15$ m (allowing for bends etc.) Entrance-dredger suction type d_s = to be determined Piping = commercial steel	$Z_d = +3$ m $L_d = 140$ m (allowing) for bends etc.) d_d = to be determined Piping = rubber hose for floating line

Solids capacity required = 200 tonnes h^{-1} (minimum)

(8) Percentage of solids by volume, $C_V = 100$ times (2) divided by (5) $\dfrac{100 \times 75.47}{542.14} = 13.9$

The size of the delivery line is selected by trial and error based upon achieving a flow rate sufficient to avoid the solids settling out from the mixture flowing, but not so high as to result in excessive pipe friction losses. The density of the slurry cannot be controlled closely in dredging service and, hence, the design must allow for short periods of higher slurry densities at reduced flow rates.

5.8.3 Selecting the pipe diameter

A spread of three alternative diameters, 200 mm, 250 mm and 300 mm is tried: The velocity of flow $V(\mathrm{m\,s^{-1}})$ in a pipeline of internal diameter d (mm) is calculated in each case from equation (5.18) re-arranged to give:

$$V = \frac{1273Q}{d^2}, \qquad \text{for } Q = 151\,\mathrm{l\,s^{-1}}$$

Pipe diameter: d (mm)	200	250	300
Pipe diameter: D (mm)	0.200	0.255	0.300
$V = \dfrac{1273Q}{d^2}$ $(\mathrm{m\,s^{-1}})$	4.81	3.08	2.14

Test for settling using the Durand equation $V_L = F_L\ 5.97\,D^{0.5}$.

Note: Figure 5.16, the plot of F_L as a function of particle diameter and concentration, is used in these calculations since they are of a preliminary nature only. The available curves are restricted to a maximum C_V of 15% and values of F_L for denser mixtures must be interpolated, noting that experience has shown that F_L increases with increasing C_V to about $C_V = 30\%$. Beyond $C_V = 30\%$, F_L decreases with increasing C_V due to increasing interferance of particles with one another.

From Fig. 5.16 for $d_{50} = 200$ and $C_v = 13.9\%$, $F_L = 1.13$ and from equation (5.17):

V_L for 200 mm pipe $= 1.13\,(5.97 \times 0.4472) = 3.02\,\mathrm{m\,s^{-1}}$
V_L for 250 mm pipe $= 1.13\,(5.97 \times 0.50) = 3.37\,\mathrm{m\,s^{-1}}$
V_L for 300 mm pipe $= 1.13\,(5.97 \times 0.5477) = 3.69\,\mathrm{m\,s^{-1}}$

The calculated flow velocities of $3.08\,\mathrm{m\,s^{-1}}$ and $2.14\,\mathrm{m\,s^{-1}}$ are below the limit deposit velocities $3.37\,\mathrm{m\,s^{-1}}$ and $3.69\,\mathrm{m\,s^{-1}}$ calculated from the Durand formula and hence the 250 mm and 300 mm pipes are obviously too large. On the other hand, although $4.81\,\mathrm{m\,s^{-1}}$ in the 200 mm diameter piping is safely higher than the corresponding limit deposit velocity of $3.08\,\mathrm{m\,s^{-1}}$ it may be excessively so and a 225 mm bore size is tried.

$$V = \frac{1273Q}{d^2} = \frac{1273 \times 151}{225 \times 225} = 3.80\,\mathrm{m\,s^{-1}}$$

$$V_L = 1.13\,(5.97 \times 0.225^{0.5}) = 3.20\,\mathrm{m\,s^{-1}}$$

A 225 mm diameter pipe provides a safe velocity differential and is selected for this duty.

5.8.4 Calculation of total head

The total delivery line length is given as 140 m of rubber hose, allowing for bends and other fittings and for the effects of flexing whilst free-ranging in the dredge pond. The suction pipe is made of commercial steel piping and has an equivalent length of 15 m. Assume suction and delivery diameters d to both equal 225 mm.

Differential column

Z_c in head of mixture is given by

$$Z_c = Z\frac{(S_m - S)}{S_m} = 3.0\,\frac{(1.23 - 1.0)}{1.23} = 0.56\,\mathrm{m}.$$

$$H_{atm} = \frac{\text{Barometric pressure head of water (m)}}{S_m}$$

$$= 10.34/1.23 = 8.4\,\mathrm{m}$$

$$H_{vap} = \frac{\text{Vapour pressure of water (m)}}{S_m}$$

$$= 0.32 \div 1.23 = 0.26\,\mathrm{m}$$

$$H_{atm} - H_{vap} = 8.4 - 0.26 = 8.14\,\mathrm{m}$$

Pipe friction factor

The Darcy friction factor f is read from Fig. 5.15 as, (a) $f = 0.0158$ for commercial steel pipe, and (b) $f = 0.0190$ for rubber hose.

Suction head and NPSH$_a$

$l_s = 15$ m; $D = 0.225$ m; $Q = 151\,\mathrm{l\,s^{-1}}$. Use Fig. 5.16 for limit deposit velocity parameter F_L and Fig. 5.19 for efficiency and head ratios ER and HR

For water:

$$V_s = \frac{1273Q}{d^2} = \frac{1273 \times 151}{225 \times 225} = 3.80\,\mathrm{m\,s^{-1}}$$

f (Fig. 5.15) $= 0.0158$

$$\frac{L_s \text{ (length of suction pipe)}}{L_d \text{ (diameter of suction pipe)}} = \frac{15}{0.225} = 66.7$$

$$H_{vs} = \frac{V^2}{2g} = \frac{3.8 \times 3.8}{2 \times 9.81} = 0.74\text{ m}$$

$$H_i = 1.0\, H_{vs} = 0.74$$

$$H_{fs} = \frac{0.0158 \times 15 \times (3.8)^2}{2 \times 9.81 \times 0.225} = 0.78$$

$$Z_s = -0.50$$

$$H_s = Z_s - H_i - H_{fs} = -2.02\text{ m}$$

For mixture:

$$Z_c = \underline{0.56}$$

$$H_s = H_s \text{ for water} - Z_c = -2.58\text{ m}$$

$$H_{atm} - H_{vap} = \underline{8.14\text{ m}}$$

$$\text{Therefore NPSH}_a = H_{atm} - H_{vap} + H_s = \underline{5.56\text{ m}}$$

Note: H_s is negative for both water and slurry in this example.

H_{fs} for water and H_{fs} for the $C_W = 30\%$ slurry are for all practical purposes the same in the short suction pipe.

The constant head loss Z_c for the slurry is significantly higher than for water because of the higher slurry density.

Discharge head and total head

$L_d = 140$ m; $D = 0.225$ m; $Q = 151\text{ l s}^{-1}$.

$$V_d = \frac{1273Q}{d^2} = \frac{1273 \times 151}{225 \times 225} = 3.80\text{ m s}^{-1}$$

$$f \text{ (Fig. 5.15)} = 0.019$$

$$L_d/D_d = \frac{140}{0.225} = 622$$

$$H_{vd} = \frac{V^2}{2g} = H_{vs} = 0.74$$

$$H_{fd} = \frac{0.019 \times 140 \times (3.8)^2}{2 \times 9.81 \times 0.225} = 8.70$$

$$Z_d = \underline{+3.00}$$

$$H_d = Z_d + H_{fd} + H_{vd} \qquad = 12.44$$
$$H_s = H_s \text{ for water} \qquad = \underline{-2.02\,\text{m}}$$
$$H_\gamma = H_d - H_s \qquad = 14.46\,\text{m}$$

For mixture:

$$Z_c \qquad = \underline{0.56}$$
$$H = H_\gamma \text{ for water} + Z_c \qquad = \underline{15.02\,\text{m}}$$

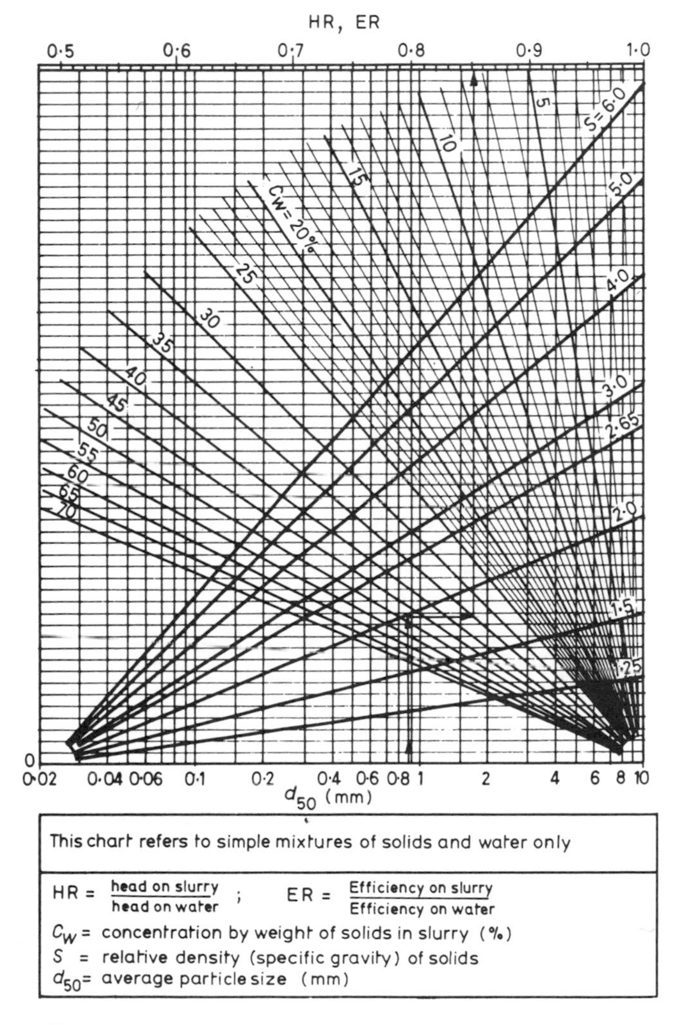

Fig. 5.19 Performance of centrifugal pumps on slurry

5.8.5 Efficiency ratio (ER) and head ratio (HR) (Fig. 5.19)

$$\mathrm{ER} = \frac{\text{Efficiency on slurry}}{\text{Efficiency on water}} \qquad \mathrm{HR} = \frac{\text{Head on slurry}}{\text{Head on water}}$$

Applying the known values $S = 2.65$; $d_{50} = 200\,\mu\mathrm{m}$ and $C_W = 30\%$

$$\mathrm{ER} = \mathrm{HR} = 0.89$$

5.8.6 Preliminary pump selection

The Warman Series 10/8 FG may be suitable for this duty, i.e. to pump $151\,\mathrm{l\,s^{-1}}$ of slurry in new condition and be capable of speeding up in worn condition.

NPSH *required*

From the performance curves (Fig. 5.20), the $\mathrm{NPSH_r}$ for this service is 3 m and the available $\mathrm{NPSH_a}$ of 5.56 m is adequate to provide cavitation free performance at the rated output and with a reasonable margin to cope with unexpected fluctuations caused by such factors as the sudden caving of bank material or partial blockages at the suction inlet.

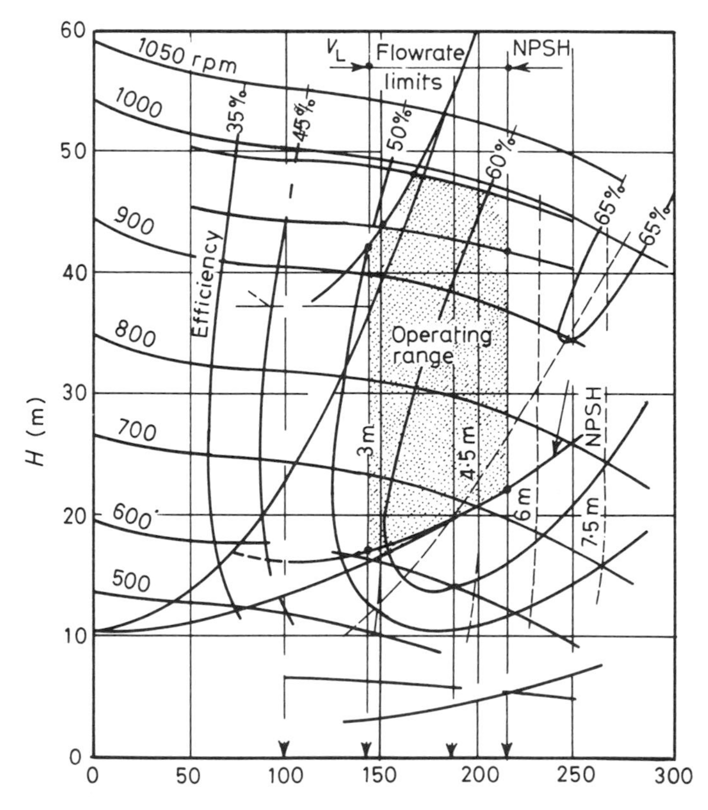

Fig. 5.20 Typical pump performance curve, Warman 10/8 FG gravel pump

Pump speed

Allowing H_m worn $= 0.90\,H_m$ new and ER = HR = 0.89,

$$H_w = \frac{H_m\,(\text{worn})}{0.90 \times \text{HR}} = \frac{15.02}{0.9 \times 0.89} = 18.75\ \text{m}$$

From the performance curve (Fig. 5.20), at $Q = 151\ \text{l s}^{-1}$ and $H_w = 18.75$ m:

$$\text{Pump speed} = 650\ \text{rpm}$$

$$\text{NPSH}_r = 3.0\ \text{m}$$

$$\text{NPSH}_a = 5.58\ \text{m}$$

Pump efficiency e

$$e_m\ (\text{worn}) = e_m\ (\text{new}) \times \text{ER} \times \frac{e_m\ (\text{worn})}{e_m\ (\text{new})}$$

$$= 59\,\% \times 0.89 \times 0.90 = \underline{47.3\,\%}$$

5.8.7 Power consumed at pump spindle (pump in worn condition)

$$\frac{Q \times H \times S_m}{1.02 \times e_m\ (\text{worn})} = \frac{151 \times 18.75 \times 1.23}{1.02 \times 47.3}$$

$$= 72.2\ \text{kW}$$

Note: If the pump is operated in new condition at 650 rpm the flow rate will be greater than $151\ \text{l s}^{-1}$. For this, as well as for flexibility in operation, dredging pumps are normally powered by variable-speed electric motors or are driven by diesel engines.

Allowing 10% loss for belt slippage, the total required power is 72.2 + 7.2 = 79.4 kW at the pump spindle and a 120 HP slip-ring motor would be a suitable drive unit.

5.8.8 Preliminary dredge-pump specifications

Based upon the Warman Series A gravel pump curves a pump would be selected similar to the Warman 10/8, Series A, Frame F, type G, gravel pump, gland water-sealed complete with 120 HP slip-ring motor and vee-belt drive to give a variable speed around 700 rpm. Details relating to impeller types, materials of construction etc. are decided in conjunction with the manufacturer.

5.9 PRIMARY HEAD-FEED PUMP

5.9.1 Duty

To pump 200 TPH of sand of average $S = 2.65$ at $C_W = 60\,\%$ of solids by weight from a densifying bin located on the treatment barge to a gravity-fed distributer feeding to the primary concentrating cones according to the arrangement described in Fig. 5.21. The duty details are shown in Table 5.8.

5.9.2 Slurry dimensions

(1) Weight of solids in slurry	200 TPH
(2) Weight of volume of water equal to volume of solids	200/2.65 = 75.47 tonnes
(3) Weight of water in slurry of $C_W = 60\%$	$200\dfrac{(100-60)}{60} = 133.33$ tonnes
(4) Total weight of slurry, add (1) and (3)	200 + 133.33 = 333.33 tonnes
(5) Total weight of equal volume of water, add (2) and (3)	75.47 + 133.33 = 208.8 tonnes

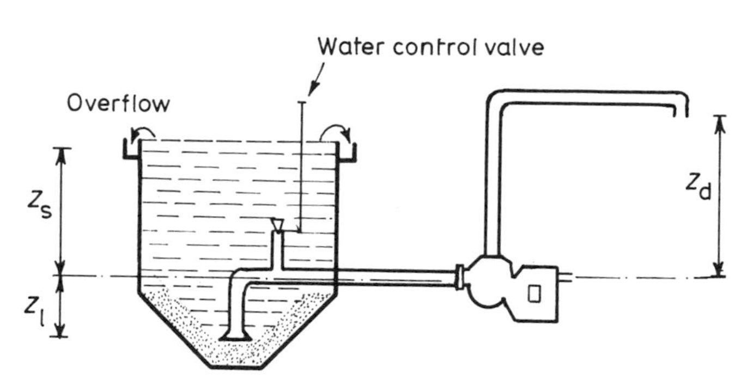

Fig. 5.21 Densifying bin pumping arrangement

Table 5.8

Material: slurry of sand and water	Pumping system	
	Suction	Discharge
$S = 2.65$ $S_l = 1.0$ $M = 200$ TPH $d_{50} = 200\,\mu$m (widely graded) $C_W = 60\%$ Water temperature = 25°C (max) Altitude at pump location–sea level	$Z_s = 1.5$ m $Z_l = 2.5$ m $L_{as} = 3.0$ m Bends = 1 × 90° long radius Tees = 1 × 'run of' Entrance – rounded Piping – commercial steel	$Z_d = 10.0$ m $L_{ad} = 20.0$ m Bends = 2 × 90° long radius Piping – commercial steel

(6) Rate of slurry flow Q, multiply (5) by 1000 and divide by 3600 $\quad \dfrac{208.8 \times 1000}{3600} = 58\,\mathrm{l\,s^{-1}}$

(7) Relative density of slurry S_m, divide (4) by (5) $\quad \dfrac{333.33}{208.8} = 1.60$

(8) Percentage of solids by volume C_V = 100 times (2) divided by (5) $\quad \dfrac{100 \times 75.47}{208.8} = 36.1\,\%$

5.9.3 Selecting delivery-pipe diameter

Try a 150 mm diameter pipe:

$$V_d = \frac{1273Q}{d^2} = \frac{1273 \times 58}{150 \times 150} = 3.28\,\mathrm{m\,s^{-1}}$$

Test for settling using the Durand formula. From Fig. 5.16, for $d_{50} = 200$ and $C_V = 36.1\,\%$: $F_L = 1.4$

$$\begin{aligned} V_L &= F_L\,[5.97\,D^{0.5}] \\ &= 1.4\,[5.97 \times 0.150^{0.5}] \\ &= 3.24\,\mathrm{m\,s^{-1}} \end{aligned}$$

This is the limit deposit velocity based upon flow in a horizontal pipe. In this head feed application, the delivery line is short and, for the most part vertical and the actual V_L would be measurably lower than $3.24\,\mathrm{m\,s^{-1}}$. The 150-mm diameter delivery pipe should be quite safe.

5.9.4 Calculation of total head

The length of the delivery line $L_{ad} = 20$ m to which must be added the equivalent lengths (see Table 5.4) of two, 90°, long-radius bends

$$L_{fd} = 2 \times 3.35 = 6.70\,\mathrm{m}$$

Therefore $L_d = L_{ad} + L_{fd} = 20 + 6.7$

$= 26.7$ m of 150-mm diameter straight pipe.

The length of the suction pipe $L_{as} = 3.0$ m to which must be added the equivalent lengths of one 90°, long-radius, bend and the 'run-through' loss of head of one standard tee for which experience has shown the loss to equal that of a long-radius 90° bend, i.e. 3.35 m. Therefore

$$L_s = L_{as} + L_{fs} = 3.0 + 2 \times 3.35 = 9.7\,\mathrm{m}$$

of 150-mm diameter straight pipe.

Conditions on suction side of pump

$$V_s = \frac{1273Q}{d^2} = V_d = 3.28\ \text{m s}^{-1}$$

$$f = 0.0170\ \text{(Fig. 5.15)}$$

$$L_s = \underline{9.7\ \text{m}}$$

$$H_{fs} = \frac{0.0170 \times 9.7 \times (3.28)^2}{2 \times 9.81 \times 0.150} = 0.60\ \text{m}$$

$$H_{vs} = \frac{(3.28)^2}{2 \times 9.81} = 0.55\ \text{m}$$

$$H_i = \frac{0.5(3.28)^2}{2 \times 9.81}\ \text{(see Table 5.4)} = 0.27\ \text{m}$$

$$Z_i = Z\frac{(S_m - S_l)}{S_m} = 2.5\frac{(1.6-1)}{1.6} = 0.94\ \text{m}$$

A conservatively low valuation of Z_s is to regard the contents of the upper portion of the hopper as simply clear water so that:

$$Z_{sm} = Z_s\frac{(S_l)}{S_m} = 1.5\frac{(1.0)}{1.6} = 0.94\ \text{m}$$

$$H_s = Z_{sm} - Z_i - H_i - H_{fs} = 0.94 - 0.94 - 0.27 - 0.6 = (-0.87\ \text{m}),\ \text{say}\ -1.0\ \text{m}.$$

Conditions on the delivery side of pump

From Fig. 5.21,

Z_d $= 10\ \text{m}$

$$H_{fs} = \frac{0.0170 \times 26.7 \times (3.28)^2}{2 \times 9.81 \times 0.150} = 1.66\ \text{m}$$

$H_{ve} = H_{vs}$ $= \underline{0.55\ \text{m}}$

Total delivery head $H_d = Z_d + H_{fd} + H_{ve}$ $= 12.21\ \text{m}$

Total dynamic head $H = H_d - H_s = 12.2 - (-1.0)$ $= 13.2\ \text{m}$

Barometric pressure at sea level

$$-760\ \text{mm of mercury} = \frac{760}{1000} \times \frac{13.6}{1} = 10.34\ \text{m (of water)}$$

$$H_{atm} = \frac{10.34}{1.60} = 6.46\ \text{m}$$

Vapour pressure of water at 25°C = 23.756 mm of mercury

$$\frac{23.756}{1000} \times \frac{13.6}{1.0} = 0.32\,\text{m}$$

$$H_{vap} = \frac{0.32}{1.60} = 0.20\,\text{m}$$

$$\text{NPSH}_a = H_{atm} - H_{vap} + H_s$$
$$= 6.46 - 0.2 + (-1.0) = \underline{5.26\,\text{m}}$$

5.9.5 Efficiency ratio (ER) and head ratio (HR) (Fig. 5.19)

Applying the known values of $S = 2.65$, $C_W = 60\,\%$ and $d_{50} = 200$

$$\text{HR} = \text{ER} = 0.79$$

5.9.6 Preliminary pump selection

From the Warman Series A slurry pumps the preliminary selection is a 6/4 Frame E, type AH, slurry pump with a five-vane, closed rubber impeller.

At $Q = 58.0\,\text{l}\,\text{s}^{-1}$ and $H = 13.2\,\text{m}$, in worn condition:

$$H_w = \frac{H_m\ (\text{worn})}{0.9 \times \text{HR}} = \frac{13.2}{0.9 \times 0.79} = 18.6\,\text{m}$$

From performance curves (Fig. 5.22)

Pump speed = 950 rpm
NPSH_r = 2.75 m
NPSH_a = 5.26 m

The reserve NPSH_a of 2.51 m is adequate

Efficiency

$$e_m\ (\text{worn}) = e_w\ (\text{new}) \times \text{ER} \times \frac{e_m\ (\text{worn})}{e_m\ (\text{new})}$$
$$= 64\,\% \times 0.79 \times 0.90 = 45.5\,\%$$

5.9.7 Power consumed at pump spindle

$$P = \frac{Q \times H \times S_m}{1.02 \times e_m\ (\text{worn})} = \frac{58 \times 13.2 \times 1.6}{1.02 \times 45.5} = 26.4\,\text{kW}$$

Allowing 10% for belt slippage the total required power at the pump spindle is 26.4 + 2.6 = 29 kW and a 45 HP TEFC synchronous motor would

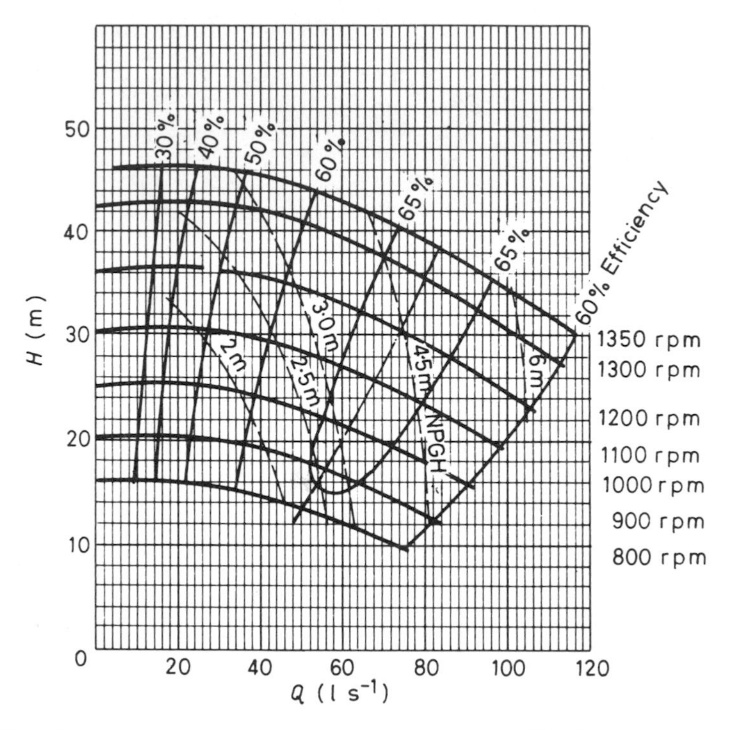

Fig. 5.22 Characteristic curves for Warman 6/4 EAH slurry pump

provide a suitable drive. The range of speeds required at various stages of impeller wear can be provided from a range of pulleys having different diameters.

5.9.8 Preliminary head-feed pump specification

Based upon the Warman Series A slurry pumps a pump would be selected similar to the Warman 6/4, Series A, Frame E, type AH, gland-sealed slurry pump with five-vane closed-type rubber impeller complete with 35 kW (45 HP) TEFC, synchronous motor and vee-belt drive for pump speed of 950 rpm in worn condition.

5.10 TAILINGS – JET STACKING

Tailings from a dredger working into a bank of material above pond level must be stacked to at least an equivalent height behind the treatment barge to safeguard against them flowing back into the pond at a faster rate than the dredger advances along its mining path. The angle of repose of the waterlogged tailings is usually around 24° as illustrated in Fig. 5.23. It is

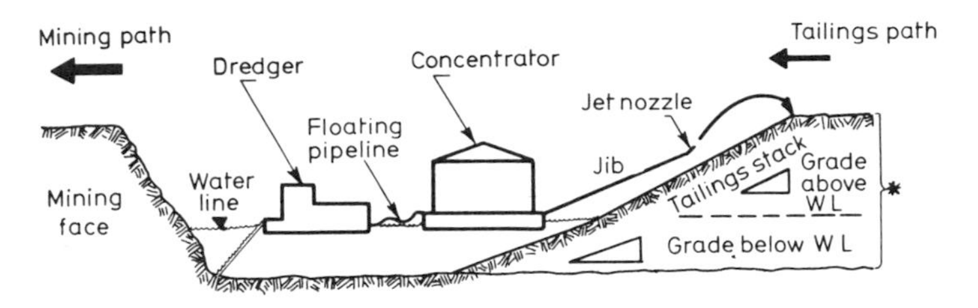

Fig. 5.23 Schematic arrangement for suction dredging

slightly steeper in the drier material above the water line and if slumping occurs too close to the barge it may be sanded in. Important jet design features illustrated in Fig. 5.24 are:

(1) The horizontal distance L from the nozzle.
(2) The vertical height H of dumping above the nozzle.
(3) The direct linear distance Y between the nozzle and the point of impact.

From Fig. 5.24:

$$L = \frac{V_e^2}{2g} \times \sin 2\theta \tag{5.22}$$

$$H = \frac{V_e^2}{2g} \times \sin^2 \theta \tag{5.23}$$

$$Y = \frac{V_e^2}{2g} \times \sin \theta (4 - 3 \sin^2 \theta)^{0.5} \tag{5.24}$$

where V_e is the nozzle exit velocity (m s^{-1}) and θ is the angle of inclination of the longitudinal axis of the nozzle above the horizontal (degrees).

The theoretical values of L, H and Y for various values of $V_e^2/2g$ may be calculated from Table 5.9.

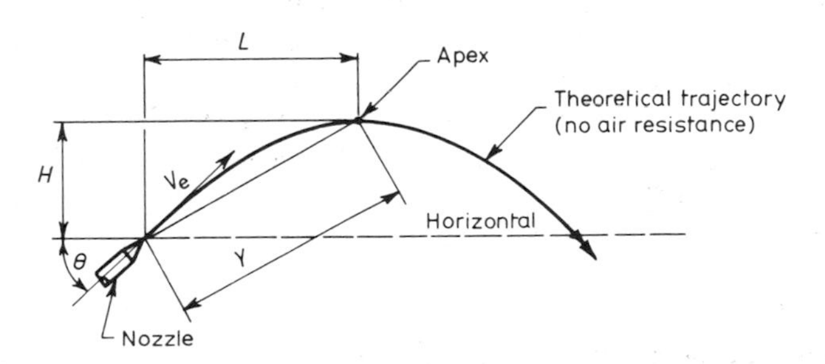

Fig. 5.24 Theoretical trajectory of stacking jet

Table 5.9 Tabulated data for calculation of theoretical values of L, H and Y

$\theta°$	For L $\sin 2\theta$	For H $\sin^2 \theta$	For Y $\sin \theta\,(4 - 3\sin^2 \theta)^{0.5}$
30	0.866	0.250	0.901
45	1.000	0.500	1.119
50	0.985	0.587	1.148
54.6	0.945	0.666	1.154
60	0.866	0.750	1.145
70	0.644	0.884	1.090
90	0	1.000	1.000

According to Warman Equipment experience, a maximum inclination of $\theta = 50°$ is usually adequate and very much steeper inclinations are seldom appropriate because of the nature of the stacked tailings and their angles of repose.

5.10.1 Practical considerations

Jet stacking is effected, normally, by running the tailings pipe along a jib extending from the stern of the treatment barge. The tailings are pumped through this pipe to give a suitable nozzle exit velocity for safe stacking. The nozzle exit velocity V_e is designed to place the tailings at a satisfactory height and distance from the nozzle and barge and the jib may be inclined at an angle just slightly greater than the upper tailings angle of repose. The use of jets avoids the need for much longer discharge pipes which would require heavier and more costly construction for both the jib and the barge super-structure and hull.

The jet

Given a well-designed nozzle and an exit velocity of up to about $30\ \mathrm{m\,s^{-1}}$, the slurry stream maintains its cohesiveness in still air for most of its rising trajectory until it reaches a point approximately 70% of the distance to its actual apex. At this point the slurry stream commences to break up into discrete droplets; it forms a wider stream which meets with more air resistance and, as illustrated in Fig. 5.25, rapidly departs from the theoretical trajectory (Fig. 5.24).

The falling trajectory beyond the apex is increasingly dominated by gravitational forces and the stream reaches a horizontal distance of only about 60% of the theoretical distance A in the horizontal plane. Theoretically $A = 2L$ in a vacuum.

Design features

Two or more nozzles can be used for stacking, as shown in Fig. 5.27, when large tonnages of tailings are to be handled or where the stacking heights

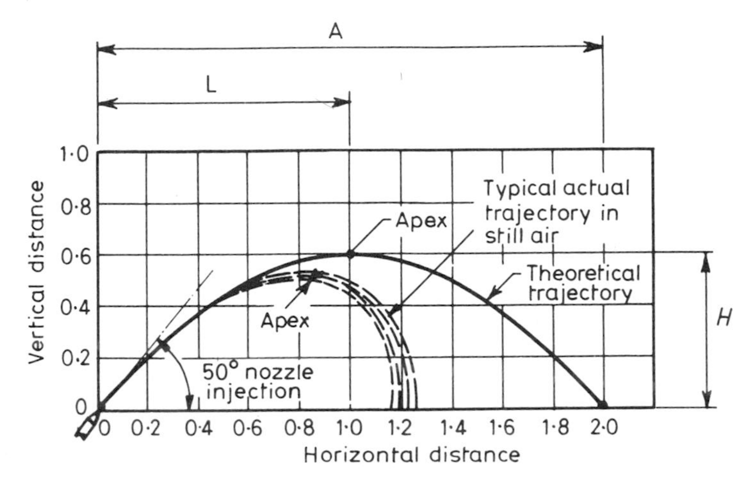

Fig. 5.25 Actual trajectory of stacking jet

are considerable. The jets are located at different elevations in order to build up the stack in layers and may also be operated in parallel at each point in order to distribute the residues from treatment over a wider area. By so doing, the frequency of the jib slewing movements is reduced and there is less danger from slumping.

Tailings bank gradients above and below the water line are dominant factors in selecting the jib angle, nozzle locations, exit velocities and jet trajectories. A sufficient margin should be allowed in each case to ensure that the bank *BC* in Fig. 5.7 will not foul the treatment barge hull at point *A* and approximately 20% in excess of the specified values of *L*, *H* and *Y* is usually allowed to compensate for:

(1) A progressive reduction in V_e as the nozzle wears and, hence, the exit diameter D_e increases.
(2) Additional V_e requirements to counter the effects of at least moderate winds unfavourable to the required trajectory, and to provide:
(3) Greater effective values of *L*, *H* and *Y* in order to reduce the frequency of jib relocation as the tailings stack advances.

Replacement tips provide easier and cheaper nozzle maintenance and alternative tips of varying diameters can be used to vary V_e if conditions change. Optimum conditions are achieved when L_d and Z_d are such that the dynamic head *H* is at a practical minimum.

Slurry density

The required value of C_W for most mineral sands tailings is around 65% of solids by weight. At this relatively high value the water contained in the mixture can be readily absorbed by, and percolate through the tailings

Fig. 5.26 Slurry stream from jet stacker showing effects of air resistance

bank, thus allowing a progressive build up of residues, without being effected unduly by the impact of the slurry stream. The disintegration and consequent spreading of this stream during its trajectory also helps to facilitate the absorption of the water and allows the system to handle, safely, a small proportion of slimes particles at high slurry concentrations. However, if C_W falls much below $C_W = 65\%$, or if the slimes content of the

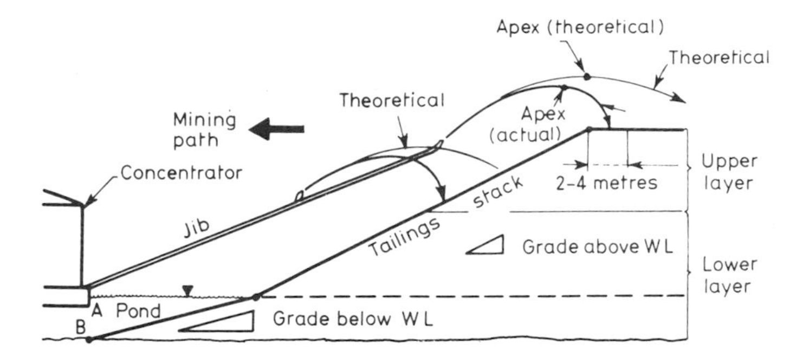

Fig. 5.27 Multi-jet stacking

slurry is excessive, the tailings bank will lack sufficient permeability and be unable to absorb the water quickly enough to avoid slumping. In some instances, provision must then be made for more efficient desliming during treatment; in most cases some preliminary test work is essential for determining the optimum value for C_W.

5.11 CONCENTRATE THICKENING

Heavy mineral concentrates produced on board a floating treatment plant are pumped ashore in the form of a slurry and thickened for stacking and air drying. A dewatering cone is sometimes used for the purpose but most installations are provided with a rubber-lined hydraulic cyclone suspended by brackets from a gantry, or attached to the end of a concentrate stacking jib. Cyclone variables (see Fig. 5.28) are the area of the feed inlet, the vortex finder diameter and length and the spigot discharge diameter. Operational variables are the feed inlet pressure, feed slurry density, pressure drop through cyclone, feed flow rates and solids characteristics.

5.11.1 Principle of operation

A dewatering cyclone relies for its action on centrifugal forces generated by the high tangential velocities of the flow. The incoming slurry forms a primary vortex around the inner cone wall and, in this zone, the tangential component increases away from the cone wall towards the centre of the cyclone, reaches a maximum and then decreases rapidly to zero. Thickened solids are discharged at the spigot and the effluent forms a smaller vortex around the axis of the cyclone.

Vertical components of the velocity act downwards near the cone walls and upwards near the axis. An intermediate zone or envelope, having zero vertical velocity, separates the coarser solids moving downwards to the

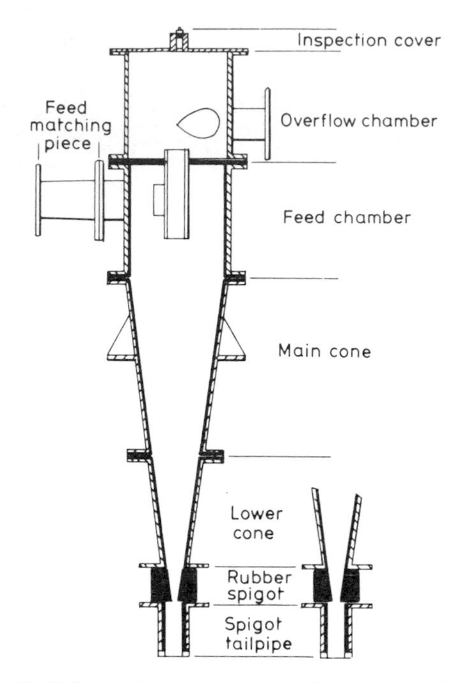

Fig. 5.28 Schematic arrangement for cyclone dewaterer

discharge spigot from the effluent containing finer solids moving upwards. Because of the reduced size of the inner vortex and its requirement to pass the bulk of the flow there is an increase in circumferential speed and higher centrifugal forces are generated resulting in a more efficient separation in the finer sizings. As a result, the rejected fine particles larger than the size of separation are returned to the primary vortex and once more have an opportunity to be discharged with the spigot product.

5.11.2 Cyclone efficiency

The efficiency of a cyclone is usually expressed in terms of a d_{50} sizing, i.e. the diameter of a particle which has an equal chance of reporting with either the underflow or the overflow. Particle density is also a factor in the dewatering of heavy mineral concentrates and, for such materials the optimum d_{50} sizings are concerned more with making high recoveries of the valuable particles than with a high level of thickening. One indication is the form of the discharge and optimum recoveries are made when the underflow leaves the spigot as a 'spray' discharge. The discharge is ropey only if the spigot has been overthrottled, thus sacrificing excessive amounts of above separation size solids into the overflow. Separation efficiencies are very sensitive to changes in pressure and to unfavourable pressure gradients and the efficiency of thickening recoveries in one pass often falls short of design predictions. For this reason it is usually prudent to return the overflow to the concentrating circuit in order to minimize losses of valuable minerals.

5.11.3 Cyclone duty

The current exercise assumes the treatment of 200 TPH of sands at, say, 2% of total heavy minerals to produce a final concentrate of, say, 80% of total heavies. The quantity of concentrates to be thickened will be:

$$200 \times \frac{2}{100} \times \frac{100}{80} = 5 \text{ TPH of heavy mineral concentrates.}$$

This quantity is to be pumped to a dewatering cyclone at a slurry density $C_W = 25\%$ for separation at $d_{50} = 25\ \mu m$

5.11.4 Cyclone selection

A preliminary selection is made by referring to the Warman Selection Chart (Fig. 5.29). For 5 TPH solids at $C_W = 25\%$, the model 6R Cyclone will give a $d_{50} = 25\ \mu m$ separation at an inlet pressure of around 140 kPa, and there is sufficient flexibility to reduce the d_{50} cut off to 17 or 18 μm by increasing the inlet pressure to 400 kPa.

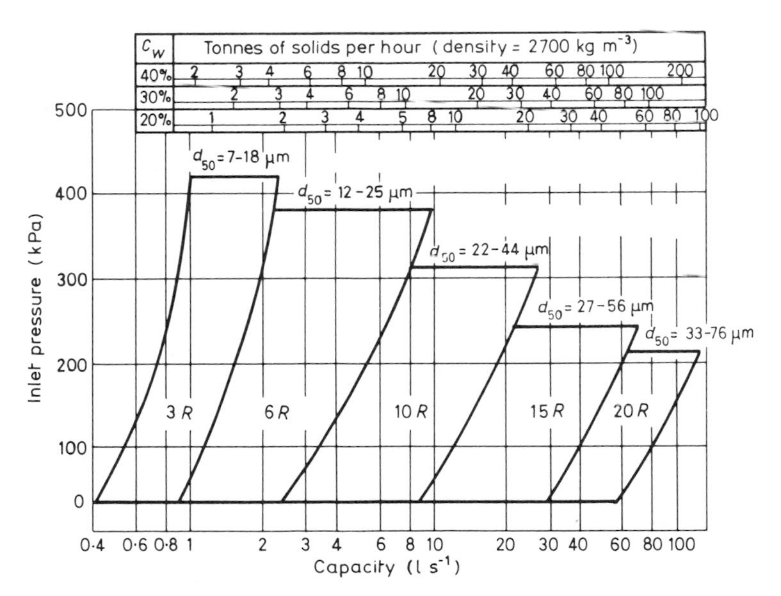

Fig. 5.29 Typical cyclone separation chart for d_{50} separations up to $C_W = 25\%$ (RD 2.7)

Manufacturers' charts for cyclone performances, as for slurry pumps, are based upon one main type of feed. For pumps it is clear water; for cyclones it is assumed that the solids in the flow will have densities closely around 2700 kg m^{-3}. In practice the valuable solids may have densities ranging from about 3500 kg m^{-3} up to more than 8000 kg m^{-3} and, consequently, such charts are useful only as a first guide for selection. The final parameters for thickening are then determined experimentally. Provided that a cyclone is chosen around the middle of the required range, the test work can usually be done when operations commence.

REFERENCES

Anon (1969) Hydraulic transport of minerals in pipelines, *Mining and Minerals Engineering*, October, 25–30.

Bonnington, R. and Denny, D. F. (1959) Research report, 526, *Brit. Hydro-Mech. Res. Assoc.*

Cabrera, V. R. (1979) Slurry pipelines: theory, design and equipment, *World mining* (part I of II), January, 56–64.

Durand, R. and Condolios, E. (1952) The hydraulic transportation of coal and solid material in pipes, *Proc. Colloq. on Hyd. Transportation*, London, p. 44.

Erickson, Ole P. (1956) Dredger pumps, *Dock and Harbour Authority*, August 133.

Hattersley, R. T. (1960) *Swirl at Suction Inlets*, Water Research Laboratory Report No. 23, University of New South Wales, Sydney.

Johnson, Glenn (1975) Dredge pump design. A modern approach, *World Dredging*, March 20–4.

Macdonald, Eoin H. (1965) *Factors Affecting the Dredging of Beach Sands and Other Unconsolidated Sediments*, University of New South Wales, ME Thesis.

Macdonald, Eoin H. (1966) The flow of slurries at suction pipe inlets, *AIMM Proc. 220* Dec. 33–47.

Pekor, Charles (1973) Advances in pump application, *World Dredging and Marine Construction*, October, 18–20.

Peterka, A. J. (1953) The effect of entrained air on cavitation pitting, in *Proc. Minnesota Int. Hyd. Convent* (ed. Hunter Rouse), 507.

Ross, W. J. (1961) Pumping slimes, *7th Comm. Min & Met. Congress, Rhodesia.*

Scheffauer, F. C. (1954) *The Hopper Dredge*, Office of the Chief of Engineers, US Army, Washington DC.

Sheehy, G. D. (1976) Submersible dredge pump – an answer for deep dredging, *World Dredging*, March, 25–7.

Smith, R. A. (1955) Experiments in the flow of sand – water slurries in horizontal pipes, *Trans. Inst. Chem. Eng.* **33**, 2, 22–95.

Wasp, E. J., Kenny, J. P., and Gandhi, R. L. (1977) Solid liquid flow, slurry pipeline transportation, *Trans. Tech. Pub.*, Claustal, Germany.

5.12 GLOSSARY OF SYMBOLS

Notes: 1. The term 'mixture' refers to a slurry mixture of solids and water.

2. The subscript m refers to a head, efficiency or other factor relating to the pumping of a mixture as distinct from the subscript w for pumping water.

3. The relative density of a mixture or slurry S_m is made up of the relative densities of the solids S and the water S_w.

Symbol	*Description*
C_g	Concentration of solids in mixture expressed as grams of solids per litre of mixture ($g\,l^{-1}$)
C_V	Concentration of solids in mixture by (true occupied) volume per cent
C_W	Concentration of solids in mixture, by weight per cent.
D	Inside diameter of pipe (m)
D_S	Inside diameter of suction pipe (m)
D_d	Inside diameter of discharge pipe (m)
d_{50}	Average particle size of solids in a dry sample. This size is equal to the screen aperture which would retain exactly 50% by weight of the total sample (mm or μm)
ER	Efficiency ratio: e_m/e_w for the same flow rate and pump speed (dimensionless)
e_w	Efficiency of pump when pumping water (per cent)
f	Darcy Friction Factor (dimensionless)
g	Gravitational acceleration 9.81 $m\,s^{-2}$
h	Head symbol for sundry purposes, each being individually defined and the units stated
H	Total dynamic head required by a system: head of mixture (m)
H_{atm}	Atmospheric pressure at pump location: expressed as head of mixture pumped (m)
H_d	Total discharge head: head of mixture (m)
H_f	Friction head loss: head of mixture (m)
H_{fd}	Friction head loss in discharge pipe: head of mixture (m)
H_{fs}	Friction head loss in suction pipe: head of mixture (m)
H_{gd}	Discharge gauge head (above atmospheric pressure) of mixture at the pump discharge tapping point, corrected to pump centre line: head of mixture (m)
H_{gs}	Suction gauge head (positive) value if above atmospheric pressure or (negative) value if below atmospheric pressure of mixture at the pump suction tapping point, corrected to the pump centre line: head of mixture (m)
H_i	Inlet head loss at junction of supply vessel and suction pipe: head of mixture (m)
H_m	Total dynamic head developed by pump when pumping mixture: head of mixture (m)
HR	Head ratio = H_m/H_w for the same flow rate and pump speed: dimensionless
H_s	Total suction head: (positive) or (negative): head of mixture (m)

H_{vap} Absolute vapour pressure head of *suspending liquid* at pumping temperature: head of mixture pumped (m)
H_v Velocity head at any point of valuation: head of mixture (m)
H_{vd} Velocity head in the pump discharge pipe at discharge tapping point: head of mixture (m)
H_{ve} Exit velocity head loss at final discharge from pipeline: head of mixture (m)
H_{vs} Velocity head in pump suction pipe at tapping point: head of mixture (m)
H_w Total dynamic head developed by the pump when pumping water: head of water (m)
L Total equivalent length of pipe $= L_a + L_f$ (m)
L_a Total actual length of pipe (m)
L_f Aggregate of equivalent lengths for all valves, bends and fittings contributing to friction head loss in pipeline: (m)
L_s L for suction pipe: $L_s = L_{as} + L_{fs}$ (m)
L_d L for discharge pipe: $L_d = L_{ad} + L_{fd}$ (m)
M Mass flow rate of dry solids (TPH)
n Pump rotational speed (rpm)
$NPSH_a$ Net positive suction head available at pump sunction flange: head of mixture (m)
$NPSH_r$ Net positive suction head required at pump suction flange: head of mixture (m)
NR Reynolds Number (dimensionless)
P Power consumed at pump shaft (kW)
P_e Absolute pressure (Pa)
P_r Pressure (Pa)
Q Mixture flow rate ($l\,s^{-1}$)
S Relative density of dry solids ρ_s/ρ_w (dimensionless)
RD Density relative to water 1.00 ρ/ρ_w (dimensionless)
S_m Relative density of mixture ρ_m/ρ_w (dimensionless)
V Average velocity of mixture in a pipe ($m\,s^{-1}$)
V_d V in discharge pipe ($m\,s^{-1}$)
V_s V in suction pipe ($m\,s^{-1}$)
Z Net static head: vertical height from end of pump discharge pipe to mixture supply surface level, positive if end is above supply surface level and negative if below (m)
Z_c Differential column head: head of mixture (m)
Z_d Static discharge head: vertical height from end of discharge pipe to pump centreline, positive if end is above pump centreline and negative if below (m)
Z_l Vertical height of suction pipe conveying slurry and surrounded by a liquid of RD lower than that of the mixture pumped (m)

Z_s Static suction head: vertical height from mixture supply surface level to pump centreline, positive if end is above pump centreline and negative if below (m)

Z_{sm} Effective positive static suction head above (positive) pump centreline: head of mixture (m)

μ Dynamic viscosity (Pa s)

ν Kinematic viscosity ($m^3 s^{-1}$)

ρ Density ($kg\,m^{-3}$) of fluid (rho)

ρ_m Density of mixture ($kg\,m^{-3}$)

ρ_s Density of dry solids ($kg\,m^{-3}$)

ρ_w Density of water ($kg\,m^{-3}$)

τ Shear stress (Pa)

γ Specific weight or weight per unit volume: $\gamma = \rho g$ (dynes)

6

Placer Mining

6.1 INTRODUCTION

Placer mining is essentially an exercise in large-scale earth moving using plant and equipment that is also employed in such other undertakings as land reclamation, the dredging of harbours and waterways, road construction and quarrying. The basic systems and methods used are also the same, only the objectives differ, and there is a need for greater precision when mining along the boundaries of deposits and in cleaning up at bedrock. The valuable constituents in a placer are rarely distributed evenly and, while the bulk of the material (overburden) can be removed as in any other undertaking, the pay materials must be taken up separately and fed to the treatment plant at a required rate and in a designated form.

As a result, some aspects of placer-mining economics are concerned more with selectivity than with the actual quantities moved per unit of time and, since the ultimate purpose of mining is to make a profit, maximum benefits accrue to operators who plan every aspect of an operation in detail while still retaining sufficient flexibility of mind and the ingenuity to deal with the unexpected. Vast sums of money have been spent on improving machine design and capabilities and the miner has an extensive array of models to choose from; the task is to select the best of the alternatives for each particular duty. Governing factors are concerned with how well a particular machine or combination of machines is likely to adjust to conditions that may in some ways be different to those for which they were designed.

Other aspects of increasing importance relate to the environmental impact of mining. There are now general requirements in many areas for the overburden to be backfilled along with the treatment-plant residues; for the topsoil to be stacked separately and replaced as required; and for the surface of the land to be re-contoured after mining is finished in order to return it to much its original form. The additional costs for environmental protection are minimal if the ecological requirements are catered for from the start; they can be very considerable if only tacked on at the end.

6.2 MINING SYSTEMS

The first consideration is whether to mine wet or dry. In some cases there may be no reasonable alternative to a system that is wholly wet (e.g. offshore dredging), or wholly dry (e.g. desert mining). At other times there is a choice and some methods may incorporate elements of both systems (e.g. dry mining followed by dredging or hydraulic sluicing). Table 6.1 compares the main features of wet- and dry-mining systems for placers.

6.3 MINING METHODS

Each placer has some distinctive features and, hence, each requires separate study to determine by what method it can be most economically mined. Basic factors influencing the choice are seen from Table 6.1 to be mainly (a) availability of water, (b) deposit size and value, (c) sediments and minerals characteristics, (d) bedrock characteristics and geometry, and (e) the total environment.

6.3.1 Availability of water

Depending largely upon the amount of water present in a deposit area, placer-mining systems feature such diverse methods as bucket-line and suction dredging, hydraulic sluicing, hand mining and a variety of cyclical methods of mining, both wet and dry. Offshore and in ground that is wholly covered or saturated by water, placers are mined by dredging and the choice lies between some form of continuous bucket line or hydraulic dredging, and a cyclical method such as clamshell dredging. Hydraulic sluicing is a logical alternative for mining deposits that are capable of being drained or otherwise dewatered, provided that ample quantities of water are located at higher levels nearby. Desert conditions call for a full range of dry-mining methods using bulldozers, scraper-loaders, hydraulic excavators, draglines, drum scrapers, articulated front-end loaders and bucket-wheel excavators. Where adequate and controllable amounts of water are present, practically all of the methods compete and the final selection is based upon economics.

Table 6.1 Dry versus wet mining systems

	Dry mining	Wet mining
Applications	Shallow surface deposits, tightly compacted or indurated sands, irregular geometry, high level dunes, desert environment.	All environments where ample water is available for mining and treatment including shallow surface deposits, high-level dunes, marine deposits.
Systems built around	Bulldozers, scraper loaders, articulated front end loaders, draglines, hydraulic excavators, bucket-wheel excavators.	Pumps and monitors, suction and bucket dredgers, bucket-wheel dredgers, clamshell dredgers, jet-lift dredgers, hydraulic excavators.
Controlling factors for selection	Proposed scale of mining, minerals distribution and value, location and physical characteristics, slope and texture of mining floor surface and bedrock geometry, availability of water.	Proposed scale of mining, deposit size and grade, location and physical characteristics, slope and texture of mining floor. Bedrock geometry, amount of water, position of water table.
Advantages	Ability to handle group of small deposits, constant feed rate under widely differing mining conditions, selective mining leads to optimization of feed-grade control, recoveries may approximate 100%.	Mining and processing can be incorporated in one unit, low unit mining costs, closer supervision and control, only possible method in excess water conditions.
Disadvantages	High unit operating costs, difficulties in handling occasional large volumes of water, requires firm base for vehicle movement, needs large on-site workshop facilities and stock of spare parts.	Mining losses sometimes high, less selectivity in mining, possible high relocation costs, high capital costs, large water requirements, ecological problems may affect large sections of environment.

6.3.2 Deposit size and value

Size and value are dominant factors in determining how much capital can be invested safely and each level of capital costs must be balanced by a proportionate level of production. It is essential that appropriate revenues flow in for a sufficient number of years to provide for the return of all capital plus a fair margin of profit (Chapter 8). Bucket-line dredgers cost most and, for economic viability, minimum volumes of average ground range from 10 million m^3 for small dredgers up to 120 million m^3 for the larger ones. Hydraulic suction-cutter and bucket-wheel dredgers, being cheaper to construct although more limited in their applications, may be viable for mining deposits with total volumes as low as 250 000 m^3 in suitable conditions. Very small deposits are usually mined by hand.

6.3.3 Sediments characteristics

Sediments characteristics are functions of the environments within which placers are laid down and, for each set of conditions, there is usually one method of mining that is superior to all others. Gold, platinum and cassiterite deposits are normally associated with long size-range sediments and bucket-line dredgers are used almost exclusively for mining large deposits of such materials. Smaller deposits may be mined by some cyclical earth-moving method; a common requirement is the ability to apply high break-out forces and to withstand heavy shock loadings.

Large deposits of fine-grained sediments such as beach sands require less effort for displacement and are more economically mined using suction-cutter or bucket-wheel dredgers. Costs are higher when other factors dictate the use of dry methods of mining in preference to dredging and attempts made to mine high-level dunes at Cape Morgan, South Africa and on Stradbroke Island Queensland by hydraulic sluicing were even more costly because of power and labour intensiveness.

Features of overburden sediment characteristics are related to the ease with which they can be ripped by heavy earth-moving equipment or cut by buckets, cutter heads etc. when dredging. The slimes content may influence the selection of a dry-mining method in some circumstances and one method might be preferred to another because it provides safer working conditions in deep pits or because of differences in the competence of footings. Overburden handling constitutes one of the most important factors in mine planning and a particular method might be chosen for its ability to cope with specific disposal or restoration problems.

Underground methods are used for mining deposits that are overlain by thick cappings of consolidated sediments that cannot be ripped or blasted economically; or where, for some other reason, the stripping ratios are uneconomic and open-cast methods cannot be applied.

6.3.4 Bedrock characteristics and geometry

The most favourable bedrock conditions are provided by flat, even surfaces sufficiently weathered to be easily dug or scraped. Valuable heavy minerals accumulate both at and just below the gravel–bedrock interface and bucket-line dredgers and other mining units equipped with positive digging mechanisms make high recoveries by cutting into the weathered rock and recovering particles lodged in cracks, open pores and joints.

Suction dredgers are less able to take an even cut across and into a soft bedrock, however, an adaptation of the hydraulic mining method, bucket-wheel dredging, may overcome this disability.

The least favourable bedrock conditions are imposed by hard, uneven surfaces covered by clusters of large boulders. All dredgers find difficulty in handling such conditions and, in some cases, diver-operated jet-suction dredgers may be used as auxilliary units for cleaning up. In extreme cases where pits can be dewatered, hydraulic sluicing may be the only feasible method. Sluicing operations can be adapted to cope with any degree of unevenness of the bedrock and bulldozers can be used in the pit for pushing the boulders aside.

6.3.5 The total environment

Climatic conditions are dominant factors in the selection of mining plant in some environments. Machinery and electrical service equipment must be able to withstand the effects of dampness and humidity in monsoonal regions. Such conditions also restrict the choice of machines because of differences in work availability in wet ground. As already observed, tracked machines are preferred to wheel-type units on very soft, uneven or wet clay surfaces; normal conditions in humid tropical climates. Machines that dig downwards from a bank rather than from the pit bottom usually give higher all round performance figures where ground-water has to be controlled and where there is danger of periodic flooding. The tropic proofing of all electrics is essential in humid conditions, regardless of ambient temperatures and a high degree of mobility is needed for machines in cyclone areas particularly along low-lying coastlines subject to inundation from storm surges (e.g. along the coastline of the Bay of Bengal).

Table 6.2 summarizes the main criteria for selecting individual mining methods.

6.4 MINE PLANNING

Mine planning covers a range of duties at two distinct levels. Planning at the first level is simply to establish broad parameters for evaluation costing and the data are sufficiently comprehensive when it is possible to determine a suitable method and scale of mining and a favourable sequence for extraction and restoration. At this level, the planner is concerned primarily with statistical probabilities and a sufficient number of samples has been taken when individual errors cancel out closely (see Chapter 8).

The second level of planning is concerned with the actual mining

Table 6.2 *Factors influencing selection of placer mining methods*

Mining System	Dredging				Hydraulic mining	Dry mining	Hand mining
	Bucket	Suction	Jet lift	Hydraulic excavation			
Minimum volume (m^3) to justify operations in average values	10 million to 120 million	250 000 land 10 million sea	100 000	500 000	100 000	1 million	Any small quantity
Preferred nature of basement	Soft and even, few hard pinnacles or bars	As for bucket but more tolerant	Less critical than other forms of dredging	Soft and even but can mine to hard bottom	Soft preferred but can handle hard	Soft and even capable of supporting heavy traffic	Hard or soft
Nature of mineralized beds	Reasonably loose with few large boulders	Unconsolidated gravels and sand	Unconsolidated gravels and sand. Boulders can be avoided	Unconsolidated gravels and sand. Boulders can be pushed aside	Can be broken and fluidized using jets	May have small degree of consolidation	Preferably soft but not critical
Preferred nature of overburden	Unconsolidated	Unconsolidated	Unconsolidated	Unconsolidated	Capable of being ripped or broken by jets	Rippable	Preferably soft but not critical
Valuable mineral type: Group 1 densities 4.6–21.0; Group 2 Densities 3.2–4.6	Group 1	Group 1 offshore Group 2 on shore	Sand and gravel offshore. Selectively mining gravels under cover of silt	Group 1 offshore Group 1 and 2 on shore	Group 2 minerals on shore Rarely Group 1	Group 1 and 2 minerals on shore	Group 1 minerals on shore
Water requirements	Large	Large	Large	Variable to large	Large	Nil	Variable
Bottom slope	Relatively flat preferably ≯ 1:40 for artificial ponds	Relatively flat	Relatively flat	Not critical	Any degree of slope but preferably around 5°	Not critical	Not critical
Ocean conditions	Maximum wave height 1.5 m	Maximum wave height 1.5 m	Depending on vessel	Not applicable	Not applicable	Not applicable	Not applicable
Environment	Marine Continental	Transitional Marine	Marine	Continental	Continental	Continental Transitional	Continental

operation and the data require to be much more complete. Differences in local features such as the abrupt widening and narrowing of channels, junctions with other channels and changes in bedrock characteristics and geometry assume much greater importance. Individual sampling errors are not cancelled out but, instead, are reflected at every sampling point and, even when pit control is handled on a day to day basis, serious errors may occur unless all such features are identified well in advance.

Some of the most important consequences of faulty or inadeqate mine planning are:

(1) Faulty evaluation. Planning should commence with the first inflow of data from prospecting and be continued throughout the valuation phase until basic issues are resolved beyond any reasonable doubt. If there is a doubt or if the data are wrong, the whole enterprise might fail when put to the test.
(2) Double handling. Every load extracted from the ground should have its ultimate destination planned in advance in order to minimize double handling. An all too frequent error is to stack overburden on ground that has not already been tested and which is later found to contain payable values.
(3) Losses of valuable minerals and excessive dilution. Both of these faults result from wrong boundary definition. Ore limits are seldom regular in either the vertical or horizontal plane and a close drilling pattern is required ahead of mining for optimum selectivity.
(4) Mechanical damage. Mechanical damage to bucket lines, cutter heads etc. is inevitable when operations extend into ground containing previously unidentified rock bars or clusters of boulders.
(5) Electrical failures. Damage and excessive downtime result from inadequate tropic proofing, faulty wiring and overload protection and lack of maintenance planning.

Such consequences can usually be avoided, or at least minimized, by efficient planning and the test is for both investment and production targets to be met while still retaining an adequate amount of reserve capacity for unscheduled stoppages and adequate spares for all normal contingencies. Emergency measures taken to compensate for a lack of planning do not solve the basic problems. For example, if because of excessive dilution, additional material must be processed to achieve the desired output of valuable products, unit operating costs will be higher and profits will be lower. If mill feed grades can be held at design levels only by mining more selectively from time to time or by avoiding difficult patches of ground, the result will be a reduction in both mine life and overall profitability. In no case is it satisfactory for one target to be met at the expense of another.

The practicalities of mine-plan design are influenced strongly in most cases by the need to provide for a quick return of capital, particularly in politically sensitive areas. Hence, it is usually desirable to establish high cash flows in the first few production years as illustrated in Chapter 8. However, national interests may, in some cases, put social values ahead of economic gain and the order and rate of mining will then be determined accordingly.

Wet-mining systems are concerned with dredge path design; dry-mining systems operate according to normal open-cast design considerations similar to those of other surface-mining operations; ore zones in high-level dunes have specific design problems associated with their spatial relationships.

6.4.1 Dredge path design

Dredgers normally work either longitudinally upstream parallel to the principal axis of the deposit or across the deposit in transverse strips and one or more dredging units may be used to cope with changes in ore grades and types, or to achieve the required production levels. Unit operating costs are generally lower for one large dredger than for a multiplicity of smaller units but many other factors such as dredging depths, overburden to ore ratios, bedrock conditions etc. also influence costs.

Dredging across a placer has advantages for mining wide deposits in which the values are likely to extend beyond the assumed boundaries. The main disadvantage, according to Popov (1971) lies in leaving wedges between successive cuts so as to minimize contamination from tailings stacked along the previous cuts. He suggests that up to 5% of the total reserves might be lost for this reason and that, for narrow placers, there is a further disadvantage of a turning cycle of 4–10 h at each end. Such problems may, in some instances, be minimized by building dredgers with a capacity for side as well as end dumping. This is one feature of the IHC Holland 85 dredger described on pp. 357–58. Apart from discharging the tailings further downstream away from the next line of cutting, the ability to convert from port-side dumping to starboard-side dumping allows the dredger to swing around at the end of the run and to start cutting in the other direction in a much shorter time. Figure 6.1 describes the transverse system of dredging in cuts across an irregularly shaped placer.

Dredgers working longitudinally upstream may do so along several adjoining faces to cover the full width of the deposit as illustrated in Fig. 6.2. Each face is advanced from 10 to 15 m at a time, the transfer from one face to the next taking only 10–15 min. Tailings are stacked far enough behind the dredger to allow free movement. No intermediate pillars are left and, hence, there are no losses for this reason. The method finds difficulties in working outside of the planned boundaries and there is more room for error in attempting to dredge to the exact limits of a deposit. A large

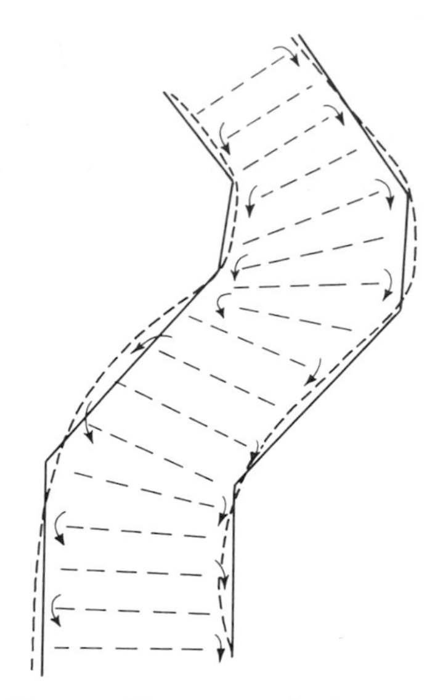

Fig. 6.1 Transverse dredge path

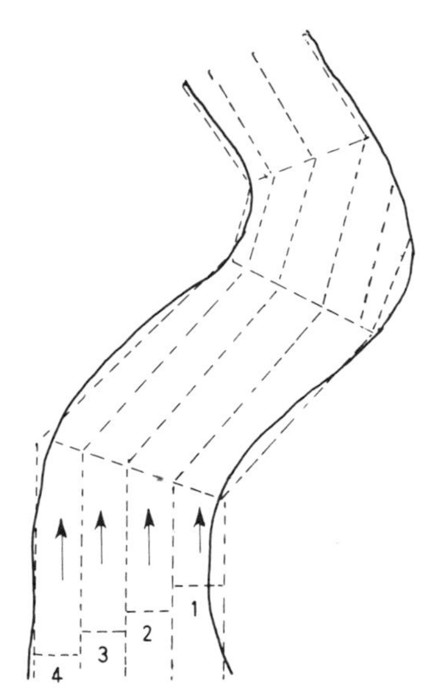

Fig. 6.2 Longitudinal dredge path in order of dredging

number of bow lines may also be needed for lateral movement (see also individual dredging systems).

6.4.2 Open-pit design

Placer dry-mining systems involve many similar features to those of other open-cast-mining systems and are subject to the same constraints. Side batters are determined by such factors as, the angle of repose of the wall material, bank height, seepage and the angles of any slip planes in the walls. The floor must be kept reasonably level and the road maintained in a trafficable condition for use by the haulage units. Deep pits in dry-mining operations are mined in benches and a variety of machines are available to cope with varying deposit characteristics.

Hydraulic sluicing is the usual method for open-pit mining in wet conditions and planning for such operations is concerned mainly with safety and mobility. For safe working conditions, water must be supplied to the monitors at a sufficiently high pressure for them to be operated effectively from a distance of at least bank height back from the face. Mobility is a major consideration because the monitors must be moved forward progressively as the face recedes and the ground races (trenches cut into the bedrock) extended to take away the spoil. Work availability is limited by the time lost in shifting pipelines and modifying the high-pressure water reticulation system, particularly where bank heights are low.

6.4.3 High-level dunes

Mine planners designing pit operations in high-level dunes face different sets of problems. One of these is to fix individual ore zones precisely within the various elements of sand migration and then to predict the spatial relationships in three-dimensional models. The choice of a mining method lies between suction dredging and some form of dry mining and success depends more upon the skill and experience of the mine planner than possibly in any other placer environment. Dry mining in high-level dunes has lost favour in recent years because of increasingly high fuel and maintenance costs, although it might still be considered for deposits in which the ore bodies are spaced widely both in the vertical and horizontal planes or where selective mining might improve the economics of the whole operation, including treatment. Costs for dredger relocation must be weighed against costs for re-locating buried loaders and pump transfer stations. Dry mining might also be preferred where major slumping problems inhibit the use of such reagents as fly ash, bentonite or aquagel to limit pond water losses through seepage.

High-level dune mine planners, consider the merits of ore zones, individually at first, and then in groups. Complex relationships between ore and waste are examined three-dimensionally and treatment plant sites are selected so that their weighted positions relative to the pits provide optimum haulage and pumping conditions at the haulage or hydraulic transfer stations and at the concentrating plant. Careful forward planning is essential and short-term compromises may be necessary for long-term benefits to be realized.

6.4.4 Overburden handling

The decision to strip overburden from a deposit brings with it the responsibility for providing a temporary safe lodgement for the material and a suitable means for backfilling when mining is complete. Methods and equipment used will probably be similar to those proposed for mining the ore but in any case they must be compatible with such methods. Particular attention is paid to the following factors.

Deposit dimensions

As well as the average dimensions of length, width and depth, varying degrees of irregularity influence the choice of one method against another. For example, back-acting excavators strip much more selectively than do bulldozers or draglines but draglines would probably be preferred for large-scale overburden removal and bulldozers might be more effective for removing and stacking topsoil in the first stages of stripping.

Nature of footings

Machines must be able to cover the ground quickly and safely. Wheel-type loaders require a firm footing at all times; tracked types are more versatile but in very wet clayey ground, or in swampy conditions, the choice might be restricted to draglines or to some form of dredging.

Rate of removal

A wider range of machine types is available for large-scale stripping than for small-scale operations and, at intermediate levels, the advantages of using big-capacity machines such as walking draglines must be weighed against the attendant higher capital costs and lower utilization. Consideration must also be given to the nature of the overburden. For example, compacted materials and flow rocks can usually, although not always, be ripped but there are various degrees of rippability. What is easy for a Caterpillar D10 tractor to rip may be much too hard for a smaller machine and deposits that are otherwise similar except in scale might use quite different methods for stripping.

Order of extraction and replacement

The order of overburden extraction and replacement is important because almost all topsoil contains some humus and thus has a potential for vegetal re-generation that is too valuable to lose. Restoration standards are generally high in the mineral-sands industry where normal procedures involve stacking the topsoil and underlying waste material separately and, when backfilling, replacing the soil cover on top of the waste. The surface is then landscaped and sown with quick-growing rye or other grasses to provide temporary surface stability while the slower-growing natural species develop once more. In Australian beach-mining practice, soil replacement is carried out as quickly as possible after mining because most of the native species are revived, or their seeds germinate, if the soil is replaced within about 2 months from the time of stacking.

A similar awareness of the need for environmental protection is becoming evident in other sectors of the placer-mining industry, although the monitoring of operations in remote areas still tends to be lax. For example, bucket dredgers working along fertile river flats in some regions still reverse the order of re-emplacement by depositing humus bearing material, stripped from the surface, underneath gravels mined from the basement of the deposit. On the other hand, many operators are actually improving the cropping capacity of some swampy environments by leaving long ridges of residues above swamp level. In such conditions, gravel surfaces provide solid footings for access and adequate drainage for some agricultural purposes (see *Restoration*, pp. 358–60).

Problems of overburden removal and replacement vary according to the type of operation and the environment. Basically, all overburden and treatment-plant tailings should always be returned to the excavations made, except where more lasting benefits can be obtained by such means as providing fill for adjacent low-lying areas or materials for road surfacing. The over-riding consideration should be to optimize the value of all overburden disposal operations both in the present and for the future.

In beach mining, where the preservation of a beach depends upon maintaining the foreshore in equilibrium with the seas washing onto it, there may be no completely satisfactory means by which the frontal dune can be mined and restored to its previous geometry and stable state. Consequently, in view of what is now known of the possible widespread damage from broaching the frontal dune, mining should not be permitted except in the hind dune areas and, even then, the ground should be restored to a safe height above sea level. Remedial measures if taken promptly may arrest the spread of erosion and minimize the worst effects, however, rarely does a beach return to its original stable state.

Offshore and river dredging operations also have problems of tailings disposal because of disruptions to the existing patterns of sediment transport and deposition. Such mechanisms, in upsetting the balance between erosion and accretion in any section of the river or sea bed, also upset the biological and bio-chemical relationships and destroy marine life and spawning grounds. Navigation hazards are created when large quantities of sediments are mobilized and re-deposited in shallow waters. Resulting changes in the contours of the sea bed affect the energy balance along shorelines and lead to conditions that call for an entirely new set of equilibria. Major changes may occur from even comparatively minor variations in critical locations such as are found along narrow spits of sand projecting from headlands.

The relatively stable measure of equilibrium attained by littoral drift along most shorelines is readily disturbed by such outside influences as the removal of material from one sector and its replacement in another. Resulting changes in the distribution of energy from changing patterns of refraction and reflection commonly introduce fresh cycles of erosion; environmental disturbances are typified by:

(1) Beach erosion and property damage.
(2) The silting up of river mouths.
(3) The introduction of navigational hazards through the formation of offshore bars and shoals.
(4) Increased turbidity, from which the main deleterious effects are: reduced rates of photosynthesis which inhibit the growth of plankton and scale fish life: and the release of toxins from the sediments, thus overloading the filter-feeding organisms that would otherwise purify the system.
(5) Anaerobic conditions which are produced when benthic organisms are buried by sediments deposited on spawning beds or feeding grounds.

Environmentalists are naturally very concerned about these matters and mine planners and governments are now giving much greater thought to the future commercial needs than they have in the past. For example, most marine environmental impact statements now include detailed studies by specialist organizations concerned with all aspects of marine animal life and coastal sedimentation and it is interesting to note that the Thai Government has placed a moratorium on dredging in the Andamen Sea, west of Phuket Island, until an environmental study, due for completion in 1983, has been considered. No new leases or lease renewals are being issued in the interim.

If such concern persists, dredging may be almost universally forbidden in fishing and spawning grounds until offshore miners learn to rehabilitate the mined-out areas so as not to cause any lasting damage to the ecology. In any case, it can be expected that all future operations offshore will have to incorporate a means for continuously monitoring environmental changes and simultaneously comparing the effects with conditions in a control area. Particular attention will be paid to dredging operations in semi-enclosed embayments; these are more likely to be affected than where dredging takes place along active shorelines or in the open sea. In all cases, however, there will be a need to take compensatory actions as soon as local conditions deviate significantly from pre-determined levels.

River dredging has seldom reflected any credit on those responsible for protecting the environment. Common sights in worked out areas are swamps and piles of gravel and trash instead of pastures and farmlands. Stream channels are choked with slimes and other residues instead of being clean and free-flowing and clearly, there can be no excuse for such things to happen in the future. Although some lives might be disrupted for a time, proper planning and attention to environmental standards can ensure both immediate and long-term benefits that far outweigh any temporary inconveniences. If a project cannot support the small additional costs of providing an adequate level of restoration it should probably not go ahead.

6.4.5 Slimes disposal

Slimes are loosely defined as suspensions of finely divided solids in water. The definition, being imprecise, has different meanings in different circumstances. In some chemical plants, the relevant size range for solids might range from about 80% of $-20\mu m$ particles down to 70% $-2\mu m$. In placer technology the definition is often broadened to include any particles whose settling qualities are inhibited significantly by the slow settling of other, finer particles so that they are all carried away in the same suspension at low velocities.

For example, particles as large as 125 μm might be considered slimes particles in a diamond-concentrating plant while, in dredge ponds and slime dams, only particles in the Stokesian size range minus 62 μm, are regarded as slimes particles.

Reasons for the slow settling of fine particles are connected with Brownian movement and associated electrical repulsion between colloids. The practical limit is reached in a slimes dam when, following a progressive slowing down in the rate of settlement, no further settling is perceptible within the space of a human life. Figure 6.3 is a typical plot of settling rate versus time for such a suspension. In this particular case the rate became asymptotic when the suspension reached a concentration of about 30% of solids by weight, only after many years of settling.

In dredge ponds where slimes-sized particles are being added constantly in great numbers, the slimes level builds up rapidly unless adequate remedial actions are taken. Dredge ponds in humid regions such as in the

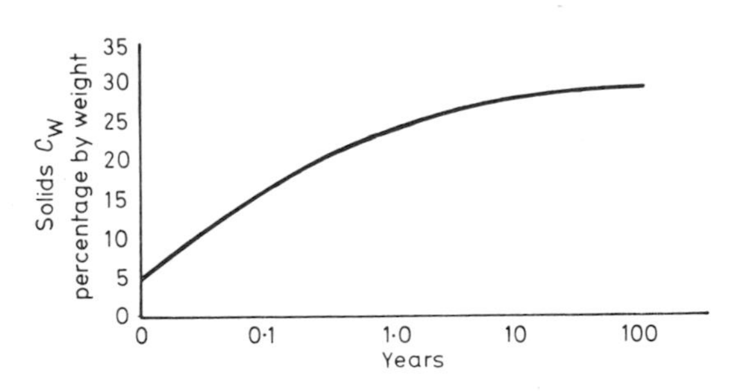

Fig. 6.3 Typical plot of settling rate for slimes

jungles of south-east Asia are particularly vulnerable to such effects and operations are quickly brought to a halt if the slimes are not removed at rates commensurate with their formation. Modern on-board treatment facilities use hydrocyclones to deslime the primary head feed to concentrators and the slimes are then pumped directly to slimes dams or into active stream channels where this is allowable, ecologically. However, some slimes-sized particles escape back into the dredge pond and, together with slimes that are formed by the flow of water over the buckets as they rise to the surface, some build up is inevitable in the pond. Normally in such cases, this build up is dealt with by continuously pumping away from the pond bottom using a slurry pump mounted on the dredger or on an independently floating barge. Equivalent volumes of fresh water are introduced into the system to take the place of the water lost in disposing of the slimes, usually as make-up water in wash-water circuits and in dressing operations that require fresh water for final cleaning.

Much research has gone into developing techniques for increasing both rates of settling and degrees of settling in attempts to minimize the basic problems and costs of slimes disposal which include pumping (sometimes over long distances), dam construction and maintenance, damage to real estate, public risk and environmental disruption. Many slimes dams remain hazardous for decades.

As a measure of the limited success achieved in improving the settling characteristics of slimes in recent years, the suspension referred to in Fig. 6.3 might now be thickened to a final slurry density of 40–45 % of solids by weight by introducing quantities of fine sand particles along with the slime. This subject is dealt with in more detail in Chapter 7.

6.5 HAND MINING

Hand methods are used widely throughout the world for mining small deposits in regions that are largely inaccessible to machinery and for cleaning up pockets of gravels remaining untouched from previous operations. Whole communities may be involved in hand-mining operations in large areas of provenance where innumerable small shallow deposits occur and labour intensiveness is a major consideration. Examples of such widespread activities are found in many parts of Africa, India, Asia and South America where regional economics have depended to a very large extent upon the hand mining of gold, tin and gemstone placers, sometimes for hundreds and even thousands of years and may continue to do so for a long time to come. Figure 6.4 shows tribesmen commencing a ground sluice for placer gold in Yakatabari Creek, Porgera, Papua New Guinea. A diamond-mining operation in the Birim River Flats, Ghana, is illustrated in Fig. 6.5.

Work output varies very much with the climate. A good strong labourer

Fig. 6.4 Commencing a ground sluicing operation for gold in Yakatabari Creek, Porgera, Papua-New Guinea

Fig. 6.5 Hand mining for diamonds, Birim River Flats, Ghana

in a temperate to cold climate can be expected to shovel between 6 m^3 and 8 m^3 of loose gravels per day if suitably motivated. This reduces to less than 1 m^3 per day in some tropical areas. Various types of sluices, rockers and pans used for recovering the valuable minerals have capacities that differ according to the ease with which the particles are liberated from their matrices of clay, sand and gravel. Some materials may break up as soon as they make contact with the water; others need a prior soaking before they can be slurried.

Pans used for concentrating valuable heavy minerals by hand vary from community to community. The European type pan is made of iron or some other metal. It is flat bottomed and circular with sides sloping outwards at about 30° from the base. Sizes range up to about 35 cm diameter at the base and 15 cm in depth. The South American batea is wooden, about 50 cm in diameter at the top, and conical in shape. The angle at the apex is around 160° and the depth is about 7–8 cm. In south-east Asia wooden pans are dish shaped; Indian pans feature a variety of shapes. Recoveries are effected in desert conditions through various combinations of screens, deck or drop boxes and bellows, known generally as dry blowers.

6.5.1 Ground sluicing

Ground-sluicing methods have been developed independently in many parts of the world over a long period of time dating back to antiquity. The usual method is to construct a dam across a watercourse above the area to be mined and to channel a stream of water along flumes leading to the pay gravels. This water is directed into a cutting in the gravels and material shovelled in from the sides is broken up and slurried manually. The valuable particles, generally of gold, are caught up in natural riffles formed by the unevenness of the bed or behind artificial riffles formed in wooden sluices built into the trench. The work proceeds in an upstream direction and the method is reasonably successful where steep slopes are available for tailings disposal.

Figure 6.6 describes the method of ground sluicing used by local miners in the Raigarth Gold Placer Field of Madhya Pradesh, India, where the banks of shallow streams are worked from a series of radiating sluices. Bunding provides a reservoir from which the water is directed along flumes to the head of the sluice. Bank material is shovelled into the sluice and puddled by hand to break up the clay. The slimes are washed away in the flowing stream and the gold is caught up in a series of separated sluice sections, the downstream sluices acting as scavengers. The flow is diverted back into the main stream during final clean up operations in which the gold is recovered by panning. When shovelling distances become excessive, a fresh sluice is dug along the base of the new bank face and the procedure is repeated.

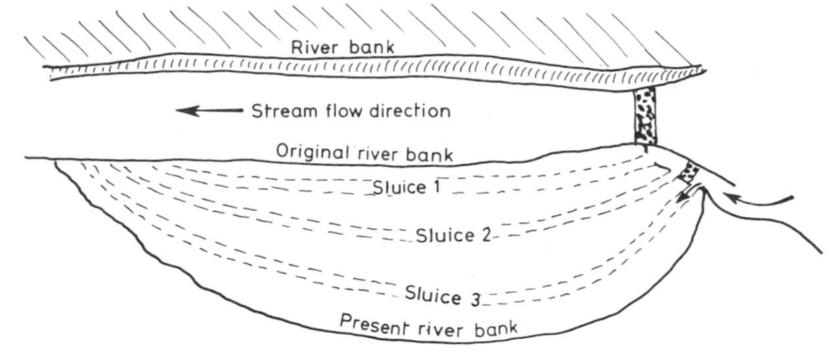

Fig. 6.6 Ground sluicing as practised in Madhya Pradesh, India

A variation of ground sluicing called 'booming' was first described by Pliny the Elder in reference to the methods used for gold washing in Spain during the first century AD. It is a generally inefficient process in terms of recovery of fine particles of gold, but has some application where the water supply is inadequate for continuous operations. The method employs a light gate, capable of being lifted or opened easily, built into the dam. Water filling the dam eventually overflows at one point into a container attached to one end of a long lever. When full, the weight of the container activates the lever and opens the dam gate, the water rushes out and scours the channel bottom. The water also spills out of the container and the dam gate then returns to its original position under its own weight. Booming has the advantage of allowing the miner freedom to work the dirt without the necessity of extending sluice boxes or tail races but it makes reasonable recoveries only of coarse gold.

Hand miners have learned, from experience, to control flume velocities below the critical erosion velocities for the particular materials over which they flow. Ground textures differ widely and wooden flumes are sometimes used when steep gradients cannot be avoided. Safe velocities for various ground conditions are generally as described in Table 6.3.

Table 6.3 Safe velocities in ditches

Lithology	Velocity ($m\,s^{-1}$)	Lithology	Velocity ($m\,s^{-1}$)
Silt	0.15	Stiff clays	1.50–1.60
Fine loam	0.20	Clayey gravels	1.60–1.70
Fine sand	0.35–0.40	Soft shale	1.80–2.00
Small gravels	0.60–0.75	Weathered granite	2.00–2.10
Large gravels	1.15–1.30	Hard rock	3.50–4.50

Earth flumes and ditches are usually constructed in trapezoidal section with sides sloping at some angle less than the angle of repose to avoid slumping. This angle may range upwards from 45° for the softer materials to 60° for hard compact ground conditions. Width should always exceed height to minimize the flow velocities and a common ratio of bottom width to height is 2:1.

Flow velocities in ground sluices should be just high enough to move the sand and gravel in traction (see Chapter 2) and to carry the non-valuable fines away in suspension. Wooden sluice boxes with riffles for treating gold alluvials are commonly given gradients of 1:12 to 1:10 or even steeper. Flat-bottomed palongs for processing cassiterite-bearing gravels are generally set on gradients of 1:30 to 1:20.

6.5.2 Miscellaneous methods of hand mining

Basic techniques are modified to suit the particular conditions in which mining takes place and a great deal of ingenuity is often displayed in utilizing available materials. At Pailin, Kamputchea, where sections of the gem fields are mined by hand from small pits, shafts and tunnels, below-surface workings are lined with woven matting and dewatered using hand-operated plunger pumps made from bamboo. Prospecting pans in most countries are made from local timbers; boulders are hollowed out and rounded stones are used as pestles to grind eluvial rock fragments and release the values. Figure 6.7 shows a bamboo and woven matting lined shaft in the gem fields near Pailin, Kamputchea.

Panning techniques differ according to custom and pan design and the techniques used along the Gold Coast of Africa have been described in Chapter 1. Basically, however, all methods rely upon a thorough puddling of the material by hand and agitation, using an oscillatory motion, to allow the heavier and denser particles to settle preferentially. A swirling motion of the pan under water continuously washes the top layer of light particles away until only a small amount remains. This rougher concentrate is a mixture of valuable and non-valuable heavy minerals with some lighter particles and no attempt is made at this stage to prepare a finished concentrate. This is done at the end of the day when the days accumulation of rougher concentrates is panned to a final product (Fig. 6.8). Only clean water is used for cleaning.

It is interesting to note that most wooden pans have one common characteristic, a large internal surface area and shallow depth. With no knowledge of boundary layer theory, but with a developed understanding of its effects, some 'primitives' may have developed hydraulically superior models to the metal prospecting pans used by European miners.

Panning for diamonds and other gemstones requires much more care than for gold because of their lower densities and rates of settling. The

Fig. 6.7 Collar of woven matting-lined shaft, Pailin gem fields, Kamputchea

Fig. 6.8 Panning of concentrates at the end of a days operations

rougher concentrates, in such operations, are not repanned but are transferred to hand-held sieves (called Crevus in Brazil) described in Fig. 6.9. The concentrates are upgraded by a combination of jigging and rotary movements so that both centrifugal and gravitational forces are employed in the concentrating process. Banks of screens, generally three in number, are used; common mesh numbers being 8, 16 and 28. The jigging action applied to the screens allows the undersize particles to fall from the centre of each screen into the next smallest. The gemstones are recovered from the central portion of each screen. Rectangular-shaped sieves are used in Ghana (Fig. 6.11) but these are less efficient.

The Australian aboriginal uses a wooden 'Yandy' hollowed out in the shape of a shield. Held in both hands and tilted slightly downwards towards one end, a double reciprocating motion is imparted that gradually works the heavies up slope while the lights pass downwards. The operation was originally developed for separating edible seeds from sand and husks but is now used widely in the desert areas for recovering gold and other valuable heavy minerals from near surface gravels (Fig. 6.12).

Many types of 'super-micro-macro' panners have evolved from the yandy; and pinched-sluice separators, now the vogue in mineral sands and other gravity concentrating applications, also date back to antiquity.

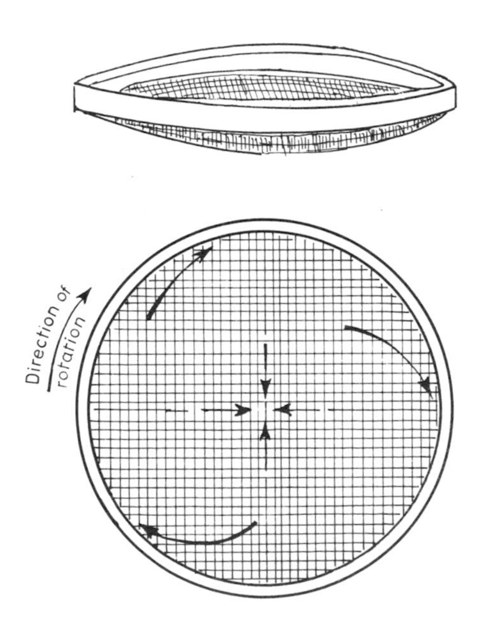

Fig. 6.9 Brazilian Crevu. Centrifugal force spins light materials to outside perimeter. Gravitational force attracts diamonds to the centre. Fines pass through the sieve

Fig. 6.10 Sieving for gemstones, Pailin Gem Field, Kampuchea

Fig. 6.11 Rectangular-shaped sieve used for panning diamonds in Ghana

Fig. 6.12 Australian prospector using a yandi for recovering cassiterite particles from eluvial deposit in Pilbara District, Western Australia

Figure 6.13 shows an Indian miner operating a pinched-sluice separator while Fig. 6.14 shows his son holding the wooden pan used for final upgrading.

6.6 MECHANICAL MINING – DRY METHODS

Dry mechanical methods of mining placers are developed around the use of various combinations of heavy earth-moving equipment for extraction and either trucks or slurry pumping transfer units for transportation. Such methods provide a means of mining selectively down to bedrock and, by exposing its surface for cleaning, to make possible high recoveries of valuable particles caught up in cracks and crevices. The various methods are positive in their actions and can be relied upon to keep stockpiles at acceptable levels for continuous mill operation, almost regardless of the mining conditions. However, most are disadvantaged against dredging wherever the dominant considerations are costs for power, labour and maintenance. Generally, therefore, if a dry-mining method is chosen ahead of a dredging method it is because conditions do not favour dredging. The following machines and combinations of machines are in current use, either alone, or in support of some other method:

Fig. 6.13 Indian prospector-miner using primitive pinched-sluice separator for gold recovery in Raigarh Goldfield, Madhya Pradesh, India

Fig. 6.14 Wooden pan used for upgrading pinched-sluice concentrates

Bulldozer – scrapers
Bulldozer – Front-end loaders
Bulldozer – Buried loaders
Land dredgers
Hydraulic excavators and trucks
Draglines
Bucket-wheel excavators
Drum scrapers and trucks

6.6.1 Bulldozer – scrapers

This combination can be used to handle large quantities of overburden if ample space is available for constructing roadways to dumping grounds away from the excavations. At Moolyella in Western Australia, scrapers were used to pick up and dump overburden with a maximum lead of around 1 km. Stanniferous gravels from the same pits were delivered to a treatment plant stockpile 3 km away by the same means. Mining rates averaged 250 $m^3 h^{-1}$ of overburden and 100 $m^3 h^{-1}$ of stanniferous gravels using power scrapers with nominal load capacities of around 20 m^3. Depending upon surface conditions in the pit at the time, one or two D9 bulldozers were used for push-loading the scrapers which then continued under their own power. After dumping their loads at either the disposal area or at the plant the scrapers returned at top speed for the next load.

Cycle time is a function of loading, running and down time and, generally a great deal depends upon the experience and temperaments of the drivers. Occasional tyre problems are unavoidable but they can be minimized by good haulage-road maintenance and frequent checks to remove sharp objects. Planned maintenance periods are allowed far away from normal working hours, and breakdowns should be infrequent. However, workshop facilities need to be extensive and all fast-moving spares should be readily at hand.

As a general rule, established earth-moving contractors are able to provide a more efficient service and operate at lower cost than mine owners. Inexperienced operators pay dearly for the experience of learning to operate and maintain heavy earth-moving equipment and few undertakings can afford a long training period.

6.6.2 Bulldozer – front end loaders

Front-end loaders, of which the articulated wheel loader is a special type, have great manoevrability when operating on a firm, level floor. They are less-well suited to following uneven ground contours at floor level and have difficulties in traversing soft patches. They are sometimes used alone, but more often in conjunction with bulldozers which prepare the more compacted and difficult ground for loading by ripping and stacking ahead of extraction. The combination is best suited to shallow surface deposits in which the water inflow can be controlled and for mining beach deposits that cannot be dredged satisfactorily.

In beach-mining applications, one or more loaders are used to pick up and carry the spoil to a semi-mobile (usually skid mounted) transfer unit where it is screened and slurried before being pumped at a controlled rate to the head feed bin at the treatment plant. Single-stage pumps are used for pumping distances of up to 400 m. A second booster stage, to increase the pumping distance, is sometimes justified depending on how the additional power costs compare with the costs of more frequent plant shifts.

A typical transfer unit comprises a feed hopper, inclined belt loader feeding to a trommel screen, densifying transfer bin and slurry pump. (Fig. 6.15). Slurry densities are controlled at around 50–55% of solids by weight by regulating both the feed rate and the flow of water into the trommel. Experience has shown that higher slurry densities result in serious sand losses in the trommel oversize fractions.

Tailings returned to the back of the excavation are contained behind a bank or levee of sand dozed up across the pit. The operation is planned so that restoration keeps pace with extraction.

The work cycle is as follows:

(1) Move the machine up to the working face or stockpile and drop the bucket. Rotate it to a flat position on the floor at ground level.
(2) Crowd the machine forward and lift the bucket slightly to fill.
(3) Back the machine off and travel with full load to the feed hopper.
(4) Raise the bucket to clear the top of the hopper and dump.
(5) Return the bucket to the running postion and travel back to the face to complete the cycle.

Cycle times vary with distances of travel (preferably less than 90 m) and the condition of the pit floor. In good conditions, loading, dumping and manoeuvring may all be accomplished for a short haul in 40 s, but both the

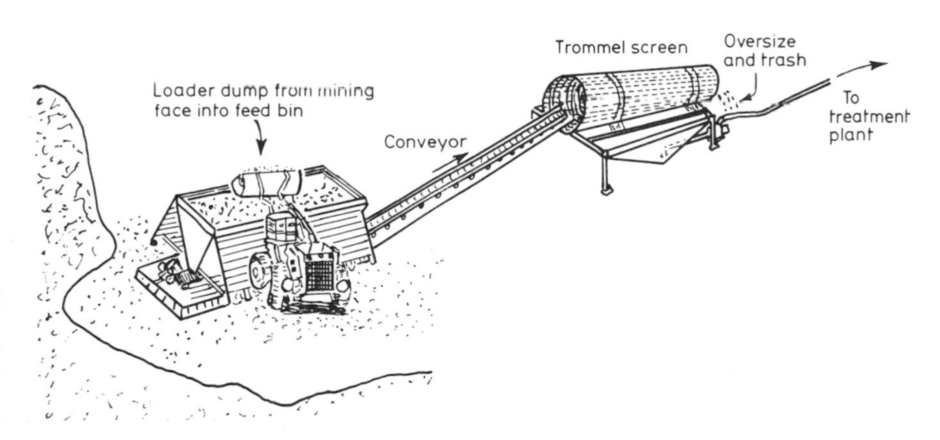

Fig. 6.15 Typical beach-mining transfer unit

height of the face and face compaction are limiting operational factors. A bucket-fill factor of 0.8–0.9 can be expected from a 2–3 m face. This rises to 0.95–1.00 with increased bank height. Slightly damp sand gives higher fill factors than dry sand because it tends to remain heaped without rilling away. Tightly compacted or indurated sands may take 10–20% longer to load than free-flowing sand.

Good bucket design is fundamental to achieving design rates of production and different types have been developed to suit different conditions. The two most commonly used buckets for beach mining are:

(1) General purpose buckets. These have straight cutting edges and are best suited for digging free-flowing sand. They lack penetration when digging into tightly compacted beds.
(2) Rock buckets. Spade-nosed cutting edges penetrate compacted materials better than do general purpose buckets but they exert higher breakout forces in dry sand because of deeper penetration. Such high breakout stresses may create excessive hydraulic operating pressures that rupture hoses and crack the welding joints unless a sufficient safety margin is allowed for in design.

Bucket teeth comb the face of the compacted sand as the bucket is lifted and it is demonstrable that loading efficiencies for all types are improved with close-teeth spacings that allow sand to be retained for much of the tooth length.

The pulping bin is skid mounted and the transfer unit consists of a feed hopper, belt conveyor and trommel screen from which the undersize passes to a densifying tank for transfer to the plant. A possible alternative, when loading from a high face of loose material, is to use large wheel loaders for both loading and haulage. Carry distances are governed by the bucket size and manufacturers claim that the Fiat-Allis 4.5 m^3 loader can load and carry economically for one way distances of 180–200 m; 7.6 m^3 loaders up to 250–275 m and 9 m^3 loaders up to 300 m.

Articulation: articulated steering provides tight turning circles in restricted places. The point of articulation is located midway between the axles in most models so that the rear wheels track in treads left by the front wheels. However, one manufacturer, 'Tenex', has placed the pivot point at approximately 40% of the distance between the front and rear wheel axles. As a result, the axles turn by different amounts and the rear wheels track inside the front wheels. It is claimed that this configuration is better suited to travel on soft surfaces.

6.6.3 Bulldozer – buried loaders

This method applies mainly to the mining of high-level dunes. A fleet of bulldozers is used for mining and plant relocation and a screening and slurrying section prepares the feed for transportation to the concentrating plant.

The bulldozing operation

Sand is dozed down from the pit to the end of a ramp where the dumping point may be as much as 25 m above the level of the buried loader draw point but, for safe operation, is preferably less. The dozing path, or windrow has a maximum decline of $7\frac{1}{2}°$ for speedy back travel and the windrow walls are maintained at around blade height to avoid side-spillage. Good operators pick the load up evenly and smoothly, thus leaving the surface in good condition for the next cycle. Provided that the machines are not over extended, efficiencies are generally good. On Stradbroke Island, Queensland, downtime averaged 16% which is a fair average performance for earth-moving equipment in any normal service. It comprised

Planned maintenance and overhauls	12%
Breakdowns	4%
Total downtime	16%

Commencing with new pins and brushes, the tracks averaged 1100 h before turning and 1300 h thereafter until replacement. Rail build up and side weld was carried out at the turning stage; segments were replaced at approximately 2000 h, carrier rollers and idlers at 3500 h and the final drive, transmission and converter at 8000 h. Motor life was between 10 000 and 11 000 h of service.

The loader–transfer complex

A typical complex consists of a buried loader, two belt conveyors, trommel screen and pumping unit. Sand rilling down the slope passes through a grizzly which screens out tree stumps and other large debris. An inclined belt-feeder system transfers the sand to the trommel screen where it is washed clear of the smaller trash and fluidized for pumping to the treatment plant at around 55% of solids by weight. A modular plant can usually be set up in a new location within as little as 4 h. If planned maintenance is carried out before the plant is brought on stream again the total shut-down period is extended to a maximum of about 10 h. Re-location is carried out in the following sequence:

(1) Doze spillage away from the loader and tow to next site using three or four D9 tractors according to the steepness of the gradients.
(2) Push the loader hard against the new face and position the intermediate feeder and trommel units. Each of these units is handled using either two or three D9 tractors depending upon the conditions.
(3) Position the pump unit and connect power, water and slurry lines; two D9 tractors are used for towing and positioning.

Note: The method is power and labour intensive with operating costs as much as three to four times those of equivalent capacity suction-dredge operations. Figure 6.16 (after Lambert, 1975, personal communication) is a schematic layout of a complete operation to the stage of carting heavy-mineral concentrates to a dry-separation plant using trucks.

6.6.4 Land dredgers

Bulldozers are also used for sand-mining duties in conjunction with skid-mounted, cutter-suction pump arrangements similar to those of floating suction dredgers. The method, illustrated in Fig. 6.17, was developed primarily for mining shallow beach placers but can be adapted for mining any other reasonably dry and shallow deposits.

In the same way as for the buried-loader system, dozing is downslope except at the end of the run where the sand is piled up to form a stockpile alongside the sump. The mining face provides both short and long pushes for an economic maximum lead of 90 m. This procedure makes it easier to control the structure of the stockpile and, hence, to regulate the flow of sand to the concentrating plant. Optimum dozer performance follows the use of a full 'U' blade and oversize grouser plates for better traction.

The dredger arrangement comprises an enclosed steel hull on which is located the power unit, pumps and cutter-drive mechanism. The cutter shaft is designed to allow the cutter to operate at a depth of about 2 m below the ground surface when the shaft is at 60° from the horizontal. The unit rests on the ground surface and picks up from a sump dug into the floor. The sump is dug just deep and large enough in cross section to contain the cutter shaft and suction pipe. In operation the pit assumes dimensions that are governed by the texture of the material and the rate of mining.

Sand is pushed up around the sump to a height of 2–5 m and rilling is assisted by monitoring. Monitoring, also provides the water for fluidization and, for sand, optimum conditions are obtained through the use of high-volume, low-pressure nozzles. The jets are directed onto the higher section of the stockpile to wash the sand down into the sump; it is fluidized and held in suspension around the suction nozzle by the action of the cutter head. Concentrations of up to 50% of solids by weight can be achieved without any overflow of water from the sump.

The main disadvantages of the system when applied to strandline deposits, are the high maintenance costs requirements for dozing. According to Everson (1973), in contrast to dozing in high-level dunes the combination of fine sand and water results in very rapid pin and bush wear. Turning may be necessary after only 400–500 hours of use; track life is similarly reduced, and a complete walking gear overhaul is usually needed after about 2000 h.

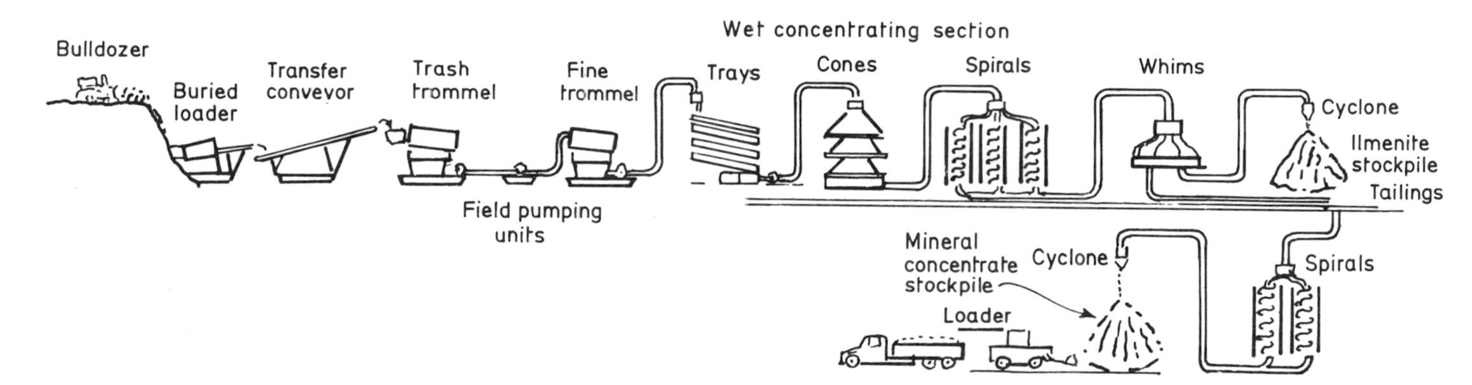

Fig. 6.16 Schematic layout of bulldozer–buried loader–primary concentration operation

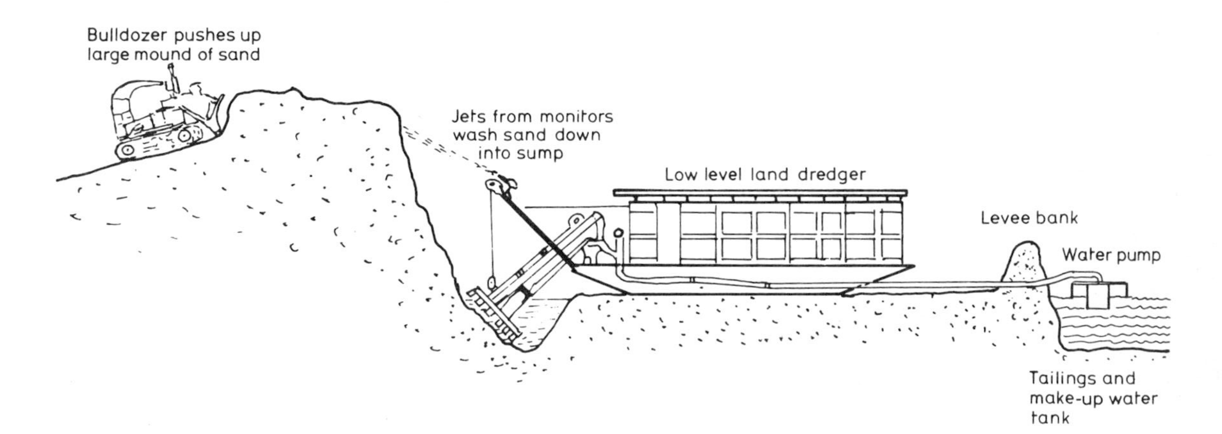

Fig. 6.17 Land-dredging operation

6.6.5 Hydraulic excavators and trucks

Hydraulic excavators, either forward or back acting, are comparatively new to placer mining but are finding increasing applications and offering serious competition to many of the older established types such as draglines and cable-operated excavators. One reason for their growing popularity is their ability to exert powerful forces at various levels in the excavation; another is their versatility. Various configurations are available to convert most models for either forward- or back-acting service. Hydraulic excavators may also be fitted with a variety of buckets and attachments for such duties as pitting, trenching, ditch digging and excavation in both wet and dry conditions. They are generally easy to manoeuvre and simple to operate.

Digging mechanisms are mounted on rotating upper structures that allow a 360° swing without change in load ratings, reach or digging depths. They normally have two-piece configurations for the forward-acting shovels and three-piece configurations for the larger back-acting shovels. Adaptors are available from some manufacturers to convert from back hoe to clamshell operation using the back-hoe sticks.

Forward-acting excavators have buckets that are either front dumping or bottom dumping. The former allow somewhat higher digging forces and greater strength; bottom dumping buckets provide more headroom for dumping. Although some models can dig below ground level, i.e. the surface they stand on, such digging depths are limited and forward acting models are used mainly on benches and floors in competition with other shovel types. Machines such as the Caterpillar 245 feature parallel crowd and can mine very selectively in good digging conditions by skimming over the top of pay gravels and other weathered rock surfaces to recover values at bedrock. A breakout force of 43 450 kg and dump height of 6.7 m is claimed for this machine and, in difficult digging conditions, the horizontal thrust exploits any weaknesses by penetrating a short distance and prying the material away from the bank. To accomplish this, it exerts a powerful wrist action using the heel of the bucket resting on the ground as a pivotting point, thus minimizing the tipping moment and relieving the machine of much of the stress.

Back-acting excavators are the better-known machines for placer operations and are used widely for both exploration (see Chapter 5) and mining. They normally dig downwards from the bank and mine by retreating along or across the deposit, loading either directly into trucks or into the feed hopper of a mobile transfer unit. Cross (1979) refers to the possibility of loading directly to floating treatment plants using grab buckets (clamshells) as is done in dragline 'doodlebug' mining. It might be feasible also to use barge-mounted back hoes for mining some shallow wet placers provided that the barges are equipped with suitably designed spuds and side lines to withstand high reactive stresses.

Buckets are adapted for digging conditions that range from very light to most difficult, using various tip- and side-cutter designs. Allowable bucket capacities vary according to the stability of the surface on which the excavator stands, the specific weight of the material being dug and the digging envelope as determined by the reach and depth of digging.

A relatively high factor of safety should be applied when using a stick that delivers the longest reach and deepest digging depth recommended by manufacturers for a particular model. Machines should be selected on the base of average all-round performance rather than on their ultimate capacities and most excavators can be fitted with a variety of front ends depending upon what is required of them. This applies to both back-acting and forward-acting shovels; typical combinations are as follows:

(1) A short backhoe front and large bucket for high-production rates.
(2) A long backhoe boom and smaller capacity bucket for long-reach work.
(3) A short shovel front and large capacity bottom dump bucket for powerful penetration forces and high-production rates.
(4) A long shovel front with forward tipping bucket for long reach.

Standard bucket capacities for forward-acting shovels range up to 15 m^3 and, for back hoes, to 10 m^3. However, although the forward-acting machines can operate using larger buckets, other important features influence the selection of back-acting machines for most placer operations:

(1) Smaller swing for digging and discharging.
(2) Less need for accurate truck positioning.
(3) Lower stresses imposed on track and running gear.
(4) Reduced danger from rockfalls.
(5) Deeper digging capacities.
(6) Ability to work more effectively from the ground-surface level.

Cycle times for hydraulic excavators are very sensitive to digging depths, dumping heights and swing, and range from 20 s (fill, turn, discharge, return and reposition) for small machines up to 30 s and more for large machines capable of peak outputs as high as 2000 $m^3 h^{-1}$. Actual production rates are considerably lower than rated capacities, and corrections must be applied for factors that relate to machine utilization, bucket fill and swell. Machine operating time is lost in machine travel and re-positioning and in waiting for trucks; an 83% utilization is normally a practical maximum. The fill factor is around 1.0 in most placer applications but the swell of the material when broken may vary considerably although it is commonly assumed to be 30% (see Chapter 2). In a typical calculation a machine operating with a 6 m^3 bucket, 28 s cycle, 1.0 fill factor and 30% swell factor would have a theoretical average hourly output of:

$$\frac{60 \times 60}{28} \times 6 \times 1.0 \times \frac{100}{130} = 593.4 \text{ m}^3 \text{ h}^{-1} \text{ (bank)}$$

This would reduce to:

$$593.4 \times 0.83 = 492.5 \text{ m}^3 \text{ h}^{-1} \text{ (bank)}$$

for an 83% utilization.

6.6.6 Draglines

Draglines are used mainly for stripping deep (20–50 m), regular thicknesses of unconsolidated overburden from large areas. The largest crawler machines have buckets (dippers) around 10 m^3 capacity and boom lengths of 50–70 m; the buckets used by walking draglines commence at about this size. Typical cycle times for the crawler types are given as:

8 m^3 dippers – 25 to 30 s
15 m^3 dippers – 25 to 40 s
20 m^3 dippers – 30 to 45 s.

Production rates approximate 65–70 bank $m^3 h^{-1}$ for each cubic metre of dipper capacity with an average work availability of around 80–85%.

Draglines found a great deal of favour during the depression days of the thirties in California for small gold-mining operations commonly referred to as 'doodle-bug' operations. Draglines were used to dig material from the bank and load it into the hoppers of small floating gold-saving plants moored alongside. A few such operations were successful but most failed due to lack of experience or faulty valuations. The concept probably developed from the crane and bucket dredgers used around the turn of the last century. Rose (1902) describes an operation in the Fraser River, British Columbia (1896–1897) in which two barges were used. One contained the dredging machinery, the other the gold-washing tables; production rates were estimated at 135 yd^3 (103 m^3) h^{-1} for a bucket capacity of 1.5 yd^3 (1.15 m^3) and a 45-s cycle. The bucket was of the Priestman grab type which opened at the bottom when it was dropped onto the gravel bed and closed automatically when the chains began to lift it.

Placer mining by dragline has been largely replaced by the use of small semi-portable dredgers, however, some operations have survived and, in both Sierra Leone and Ghana, large-scale diamond-mining operations are conducted using draglines almost exclusively for mining in very swampy ground. Crawler draglines are used for excavation and ore trucks for haulage; the main difficulties are associated with drainage and road building. Generally the ground must be prepared for mining at least 1 year ahead of extraction and the first task is to drain each block that is to be mined. One procedure is to make a dragline cut around the block, stacking the gravels and overburden separately. The cut is then backfilled with lateritic soil to the top of the gravel horizon and consolidated by dropping the bucket onto it to make a relatively impermeable seal against inflow. The topsoil is replaced and pump wells are sunk in the deepest parts. The block is pumped dry and kept in that condition until mined. Haulage roads require at least 1 year to consolidate and it is difficult to see the method competing economically with bucket-line dredging in very wet ground.

6.6.7 Bucket-wheel excavators

Bucket-wheel excavators have found occasional favour for mining placers in some environments. A typical arrangement comprises a standard bucket-wheel excavator with an extended belt conveyor loading into a pulping bin. Spillage is returned to the loading face and a bulldozer is used to rip the harder materials. The spoil is washed in a skid-mounted trommel screen from which the trash is discharged onto the ground. The undersize passes to the pulping or transfer bin and is pumped in slurry form to the concentrating plant.

A bucket-wheel excavator was used successfully at Jerusalem Creek, northern New South Wales, Australia to mine relatively hard, indurated sands containing rutile and zircon. The sand was mined on a 30-m wide face and the concentrator plant was floated on pontoons in a dammed-off section of the mined out area. Periodically, a new dam wall was constructed closer to the face and the water level was brought up to the level of the existing pond. The old wall was then broken down and the concentrating plant brought forward to its new position. Tailings were directed into the back of the mined out area and the surface was continually contoured and restored in the direction of mining.

For given capacities bucket-wheel excavators cost less than bucket-line dredgers and generally more than suction-cutter dredgers. They offer some advantages where deposits are heavily indurated and because they provide a regular rate of feed material to the concentrating units, thus allowing close metallurgical control and recording. Bucket-wheel excavators are not suitable for sticky clays which clog bucket discharge and conveyor systems and are normally suited only to dry ground with firm, level footings. There were good reasons for selecting a bucket-wheel excavator instead of a suction-cutter dredger at Jerusalem Creek more than a decade ago. The same machine might not now be selected ahead of a bucket-wheel suction dredger in a similar application because bucket-wheel dredgers can also adapt to hard digging conditions and a more compact digging–treatment operation can be put together using them.

Bucket-wheel excavators are best suited to the large-scale removal of overburden. Hodgeson (1978) describes the stripping operation on a diamond mining property in Nambibia where an O & K SH400 is used to strip 140 000 tonnes of overburden per day to expose 40 000 m^3 of diamond-bearing gravels per month. The gravels are mined using bulldozers, hydraulic excavators and front-end loaders and are hauled to the treatment plant in trucks. Bucket-wheel excavators are less suitable for small-scale operations. Parkianathan and Simpson (1965) described how a small unit was used to remove about 60–80 ft (18–25 m) of sand from an area which had previously been dredged. From their observations, the stripping rate approximated 43 000 m^3 of sand per month, however, the authors were not impressed by the economics of the operation and concluded that 'the method is limited to larger mines where ample reserves justify the large capital outlay'.

6.6.8 Drag-scrapers and trucks

The increasing costs of operating wheeled and tracked equipment have led operators to re-examine some of the older mining methods and double-drum drag-scrapers have been upgraded to meet many of the requirements of placer mining. Depending upon the duty and local conditions the basic equipment may be transported on rails laid transversely to the direction of scraping (transverse running gear); on circular rails swung around the delivery point (circular running gear); or it can be skid mounted. Drag scrapers on rails are usually moved by means of a hand winch but can if necessary be provided with a drive system. The skid-mounted drag-scraper may be rotated on the spot or hauled by tractor to a new position.

Figure 6.18 illustrates the method of mining by drag-scraper. Depending upon local conditions the tail sheave, firmly anchored, is either mounted on land or on floats in the pond in which case two warping winches are set up on land for manoeuvring. The distance between warping winches can

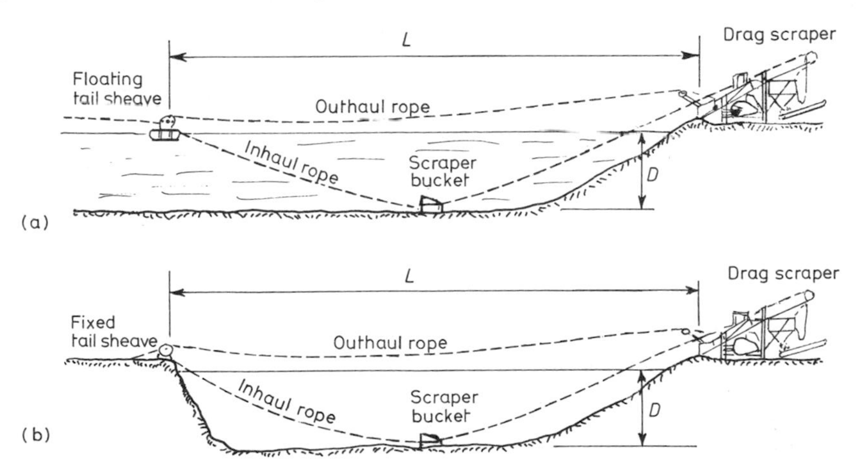

Fig. 6.18 Drag scraper operator: (a) wet application, (b) dry application

be up to 100 m depending upon the distance from the winches to the float. The warping winches are normally operated by hand but can also be electrically driven and remotely controlled. Digging depths are functions of the horizontal distance between the drag-scraper and the tail sheave.

For dry conditions the maximum attainable depth $D = L/4$.
For wet conditions the maximum attainable depth $D = L/5$ to $L/6$.

Double-drum drag-scraper winches have an outhaul drum and an inhaul drum. Gravels dredged from the pit are hauled up to the dumping point using the inhaul rope driven at a speed of about 1.2 m s^{-1}. The empty bucket is retracted via the tail sheave by means of the outhaul rope.

Drag-scrapers load either directly into trucks, onto a conveyor-belt system for feeding stockpiles up to a height of about 8 m or into a slurry transfer system for pumping away to the treatment plant. Units can be equipped with attached vibrating screens or trommels to screen out the oversize material and trash; where applicable, a vibrating grizzley is used to reject stones larger than about 100 mm diameter.

Fritz Stichweh GMBH–Maschinenfabrik, West Germany, produce drag scraper units with bucket capacities of 2–10 m^3 for extraction rates of 50 m^3/h up to 350 m^3/h in average conditions. All units are equipped with torque converters to enable them to be operated in difficult conditions. It is claimed that:

(1) The drag speed and cables tractive force can be precision regulated by hand and are automatically and immediately adjusted to varying loads, the tractive force being capable of tripling where necessary.
(2) The converter needs no maintenance apart from oil changes every 1000 h.
(3) A limit switch automatically switches the clutches off when the drag bucket is in either the front or the rear position.
(4) Operation is simple using only a single lever control.
(5) Gravel banks up to a height of 25 m can be excavated safely.

Table 6.4 indicates the output of various Stichweh drag-scrapers relative to the length of haul in favourable conditions. The company stresses, however, that the output figures assume continuous operation and in no circumstances can be regarded as guaranteed outputs. Details of the various units are shown in Table 6.4.

6.9 MECHANICAL MINING – WET METHODS

6.9.1 Hydraulic sluicing

Hydraulic sluicing is one of the oldest systems for mining placers using mechanical means. It is sometimes referred to as gravel pump mining and,

Table 6.4 *Performance data – Stichweh Drag Scrapers*

Type	KS200	KS300 KS300 S	KS400 S KS400 RA	KS600 S KS600 AA	KS1000 S
Power required	65 kW (90 HP)	100 kW (135 HP)	125 kW (170 HP)	205 kW (280 HP)	250 kW (340 HP)
Bucket capacity	2.0 m^3/2.6 yd^3	3.0 m^3/3.9 yd^3	4.0 m^3/5.2 yd^3	6.0 m^3/7.8 yd^3	10.0 m^3/13.0 yd^3
Rope diameter	20/22 mm	22/26 mm	22/29 mm	26/36 mm	29/40 mm
Rope speeds					
Inhaul up to	1.2 m s^{-1}	1.2 m s^{-1}	1.2 m s^{-1}	1.2 m s^{-1}	1.2 m s^{-1}
Outhaul up to	1.8 m s^{-1}	1.8 m s^{-1}	1.8 m s^{-1}	1.8 m s^{-1}	1.8 m s^{-1}
Weight including skids (approx.)	18 tons/20 sh tn	26 tons/29 sh tn			
Weight with walking equipment		32 tons/35 sh tn	34 tons/37 sh tn	70 tons/77 sh tn	110 tons/121 sh tn
Weight with track-type crawlers and integral boon (approx.)			76 tons/84 sh tn	135 tons/149 sh tn	
Hopper capacity	3.0 m^3/3.9 yd^3	5.0 m^3/6.5 yd^3	6.0 m^3/7.8 yd^3	10.0 m^3/13.0 yd^3	15.0 m^3/19.5 yd^3
Maximum drag distance	150 m/495 ft	180 m/595 ft	180 m/595 ft	200 m/660 ft	220 m/730 ft
Maximum drag distance with integral boom			38 m/125 ft	46 m/152 ft	
Traction at rope	6 tons/6.6 sh tn	9 tons/9.9 sh tn	10 tons/11 sh tn	18 tons/19.8 sh tn	24 tons/26.5 sh tn

in Malaysia, more than one-half of the tin production is contributed by sluicing. The method was first used in mountainous regions where a natural head of water could be drawn upon to provide high-pressure water at the face without the need for power machinery. Steep gradients at bedrock allowed the tailings to be discharged by gravity and virtually the only operating costs were for labour and pipeline maintenance.

With the invention of the venturi hydraulic elevator flatter gradients could be mined hydraulically using the hydrostatic head to elevate the slurry to sluices at a higher level, thus providing steeper gradients for the tails races and for stacking the tails. A typical layout is illustrated in Fig. 6.19. With full mechanization, the method is simple and practicable for a wide range of the more primitive placer deposits (see Chapter 3) although it becomes very power intensive when the nozzle water has to be pumped against high dynamic heads (see Chapter 5) or where the energy requirements for slurrying at the face are high. The main items of equipment for a fully operational plant are, (a) one only high-pressure water pump, (b) one or more hydraulic giants (monitors) with accessory valves pipelines and fittings, (c) one only low- to medium-head gravel pump (preferably the vertical submersible type), and (d) sundry concentrating units.

6.9.2 Sluicing practice

Water from a pressure tank or pump is directed at high velocity against the bottom of the pit face using a monitor to direct the flow. The face is undermined and sections of the bank collapse. The jet is played onto the broken material to disintegrate and wash it back through races cut into the bedrock to the gravel-pump sump. The larger stones are forked out of the races and stacked along the sides of the excavation. The slurry is elevated from the sump into a feed preparation and desliming section and thence to jigs, palongs, sluice boxes or other devices to recover the valuable heavy-mineral concentrates. Concentrates are 'dressed' in a 'dressing shed' to separate the valuable minerals from the non-valuable heavies and to recover them in a marketable form (see Chapter 7).

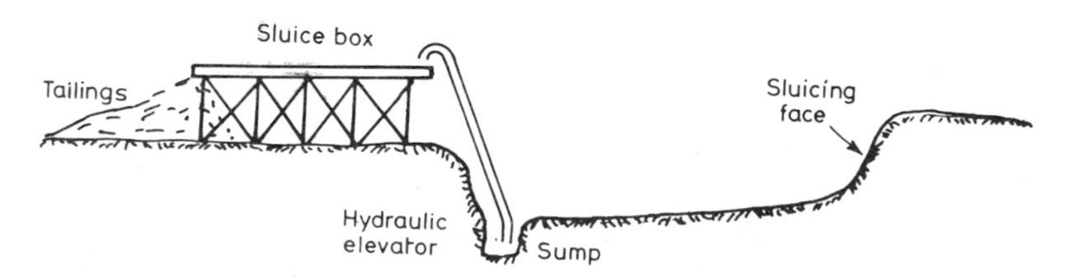

Fig. 6.19 Hydraulic sluicing operation using hydraulic elevator

The method makes poor use of the available energy because of hydraulic line losses and because the jet momentum is utilized for only part of the time doing useful work in breaking down the face. There are practical difficulties in being able to direct the jet continually against unbroken or unslurried material and excessive quantities of water brought into the pit may have to be elevated out and away from it, thus increasing pumping costs. The inefficient use of hydraulic power is not critical where an adequate natural head of water is available but some mechanical operations now face crippling costs for power because of high fuel costs.

Possibly the greatest inefficiencies in hydraulic sluicing stem from the need to locate the monitor at least bank height away from the face for safe working. The kinetic energy in a jet of water is used up rapidly in its passage through the air (see Chapter 5) and the extent of the problem is well illustrated in the chart of performances shown in Table 6.5 which shows that the productivity of a jet is directly related to the distance traveled. Table 6.6 relates the effective range of a jet to the head of water at the jet nozzle.

Nozzle diameters vary normally between 2.5 and 12.5 cm with jet velocities from 20 to 50 m s^{-1}. Velocities are related to pressures through the expression for velocity, i.e.

$$V = C(2gh)^{0.5}$$

where, V is the velocity of flow at the nozzle outlet (m s^{-1}), h the head of water to the nozzle (m), g the acceleration due to gravity (9.81 m s^{-2}), and C is a nozzle constant (0.95 approx.).

Table 6.5 *Hydraulic giant performance and water consumption per unit of material washed (after Shevyakov, 1970)*

Distance between nozzle and working face (m)	5	10	15	20	25
Monitor performance (m^3) of ground per hour	100	93	74	48	18
Water consumption (m^3) per m^3 of ground	8	8.6	10.8	16.7	44.5

Based upon all considerations, safety and otherwise, the effective ranges of jets are given by Shevyakov as listed in Table 6.6.

Table 6.6 *Effective range of jet (m) versus head (m) (after Shevyakov, 1970)*

Head (m)	20	40	60	80	100	120	150
Effective range (m)	10	21	31	41	52	62	78

Example: To obtain a jet velocity of $50\,\mathrm{m\,s^{-1}}$:

$$h = \frac{V^2}{2gC^2} = \frac{50^2}{2 \times 9.81 \times 0.95^2}$$
$$= 141.2\,\mathrm{m}$$

to which must be added sufficient extra head to compensate for line friction and other losses (see Chapter 5) and possibly a total head of at least 160 m might be required.

Productivity also varies with the characteristics of ground treatment. Easily dispersed clays and other materials are quickly fluidized and form high slurry concentrations. Puggy clays are the most troublesome and such materials require additional energy and time for puddling. Combinations of methods often give much better results than hydraulic sluicing alone. Bulldozers, draglines and other excavators are in frequent use for removing overburden and for clearing topsoil and vegetation. Very tough clays and partly cemented gravels respond better to jetting if broken initially by some mechanical means. Bulldozers are favoured above most other mechanic types because they are usually the most suitable machines for restoring the ground and its contours after mining.

A particular hazard in mountainous country is the possibility of initiating a massive slumping of material into the pit. The essential ingredients for a major landslide are found in a lack of support across a wide face, a slip surface between water-saturated sediments and bedrock and sediments that contain considerable quantities of clay or mud. In one operation at Porgera in the highlands of Papua New Guinea a trial mining programme was implemented to determine if a sluicing cut could be made safely in bank heights up to and exceeding 12 m. There was some evidence of previous massive slides in the valley. Pit safety precautions were planned and taken as follows.

(1) Monitors were positioned to cut from the sides and not from directly in front of the face. This was to avoid being smothered by a major collapse.
(2) A bulldozer pushed bank material down to the monitors at a slope of 45° or flatter depending upon the material. It was not allowed at any time to undercut the face.
(3) Pit dimensions were maintained at a satisfactory level to allow ready manoeuvring. Apart from the bulldozer, a traxcavator was also in constant service picking up and stacking the larger stones and small boulders. The width of the cut was held to a practical minimum to minimize the risk of block flow.
(4) The daily and instantaneous movements of the upstream materials were monitored using an integrated system of surface markers tied in

by theodolite survey over the whole deposit. Experience has shown that general failure is usually preceded by rapidly accelerating rates of deformation and that the periods of greatest danger are predictable provided that the readings are taken at short intervals. Failure did occur at Porgera, in spite of all the precautions taken in mining, but there was no loss of life and only the sluice boxes were buried under the debris because early warning gave time to get the earth moving equipment to safety.

Block failures can also be predicted using borehole slope indicators to register instantaneous changes in the angles of boreholes drilled back from the face. The subsidiary use of the method at Porgera allowed extra care to be taken just prior to the total collapse.

Most mechanical sluicing operations rely upon gravel pumps for elevating the spoil. The older pump types were single stage, open impeller centrifugal pumps, belt driven from a diesel engine or slip-ring electric motor. They often proved difficult to prime and submersible pumps are now more common. Vertical slurry pumps can be raised or lowered in the sump to increase or decrease the density of the slurry, thus controlling the output according to fluctuations in the inflow. Simple tripod and pulley arrangements constructed from materials on site are set up over the deepest part of the sump.

The capacities of a range of gravel pumps having inlet diameters from 5 in (13.7 cm) to 10 in (25.2 cm) as described in Table 6.7 (adapted from Pakianathan and Simpson, 1965) can be used as a guide to what might be expected in average conditions. However, it will be noted that the capacities given, vary closely with the square of the diameter thus assuming identical flow velocities and slurry concentrations throughout. See, however, the variation in limit deposit velocities in pipelines of different sizes according to differences in solids characteristics (Chapter 5).

Table 6.7 *Capacities of gravel pumps (adapted from Pakianathan and Simpson, 1965)*

Size of pump inlet diameter		Solids pumped/month, 16 h/day	
inches	cm	yd^3	m^3
5	12.70	8 000	6 116
6	15.24	11 000	8 410
7	17.78	15 000	11 468
8	20.32	20 000	15 291
9	22.86	25 000	19 114
10	25.40	31 000	23 701

6.10 DREDGING

Bucket-line and suction-cutter dredgers are the main dredger types used in placer-mining operations. Each has a particular field of operation in which it is clearly superior to the other; there are relatively few applications where the issue might be in doubt. Of the other alternatives, clamshell dredgers have a limited scope for mining small deposits of loose gravels underwater or at sea where drag heads and jet lifts are usually preferred. Bucket-wheel dredgers currently being developed to compete with suction-cutter dredgers have some distinct advantages, but suction cutter dredgers can still mine a wider range of materials at higher production rates and lower costs.

6.11 BUCKET-LINE DREDGERS

6.11.1 Description

A bucket-line dredger (Fig. 6.20) consists of a large pontoon onto which are built the supporting structures, digging mechanism and treatment facilities. The digging mechanism consists of a line of heavy steel buckets on an endless chain supported by a dredging ladder made from steel girders

Fig. 6.20 Bucket-line dredger working on gold lead, Ghana

held together by cross members and pivotted from a central framework which also holds the drive. The pontoon is boxlike in shape, with a slot-like opening at the forward end for the ladder. Modern dredger hulls are all steel welded to economise on weight and are partitioned off for strength and safety. The bucket line provides the mechanism for excavating the spoil and feeding it into the treatment plant. It runs between two tumblers, one at each end of the ladder and is supported by closely pitched rollers. Bow and stern gantries carry the pulleys and steel cables used for tilting the ladder and for supporting the tailings disposal sluices and conveyors. The aft gantry also supports the spuds and their lifting gear.

The first dredgers were single-bucket types but the advantages of multi-bucket operation were quickly realized and dredgers have since been developed to handle increasingly difficult conditions both on- and offshore. All-electric dredgers are the most popular although direct diesel power is used also and some of the older units in remote areas may still be powered by steam. Great strength is needed and bucket-line dredgers are massive structures of considerable weight. The weight per $m^3 h^{-1}$ capacity ranges from about 1.5 tonnes/$m^3 h^{-1}$ for small modern dredgers to about 8 tonnes/$m^3 h^{-1}$ for large deep-digging dredgers. Typically, a 130 $m^3 h^{-1}$ dredger weighs around 200 tonnes whereas a 1000 $m^3 h^{-1}$ unit might weigh as much as 7800 tonnes. Popov (1971) gives the weight of Soviet-manufactured dredgers in Table 6.8.

The second of the two Orkutsk No. 1 dredgers was apparently designed for much simpler dredging conditions because of its larger bucket capacity, greater weight and higher bucket speeds.

On-board treatment processes have changed little in past decades. A large trommel screen with lifters and high pressure sprays is used to break up the wash and to eliminate trash and oversize material. A two- or three-stage jigging plant makes a rough concentrate; desliming cyclones, pumps and stackers handle the slimes and tailings (see Chapters 5 and 7 for a more detailed treatment of the subject). Sluice boxes for gold and palongs for tin are still used on some of the older dredgers but they are much less efficient for recovering valuable particles in the finer sizings. The tailings are discharged over the stern or sides of the dredgers or are stacked some

Table 6.8 Dredger bucket capacities, digging rates and weights – USSR

Plant	Bucket capacity (litres)	Digging rate/min	Weight of dredger (tons)
Irkutsk (1)	150	22	720
Irkutsk (1)	250	20–34	124
Perm	380	22	1950
Perm	380	22	3370
Irkutsk (2)	600	20	7500

distance away from it using pumps for slimes and conveyors for the coarse materials. Rough concentrates are normally taken to an onshore dressing plant for final separation.

A typical land-based bucket line dredger is illustrated in Fig. 6.21. Offshore dredgers have similar working arrangements but are equipped with their own means of propulsion and are built to maritime standards.

6.11.2 Applications of bucket-line dredgers

Bucket-line dredgers were developed initially for mining deposits in rivers and swamps in which the digging conditions were reasonably easy to handle. Their uses have been extended and dredgers have been modified to cope with the problems of mining in a range of conditions. Dredging in higher ground is made possible by raising the banks or building levees to form artitifical ponds for floatation. Recent developments affecting offshore dredgers are related to systems providing more exact positioning, the development of hull-stabilizing devices, swell-compensating factors, increased capacities and digging depths. Bucket-line dredging is normally preferred to other systems for dredging and mining where:

Fig. 6.21 Suction-cutter dredger mining into a high bank of sand on Stradbroke Island, Queensland, Australia

(1) The reserves are large enough to justify the high capital requirements. Capital costs (US$) in 1980 ranged from $9000/m^3 of rated hourly capacity for small dredgers to as much as $21 000/m^3 for the larger ones. Thus a 700 000 m^3/annum dredger might have cost $1 050 000 and one for 6 000 000 m^3/annum of the order of $21 000 000. Because of the present high cost of money, total reserves of not less than 10 000 000 m^3 are needed for small-scale dredging and at least 120 000 000 m^3 of average-grade material for the larger operations.

(2) The supply of fresh pond water is sufficient for all purposes connected with mining and treatment. According to Popov (1971) the water availability in a placer valley should not be less than 100–150 l/s for large- and medium-sized dredgers and that, where this amount is not available special measures are required. In the USSR, these measures include the provision of dredge pond make-up water from other sources, sometimes as much as 40–50 km distant or from the recovery of water from slimes and other tailings in dams nearby. Conditions which result in the sliming up of dredge ponds due to insufficient make-up water, the failure to dispose of excessive slimes away from the pond, inadequate settling areas, and so on, cannot be tolerated.

Favourable dredging conditions require a gently sloping bedrock and the absence of many large boulders. Although it is technically feasible to build high banks and to step dredgers up quite steep slopes the procedure is very costly. Slopes should preferably be not much greater than 1:40 for artificial ponds. Greater tolerances are allowed for natural ponds and river beds but flat surfaces to dredge onto are always an advantage. Large quantities of boulders are very difficult to handle; the dredge-master may by-pass an occasional one or perhaps shatter it with explosives, however, the operation becomes excessively costly if clusters of large boulders have to be dealt with.

Also favourable for bucket-line dredging is a dredge-pond floor that can be cut by the buckets without throwing undue stress on the digging mechanism. Valuable heavy minerals tend to collect and accumulate at bedrock where they are caught in crevices, potholes and other natural traps. Serious losses may occur if the surface is not scraped down to at least 20–50 cm, depending upon local conditions, but it must be possible to do this without damage. Large dredgers clean up the bottom more efficiently because of their greater breakout strengths and robustness.

Of at least 130 bucket-line dredgers in current use around the world, Cross (1979) places three-quarters of the total in tin mining, the remainder in gold, platinum and gemstone operations. Dredgers in the tin fields of south-east Asia are the most advanced. Throughputs for the older models range from 700 000 m^3/annum to 5 million m^3/annum and recently commissioned units operating to depths of 45 m below the pond surface handle as much as ten times those of the smaller units. The largest Soviet dredgers have digging capabilities down to at least 50 m; the smallest are constructed from wood and, with 50-l buckets, are a feature of many small-scale operations.

A need in the foreseeable future is for dredgers that will dredge to greater depths and considerable thought is being given to the design of ladder arrangements that will give the additional digging depths without incurring excessive costs. Bucket-dredger capital and operating costs are closely related to digging depths because increased ladder dimensions and stresses must be compensated for by enlarging and strengthening the superstructure and by increasing the hull size to obtain the required buoyancy.

6.11.3 Construction details

Dredging ladders

Dredging ladders are heavy steel frames designed to carry the bucket line down to the required digging depths. Top and bottom tumblers and closely spaced rollers on the upper side of the framework provide support for the buckets, regardless of the angle of the ladder. Pivotted at the top, the ladder is raised and lowered either hydraulically or by winching, using steel cables passing over the sheaves on the main gantry.

Various proposals, for the future, include the call for more sophisticated structural designs with individual steel ladder sections being constructed from different sections according to the actual bending, shear and impact stresses that they have to withstand. Such proposals postulate the building of longer ladders without a reduction in strength and deeper digging depths without increasing the ladder weight. Another suggestion is to provide ladder support at the lower ladder points by the bow lines. According to Cleaveland (1976) this should make it possible to lower the top tumbler to some 20 ft (6.1 m) above the deck, thus adding 30 or more feet (approx. 9 m) to the digging depth of the larger dredgers. A smaller and lighter bucket elevator could then be used to elevate the spoil to the treatment-plant level.

Bucket lines and buckets

Regardless of the methods used to achieve increased digging depths, additional costs are involved in both the construction of the dredger and in the elevation of the spoil over greater distances. Accordingly, there must also be improvements in recovery and increased throughputs in order to hold unit operating costs at economic levels. Economic considerations suggest a minimum bucket capacity of 600 l for depths exceeding 35 m but this has not always been fully appreciated and it is only too common to see dredgers employed in defiance of good economics. For example, the two dredgers at the Perm Plant, USSR, described in Table 6.7 have equal bucket capacities and digging rates (380-l buckets at a rate of 22 buckets per minute) but different digging depths of 11 m and 30 m, and overall weights of 1950 and 3370 tonnes respectively. Clearly the larger plant would need to increase its output to be competitive with the smaller unit and normally this would be done by increasing either the bucket size or the bucket-line speed. Perhaps other factors are involved, that have not been stated, however, practical considerations affecting the design of any digging system clearly relate to the amount of material that can be safely and economically dug and treated per unit of time. Determining factors are bucket sizes and bucket-line speeds; the ability of the bucket line to withstand shock loadings and to resist wear is a factor determining the percentage availability of dredging over extended periods.

Bucket construction

Dredger buckets are made from high-grade manganese steels and are provided with specially hardened lips. The thickness of steel used is generally around 6–10 mm; the lip is as much as 30 mm in thickness depending upon the size of the bucket. Buckets are either attached to one another to form a continuous chain or are separated by idler links. The continuous-chain type is more adaptable for a variety of working conditions and usually cuts more effectively into a weathered bedrock.

Buckets have lives of up to 10 years depending upon the type of formation being dug. Sandy materials are very abrasive and give shorter lives. Romanowitz and Cruickshank (1968) refer to dredging operations in unthawed permafrost formations, mostly sand, in Alaska where the bucket lips wore as much as 1 in (2.54 cm) in a month thus giving a lip life of only a few months. An inside–outside rivetted lip was developed that gave a longer life to the fastening which finally developed into a two-bolted design. This enabled the complete changing of lips on a long bucket line in the space of about 4 h. Current practice favours casting the lip integrally with the hood and base; wear is compensated for by welding inserts into the lip portion.

Bucket pins require to be very tough and strong because if one breaks, the whole of the bucket line will collapse into the pond. Pins are constructed from nickel chrome, molybdenum and other high-quality alloy steels and should be fitted accurately and be well seated. Deep digging mechanisms impose high stresses due to the catenary pull on the underside of the ladder; idlers are installed to reduce the effect in most deep-digging dredgers.

Bucket size

Buckets are sized very largely according to the dimensions of the area to be dredged and the ease of dredging. Hence, although increased bucket sizes lead to increased production, there are practical considerations that limit the size. For example, large boulders, cemented wash and rocky bars impose shock stresses that must be compensated for by providing extra strength in both structure and drive units. This necessitates smaller buckets for a given size of hull and weight. Small buckets are also used for dredging shallow deposits because of limitations in hull sizes and manoeuvrability. Convention also appears to have a significant influence in determining bucket sizes; in the USA, 510-l buckets are are the largest for deep (> 40 m) dredging whereas in Malaysia 680-l buckets are common. At the Urkutsk No. 2 plant in the USSR 600-l buckets are used for a 50 m maximum depth. In Columbia the usual bucket size is 400 litres; the larger buckets are usually cast in three sections, base, hood and lip.

Bucket speeds

The speed of a bucket line is also a determining factor governing dredger output and obviously the speed can be safely higher in good ground than in more difficult conditions. In Columbia where conditions are excellent the bucket speeds range from 32 to 35 buckets/min. However, again, convention has influenced decisions to some degree and most dredgers around the world work with speeds of around 20–24 buckets/min regardless of digging conditions. It is interesting to note in this respect that at the Irkutsk No. 1 plant in the USSR, dredgers with 250-l buckets have line speeds ranging from 20 to 34 buckets/min. Obviously this variation in speed has to do with variations in digging conditions and the dredger has been designed for operational flexibility.

The control of bucket-line speeds and, thus, the ability to achieve optimum throughputs in modern dredgers is obtained by using the Static Stepless Control System for both AC and DC motors. This system has, to a large extent, taken over from the earlier Ward Leonard System. The SSC system for dredgers incorporates features to compensate for overload and vibration and can be highly sophisticated or relatively simple for operation in remote regions.

6.11.4 Operation

Manoeuvring

Manoeuvring the dredger and holding it in its digging position is effected using winches and land lines with or without spuds. In good digging conditions, head, bow and side lines are anchored to the banks or pond bottom and tightened or slackened independently, using multi-barrel winches, to move the dredger about and hold it in its working position against the face.

In more difficult digging conditions, the reactive thrusts are taken up by rivetted steel columns or boxworks (spuds) which are raised or lowered from the bed using steel cables running over pulley blocks at the aft gantry end. With one spud raised, the side lines are manipulated to swing the dredger about the other. Figure 6.22(a), (b) demonstrate the main features of each method.

Dredgers that use a headline system with a bow anchor for holding the dredger in position are known as 'New Zealand' type dredgers and these were the first types developed. Suitable only for good digging conditions, the system was soon modified by the introduction of box girder spuds to increase the digging capacity in tighter formations with more boulders and at greater depths. This is the California-type dredger.

Digging order

Most deep dredging operations provide for the selective removal of the upper barren layers by digging from the surface downwards. The commencing angle for the ladder is usually about 15° from the horizontal and the overburden is removed in strips down to the payable horizon. Provision is made for passing the valueless material directly through to waste, thus by-passing the recovery system. Only the pay gravels are treated and the treatment plant is used only for part of the available time.

Where conditions permit, two dredgers may be used; a suction cutter for stripping and a bucket line for mining and treatment. This has the advantage of providing for more effective process control on an around-the-clock basis using a smaller treatment facility. However, although the capital costs are reduced, there may be other considerations. In difficult dredging ground, one large dredger will have greater strength and be able to apply higher break out forces and this might be a deciding factor. Labour costs are higher for operating two dredgers than for one and there are likely to be increased costs for handling and disposing of the material stripped. Each case must be considered individually, balancing capital

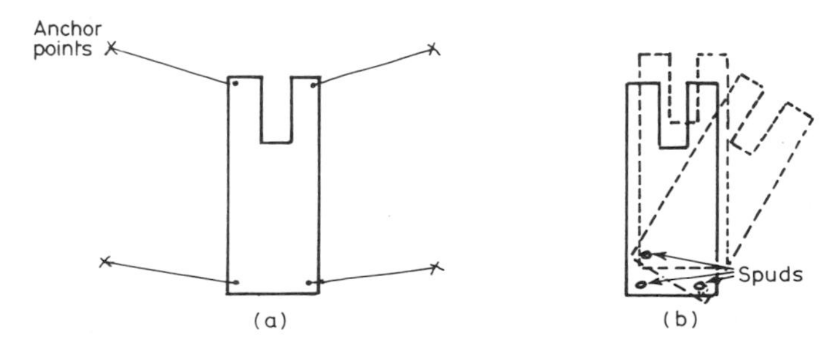

Fig. 6.22 Diagram showing main alternative systems for manoeuvring: (a) using side lines, (b) using spuds

costs against operating costs on the one hand, and the relative effectiveness of the two systems on the other.

Shallow deposits, or deposits in which the values are distributed from the surface downwards are usually dredged over the full depth of the face. All of the bank material is mined and treated and the treatment plant runs continuously. Full-face cuts may also be taken in deeper ground if there is a mineral-processing advantage to be gained from mixing the upper and lower sections.

Tailings disposal

Dredgers floating in artificial ponds advance by excavating spoil at the front end and dumping it at the back or to the side. The waste, which comprises both overburden and treatment plant rejects, must be dumped or stacked at a sufficient distance from the dredger to allow it to float freely and be fully manoeuvrable at all times. Slimes must not be allowed to build up in the dredge pond and should, preferably, be kept out of the treatment circuit. The more granular tailings should be replaced in such fashion as to yield the greatest future benefit for agricultural or other community purposes.

Slimes

Small quantities of slimes are sometimes useful in helping to minimize seepage through the pond walls but large quantities must be isolated or removed from the system. This is done on board using desliming cyclones ahead of the digging section and, in the pond, using a slimes pump sucking from near the bottom of the pond. Most modern dredgers incorporate a submersible slimes pump mounted near the bottom of the ladder. A floating pipeline is used to transport the slimes to the shore and thence to the slimes disposal area. A slimes pump having a similar discharge line arrangement can also be mounted on a separate barge for free ranging in the dredge pond.

In extreme cases, where slimes are allowed to build up in a pond, the trash held in suspension due to the increased density and viscosity of the slurry eventually rises above the hull bottom, giving it increased free board and imposing additional stresses on motors and drives for swinging. Cleaveland (1976) notes that dredger operators in Malay's jungle areas keep track of the paddock density by probing for the 'trash level', a mat of slurry-cum-sludge-cum-mud, and do something about removing it only when there is a danger of the pump intakes becoming choked.

Solids

The trommel oversize normally passes directly onto a stacking belt for disposal but some dredgers, particularly the older ones, return all tailings to the pond using open chutes and launders. At shallow digging depths this may not matter too much but because of the swell factor, usually around 30%, and the flat angle of repose of the fine sediments, about 24°, deep-digging dredgers require the tailings to be deposited further away from the dredger, thus necessitating longer chutes and higher starting points. The larger super-structure then requires an increase in dredger hull size to compensate.

Alternatively, some form of conveyor belt stacking system may be used with a high-speed belt section at the end to act as a 'thrower'. In such an arrangement, the tailings are dewatered using sieve bends or classifiers ahead of the belt. Some of the tailings may also be pumped away from the dredger if the additional costs for doing so are justified by the resulting larger pond area for manoeuvring.

Side-stacking dredger

The side stacking of tailings allows systematic dredging backwards and forwards across a deposit without having to re-dredge previously stacked tailings or, alternatively, leaving narrow strips unmined between cuts as described in Fig. 6.23. The chute arrangements are reversible and allow the dredger to swing around in the opposite direction at the end of each cut.

As an example of the additional thought now being given to designing

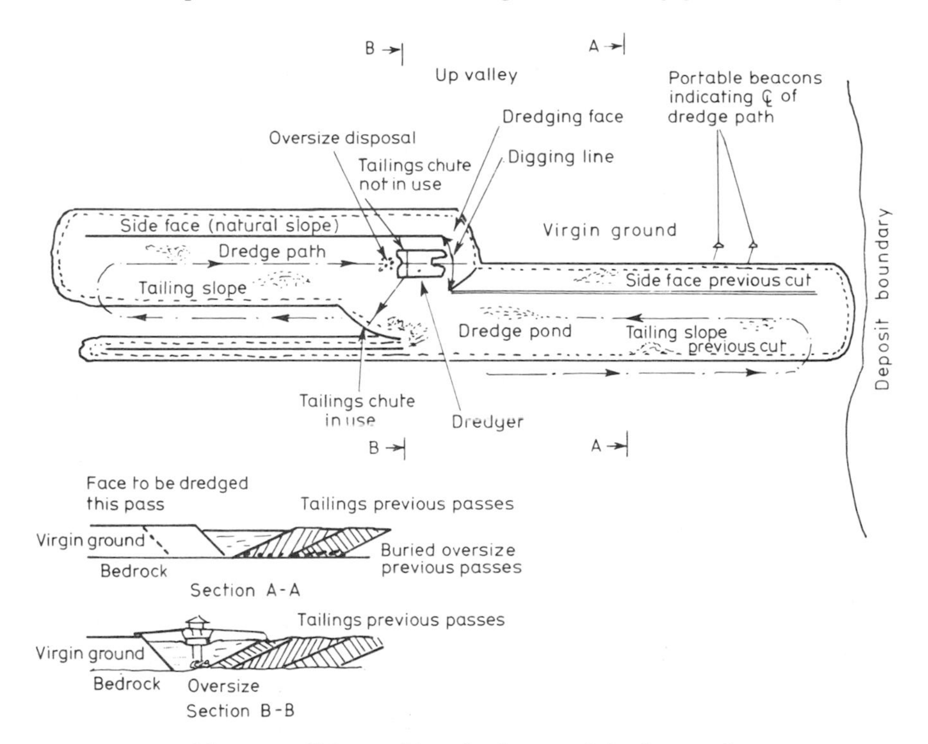

Fig. 6.23 Side stacking dredger and dredger path

dredgers specifically for the conditions under which they will have to operate, the IHC Holland Group has designed an 85-litre bucket-line dredger weighing only 200 tons for service in small and remote areas. The dredger has the following basic specifications:

Dimensions:	Hull length 19.25 m; width 10.0 m; draft 1.5 m, dismountable into sections of maximum weight 12 tonnes and dimensions $10 \times 3.5 \times 2.5$ m.
Capacity:	$135\,m^3\,h^{-1}$ at 75% bucket fill to maximum dredging depth 10 m.
Dredging gear:	Bucket speed 8–27–35/min.
Treatment plant:	Trommel screen 4.5 m long × 1.2 m diameter. Sparge nozzles and pump $400\,m^3\,h^{-1} \times 30$ m head; IHC primary jigs 42 in × 42 in cells, six lines of three cells; IHC secondary jigs 42 in × 42 in cells, four lines of two cells; all jigs variable stroke and variable feed, jig feed pump $590\,m^3\,h^{-1}$; pump for hutch, make up and sluicingwater $280\,m^3\,h^{-1}$; bucket monitor pump $40\,m^3\,h^{-1}$.
Power plant:	Diesel generating set 214 kW
Optionals:	Tailings pump, desliming cyclones, slimes discharge pump, tertiary jig etc. require additional power facilities.

Tailings stacked to the side of the dredger as shown in Fig. 6.23 are expected to assume a natural tailings slope of 24°. The dredger uses two spuds for lateral movement and advances while swinging. The ladder is hoisted and lowered using hydraulic rams. Concentrates are either transported ashore in drums or pumped through a small floating pipeline to a stockpile area for final upgrading.

The manufacturers state that one such dredger was transported on trucks to a remote mining prospect and assembled on site by the mine personnel with only primitive means in 5 weeks. The equipment is such that most maintenance requirements are readily handled by ordinary skilled mine personnel.

Restoration

Fundamental differences in stacking procedures distinguish 'New Zealand' from 'California' dredger types. The 'New Zealand' dredger deposits the spoil in somewhat the same stratigraphic sequence as it was in before mining by providing long tailings flumes and short stackers off the stern for the gravels. The basic Californian type does the opposite, thus reversing the order of deposition and leaving the gravels on the surface. Modifications have resulted in the California resoiling dredger design which leaves the ground in a suitable state for agriculture.

Considerable importance is now attached to environmental impact statements and experience in dredging in many different environments has shown that good restoration practice in one area is not necessarily good practice in another. As distinct from restoration in the transitional placer environment where the requirement is for separate stacking and replacement of the topsoil and for the surface of the ground to be returned to its earlier contours, good practice in the continental placer environment may sometimes require quite the reverse order of replacement as shown by studies of dredge-tailing agriculture on the Rio Nechi, Colombia, by Schlemon and Phelps (1971).

Dredgers have operated on the lower Rio Nechi in the Department of Antioquia for at least 60 years and in recent decades the tailings have become a favoured material for agriculture. The local people, Colonos, follow the dredge path, planting crops in the debris piles and, over the years, the once barren tailings have become dotted with plantain groves. A similar use of tailings has been made in the diamond fields of Akwatia in Ghana. Ridges of tailings are now cultivated by the local Ghanaians where previously the ground supported only swamp and jungle plants.

It has been found, however, that not all tailings heaps are equally productive as farming land nor are they all suitable for the same crops. Schlemon and Phelps recognized six distinct alluvial units along the lower Rio Nechi that, when dredged, provided various combinations of sediments and similarly various tailings types. Figure 6.24 illustrates the alluvial morphological units of a typical bucket-type gold dredger on the Rio Nechi. The basal channel and buried terrace gravels give rise to tailings most suitable for plantain cultivation whereas the leached bench gravels almost invariably yield infertile tailings. According to the authors, the unique use of the Rio Nechi tailings stems from their physical properties, in particular the clay: gravel ratios and to the inversion of the topography. The ideal ratio is 1:1 and, in contrast to the highly acidic (4.5–5.0 pH) soils of the surrounding area, the tailings measure about 6.5 pH. This value has changed little over the past 10 years of cultivation.

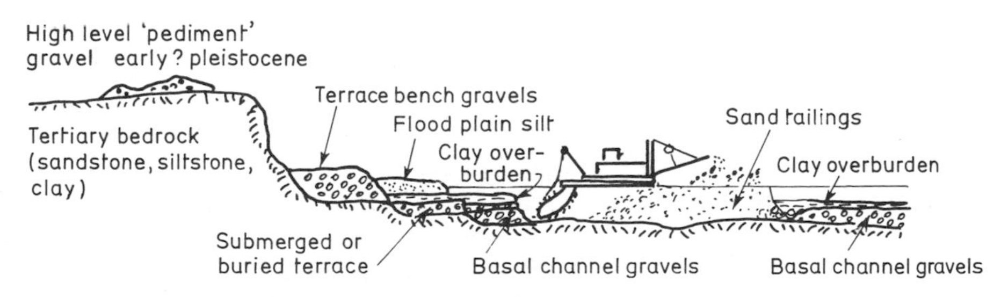

Fig. 6.24 Alluvial morphological units and diagram of typical bucket dredging operation on Rio Nechi. The basal channel and buried terrace gravels give rise to tailings most suitable for plantain cultivation

Such effects are likely to apply in many other humid, tropical regions and although restoration patterns vary according to the local environment many previously marshy waste areas can be turned into arable ground by depositing the dredger tailings in ridges or heaps sufficiently high above flood level for the crops to be safe. Many settlements, both small and large, have sprung up in Colombia over the past decade, supported entirely by farming on tailings.

6.12 SUCTION-CUTTER DREDGERS

6.12.1 Description

Primitive suction dredgers were made from rafts of timber and empty oil drums on which were positioned sand suction pumps and small water pumps for jetting and priming. Dredging was carried out by lowering the suction pipe onto the bottom of the pond and jetting around it with water to fluidize the sand so that it could be sucked into the nozzle. The sand was pumped away through floating lines to a land-based treatment plant. The arrangement was workable only in free-flowing sands and heavy mineral losses were high (Macdonald, 1962). The main advantages of such dredgers were simplicity, low cost and ready portability for mining small, isolated high-grade deposits.

Cutter heads were introduced to break up the more compacted and indurated materials and to increase the slurry densities by directing the broken material towards the nozzle opening. Cutter heads composed of curved steel blades, drive shaft and drive machinery were mounted on a ladder with the dredge suction pipe. The ladder was pivotted downwards at an angle from the pontoon and raised or lowered as required using small hydraulic motors and a gantry-pulley system.

Improvements in modern dredgers relate mainly to cutter-head design, gravel- and slurry-pump design and to materials of construction and methods of control. Dredgers for offshore work have both dredging and treatment facilities mounted in the one hull. The concept was also applied to some early dredgers on Stradbroke Island, Queensland, Australia but they lacked mobility and manoeuvrability. Modern onshore dredgers are preferably self contained so that they can range freely in the pond without the need for the plant to follow every movement. The slurry is pumped to a surge bin at the treatment facility through a flexible floating pipeline supported on floats. In some small land-based operations the concentrating units are skid mounted or in some other way made mobile but major economics can usually be effected by floating all plant on a pontoon directly behind the dredger (Fig. 6.21). Manoeuvring in the dredge pond may be effected through a combination of spuds and anchor lines although, in most operations, only crossed bow lines are used.

Although practical design considerations limit the suction lift of hull-mounted pumps to about 5.5 m, single-stage slurry pumps can pump efficiently against heads of as much as 60 m of water; delivery lines of 300 m and more in length are common. Suction-cutter dredgers are more power intensive than bucket-line dredgers, some of the larger dredgers have 6000–7000 HP connected to pumps and cutters, however, the stresses imposed upon ladders and hulls are much lower and they are constructed with overall weights of the order of 1.6 tonnes $m^3 h^{-1}$ of capacity in free-flowing sands, rising to 2.0 tonnes $m^3 h^{-1}$ for hard digging. As a result, capital costs are lower and operational costs higher compared with bucket-line dredgers of the same rated capacity provided that each is working in its particular field of application. Davis and McKay (1969) have illustrated this point graphically by comparing the capacities and costs of building and operating bucket-line and suction-cutter dredgers under a range of conditions. Figure 6.25 presents the data according to the economics of those times. Although the actual costs have increased considerably since then, the ratio of costs and performances illustrates the clear distinction between conditions that are suited to the one and those

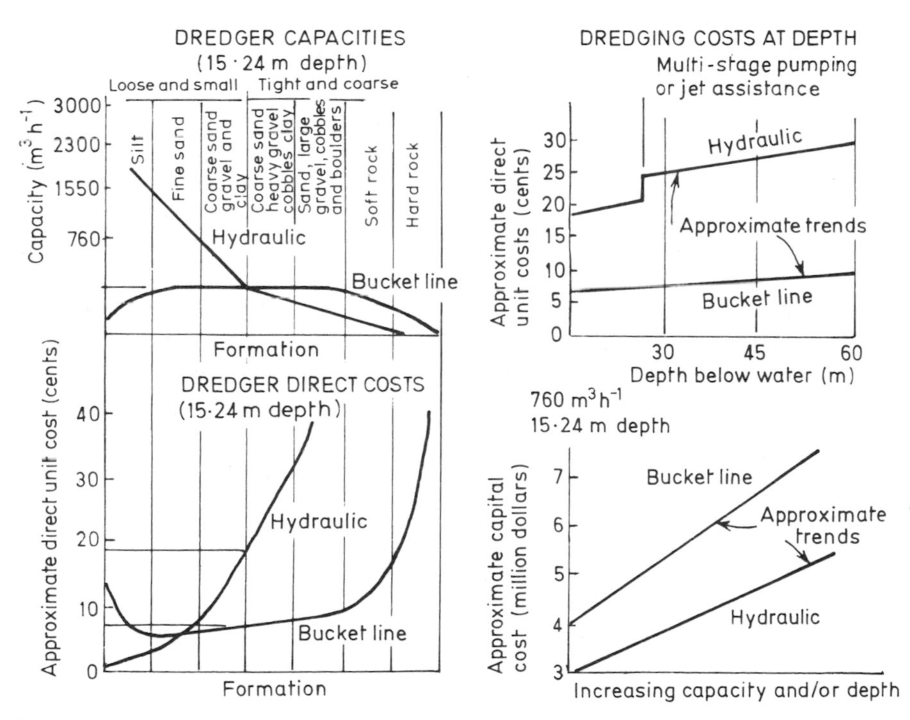

Fig. 6.25 Capacities and costs of building and operating bucket-line and suction-cutter dredgers in varying conditions in 1969 (from Davis and McKay, 1969)

that are suited to the other. The authors point to cases in which features of both systems can be used to advantage in the one project.

6.12.2 Applications of suction-cutter dredgers

Suction-cutter dredgers like bucket-line dredgers are high-capacity, continuous-feed machines with applications in placer mining that include both the digging of ore and stripping of overburden. They differ in their respective applications largely according to the nature of the ground to be dug and their cutting actions and, whereas bucket-line dredgers can dig successfully into tight formations containing cobbles and boulders, suction-cutter dredgers work to full capacity only when dredging fine unconsolidated materials. Because they undercut the mining face and rely upon the sand rilling freely towards the nozzle, suction-cutter dredgers are limited to a non-sloughing above-water bank height of 2 m to avoid dangerous shock waves from the bank collapsing. Bucket-line dredgers can carry a bank of up to 5 m in height because of their ability to cut downwards from the surface at some distance above pond level.

One of the main features of suction-cutter dredgers favouring their selection for many dredging duties, e.g. large-scale overburden stripping and land reclamation, is their ability to mine and transport large quantities of sand over considerable distances in the one operation. Dredgers can be constructed with capacities in excess of 2000 $m^3 h^{-1}$ of solids using only one pump and deliver the spoil through pipelines for distances of 400 m and more. In mining high-level dunes along the east coast of Australia, dredgers with pumps mounted in the hull have handled, successfully, faces of free-flowing sand rising to as high as 45 m above pond level.

Suction-cutter dredgers working offshore (e.g. near Phuket, Thailand) already operate at depths of up to 35 m below waterline using specially designed pumps and drives installed close to the bottom of the ladder. Future dredgers onshore will eventually be expected to handle faces up to 100 m above pond level and offshore, depth limitations will probably depend entirely upon mechanical design considerations and economics; there are no hydraulic limitations to the heights to which slurries can be elevated by pumping.

6.12.3 Construction details

Cutter heads

Rotating cutter heads are driven through suitable gearing at between 15 and 45 rpm depending upon the nature of the materials and the size of the cutter. The cutting devices are divided into two main classes:

(1) Those that rotate on a shaft running approximately parallel to the suction line. Typical designs are crown, basket and helicoidal.
(2) Those that rotate on a shaft running at right-angles to the line of the suction pipe and appear as mirrored twins on either side of the suction opening (bucket wheels).

Although designed primarily to fluidize beds of relatively clean free-flowing sand, cutters must be capable, also, of resisting the high stresses and wear involved in digging more consolidated materials such as indurated sands and sands that are compacted with clay and fine gravel. They should also be capable of dealing with such trash as tree roots and vines because, regardless of how well the ground surfaces are cleared ahead of dredging, some of the deeper roots will remain in place and will, if not handled effectively by the cutter, cause pump blockages that may require the whole operation to be stopped while they are being cleared. Cutter-head assemblies deal with such trash by chopping the fibrous materials into pieces small enough to pass freely through the pump without interrupting the flow.

Cutter-head wear takes place by scratching and abrasion and by impact when striking rocks or other hard solids. Consequently, the materials of construction must be able to resist a combination of wear factors and the necessary parts are furnished to minimum high-tensile strength levels of, (a) 8000 kg cm^{-2} for the arms and rings, (b) 14 000 kg cm^{-2} for the adaptors, and (c) 17 500 kg cm^{-2} for the tips and points. Venner (1973), emphasizes that the steels used must be tough as well as hard in order to minimize breakage; the wearing parts should also have the lowest carbon level commensurate with strength requirements to provide good weldability and allow hard surfacing materials to be applied.

Dredge pumps

Dredge pumps, as the most important single units in suction-cutter dredgers have been discussed in some detail in Chapter 5 in which a typical design calculation demonstrates the method of pump selection for specific duties.

6.12.4 Operation

Manoeuvring

Manoeuvring may be effected through the use of head, side and bow lines as for bucket dredging, however, crossed bow line manoeuvring is the most common system for hydraulic dredgers. Where hard digging conditions call for a spud arrangement to take up the reactive stresses the three most favoured systems are as follows.

Independent twin spuds
Two spuds located in sockets aft in the dredger govern the swing. The dredger is swung around the working spud, with the other spud raised, by means of anchor lines set behind the bow of the dredger. The cutter works progressively downwards through the arc to the required dredging depth. It is advanced by means of another set of anchor lines set at an angle on either side of the centre line of the channel after lowering the auxilliary spud, raising the working spud and hoisting the ladder.

Spud-carriage system
The working spud is located in a carriage travelling in a well in the afterpart of the dredger. An auxilliary spud is mounted in a guiding device on the pontoon for temporary anchorage. The distance of travel is usually around 6–7 m but can be increased if required.

Swinging by moving the ladder in an arc
Dredgers constructed for this purpose are anchored by three or more spuds which are then used to walk the dredger forward or to the side. Such dredgers may also be made fully manoeuvrable using mooring cables, anchors and winches but more effective cutting is achieved with spuds taking up much of the thrust.

Digging

Suction-cutter dredgers cannot dig very selectively and their bottom-digging characteristics are quite unsuitable for some conditions. In comparison with bucket lines which have similar cutting characteristics on both port and starboard swings, suction cutters, rotating in one direction only, cut evenly on one swing but tend, on the other, to ride up and over any harder parts and obstructions. In softer digging conditions where bucket lines leave behind a relatively even floor, suction cutters form potholes into which valuable particles sink and are lost.

Dredging efficiency and control

The efficiency of a suction-dredging operation is measured by the output of the dredger pump. This, in turn, depends upon the percentage of solids in the slurry pumped, the rate of flow and the actual time spent in dredging. Density control is exercised by varying the pressure of the cutter against the face, the aim being to hold the pump suction pressure just below cavitation point and the pump discharge pressure just below plugging point. Flow rates and time availabilities are governed by how well the individual units are maintained and how closely the dredging parameters are measured and controlled.

The extent to which the optimum pressures can be sustained in a manual dredging operation rests with the skill, knowledge and experience of the operators and their motivation. The overall control of dredging is, thus, a measure of how individual operators perform their duties and cope with a variety of different digging conditions. Experience has shown that even the best operators can rarely maintain slurry densities higher than 25 % of solids by weight; the average for manual operations is generally closer to $C_W = 20\%$.

On the other hand, large modern hydraulic dredgers equipped with nuclear, magnetic and differential pressure meters for flow measurement; gyro compasses to measure and record swing angles; pneumatic depth recorders, closed circuit television, sonar profiling and other automative devices can regulate their performances in accordance with optimum standards. There is less emphasis on operator skill and dredgers, automatically controlled, may consistantly average slurry densities of $C_W = 30\%$ in normal ground conditions.

Automatic control equipment integrates the various pressure and other readings and compensates automatically using an electronic system to evaluate the data and issue the relevant commands. The slave mechanism pushes the cutter up against the face and holds it at constant pressure. Some systems also provide visual indication of the cutter head position relative to the face and compensate automatically for changes in the dredger position to give a linear indication on a graphic scale. Any desired channel configuration can be programmed in advance and the operator has only to follow the chart.

The relative efficiencies of dredging controls flow on also to the concentrating plants where individual units require to be fed under closely controlled conditions. Primary concentrating units operate satisfactorily only when the feed rate is held constant and the slurry density is maintained at around 60–65 % of solids by weight (see pinched sluice separators, Chapter 7). Such conditions require a surge bin of adequate capacity between the dredger pump and the concentrator. Even when dredging in ideal conditions, the output from a dredger pump fluctuates widely and cannot provide a consistantly dense slurry feed. To achieve such conditions, the surge bin is equipped with a densifying mechanism on the suction side of the pump to regulate the proportions of solids and water entering the pump; sufficient surge capacity is usually provided for about 10 min of continuous operation. A simple but effective densifying arrangement has already been illustrated in Chapter 5; there are many variations including jet-assisted suction inlets.

Most dredging operations require surge-bin capacities of up to 10 min to allow for dredging time lost in manoeuvring and for shut-down periods when removing trash from around the inlet nozzle basket. A constant-level alarm system warns the dredger operator when sand levels in the bin are either too high or too low so that he can regulate the density of the dredger pump feed accordingly. He will either raise the ladder slightly to reduce the sand intake or bury the nozzle deeper into the sand to increase it. Excess water flowing back into the dredger pond is a source of significant losses in many installations using inadequately sized bins.

Surge bins with densifying controls are either situated on separate pontoons or, as is more usual, incorporated with the concentrating units on the treatment pontoon.

Land-based operations require the bin to be located alongside the concentrating plant and provided with the same type of mobility as the plant. Offshore, where the dredging plant, screens, surge bin and concentrating units are all contained in a single, self-propelled vessel built to maritime standards, one of the most severe constraints is the need to compensate for high waves and swell. The present tolerance is for waves less than 1.5 m high. In one offshore application near Phuket, Thailand, a sensing device was located further out to sea to give warning when large waves were coming towards the dredger. On receiving the signal the ladder would be raised to avoid damage, and lowered again when the wave had passed. This procedure helps to avoid damage but dredging time is lost and the basic problems of coping with greater stresses from larger waves have still to be solved.

However, new methods for compensation and control are being tried. One, described in *Mining Magazine*, July 1972, proposes maintaining continual digging contact at the dredging face throughout the wave and swell cycles by paying out and taking in the ladder hoist lines while manipulating the head, stern and quarter lines. The system is automatic and is controlled electronically. It is stated to compensate for any fore and aft trim variation in the dredge hull and to be applicable to either one-piece ladders or to multiple-section articulated ladders provided the dredger works on lines and not spuds.

6.13 BUCKET-WHEEL DREDGERS

Manufacturers have, for some years, taken much interest in combining underwater bucket-wheel excavating mechanisms and pumps for dredging. There were many early problems and a prototype used for mining beach sands near Wyong, New South Wales, Australia a decade or so ago was later replaced by a suction-cutter dredger because of high maintenance requirements and frequent stoppages due to an inability to deal with plant roots and other buried debris.

These problems have now largely been overcome and models in standard sizes are now coming on the market in direct competition with suction cutter dredgers. Advantages claimed for the method are:

(1) A closer control of slurry densities because of the more positive breaking out of bank material. The cutting forces are concentrated on

small cutting edges which allow forces as high as 140–200 kg cm^{-2} to be imposed on the surface to be cut.

(2) The ability to dig equally effectively in either swing direction thus allowing it to clean up efficiently along a soft basement.

(3) Fewer losses around the intake nozzle because the material, as it is dug is channelled directly into the suction inlet.

(4) A lesser degree of either overdigging or dilution because of the ability to clean up along inclined beds as well as horizontal beds.

(5) A closer control of flow rates and solids output because of the ability to regulate both the pressure against the face and the speed of rotation at the same time as the spoil is channelled away from the face.

According to Turner (1976) bucket-wheel dredgers also have an 'amazing ability' to deal with submerged stumps. He cites, as an example, a rig operating successfully in hard pan material in Florida where dredging is complicated by a heavy concentration of cyprus stumps.

Bucket-wheel dredgers manufactured by Ellicott range from units with small buckets and 12 in (28.8 cm) pumps that are claimed to handle 300–350 m^3 of sand per hour, up to larger wheels designed for 24 in (57.6 cm) and 30 in (76.2 cm) pumps which can dredge four or five times this quantity of solids. All units are capable of passing occasional stones up to 60% of the suction pipe diameter. Ellicott has also developed a new spud system that is stated to offer many advantages over conventional systems.

Orenstein and Koppel, A. G. (O & K) have introduced a prototype model UWS250 featuring a 5 m diameter bucket wheel fitted with ten 250-l buckets discharging at the rate of 1200 $m^3 h^{-1}$ up to peak outputs of 1900 $m^3 h^{-1}$. The drive power of 380 kW compares very favourably with the requirements of suction-cutter dredgers of comparable size.

The range of applications for bucket-wheel dredgers appears to be extending into areas formerly believed suitable only for bucket-line dredgers. This is largely because, like the bucket-line dredgers, they can clean up along a weathered bedrock leaving a relatively flat surface behind instead of one that is scarred with potholes. It has still to be shown that large bucket-wheel dredgers can successfully challenge bucket-line dredgers in typical placer tin and gold dredging applications, however, the method is comparatively new and the competition is good for both designers and operators. Figure 6.26 illustrates the main features of the bucket wheel in operation.

6.14 JET-SUCTION DREDGERS

This type of dredger has been used mainly for dredging sand and gravels for the construction industry and for small-scale, diver-controlled operations as described in Chapter 5. It utilizes a jet-ejector system in which

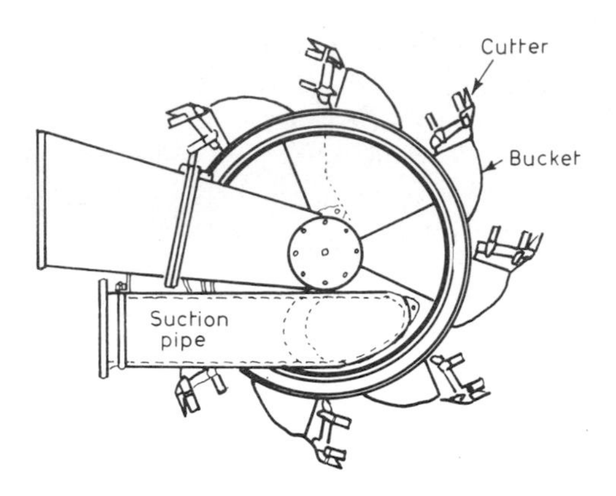

Fig. 6.26 Ellicott bucket wheel for dredging

flow velocities and pressures are derived from the output of a conventional high-pressure centrifugal water pump. A pressure drop is caused when high pressure water is injected into a mixing chamber in the digging device located at the bottom of a long, flexible delivery hose. Solids are drawn through the suction inlet into the vessel and the mixture is elevated to the surface. There are no mechanical parts in contact with the slurry and large stones may be dredged without causing obstructions. A 90 $l\ s^{-1}$ jet-lift pump will transport gravels containing stones up to 15 cm diameter without difficulty.

In the mining of gravel beds offshore, the digging device is able to penetrate any overlying fine sediments by jetting and only when the gravels are reached, is the suction bell mouth opened out for dredging Since there is no suction head, but only a few entrance losses to overcome, the limit of elevation for the spoil is determined only by how much power is available in the form of pressure and velocity energy. Current units raise materials to a height of around 100 m but additional lifts are purely a matter of increasing the power availability and staging.

Diver-operated, jet-suction dredgers, although comparatively new to the placer-mining industry are beginning to fill a very important role in cleaning up operations around rocky bars and potholes in the beds of streams where dredgers cannot dig satisfactorily and for small-scale river operations. They range in size from models small enough to be carried by one person up to units capable of passing rocks as large as 150 mm in diameter. A jet pump manufactured by Alluvial Mining (UK) Ltd has a throughpass clearance of 152.4 mm and is powered by a fully submersible, electrically driven, high-pressure water pump which also provides power for the remote jetting lance operated by the divers. The unit weighs approximately 1.5 tonnes and is readily handled by a 5 tonne derrick. It can be either suspended or free-standing when operated. The water pump is a three-stage 96 HP, 440/460 V, three-phase 60 Hz unit with an output of 1000 gpm (75.8 l/s) at 240 f (73.15 m) head. Power requirements are 130 amps FL currents and 450/500 amps starting current. A solids output of up to 60 tonnes h^{-1} is claimed for pumping free-flowing sand through 30 m of suction pipe at zero elevation and 30 m of delivery.

One such operation developed in Venezuela for dredging gold and diamond river placers is now operated by divers wearing conventional diving gear instead of the scuba gear used generally elsewhere. Divers clad in this manner have greater stability when working along the bed of a stream and are able to direct the high-pressure water jets more effectively to break up compacted materials.

Wear in jet-lift pumps is confined to the mixing chamber where velocities are high; otherwise the pump maintenance and replacement costs are minimal.

6.15 DRAG-SUCTION DREDGERS

Drag-suction dredgers, or hopper dredgers as they are also known operate by trailing a drag head in contact with the bed material by vessels which move under their own power. The material is sucked on board and stored in hoppers for disposal elsewhere. Drag suction dredger pumps are high-lift, low-head types developed specifically for this duty (see Chapter 5).

Drag-suction dredgers are the only types of dredgers able to work effectively, safely and economically in the open seas and are responsible, offshore the UK alone, for the production of some 13 million tonnes of sand and gravel annually. Fine silts are a problem in some applications because of difficulties of settling in the hoppers and reduced outputs and ecological problems resulting from the overflow, nevertheless the method has many attractive features:

(1) Dredgers can be moved quickly and cheaply to other locations under their own power.
(2) By excavating a shallow cut during each pass over a gravel bed the dredger accomplishes a progressive increase in digging depth as the work proceeds, thus taking advantage of any pronounced stratification in the beds.
(3) The method can be used for agitation dredging whereby material that settles slowly is sucked up and discharged overboard to be carried away by direct channel currents or cross currents.
(4) The ripping action of the drag head allows it to dredge, without blasting, into more compacted types of underwater beds than any other type of suction dredger.
(5) The method can be adapted for stationary operation during which manoeuvring can be achieved by means of an anchor winch.

An inherent disadvantage is the intermittent nature of the operation. Part of the working time is taken up with filling the hoppers, the remainder in transporting the material to shore. Actual dredging times may often be as low as 20% of the total operational time and direct loading into barges is preferable in calm waters.

REFERENCES

Cleaveland, N. (1976) Dirt, diamonds and gold dust, in *Placer Exploration and Mining, A Short Course*, Mackay School of Mines, Nevada.

Cross, D. J. (1979) Dredges in the alluvial mining industry, *Mine and Quarry*, October, 58–79.

Davis, C. and McKay C. E. (1969) Mine planners see wider use of dredging, *Engineering and Mining Journal.*

Hodgeson, D. L. (1978) Mining the beach for diamonds at CDM, in *E/MJ. Operating Handbook of Surface Mining*, Vol. 2., McGraw Hill, New York.

Macdonald, Eoin H. (1962) Some aspects of suction dredge design, *Proc. Aust. Inst. Min. Met.*, 204, 59–77.

Pakianathan, S. and Simpson, P. (1965) Open cast mining of alluvial tin deposits in Malaysia, *Eight Comm Min & Met Congress*, AIMM, Melbourne.

Popov, G. (1971) *The Working of Mineral Deposits* (Trans. from Russian by V. Schiffer), MIR, Moscow.

Rowmanowitz, C. M. and Cruikshank, M. J. (1968) The evolution of floating dredges for mining operations, *Proc. Inst. Surface Mining Conference*, Minneapolis, Minnesota.

Rose, T. K. (1902) *The Metallurgy of Gold*, Chas. Griffin & Company Limited, London.

Schlemon, R. J. and Phelps, L. B. (1971) Dredge tailing agriculture on the Rio Nechi, Colombia, *The Geographical Review, California*, LXI, 3, 396–414.

Shekyakov, L. (1970) *Mining for Mineral Deposits*, Foreign Languages Publishing House, Moscow.

Turner, T. N. (1976) The bucket wheel hydraulic dredge – the modern mining tool, in *Placer Exploration and Mining; Short Course*, Mackay School of Mines, Nevada.

Venner, L. J. (1973) Quality control for cutters, *World Dredging and Marine Construction*, May, 21–5.

7

Placer Minerals Processing

7.1 INTRODUCTION

Having extracted placer material from the ground, the next task is to reduce its bulk and recover the valuable minerals for sale. Where only one product is concerned, e.g. gold, and the ores are simple accumulations of readily dispersed particles with qualities that lend themselves to easy separation and concentration, the recovery processes also tend to be simple and some highly profitable operations require only pumps and sluice boxes or jigs for making finished products. Others are more complex and pose problems that only extensive test work can solve. For some ores it is a matter of determining how to liberate the valuable particles from tenaceous clays or cemented wash materials; for others, it involves learning how to deal with persistent coatings of organic and inorganic materials that interfere with the normal responses of individual particles to electrical and magnetic influences. Mineral-sands placers may contain as many as four or five different mineral types and require complicated circuitries for separation involving various forms of gravity concentration, attritioning, high tension, electrostatic and magnetic separation; size relationships are important parameters at all stages of the recovery processes and for subsequent marketing.

7.2 PROCESS DEVELOPMENT

Existing prototype operations offer little opportunity for scientific observation because many of the operating variables cannot be effectively isolated and measured. Programmes aimed at establishing design data for other commercial operations normally employ small-scale experiments as a guide to predicting the large-scale results. Tests are conducted, first on a bench scale in the laboratory and then in a pilot plant designed to simulate prototype-plant conditions. Model theory is applied to predict the scale effects and give a reliable forecast of prototype performance.

7.2.1 Bench-scale testing

The initial steps towards designing a commercial processing plant are taken in the laboratory where confirmatory evidence is sought of first impressions and direct tests are made to resolve important features of technical and economic interest. The apparatus for testing single product ores, such as auriferous gravels, may be relatively crude and tests commonly relate only to liberation, sizing and sedimentation. However, much greater complexities are associated with most transitional placer ores and a modern mineral sands laboratory will provide a comprehensive range of facilities covering every phase of project development.

Problems of liberation involve comparing mechanical and hydraulic scrubbing systems and, for experienced operators, hand puddling exercises are a valuable guide to predicting the required amount of energy. The relative effectiveness of various chemicals used for removing organic or metallic oxide coatings can be compared using only a few grammes of the affected particles. Screen sizing analyses provide useful data on fractionation for planning gravity concentration experiments using laboratory-sized shaking tables, jigs, spirals, heavy media separators and sluices.

Particles are examined mineralogically and mineragraphically to distinguish the individual mineral types and to examine the micro textures of individual particles. The transparent minerals are studied in respect of their optical properties, refractive indices, fluorescence etc. The opaque minerals are viewed in polished section to study the particle textures and to determine the levels and natures of impurities that may be present in the form of intergrowths, exsolution lamellae or inclusions. Valueless particles having somewhat similar physical properties and responses to those of the valuable particles, and tending to report with them at various stages of treatment, are identified. In specific instances doubts are resolved and individual particles may have their compositions confirmed by drilling out minute quantities of the mineral under the microscope and using X-ray diffraction and micron-probe methods of analysis on both the powder and the drilled surface. Lawrence (1965) suggests the use of fine dental burrs to drill particles of 0.25 mm diameter and smaller burrs can be specially made. The sequence for testing placer samples in the laboratory is generally as described in Table 7.1 for continental- and transitional-type ores.

Clearly not all placer analyses will follow the above sequences slavishly; steps are omitted in examining some materials; other procedures may be modified to suit unique problems. Essentially, however, the work will give the investigator an initial feel for problems affecting both separation and

Table 7.1 *Generalized model bench-scale testing*

Visual examination of sample – note any unusual features record all details and prepare programme		
Continental placer ores		Transitional placer ores
Slurry with water and decant slimes		Screen into appropriate size fractions. Analyse and prepare frequency distributions and histograms
Solids measure	Slimes measure	
Gravity separating tests Jigs, tables, micropanners etc. Analyse all fractions, and prepare curves, calculations etc.	Slimes settling tests Analyse and prepare curves	Magnetic and electromagnetic separation using isodynamic, belt and roll type separators to prepare fractions based upon magnetic susceptibility
		Attritioning, high tension and electrostatic separation; co-ordinatewith magnetic separation to prepare magnetic conductor, non-magnetic conductor, non-magnetic non-conductor and magnetic non-conductor fractions for analysis.
Detailed microscopic analysis all relevant fractions, fluorescence tests, radiometric tests, chemical and instrument analysis. Prepare full report		

marketing and, when the studies are completed there will be a reasonable understanding and knowledge of such ore characteristics as:

(1) The average particle size and distribution of both valuable and non-valuable minerals.
(2) The degree to which some particles are contaminated by foreign elements and other minerals, and the likely purity of concentrates prepared for marketing.
(3) The properties of valueless minerals tending to report with specific valuable mineral types because of similar properties, e.g. garnets reporting with monazite in high tension and some magnetic circuits.
(4) The extent to which the electrical properties of particles of some mineral types have been affected by staining, coating or alteration of the mineral surfaces.
(5) The extent to which similar coatings on particles of gold interfere with their amalgamation with mercury.
(6) The extent to which, as a result of these things, individual particles may react anomalously to the forces that will subsequently be imposed upon them in future processing operations.

It must be stressed, however, that, while bench-scale testing is of great value in defining the problems and in helping to obtain some of the answers it does not of itself provide all of the solutions when treating complex placer ores but merely points to the course to be taken for further experimentation on a scale intermediate between that of the laboratory and full commercial-scale operations. It is, of course, technically feasible to apply the bench-scale findings directly to prototype design, but only if unlimited funds are available for the designer to incorporate huge safety factors and for the operator to meet the costs of overcoming massive teething troubles. The enormous losses incurred by some operators a decade or so ago have shown the dangers of such a course of action and the safest method is to regard bench-scale test work as a preliminary fact-finding exercise from which the final project dimensions will evolve through a succession of studies conducted on a larger, pilot-plant scale.

In this regard there is no formula for deciding how much money can justifiably be spent on a pilot plant, or to what lengths the testwork should be carried; this is a matter for individual judgement and, at every point, the anticipated costs for continued experimentation must be weighed against the anticipated costs of ironing out any imperfections remaining in the process at the commissioning stage. It is to be emphasized that quite subtle differences in a range of operating parameters may sometimes result in very significant departures from planned prototype performance and such decisions should not be taken lightly.

7.2.2 Pilot-plant testing

A pilot plant is, essentially, a scaled-down version of the prototype with such additional in-built flexibility as needed to allow individual units to be superimposed on others or be taken out, and for flowlines to be changed or modified at will. Its purpose is to overcome technical problems for which there is no adequate theory and to test assumptions for which the supporting empirical data are inadequate. When such data are re-examined and enlarged upon there are likely to be fewer troubles on start up and smaller margins can be allowed for error. For some ores from which the valuable constituents, when extracted, will be used as raw materials for other processes (e.g. ilmenite concentrates for the direct chlorination route in pigment manufacture) the pilot plant will also provide samples for market research.

The scale of testing

Model theory can be applied directly to design scale-up provided that all of the pilot-plant dimensions bear a direct size relationship to those of the prototype. However, decisions on scale are also influenced by the fact that already small sedimentary particles cannot be scaled down in size without causing changes in rheological patterns of behaviour, some of which are unsupportable in the context of process solids–fluids flow. For example the behaviour of 20 μm-sized particles cannot be related by any known formula to that of a mixed population of plus 100 μm particles either in pipeline flow or as feed to separators in a dry milling process. For this reason, most pilot-plant units are single commercial-sized units or are scaled down in a dimension that does not also require a scaling down of the feed particle size, only its rate of flow. For dry-milling operations in high-tension and induced-roll magnetic separation a suitable reduction in feed rate is accomplished by reducing the roll lengths to 100–150 mm to limit the unit capacities. All other process conditions can then be simulated closely by holding constant all of the other dimensions of roll diameter, roll speed, ionizing field strengths, magnetic flux densities and so on. Scale up is then a simple matter of the multiplication of roll lengths.

Because of the varying complexities of placer minerals circuits, pilot plants are designed in either open or closed circuit, and this has considerable influence upon the sizes of samples to be tested. An essential pre-requisite is to have sufficient feed material available and the quantities involved may range from a few kilogrammes in weight up to several hundred tonnes depending upon the scale of the plant and the purpose of the study. For wet shaking tables in closed circuit a few kilogrammes only of feed material are needed. Pinched-sluice separators may be run satisfactorily on as little as 5 tonnes of feed in closed circuit. Full-scale tests in open circuit require quantities of feed materials appropriate to the rated capacity of the plant and to the duration of the tests. Such samples may amount to several hundreds of tonnes. Figure 7.1 illustrates the use of a Reichert Cone for closed-circuit test work at Mineral Deposits Ltd Laboratories, Southport, Queensland.

Selecting the samples

Bearing in mind that the ultimate purpose of pilot-plant testing is the design of a commercial operation based upon the ore reserves as a whole, the investigations must cover the full range of possible feed materials and samples must adequately represent the deposit both in size and in type. Important considerations are as follows:

(1) If the pilot plant studies are designed around only the most difficult of the possible range of feed materials the resulting prototype installation will be unnecessarily over designed and costly.
(2) On the other hand, designing for the best conditions only, may result in a prototype that is in some circumstances, inadequate to the point of failure.
(3) Designing for average conditions, in the absence of blending, will almost certainly result in a prototype that is satisfactory for part of the

Fig. 7.1 Reichert Cone in closed-circuit testing

time but with periods of inadequacy when both output and product quality is reduced.

Sample selection for testing normally considers how best to blend the material from different parts of the deposit to avoid extremes in the quality of feed entering into the mill. Each of the above alternatives represents a possible source of error when considered alone and the final tests should always be conducted using blended materials so that the prototype design can be concentrated about the practical mean.

7.2.3 Open versus closed-circuit testing

There can be little argument that it is much simpler to forecast prototype performance from open-circuit test results than from the results of closed-circuit testing, however, it is not always technically or economically feasible to build an open-circuit rig at the outset because of the large numbers of uncontrolled variables that may have to be evaluated. Fundamental differences in the two methods of testing are as follows.

Open-circuit testing

In open-circuit testing the supply of new feed is continuous and individual samples taken with the plant in equilibrium reflect the instantaneous characteristics of the flow materials at the point and time of sampling. Automatic sampling at pre-determined intervals provides composite samples that describe the overall characteristics of the flow. The main advantages of the system derive from the continuous recirculation of middlings fractions so that each can apply its full weight to the total flow.

Closed-circuit testing

In closed-circuit testing all products, middlings and tails fractions are returned continuously to the head feed bin and the total flow is in equilibrium when the materials in circulation have the same bulk properties as the bin material. The accuracy of sampling depends upon taking timed samples simultaneously across the total flow at the required number of sampling points, either mechanically or manually, using a sufficient number of samplers. The test is that the plant remains in the same state of equilibrium after sampling as it was before the samples were taken.

7.2.4 Scale up

Interpreting the results from batch-scale test work is much more difficult than for open-circuit studies. Mineral-sands, dry-mill circuits, in particular, are complicated by repeated cleaning, recleaning and scavenging operations requiring a multiplicity of individual units and flow lines. Only by mounting a full-scale operation with a massive degree of over-design could all of the required units be incorporated in a pilot plant; apart from creating chaos, this would defeat the purpose of the exercise. It is usual in such cases to have sufficient items of standard equipment available for grouping and to test each main part of the proposed circuit by batch testing, using the same equipment over and again. The individual results are then integrated and interpreted in the knowledge that some allowances must be made in final design for deficiencies in the data occasioned by the inability to recirculate middlings in the batch tests.

Technical aspects of treatment plant and equipment discussed later in this chapter apply equally to small-size and full-scale units and, if only one product is to be recovered, scale up is generally not too difficult, regardless of the system of testing. The additional recovery of bi-product minerals such as are contained in the 'amang' of south-east Asia introduces further complexities and possibly the greatest problems are encountered in the treatment of indurated fossil-dune deposits where, for example, a predominantly rich ilmenite feed might contain small but economically important quantities of rutile, leucoxene, zircon and monazite. Experience has shown that the final circuitry will probably be developed from elements of both open- and closed-circuit testing plus final adjustments that can only be made with the prototype in its commissioning phase.

Principles of similarity

Accurate scale up presupposes a reasonable correlation between the behaviour of the same class of materials in both pilot plant and prototype. The validity of the relationship depends upon satisfying or adjusting to the following similarity criteria between the model (pilot plant) and prototype, (a) geometrical similarity, (b) kinematic similarity, (c) dynamic similarity, and (d) operational similarity.

Geometrical similarity

This condition is satisfied when the ratios of all corresponding linear dimensions of two systems are equal. It can be achieved in primary wet concentrating plant by using one full-scale model such as a sluice, to represent a multiplicity of entirely similar units. If the feed materials are the same and the sampling is accurate, the slurries and the slurry system used in the pilot plant will be geometrically similar to those of the prototype and scale up is obtainable by simple arithmetic means.

Reference has already been made to geometrical similarity in dry-mill separation using high tension and magnetic equipment that may differ only in the length of the rolls, all other variables such as roll diameter and speed, flux density and electrode configurations being identical. Since the roll lengths only influence total and not average feed rates the scale up is again arithmetic.

Kinematic similarity

Kinematic similiarity depends upon the time and space relationships between identical particles flowing through the respective systems. Two geometrically similar systems are also kinematically similar when the ratios of the velocity components at all homologous points are equal. In other words, geometrically similar moving systems are kinematically similar when corresponding particles trace out geometrically similar paths in corresponding intervals of time (Johnson and Thring, 1957).

Dynamic similarity
Dynamic similarity is concerned with the rate of change of motion of particles in dynamic systems. The principle forces are those of gravity, inertia, pressure, impact and viscosity. Centrifugal forces are important factors affecting dynamic similarity in high-tension and induced-roll magnetic separation. In fluid flow systems, dynamic similarity has a direct bearing on establishing power consumption and head loss predictions.

Operational similarity
This is a function of the human equation. In a succession of plant shifts, the operators vary in their abilities and application to the task. The amount produced in any one period is the sum of their respective efforts and average performance figures are time-related means. On the other hand, pilot plant exercises are normally conducted by skilled technicians whose performance can be expected to be above the norm for plant operation. The allowance made for this in scale up is a matter for individual judgement having regard to the circumstances of each case.

7.3 PROCESS EQUIPMENT AND DESIGN

With the completion of pilot-plant studies, flow sheets are prepared to co-ordinate the mining and milling activities and to provide a metallurgical balance for both materials and services. The mill depends upon the mine for a balanced flow of feed materials and upon the various service departments for power, maintenance and supplies; procedures and equipment for processing vary according to the characteristics of the feed and its valuable constituents and are considered under four main headings, (a) feed preparation and control, (b) gravity concentration, (c) magnetic separation, and (d) high-tension and electrostatic separation.

Auxilliary items of equipment such as screens and dryers are discussed in the appropriate sections.

7.3.1 Feed preparation and control

Recovery efficiencies depend primarily upon how well the feed is prepared and controlled at each stage of processing. It is surprising therefore, how little attention has been paid to this aspect of design both in the past and in many cases, even in the present. As recently as June, 1980, Horst and Enochs, in discussing the benefits of modern instrumentation and process control observed that 'little process control is applied in (placer) circuits other than slurry density, mass flow and wash water addition and even these variables are usually monitored visually and adjusted manually'. Harris (1976) referred to a speech made in 1858 to the Royal Institute of Mining and Metallurgical Engineers in the UK in which A. Dalton-Browne advised that body that the alluvial tin dredging industry of Malaysia was achieving a cassiterite recovery of only 46%. According to Harris, this was because the feed and recovery plant characteristics did not match. In his own observations at that time the jigs were unsuited for fine tin recovery and probably captured no cassiterite smaller than 100 mesh (149 μm).

Particularly in gold and tin dredging, initial losses between 40 and 60% are still common and many operators consider the redredging of previously worked ground as a normal extension of their present operations. The carry-over of masses of unslurried clay from the trommels is accepted, slimes constitute part of the feed materials and, in most bucket-line dredgers, the only feed-rate control is that exercised by the digger. The mineral-sands industry is generally more efficient but even in the most modern plant arrangements there is room for improvement.

The adequacy of feed preparation is a critical factor in the operation of a heavy-media separating plant for the recovery of gemstones, specifically in the removal of the fines (−0.5 mm) fraction by wet screening, using high-pressure water sprays. Failure to do this well results in medium contamination and an increase in viscosity which adversely affects separation. A rotary scrubbing treatment is used in conjunction with wet screening when the ore requires additional energy to break down clayey agglomerates or where there is a need to remove clay coatings from particle surfaces.

The main elements of feed preparation are, (a) slurrying, (b) sizing, (c) densification, and (d) distribution.

7.3.2 Slurrying

Placer sediments are fluidized and slurried with water to ensure that all of the valuable particles are liberated from the materials with which they occur and are able to settle freely if allowed. Such materials are notoriously variable in character because of environmental factors already discussed in earlier chapters, and feed from the one deposit may change dramatically in texture in a very short distance both laterally and in the vertical plane.

The sediments of continental-type placers feature mainly the higher-density, high-value minerals, gold, platinum, cassiterite, tantalum-group minerals and gemstones. Compared with the quantities of associated non-valuable particles, the concentration of valuable particles is very small.

The particles range from micron-sized fragments to very coarse nuggets and crystals in partly weathered mixtures of clay, sand and gravel, known as wash.

Transitional placer sediments have undergone more intensive weathering and erosion and contain mainly the lower density, less-valuable minerals, rutile, zircon, ilmenite, magnetite and monazite, although some beach deposits also contain gold, platinum, cassiterite and diamonds. Particles in recently formed deposits have generally clean surfaces and are readily dispersed in a fluid, however, many of the older deposits exhibit varying degrees of induration due to the deposition of humic and metallic oxide materials which have acted as a cement. Additional problems in liberation and sedimentation are the result of percolating waters that have deposited slime-sized particles in the voids.

Marine placers are drowned continental or transitional type deposits and hence may contain sediments of either type modified according to both sub-aerial processes and the subsequent action of marine chemical and biological agents. Generally, also, they are overlain by later marine sediments of no commercial value.

Machines for slurrying

Free-flowing sands present few difficulties in slurrying and the fact of being screened and pumped is sufficient to liberate all of the values. Indurated materials and most continental-type placer ores require much greater energy for slurrying and various techniques include the use of high-pressure jets of water in conjunction with vibrating screens and trommels and rotating drum washers. Log washers and other mechanical puddlers much favoured in the past are now rarely used because of their low mechanical efficiencies, high-power requirements and high-wear characteristics. Hammer mills and similar acting impact breakers are used in some installations to break down aggregations of indurated sands and cemented gravels and, if run at reduced speeds, may serve a useful function without incurring excessive wear.

The classical example of slurrying continental placer sediments using high-pressure water jets is in hydraulic sluicing (see Chapter 6, pp. 344–47). Monitors (hydraulic giants) are used to direct streams of water against an alluvial bank to break it down and to direct the spoil to a feed sump in the form of a slurry. A gravel pump elevates the slurry to the recovery units, thus providing additional energy for breaking down lumps of clay and other unslurried material. However, the method is wasteful of both power and water and the degree to which the more difficult materials are fluidized depends almost entirely upon the skill and judgement of the operators. Surging is common as, also, is the carry over of valuable particles caught up in clay balls or in slime-laden mixtures.

A common fault in both the past and many recent installations has been to design the slurrying plant for assumed conditions and to maintain a fixed rate of flow regardless of differences in the tenacity of the clay-gravel matrices from different areas or zones. The resulting losses from the carry over of clay balls and masses of unbroken wash would have paid the cost, many times over, of dealing more adequately with the problems from the start.

Modern practice is to incorporate an autogenous scrubbing section in trommel screens in order to provide additional energy and retention time and to increase the effects of the water jets and rumbling by removing the fine particles as soon as they have been released. Three main factors govern the success of scrubbing plant and equipment:

(1) A sufficient residence time for all of the valuable particles to be released and for the oversize waste materials to be discharged with clean surfaces. For rotary-drum scrubbers, the residence time is governed by the length of the scrubber, its slope, internal arrangements and speed of rotation.
(2) The removal of all fines from the system as soon as they are freed. Slimes and fine particles provide a cushioning effect against the impact forces, thus wasting some of the energy applied to lifting the particles and setting them in motion, one against the other.
(3) Sufficient capacity for slurrying a blend of materials in which the most tenacious clays are fairly represented. It should be noted in this respect that some clays have tempering qualities, well known to brickmakers, in that changes in cohesion follow exposure to the atmosphere or soaking in water. Some clay-bound materials harden quickly on exposure to the atmosphere and stockpiling makes them difficult to treat. Certain others, which display high degrees of plasticity, thus allowing ready deformation and resisting dispersion, are amongst the most difficult to slurry when freshly mined. Typical examples are found in the Moolyella tin fields of Western Australia where some of the clayey gravels respond readily to treatment only after spreading and sun drying.

In a normal operation the unslurried feed materials are delivered into a head feed bin and washed down into the scrubber through fixed or vibrating grizzleys as described in Fig. 7.2. Clay-bound material is broken up using hydraulically controlled monitors operating at up to 7 atmospheres (100 psi) water pressure. Large lumps on fixed grizzleys may either be reduced in size to pass between the bars using pneumatic hammers or 'cannons', or be removed using a small mechanical grab. Vibrating grizzleys provide a continual discharge of generally plus 100 mm stones and other materials away from the bin. Terrill (1980) makes the point that a fixed grizzley system is only viable for feed rates up to approximately 100 TPH if the ore contains a high proportion of rocks, clay and timber and

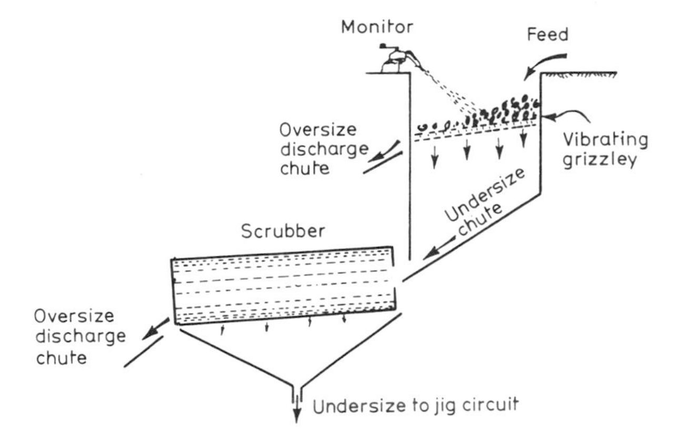

Fig. 7.2 Schematic arrangement of feed preparation scrubbing section

that a large plate feeder or vibrating grizzley is essential for feed control in large operations. Water requirements for the monitors used to break up the large clay lumps in such operations vary between 0.5 and 1.5 $m^3 h^{-1}$ of water per tonne of feed, depending on its clay content primarily. Terrill described as a typical example, a clay-bound barytes operation for which the water usage was 1.3 $m^3 h^{-1} tonne^{-1}$ of new feed containing significant quantities of timber and 15–20% of plus 100 mm oversize.

Most modern scrubbers are of the revolving-drum type, either in the form of a conventional rotating trommel with lifters and baffles, or with separate compartments for autogenous washing. They are preferred to most other types of scrubbers because they are more adaptable and are easier to operate, control and maintain. Essentially large, open-ended cylinders, they usually incorporate both scrubbing and screening sections. They are set in motion using friction rollers, or chain and sprocket mechanisms through reduction boxes. The correct speed of rotation is based upon the diameter of the larger, autogenous section. The optimum speed of rotation is some degree lower than the critical speed of rotation and is aimed at providing maximum rubbing and impact forces on individual particles and their neighbours. Scrubbing compartments are usually equipped with lifters in the form of bars and chains to carry the load higher up the sides of the drum before dropping it. This action allows for lower speeds of rotation and, hence, savings in power. A scrubber used in the Moolyella tin field of Western Australia is illustrated in Fig. 7.3.

Experience has shown the advantages of maintaining a reasonable proportion, say 20%, of the total circulating load of plus 25 mm rocks in the scrubbing compartment to act as autogenous grinding media. Baffle plates help to contain the media and, by holding back the flow, provide a longer retention time for unslurried materials. Since this leads to increased wear,

Fig. 7.3 Scrubber used on cassiterite placer ores, Moolyella, Western Australia

heavy-duty scrubbers are fitted with renewable steel liners and screen sections punched from manganese steel plate and bolted to the frame.

High-pressure water jets are directed onto the mass of the material through sparge pipes set with nozzles spaced at various angles around the effective screening arc. Considerable success has been achieved by a system of counter current scrubbing in which − 3 mm particles are washed back to the feed end of the drum and there discharged through fine peripheral slots. The counter-current flow is achieved through differences in discharge heights along the scrubber section and by injecting sparge water through the coarse material discharge opening.

A replaceable trommel section may sometimes be mounted advantageously onto the coarse discharge end of the washer for the delivery of timber and coarse rock to disposal conveyors or crushers. The − 25–50 mm trommel undersize is usually rescreened on a single- or double-deck vibrating unit to separate the finer rock into appropriate fractions for further treatment. Either, or both of the trommel and vibrating screen oversize material can be crushed further if required and recycled to the scrubber to maintain an optimum load of scrubbing media. Provision is easily made to by-pass excess rock when the clay/sand content of the feed is such that there is no need to recycle the media.

Energy requirements for effective scrubbing range from 0.25 to 1.0 kW per hourly tonne of capacity. For example, a 100 TPH scrubber system with a full load of sand, slime and recirculated rock media would be powered adequately by a 100 kW drive and would probably not exceed 65 % demand with the system rolling. Optimum sand/slimes residence times are achieved by controlling the amount of water introduced into the system and the counter-current washing arrangement is, thus, an added advantage.

7.3.3 Sizing

Natural sorting processes result in the segregation of high-density minerals together with larger, lower-density particles of similar hydraulic equivalence (see Chapter 3). Other particles, both larger and smaller, are present because of varying hydrodynamic relationships at the time of deposition. Sizing control is exercised closely in placer processing plant along with other operating variables in order to achieve what nature has been unable to do, that is, to differentiate completely between the valuable heavy minerals and the waste. Sizing has three main functions in placer processing operations:

(1) To reject a coarse fraction of valueless materials (primary screening).
(2) To reject a slimes fraction (desliming).
(3) To prepare closely sized fractions of the remaining constituents of the feed for presentation to the various treatment units to enhance their operational capabilities (fine-screen sizing and classification).

Primary screening

In addition to large rock fragments, placer sediments commonly contain quantities of such trash as plant fibres, buried timbers and shells that must be removed for the proper functioning of the primary concentrators and to avoid pump blockages. The process is carried out simply in hydraulic sluicing operations using pitch forks with tines spaced 5–8 cm apart to fork stones manually out of the races; vertical grizzleys with similarly spaced bars are installed in the head boxes of sluices and palongs.

Rotating trommels and vibrating screens are used in mineral-sands operations to reject particles larger than 3–5 mm diameter and trash. The valuable constituents are seldom larger than 300 μm in most deposits of this type and screens with much smaller apertures than those in current use would be used if it was possible to do so on a large scale without incurring excessive losses of the valuable particles. No current technology is capable of meeting the required conditions of high capacity, low floor space, low cost and high recovery and none is foreseen. In making the same point almost two decades ago, Pullar (1965) drew attention to the advantages that would be gained from rejecting all beach-sand particles larger than around 72 mesh (210 μm) in the northern areas of New South

Wales and 44 mesh (367 μm) for sands in the Belmont area of the central coast. According to this author, 'there does not appear to be any commercial screen capable of handling the required tonnage and making an efficient separation – large capacity DSM sieve bend screens were tried but failed because of excessive losses at fine sizings'.

Most land-based continental placer operations employ vibrating grizzleys assisted by high-pressure jets of water for primary screening as already described in the notes on scrubbing. There seems no reason to believe that similar techniques would not be equally effective in bucket line dredging practice but so far they do not appear to have been recognized by many dredger design groups. The possible benefits are obvious to any who have observed the large quantities of unslurried materials being disgorged by over-loaded trommel screens or have sampled the tailings from such operations. A possible modification to some existing installations would be to shorten the trommel and to scrub the rejects on a vibrating grizzley using high-pressure water jets in a fully automated operation. A small secondary screening unit could then make a final rejection of the clean oversize, and direct the fines to the recovery circuit.

Trommels used in current dredgers are massive affairs, typically up to 20 m in length, 3 m in diameter and sloping downwards to the discharge end at around 6° from the horizontal. The walls are fabricated from metal sheets perforated with round tapered holes that increase from about 8 mm at the upper end to 15–18 mm at the lower end.

Trommels for land-based screening operations are commonly fitted with an internal cylindrical bar grizzley that is slightly longer than the outer screen cylinder and approximately one-half of its diameter. It is fixed into position along the longitudinal axis of the trommel and its purpose is to scalp off the larger-sized rocks, thus reducing wear on the outer screen sections which may then be constructed from lighter materials.

A recently developed land-based system known as a 'Wobbler' combines primary screening with feeding. It derives its name from the wobbling motion imparted to the feed materials which are kept agitated at the same time as they are moved forward. The motion is obtained by means of rotating eliptical bars set on shafts with each elipse positioned at 90° to its neighbour so that while the long dimension of one is in the up position the succeeding one is lying flat. The undersize passes through the spaces between rollers and high-pressure water jets are used to wash the material and carry it through to the next treatment unit. Wolfe (1976) describes an installation in Alaska in which water was supplied by a 4-in pump, the undersize material was 25–35 mm diameter and, for a capacity in excess of 75 m^3h^{-1}, the total power was 2.24 kW. The wobblers rotated at 36 rpm.

The particle rejection size in primary screening circuits is governed by the range of sizes of the valuable mineral particles and by the difficulties associated with fine screening at high rates of flow. In relation to the overall process economics the upper limit for rejection is obtained with screen apertures such that the larger valuable particles are rejected if they cannot be recovered economically. In other words, when, by sacrificing only a few of the valuable particles, a large volume of worthless material is rejected that would otherwise have to be treated at a loss. Obviously there can be no justification for going to great lengths to recover the occasional specimen-sized particle. This is one of the reasons why, periodically, large nuggets of gold are found in the waste dumps of abandoned mines.

Desliming

At the other end of the size range there are very fine particles of valuable minerals that are uneconomical to recover, either because of deficient technologies or because of the low grade of the fraction they report in. Desliming is, in any case, an essential feature of all placer-feed preparation systems where appreciable quantities of slimes are to be handled and it is convenient to set the rejection size according to the same economic considerations as for large particle-size rejection. For example, slimes rejection may take place at a cut off size of 120 μm for a diamond placer, whereas for gold it may be within the Stokesean range for quartz spheres settling freely in water (see Chapter 2).

Quite apart from the elimination of very fine-sized particles in order to reduce the quantity of feed material, such particles are rejected also because their presence in a slurry alters its rheology through changes in apparent viscosities and densities (see Chapters 2 and 5). The allowable tolerance of slimes in feed materials varies according to the concentrator type. Thus, while most pinched sluices and spirals perform reasonably well with feeds that contain up to 5 or 6 % of slimes, shaking tables and jigs are less tolerant and, basically, all wet gravity concentrators operate more efficiently with a deslimed feed. Desliming is carried out in a number of scavenging stages if significant quantities of valuable particles appear around the separation size.

Final heavy-mineral concentrates are generally free of discrete slimes-sized particles but there may be a significant carry over due to such particles adhering to the surfaces of larger particles. This has important consequences in mineral-sands dry separating processes. There is a tendency for such particles to become baked onto the larger particles during the drying process thus affecting individual particle responses to subsequent electrical process elements. Slimes particles also become coated onto high-tension rolls seriously impairing their separating qualities; those that do not form coatings create a dust problem that may be quite severe.

Such problems are avoided by interposing an additional step of attritioning and desliming ahead of drying and dry separation.

Attritioning

Attritioners are mechanical scrubbing devices used to remove the surface coatings from affected particles either with or without the supplementary use of chemicals. A favoured design uses oppositely pitched impellers mounted on a vertical spindle to agitate the pulp and to project the particles towards a common central zone, the top layers being forced downwards and the bottom layers upwards. The density of the pulp is held at around 70–80% of solids by weight to ensure close rubbing contact between the particles. Attritioners are single-stage or multi-stage units depending upon how difficult the films are to remove.

Some surface coatings are not removed by any of the recognized mechanical attritioners and various chemicals are used to dissolve or soften the films. Simple organic coatings usually respond well to caustic soda solutions and to certain commercial reagents such as calgon. Metallic oxides may be more difficult to treat and at Bunbury, Western Australia, hydrofluoric acid has been found effective although somewhat dangerous to handle. Typical reagent usages at ambient temperatures are 0.86 kg each of HF and $Na_2S_2O_5$ per tonne of solids treated.

Attritioner performances are difficult to evaluate from small-scale experiments because of the large number of uncontrolled variables; normally, the results of changes made in the attritioning section of a mill become evident only after an extended production period and following a comprehensive metallurgical balance. The main value of experimental attritioning trials is in indicating possible benefits from the actions taken. For example, observations based upon the rate of slimes removal in experiments at Bunbury showed that more slimes were removed during the second stage of attritioning than during the first stage. This was attributed to a number of possible reasons including:

(1) Longer contact with the chemicals.
(2) A heat rise with correspondingly higher reaction rates due to the progressive conversion of kinetic energy to heat energy through friction.
(3) The number of collisions experienced by each particle. It has been suggested that the initial collisions serve only to loosen the coatings and facilitate attack by the chemicals.

Possibly all three reasons are involved in some way but certainly the number of collisions experienced by each particle is a most important factor in film removal because of the observed abrupt increase in the slimes build up after a certain critical time and, presumably, after a critical number of collisions. Opposed hydraulic jet-type attritioners rely upon only a few high-speed collisions for each particle and have failed completely in mineral-sands attritioning service.

Fine screening

Following the large-scale rejection of worthless oversize material using high-capacity primary screening devices and the removal of worthless slimes-sized particles and other fines, the remaining fraction is treated to separate the valuable particles away from the non-valuable particles using various types of concentrating and separating devices that operate best within certain specific feed-size limitations. Table 7.2 describes the typical particle-size limitations for standard plant items. Note, however, that although jigs are able to process feed materials at relatively coarse sizings their performances are improved by first screening the trommel undersize on single- or double-deck vibrating screens to obtain coarse and fine fractions for separate treatment. Further sizing of the rougher concentrates prepares them for final concentration by other means.

Static screens for wet screening are represented by the DSM screen deck which is constructed with profiled wedge bars fitted across a concave surface. The curved surface creates centrifugal forces to assist in the screening action, the normal separating range being between 0.3 and 4.7 mm. Screen apertures are made approximately 10% larger than the desired sizings for slow feed rates and are increased for faster flows. Other static screens utilize round holes punched out as much as 25% larger than the maximum size of particles passing, unless all of the particles are well rounded. Screening efficiencies are generally low in the finer sizings as already noted.

Screen sizing as an adjunct to separation in dry-treatment processes is limited in its applications largely because vibrating screens have increas-

Table 7.2 Typical particle size limitations – mineral processing plant

Unit	Particle size range
Jigs	75 μm–25 mm
Shaking tables	15 μm–3.0 mm
Spiral separators	75 μm–3.0 mm
Pinched sluices and cones	30 μm–3.0 mm
Buddles, vanners etc.	7 μm–30 μm
Strakes and riffles	70 μm–25 mm
Hydrocyclones	40 μm–3 mm
Amalgamation	70 μm–1.5 mm
Cyanidation	–200 μm
Magnetic separators	
Wet	5 μm–2 mm
Dry	70 μm–2 mm
High-tension and electrostatic separators	70 μm–0.6 mm

ingly poor performance characteristics at smaller apertures and the flow rates are sometimes quite large. Usually it is much more economical to provide additional capacity to handle large circulating loads in the primary concentrating and separating units and to screen only those fractions that can be dealt with on a relatively small scale. At this level, screen sizing has a number of important functions to perform in dry-milling practice, particularly in the final dressing of concentrates. One application is in the removal of large particles of weakly magnetic non-conductor minerals from rough concentrates of the smaller more magnetic non-conductor mineral monazite. Magnetic flux densities are functions of the width of the air-gap in magnetic separators and the larger particles are attracted preferentially even though they may have lower magnetic susceptibilities. Another application is in removing coarse zircon particles from a rutile product. Although zircon is a non-conductor mineral and rutile is a conductor mineral, coarse zircon tends to be thrown from high-tension rolls along with the smaller conductor minerals.

Apart from standard vibrating screen types, two main systems of fine screening in the dry mill are typified by the 'Rotex' and 'Sweco' screens.

The Rotex screen

Rotex-type screens are designed to spread the feed material quickly over the whole width of the screen cloth and to discharge the oversize particles at the far end. Basically conventional screen types, blinding is inhibited by the action of hard rubber balls bouncing up and down against bevel strips on the underside of the screen mesh. The motion is basically gyratory and a recent development is a 'quicksnap spring tension clip' for ensuring constant tension on the screen surface and to promote quick replacement. Trelleborg offers egg-shaped balls of wear-resistant, highly elastic rubber and claim these to be twice as effective as round balls.

The Sweco screen

The principle of the Sweco screening action is vibration about the centre of mass. The feed flows outward from the centre of the screen and eccentric weights on the motor shaft can be adjusted to provide different screening flow patterns. Figure 7.4 illustrates the action of a Sweco screen in producing a monazite-rich concentrate from a mixture containing mainly valueless minerals in the larger sizings.

In general, the capacity of a screen is related to its width and separating efficiencies depend upon the length. Screen performances are concerned also with the sediment properties of abrasiveness, cohesiveness, hygroscopicity, shape and content of very fine particles such as clay. Surface moisture is considered to exceed allowable screen standards when it rises above 3%.

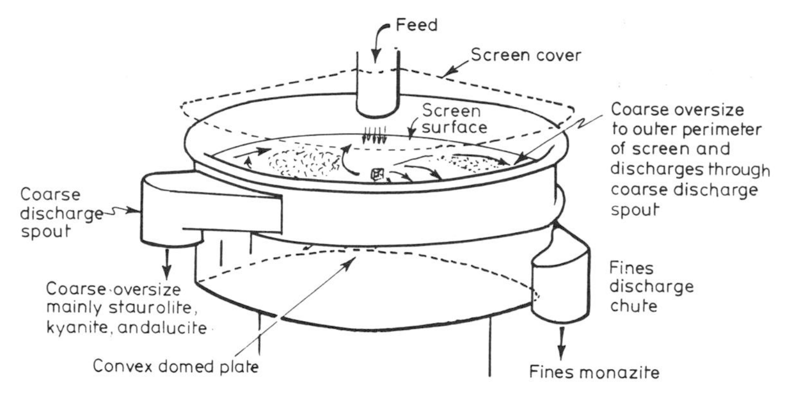

Fig. 7.4 Schematic arrangement of Sweco screen in monazite cleaning service

7.3.4 Classification

For most practical purposes in placer-processing plant, mechanically operated screens are restricted to sizing at apertures larger than 2 mm and some form of hydraulic classification is used for the finer sizings. Alternatives for hydraulic classification include hydrocyclones, hydrosizers and spiral classifiers.

Hydrocyclones

Hydrocyclones are preferred units in modern gravity-treatment processes for sizing or desliming large flow volumes cheaply and because they occupy very little floor space or head room. They are sized individually according to the particle size split and a multiplicity of cyclones may be used to obtain desired total capacities at required splits. For example a bank of five to ten RSW159 hydrocyclones would be used to handle $40\ m^3\ h^{-1}$ of slurry at a cut off of 15 μm rather than one RWN6118 which could easily handle the quantity but not the fine cut off (see Table 7.3).

Depending upon the required accuracy of the split, one or two stages of hydrocycloning will be needed and practical tests are carried out to resolve the variables. The spigot discharge characteristics are determined by throttling. A rope discharge provides a high-density underflow with some losses in the overflow. An umbrella or spray discharge makes higher recoveries but passes additional fines into the underflow. Optimum performances depend, in all cases, upon constant feed conditions in terms of volumetric flow rates and slurry densities. Such conditions are obtainable only with constant sump levels to keep a steady head and to avoid vortexing, thus sucking air into the system (see Chapter 5).

Specific cyclone types such as the circulating bed classifier developed by Amberger Kaolinwerke GmbH in West Germany can handle feed materials up to 5 mm in diameter and obtain a consistantly high density jig feed from the underflow. This system separates at approximately 150 μm based upon silica of density 2.65. According to the experience of Minerals Deposits Limited of Southport, Queensland (Terrill, 1980), the CBC cyclone underflow has proven to be particularly amenable to jig separation. Because of its relatively smooth size distribution curve and the almost complete removal of fine waste particles, the system often yields a high-grade concentrate in one pass from a relatively long-range feed.

Hydrocyclones are also used for thickening, desliming and selective classification utilizing differences in density as well as size. Table 7.3 gives the technical data for AKW hydrocyclones for cut off points between 12 and 100 μm.

Hydrosizers

When more than two sized fractions are to be prepared for separation on jigs, shaking tables, vanners etc., it is sometimes convenient to perform the sizing operation in a hydrosizer which consists in its simplest form of a series of compartments arranged in ascending order of size in the direction of flow. The largest and heaviest particles sink to the bottom of the first compartment and as the flow velocities are reduced in each successive box so are the remaining particles sorted according to their sedimentation rates. Each compartment has a spigot discharge that can be a simple gooseneck, with an adjustable constriction, or be operated automatically from a solenoid to regulate the flow.

Spiral classifiers

Long size-range feeds that are not handled efficiently by hydrocyclones may be sorted in a classifier of which the spiral type is favoured. Such classifiers are engineered to provide an effective pool area and overflow velocity for settling suitable for the particular size separation requirements. Spirals, which may be single or duplex, are sectionalized steel flites which form a continuous helix with a high pitch for optimum raking capacity. The classifier sands are moved up the incline to the discharge point where they may be screened to provide a well-classified jig feed. The

Table 7.3 Technical data for AKW hydrocyclones

Type	Diameter (mm)	Pressure drop	Capacity		Cut point (μm)
			Slurry $m^3 h^{-1}$	Solids $T h^{-1}$ max	
RWS159	75	1–2.5	4–12	2	12–20
KRS2118	100	1–2.5	9–27	5	30–50
RWS2515	125	1–2.5	11–30	6	25–45
RWS4118	200	0.7–2.0	18–60	15	40–60
RWN6118	300	0.5–1.5	40–140	40	50–100

dilute overflows can be freed of floating trash by means of sieve bends or trommel screens and hydrocycloned to yield closely sized fractions for feeding to sluices or tables.

7.3.5 Densification

Many of the advantages to be gained from closely controlling feed-slurry densities were not fully recognized until the advent of pinched sluices and cone concentrators for which optimum values lie between 57 and 65% of solids by weight. Apart from the substantial power savings referred to in Chapter 5, experience has shown that, although most other types of gravity concentrators operate reasonably well over a wider range of slurry densities, both throughputs and recoveries are improved measurably by exercising density control.

A simple densifying unit for head-feed control from a surge tank of suitable capacity has already been illustrated in Chapter 5 as Fig. 5.21. Valve A controls the pump-water inflow and with the valve fully open, only water enters into the pump. By closing it gradually, the negative pressure builds up in the suction line and sand is lifted into the flowing stream. A point is eventually reached at which the inflows of sand and water balance out at the required pulp density and this density will be maintained for as long as sufficient sand is available for pumping. Signals from sand level indicators installed in the surge tank alert the operators to increase or decrease the flow of fresh quantities of sand as required. Other forms of densification include dewatering cyclones and DSM type screens.

7.3.6 Distribution

One of the most important requirements in process plant design is to provide all concentrating units with feed that is controlled within close limits in terms of quantities and grades and, for wet processing circuits, in terms also of slurry concentration. The problem is compounded when a multiplicity of machines is to be fed from a common source and the total feed is to be split evenly between a line of machines.

Accurate feed distribution to wet processing units is normally difficult and some standard distributer types are less than satisfactory. The main problems arise from the heterogeneity of the mixtures flowing and because of the tendency for the higher-density solids to segregate along the lower parts of pipelines. Self-driven turbine distributers, even when equipped with rotating baffles, suffer from a certain amount of differential settling. Simple steady heads with an overflow pipe have no means of avoiding segregation and the resulting splits may vary considerably in both quantities and grades.

McSweeney (1975, personal communication) describes a reasonably satisfactory steady-head feed bin in which the incoming feed enters a baffle box where the kinetic energy in the flow is expended in mixing. The stream of pulp flows from this box to a shallow conical distributing plate under a rubber flap to guard against splashing. The plate is rotated at 12 rpm to spread the pulp evenly over the full 360° and is raised or lowered to control the rate of flow to the individual units. In the raised position all of the pulp is returned to the feed bin. In the fully lowered position all of the feed is distributed to the separators. Measured portions of the pulp can be taken either way at intermediate positions of the plate and by adjusting the position of an inner cylinder, distribution is controlled in all directions without segregation. No valves or plugs are used and all outlets are fully opened and without restrictions.

A patented system used by Mineral Deposits Limited provides for either bottom or top pressure feeding or top gravity feeding. It is made of polyurethane elastomer to resist wear and is stated to be highly effective.

Feed distribution in the dry mill is concerned largely with maintaining an even rate of flow of particles one layer deep over the full widths of rolls and plates. One means of achieving this is with adjustable gates that can be raised or lowered to control the rate of flow. Flow to machines for which the flow rates have been predetermined is regulated more effectively using calibrated, removable control slides with holes drilled along their lengths to distribute the flow evenly across the required width of roll or plate. The diameter of the holes controls the rate of feed and a number of such strips can be drilled for a range of feed rates.

Maintaining correct feed temperatures is also important and supplementary heating strips, infra red lamps etc. are incorporated into the feed distribution systems to replace heat lost in feed passing to high tension and electrostatic separators. One method is to use a vibrating feeder, with the tray enclosed in an insulated box. Infra red lamps, mounted in the box, direct heat rays over the tray and, because of their motion, the individual particles tend to become evenly heated throughout.

Figure 7.5 after Terrill (1980) illustrates a typical feed preparation circuit for a claybound ore.

7.4 GRAVITY CONCENTRATION

Gravity concentrators are used to separate particles of valuable high density minerals from the lighter non-valuable minerals in solids–fluids mixtures, utilizing differences in particle sedimentation rates. As already discussed in Chapter 2, the sedimentation rates of individual particles in free fall are governed by their size, shape and density but such rates are modified by the presence of other particles and by conditions of hindered settling either naturally or artificially induced. Concentrators are of six

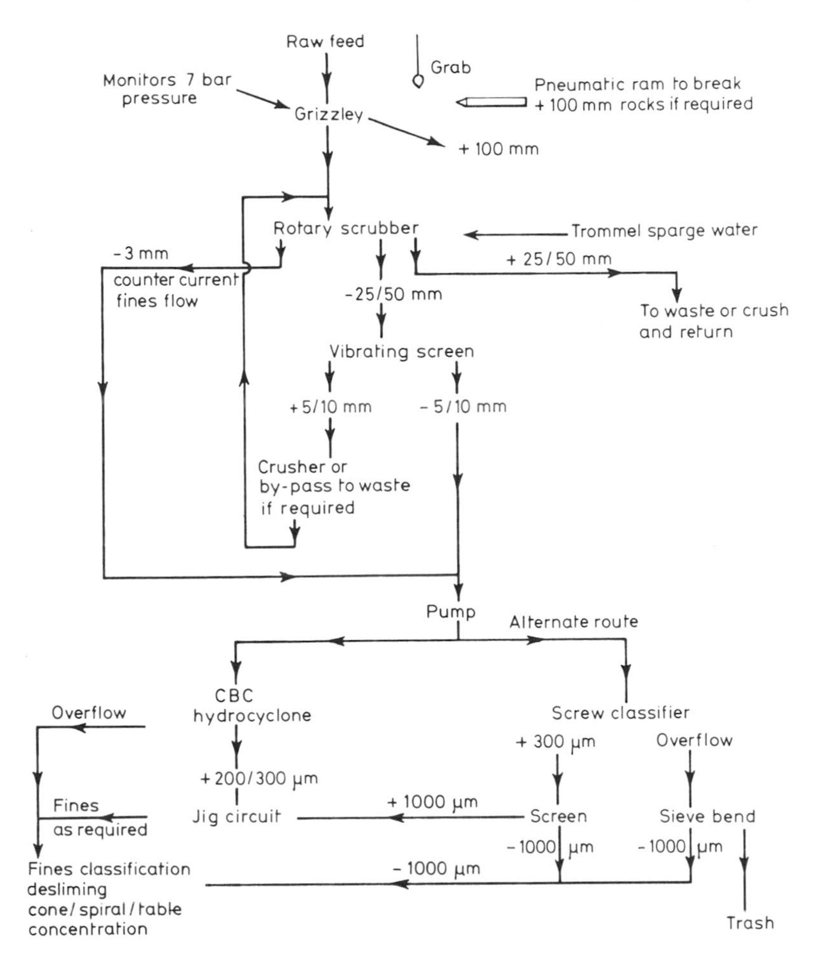

Fig. 7.5 Typical placer feed preparation circuit (adapted from Terrill, 1980)

main types, (a) riffled sluices, (b) pinched sluices, (c) spiral concentrators, (d) jigs, (e) shaking tables, (f) rotating cones, and (g) heavy medium separators.

7.4.1 Riffled sluices

Riffled sluices, cradles and long toms used by both prospectors and miners have been variously described by a number of authorities according to the particular mining fields in which they were used. Production units, primarily for gold are constructed as open launders, some 200–600 mm wide and about 200 mm high in sectional lengths of 3–4 m. Each box tapers sufficiently to allow it to fit into the succeeding box; the total box length is long or comparatively short depending upon the consistency of the gravel wash and the particle size range of the gold.

Opinions differ widely in respect of sluice-box capacities and performances thus emphasizing the dangers inherent in using the performance data from one mine to predict performances in another. For example, according to one authority, a 600 mm wide × 15 m long sluice box can process about 6 tonnes/h of −10 mm gravels using between 10 and 25 tonnes of water per tonne of ore. Another suggests that a 300 mm wide sluice box will handle 7 tonnes/h of gravels using 10 tonnes of water per hour and that a 360-mm sluice should about double those quantities. In practice, sluice capacities and performances vary according to the slope and width of the boxes, the particle size range of the gold and accompanying sediments, the slimes content and the content of heavy amang minerals and the adequacy of water for sluicing. Tough clayey gravels and cemented wash materials require flatter gradients and more work done on them to effect disintegration than do loose materials and, where much fine gold is present, the use of undercurrent sluices will allow the coarser fractions to be treated at higher flow rates than if all recoveries are made in the one sluice.

Opinions differ also on the design of riffles and their spacings and again the optimum parameters depend upon the feed characteristics and availability of water. Little notice can be taken of designs illustrated in some reference books, except to appreciate the wide variety of types, all of which have had their fervent supporters in the past.

Peele (1927) gives some details of the riffle types and spacings used in various hydraulic mines in Western USA in 1932 as shown in Table 7.4.

Basically, any form of riffling is satisfactory provided that, in the spaces between the riffles, conditions of turbulence are such that the gold particles can sink to the bottom and not be disturbed by eddies having greater components of velocity in the vertical plane than the settling velocities of the gold. Taking the Hungarian riffle as a model, trial and error as to their spacings will soon determine the optimum spacing and

Table 7.4 Typical riffle types and dimensions

Unit	Type of riffle	Width (in)	Height (in)	Spacing (in)
1	Wood cross	2	6	4
2	Wood block	12	12	12
3	Hungarian	2	4	$4\frac{1}{2}$
4	Angle iron	2	2	4
5	CI bars 4 ft long	3	$1\frac{1}{4}$	5

Note: The Hungarian riffles are commonly fabricated using 2 in × 4 in scantlings.

height for the particular material being processed. The development of longitudinal log riffles is interesting in that it may have stemmed from observations of preferential gold deposition in longitudinal depressions in auriferous river beds. The longitudinal riffle is not generally a recommended form.

Undercurrent sluices

Undercurrent sluices are installed at the bottom end of a sluice box to scavenge fine gold from the fines passing over the riffles. A grizzley with closely spaced bars rejects the coarse material and allows most of the water and fine material to pass through and fall into the head box of the undercurrent sluice. This sluice is placed at right angles to the longitudinal axis of the main sluice and is generally short and wide. Check boards distribute the flow evenly over the whole width of the box. The gradient is generally steeper than the main box, usually about 100–200 mm/m of length so that a broad shallow stream of pulp flows over the surface. The surface is usually riffled for about one half of the length and then decked with corduroy or coarse sacking. Mercury traps have been used in place of undercurrent sluices in the past but should not be considered today for ecological reasons.

Sluice boxes designed for tin alluvials are called palongs in south east Asia where they are commonly 2–4 m in width, 1 m in height and up to 50 m in length. Slopes are generally flatter than for gold, usually around 1 : 30 and instead of riffles, baffles are placed at intervals of 2 to 4 m for the full length of the boxes. The baffles are built up in height as soon as cassiterite is observed to flow over into the next compartment. A 'boxman' continually agitates the material in the boxes with a shovel to allow the heavier particles to sink. He is normally in control of the whole operation and, by means of signals to the pump and nozzle men, indicates the required density of flow at any one time.

7.4.2 Pinched sluices

The use of constricted sluice configurations dates back to antiquity and even in modern times both parallel and pinched sluices were used for coal cleaning well before the minerals mining industry recognized their potential for heavy-minerals concentration. According to Pullar (1965) one of the first pinched sluices to be developed for minerals with some measure of success was the 'Fanning concentrator' designed by Carpco Research and Engineering Inc., Florida. The purpose was to find an alternative to the heavy cast iron Humpreys spirals, for use on a floating barge. The Fanning sluice was approximately 70 cm long, tapering from 23 cm at the feed end down to 1.3 cm at the discharge end. The slurry, in discharging, fell onto a plate set at a slight angle to the vertical and almost

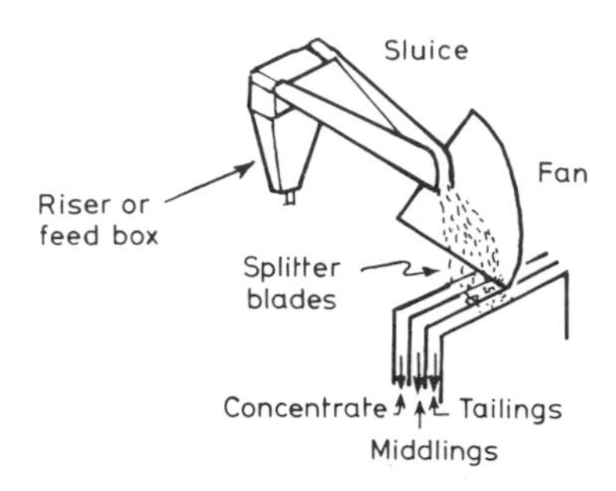

Fig. 7.6 Fanning concentrator

parallel to one side of the sluice, and spread out as a fan. Adjustable splitters directed the various splits to collecting launders as described in Fig. 7.6. The feed density was controlled between 55 and 65% of solids by weight and the slope of the launder was adjusted to an angle at which the solids were about to settle, usually between 14° and 20° from the horizontal.

Fanning separators suffered the disadvantages of excessive circulating loads (up to 1000% of the original feed) and a partial remixing of the stratified slurry due to wall effects. Multi-stage operation was necessary because the concentration criteria for each stage of upgrading had to be held at very low levels in order to achieve high recoveries and high concentrate grades.

York Bros. Pty. Ltd. of Swan Bay, New South Wales, Australia developed a modified form of pinched sluice called a 'Star separator' which was named for its configuration. Some advantages were obtained with certain arrangements of the upgrading and treatment trays which allowed the tailings and concentrates to be retreated on scavenging and cleaner sluices. The unit was soon superceded by the RZM cone concentrator, later to be known as the Reichert cone concentrator, after its developer Ernst Reichert. Originally made operational around 1964, the Reichert cone has undergone intensive experimentation by the manufacturers, Mineral Deposits Limited of Southport Queensland and may now be the most versatile and sophisticated of all the pinched sluice types.

Reichert Cone and other pinched sluice and tray concentrators are essentially sloping launders in which the flow conditions are so controlled that the lower layer of a stream of flowing pulp contains the bulk of the heavy mineral particles within its boundary layer. These are taken away through a throat or slot as described in Fig. 7.7. The launder length, taper, slope, slurry density and flow rates are adjusted for optimum stratification at the splitting or take-off point according to the properties of the solids and their percentage distributions. The inter-relationship between the various factors is such that size effects are more important in short-sluice sections and density effects predominate when sluice lengths are extended.

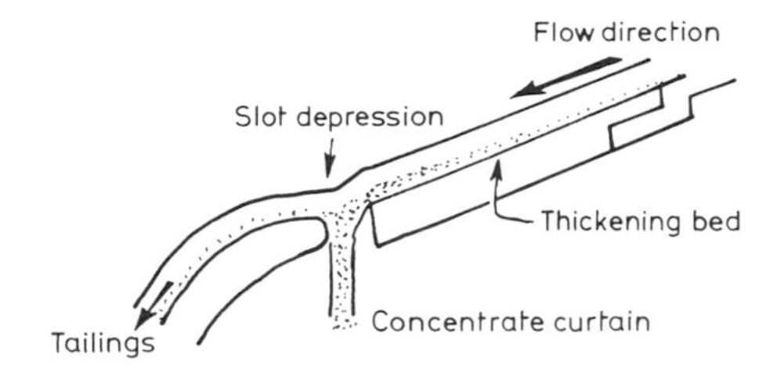

Fig. 7.7 Stratification on Reichert Cone Concentrator

Reichert Cones

According to Ferree and Terrill (1978) well-defined stratification on Reichert Cone Concentrators is usually evident after approximately 0.5 m of inward movement. They point out that the effectiveness of the operation is a function of many variables including loading, pulp density and the relative size distribution and densities of waste and mineral particles. Cone units are predominantly 2 m in diameter but 3.5 m diameter cones, now being developed in a new generation of cone systems, are expected to provide further operational simplicity and higher separation capacity at lower overall plant costs.

Unit capacities vary between 60 and 100 tonnes h^{-1} and recoveries are made down to particle sizes of as little as 40 μm for heavy minerals. Mathematical modelling programs for computer simulation of complete circuits have been developed over a considerable period to establish optimum settings and maximize the output while still increasing the tolerance of the unit to varying conditions. According to Ferree and Terrill the development of the matrix configuration system in recent years has permitted the design of compact cone units capable of treating up to 300 TPH. Figure 7.8 shows a general form of the upper three assemblies of a 3 m diameter matrix cone concentrator.

Tray separators

As an alternative to cones, pinched sluice or tray separators have been developed in a number of different designs for different duties. For example, the Wright impact plate gravity concentrator is particularly suited to upgrading low-grade ores. The Wright design is based upon the premise of a direct correlation between the percentage of solids taken into concentrates and the percentage recovery of the heavy minerals for any given material. It is claimed that when the impact plate has been set to separate the light and heavy fractions in a pre-selected ratio, it will maintain that ratio irrespective of operating variables in the flow rate of the fluid bed and the feed rate of the solids. Figure 7.9 illustrates the product distribution pattern in one start of a four-tray configuration. Figure 7.10 is a graph showing the relationship between the percentage

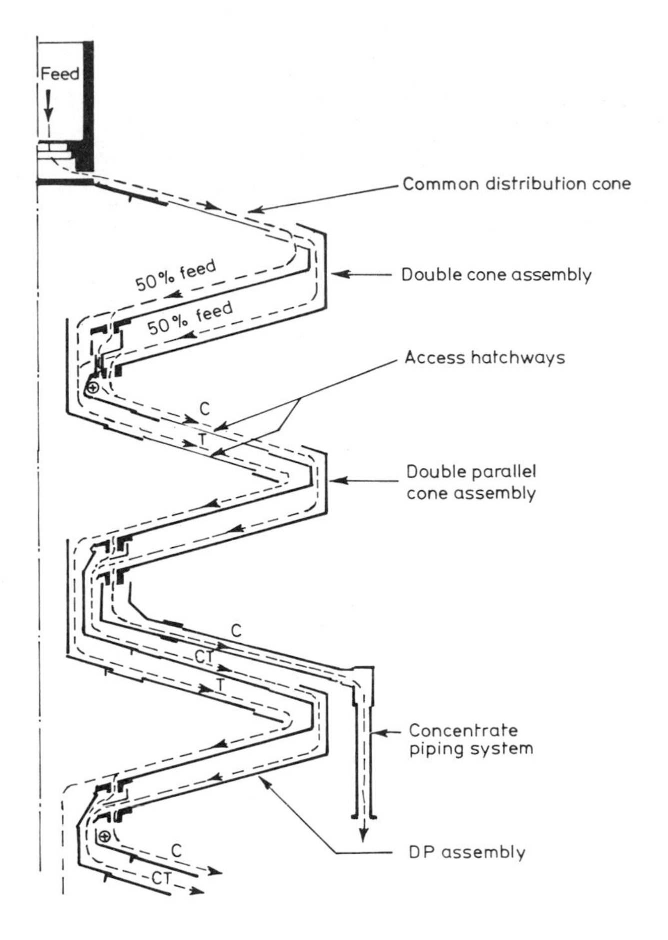

Fig. 7.8 Upper three assemblies of Reichert 3-m diameter matrix cone concentrator

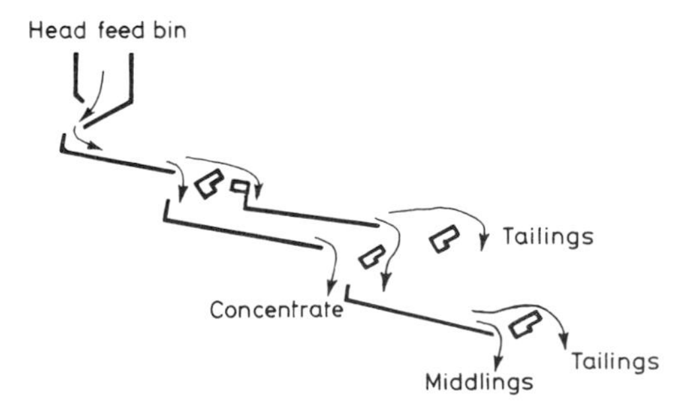

Fig. 7.9 Product distributor pattern in one start of Wright tray separator

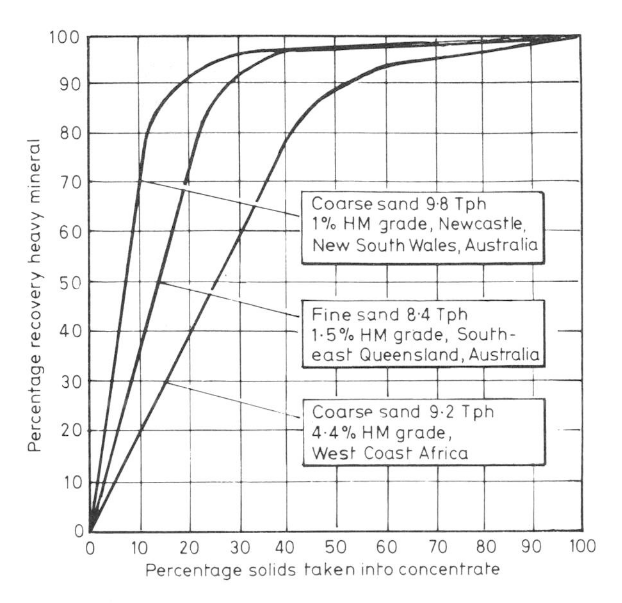

Fig. 7.10 Plot of percentage solids taken into concentrate versus percentage recovery for Wright impact plate gravity concentrator. Typical results for fine and coarse sands are shown

solids taken into the concentrate and the percentage recovery of heavy minerals in a four-tray configuration.

Determining which of the many types of concentrators is best for a given ore can involve an enormous amount of detailed test work. Major considerations are recovery rates, upgrading ratios, the ability to achieve planned production targets at the lowest practicable cost for plant and equipment, maintenance and production costs. There are, however, economic limits to the degree to which individual sluice performance characteristics can be determined because of the number of variables and perhaps the intending purchaser should also consider very carefully the experience of other operating plants before making his choice. Much of the present test work still relies upon trial and error and, with an enormous scope for varying the settings in individual trials the results, if not properly evaluated, may lead to wrong conclusions in comparing one type of separator with another. Thus while a number of sluices might appear to be similarly satisfactory in a given set of conditions, they may later be found to vary in their responses to other sets of conditions. One may be more prone to troubles and another less flexible in its ability to handle a range of different head feed grades. A separator that gives all round satisfactory, trouble-free performance will generally be preferred to a higher performance model less able to handle a range of operating variables.

7.4.3 Spiral concentrators

Spiral concentrators are essentially low feed-rate, low feed-density devices capable of considerable selectivity and control if used in the correct application. Operating variables are the diameter and pitch of the spiral, the density of the pulp, the location of splitters and take-off points and the volume and pressure of wash water introduced into the system. In order to save floor space most spiral separators are constructed as two or three start units around a common vertical pipe by reducing the height of the spiral walls and increasing the pitch. All standard spiral types have some form of adjustable splitter to divert the concentrate stream away from the main flow and direct it into a pipe leading to a collecting box. Fixed splitters have been tested and rejected because of their inability to handle large fluctuations in concentrate production. Fig. 7.11 is a schematic arrangement of a twin spiral concentrator.

The first commercially applied spirals were the Humphreys cast-iron spirals. They were simple to operate but suffered the disadvantages of being very heavy and difficult to control. The standard pitch of about 345 mm was not suitable for all classes of feed, the joints between each 120° spiral section were rough and caused turbulence, wear was rapid and the wash water distribution was irregular. Although still used by some operators the Humphreys cast-iron spiral has been largely superceded by a variety of other types made from concrete, fibreglass and moulded plastics.

Pullar (1965) describes early concrete spirals made by various companies in Australia. Individually designed spirals were cheaper and lighter than the cast-iron spirals and attempts were made to provide better wash-water control utilizing a series of small plastic tubes. However, these tended to block and many of the other disadvantages of the Humphreys remained uncorrected.

Experimenters then commenced working with moulded plastic, using

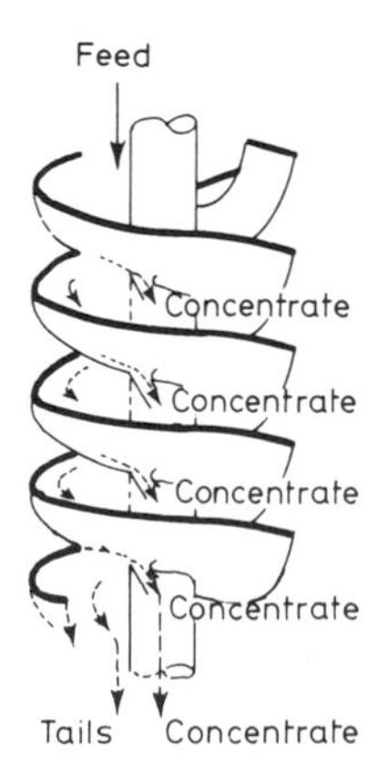

Fig. 7.11 Schematic arrangement of twin spiral concentrator of capacity 2–4 T h^{-1} depending upon application

fibreglass as reinforcing. The original Reichert spiral was a 610-mm, five-turn spiral moulded in one piece and lined with linatex to provide a tough wearing surface. Concentrate take off points were located at 180° intervals and a pitch of 343 mm was increased later to 394 mm for both roughing and cleaning service. Being extremely flexible the increase in pitch was obtained simply by stretching the spiral out, the steeper pitch allowing individual spirals to be mounted in two start configurations. Wash water was injected into the flow at two different locations between each pair of take-off points, fed from a central hollow fibreglass column. As well as providing a mounting for the spirals the column was fitted with a series of cascading water pockets from which the water was delivered to the spirals through plastic tubes. In current models, the rubber linings have been replaced by polyurethane elastomer for increased resistance to wear and with the exception of some stainless steel screws used during assembly the whole spiral is non-metallic and lightweight. For example, a double-start, five-turn spiral weighs approximately 65 kg.

Another early spiral type was that developed by National Minerals Pty. Ltd. for mining beach sands in the Belmont area of New South Wales, Australia. Spirals of $2\frac{1}{2}$ turns were formed from half sections of motor tyres and three start units were constructed by mounting three spirals, each approximately 1.1 m diameter on a common stand. This was followed by the development of a $3\frac{1}{2}$-turn unit of moulded plastic and fibreglass by another company, Wyong Minerals, working in the same area. This unit had a diameter of about 1.07 m and a pitch of 953 mm and was in the form of a four-start screw.

Its operation was described by Pullar as follows:

> Concentrate is directed by fixed moulded splitters at intervals down the length of the spiral from the outer to the second channel where it is reconcentrated. Higher grade concentrate is cut successively into the third and fourth channels. The object is to obtain a high grade concentrate and low grade tailing in one pass, with a small fraction for recirculation or separate cleaning.

The unit operated without wash water and, according to Pullar, appeared to be capable of development as a useful concentrator. However, it never lived up to its expectations and failed to impress observers when tested against conventional spirals treating the finer sands of other parts of the New South Wales coastline. Figure 7.12 (after Pullar, 1965) demonstrates the early range of spiral profiles.

Spiral concentrators offer an effective means of dealing with feed particles in the size range 3 mm down to 75 μm and most types can tolerate a slimes particle content of up to 5% provided that the wash water is plentiful and clean. Feed rates vary according to the particle properties of size, shape and density from a maximum of 1.5–2.0 tonnes of solids/h/spiral in roughing service down to 0.5–0.75 in cleaning service. Feed densities are

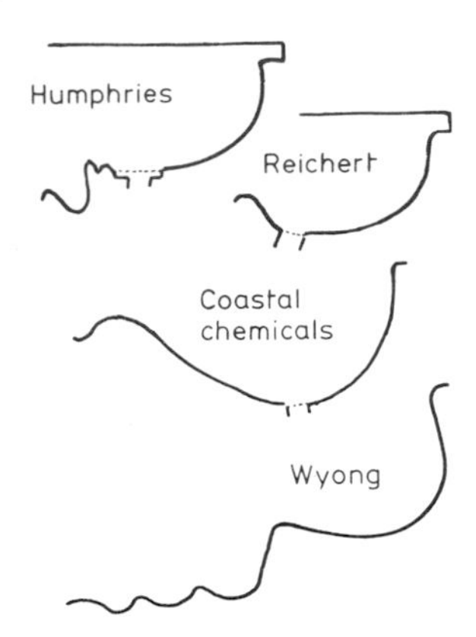

Fig. 7.12 Separating launder profiles of various spiral types

much lower than for cones and trays, usually less than 25% of solids by weight and consequently the pumping circuits are power and capital-cost intensive. Individual spirals are controlled easily but a large number of such units is required for primary concentrating service and a bank of spirals requires constant attention. Large concentrating plants treat massive quantities of sand and require complex feed distribution systems and product launder systems which are costly to build and maintain. Their main application in modern plants is in final cleaning service where a high degree of selectivity is a major benefit.

Wash-water handling and control has always provided problems in spirals operation and it is interesting to note that another manufacturer, Readings of Lismore, now offers a spiral configuration called a Cyclone Sluice with one, two or three starts per unit. The sluice requires no wash water and has only one each of concentrate, middlings and tailings take-off points, located at the bottom of the spiral. They claim capacities up to 2.5 tonnes per spiral start and the rejection of up to 95% of the silica in one pass from grades as low as 0.5% HM.

7.4.4 Jigs

Jigs are hindered settling devices used in the placer mining industry for concentrating valuable heavy minerals from feed materials in the general size range of 105 μm up to a maximum of about 25 mm. They consist essentially of shallow, flat trays with perforated bottoms through which water is caused to pulsate up and down. Jigs are sometimes allowed to develop their own beds from the material passing over; normally, however, a bed, several layers thick, of sized pieces of haematite, steel punchings or some other form of 'ragging' is placed on top of the screen as a means of promoting the strong directional currents of water needed to effect separation. The jig bed is dilated on the forward stroke of a plunger and compacted on the backward stroke in a series of rapid pulsations in which the direction of flow through the bed is reversed sharply several times each second. The feed slurry flowing across the bed is thus subjected to the action of upward and downward currents that tend to preferentially draw the heavier particles downwards and keep the lighter particles in suspension. Under such conditions of hindered settling the heavier particles either sink down through the bed to be taken off from spigots at the bottom of each box or compartment, or are removed by an end draw off if too large to pass through the screen. The lighter particles continue on and over the end of the jig as tailings. The principles of jigging are described in great detail in Taggart and other standard texts.

Jigs are used principally for the recovery of gold, platinum and other continental-type minerals and gemstones and in roughing service may consist of two, three or four cells in series. Standard units with cells each having about 1 m^2 jig bed area are usually rated at 18–20 m^3h^{-1} per unit. The feed water: solid ratio has been found to be most effective at around 1:1, the main criterion being that the pulsations should not dilate the materials excessively. Make-up water varies between 3 and 7.5 tonnes per tonne of solids in the feed and the best results are obtained with feed materials that have been deslimed and pre-sized into a number of fractions for separate treatment.

The main change in new jig design has been the development of the circular Cleaveland jig marketed by IHC Holland. Circular jigs are fed at the centre and discharge at the periphery, the most distinctive feature being a skimming device which contours the jig-bed surface and prevents both pyramiding and channelling of the pulp. The skimming action ensures a reasonably uniform jigging action over the entire jig bed and the circular shape of the bed results in deceleration of the pulp as it moves towards the discharge at the periphery. The deceleration facilitates concentration by increasing the retention time for particles of dissimilar density allowing them to move further away from one another, thus substantially increasing capacities.

From an initial 5 ft (1.524 m) diameter, the sizes of circular jigs have been increased to 22 ft (6.7 m) with a capacity of about 230 m^3 h^{-1}. According to the manufacturers much larger jigs of this type appear practicable and, because of the consequent reduction in floor space: throughput ratios, dredger treatment capacities up to ten times the above amount are within the range of possibility.

7.4.5 Shaking tables

Shaking tables are designed primarily for the wet gravity treatment of fine granular materials, however, air tables have certain specific applications in some dry separation process plants.

Wet shaking tables

Wet shaking tables have been used extensively in minerals-processing operations since they were first introduced around the turn of the century. As described by Rose (1902) the first tables were made of wood or sheet iron, the surface being as smooth as possible and the sides being flanged. They were hung by chains or similar so as to be capable of limited movement and received a number of blows delivered at their upper ends. The blows were given by cams acting through rods or, alternatively, the tables were pushed forward against the action of strong springs by cams on a revolving shaft. Being suddenly released the tables were thrown back violently by the springs against a fixed horizontal beam. As a result of this differential movement and the inertia differences of individual particles the larger and heavier particles were made to move up the table in the direction of travel of the blow and by regulating the flow of wash water the lighter particles were carried down. Only coarse particles were recovered because the fines were washed away with the slimes.

Shaking tables in current use operate on the same general principles although riffles have been introduced to provide pools for separating out and trapping valuable heavy particles. They are used essentially in cleaning service although Dunkin (1953) refers to special forms of riffling that allowed tables to be used in the beach-mining industry at feed rates of up to 8 tonnes/h/table in roughing service. Compared with other primary concentrator types, shaking tables occupy an excessive amount of headroom regardless of capacity and are now used only for cleaning up the small quantities of concentrate produced in the roughing circuit. As cleaners, they have the advantage of being readily adjustable by means of tilting and wash-water control with results that are apparent at all times. This visibility allows fine adjustments to be made to the splitter positions in order to produce high-grade concentrates, low-grade tailings and a middlings that can be returned to head feed or retreated separately.

Table feed slurry concentrations, averaging about 25% of solids by weight, flow across the table decks which are supported in the horizontal plane but can be tilted to direct the flow as required.

The deck is given a reciprocating motion in a longitudinal plane by means of a vibrator mechanism or an eccentric head motion. Speeds of 280–325 strokes min^{-1} cover the normal range for most materials and the stroke length is adjusted for the particular service. Coarse feeds require long strokes at lower speeds; fine feeds require shorter strokes at faster speeds. The inclination of the deck is adjusted when operating.

Because of the reciprocating action of the table and the transverse flow of water, the pulp fans out immediately upon contacting the table surface. When it reaches the riffles the lighter particles tend to be washed over the riffles while the heavier particles are held back and moved along the riffles by the reciprocating movement to be discharged over the end of the deck.

Tailings are washed over the lower edge and a middlings fraction is taken off, as required, between the concentrates and the tailings.

It is customary in mineral-sands, dry-milling operations to divert the rougher non-conductors to a wet tabling section to remove the greater part of such troublesome non-conductors as quartz, kyanite and leucoxene. The concentrates are then redried and returned to the dry-milling circuit electrical units. Most full-sized tables have capacities around 0.75–1 tonne/table/h in final cleaning operations.

Air tables

Air tables have a somewhat similar motion to wet shaking tables but instead of water, air is used as the fluidizing and transporting medium. The table deck is covered with a porous material and air under pressure is blown up through this material from a plenum chamber situated beneath the deck. The purpose of the plenum chamber is to equalize the pressure and hence the flow over the whole of the deck.

Dry feed materials are presented at one corner of the deck as illustrated in Fig. 7.13. The deck is shaken laterally and air pressures are regulated to keep the lighter particles in a state of fluid suspension, thus allowing them to gravitate down the slope of the table by the shortest route. The heavier particles sink to the bed and because of the reciprocating sideways movement of the table are moved up slope and discharged on the high side of the table. Splitters at the end of the deck allow a middlings fraction to be taken as well as concentrates and tailings.

The sizing effects of air tabling are opposite to those of wet tabling and this makes them very useful in specific applications where wet tables are ineffective or inefficient. For example, fine silica particles have a tendency to report with zircon particles on wet tables but are easily removed by air tabling. On the other hand coarse waste minerals report in air-table concentrates and are better removed at that stage by screening. Feed rates, deck angles and slopes are all adjustable. The capacity of a 3.1 m × 2 m air table is about 1.1 tonnes h^{-1}.

7.4.6 Rotating cones

Although not very well known in the placer-mining industry rotating cones have an application in some arid and semi-arid areas because, of all

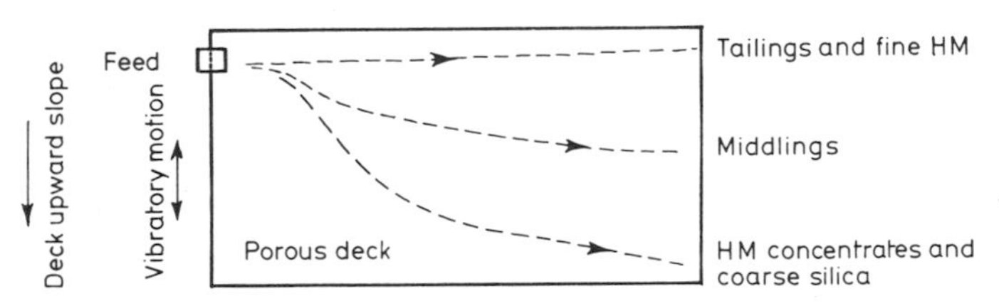

Fig. 7.13 Stratification on air table

wet primary concentrators, they use the least amount of water and make a high-grade concentrate in one pass. They have a very high tolerance to slime and, in fact, depend upon a reasonably high slime content in the slurry for adequate fluidity when treating a long size-range feed such as is found in most continental-type deposits. However, because of the high slime content, small particles are more easily held in suspension and, as a result, recoveries of valuable heavy minerals in the finer sizings are low. Rotating cone concentrators in service in the Moolyella tin field are shown in Fig. 7.14.

Rotating cones are fabricated with diameters of 3–5 m with included angles ranging from 105° to 115°. They are mounted singly or in multiples on a hollow shaft that is tilted and adjustable between about 50° and 60° from the horizontal. Single cones are more common in the tin fields because of the high unit value of the minerals. Problems of distribution and the critical nature of the cone adjustments make multiple-mounted units difficult to control. On the other hand multiple units have been used most in the mineral sands industry because of their higher capacities in terms of floor-space utilization.

In operation, high-density pulp at the rate of 3–15 tonnes h^{-1} is fed into the cone near its periphery on the trailing side and wash water is added on the leading side. The heavy minerals move gradually to the centre of the cone where they become concentrated; the lighter minerals overflow at the lowest edge of the cone. The concentrates are drawn off periodically

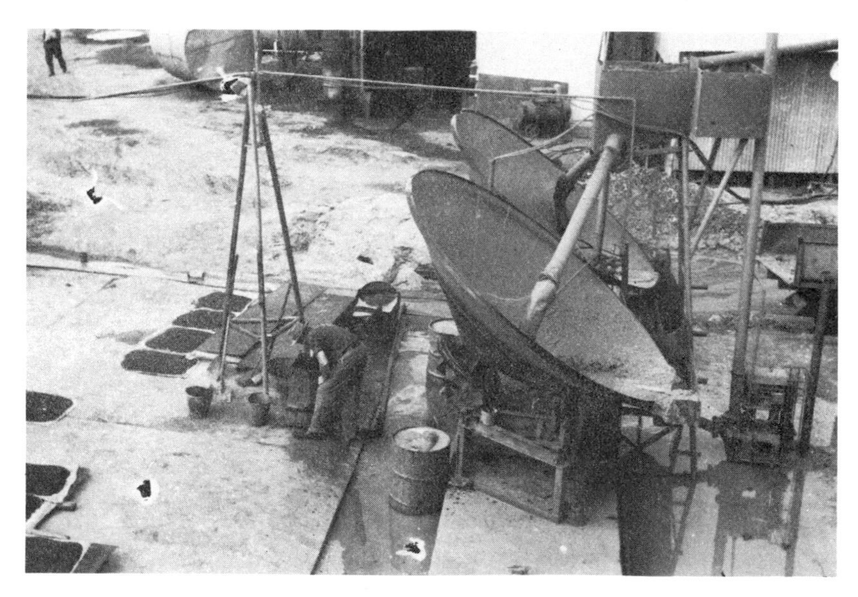

Fig. 7.14 Rotating cone concentrators used for coarse cassiterite recovery in a semi-arid environment

through the hollow shaft. The amount and grade of the concentrate is determined by the slope of the cone, the grade of the ore and time. Separation in beach-sand service is improved by artificially roughening the cone surface by some such means as spreading coarse sand on a mastic base. Pullar (1965) quotes production figures from Ilmenite Minerals Pty. Ltd. of Wonnerup, W.A. and Cable (1956) Ltd. as follows:

Operator	Feed rate (tonnes h^{-1})	Feed grade (%HM)	Concentrate grade (%HM)
IMP	3	10–20	92–93
Cable	15	10–25	50–70

In both cases the recoveries were low and at Cable, the tailings were scavenged using spirals.

In the tin fields at Moolyella in Western Australia, tin-bearing gravels containing an average of about 0.2% SnO_2 are upgraded in one pass to between 60 and 65% SnO_2 at feed rates of up to 15 tonnes/h/cone. In one plant, the fine tailings were scavenged using a Reichert Cone Concentrator followed by a Yuba four-compartment, clean-up jig.

7.4.7 Heavy-medium separators

The process of heavy-medium separation relies upon sustaining a medium consisting of finely divided solids and water in a highly fluid condition at a relative density intermediate between that of the mineral to be saved and those of the minerals to be rejected. For example, a ferrosilicon–water mix might be maintained at a relative slurry density of 2.9 to separate diamonds (RD 3.51) from gravels (RD 2.65).

According to Mitchell Cotts (1971) the principle was first applied in 1921 to the separation of coal from shale, using a fine silica-sand suspension in water. Because of the low relative density of the silica the maximum separation density was approximately RD 1.58 obtained at a concentration of 60% of solids by weight or 36.5% by volume. A 9 ft (2.74 m) cone-type vessel was successfully commissioned by the American Zinc, Lead and Smelting Company at Mascot, Tennessee, using a galena suspension as the medium in 1939 and, since then, there have been many improvements in machines, media and techniques.

The application of heavy-medium separation to placer-mining technology is mainly in the recovery of gemstones using heavy-medium cyclones in place of jigs and diamond pans. Heavy-medium cyclones have been developed to overcome the floor space and cost problems of drum, bath and cone type HM separators which rely upon gravitational forces alone for the settlement of particles in the medium provided. The centrifugal forces generated within a cyclone are many times greater than the gravitational force and with careful density control and selection of medium, accurate and economical separations are common place down to particle sizes much smaller than the minimum economic sizes of gemstones. A typical HMS flowsheet is described in Fig. 7.15.

Media solids

Preferred solids for mixture with water as HM materials are magnetite, ground ferrosilicon and atomized ferrosilicon because of the simple magnetic methods that can be applied to their recovery. According to Dreissen (1980) ground ferrosilicon costs about four times as much as magnetite and about one-half that of atomized ferrosilicon but each is worthy of careful consideration for any new application because of

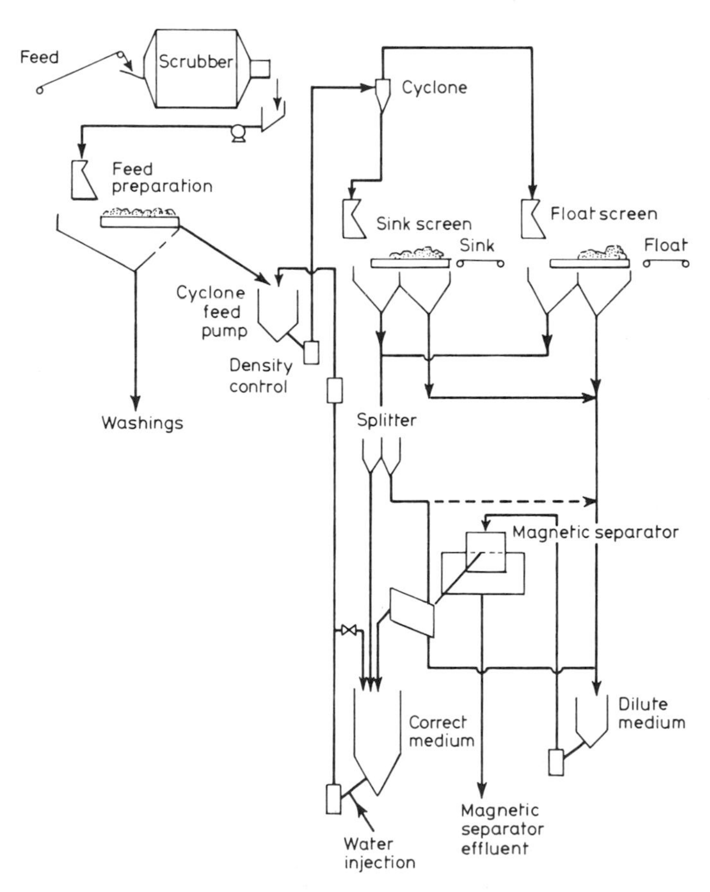

Fig. 7.15 Typical flowsheet heavy-medium separation

individually unique advantages and disadvantages. For example, medium losses are somewhat lower when atomized ferrosilicon is used and, within certain limits of relative density, magnetite can be mixed with atomized ferrosilicon to reduce the costs of medium consumption.

The size distribution of the media solids is an important factor limiting the allowable concentration and, hence, the upper limit of slurry density. Solids ground to 0.05 mm allow a suspension concentration of maximum 34% by volume above which the slurry viscosity is unacceptably high. For magnetite with RD 5.0 the maximum slurry RD is 2.2 and for ferrosilicon of RD 6.8 the maximum slurry RD is 3.0. Atomized ferrosilicon, by virtue of its truly spherical shape, can be used to a volume concentration of as high as 45% corresponding to a slurry density of 3.5 while still limiting the viscosity to a workable level.

Bath (1978) lists the main factors determining the selection of the most suitable medium for a particular HM application as: (a) RD of required separation, (b) required sharpness of separation, (c) whether a static or dynamic separatory vessel is to be used, and (d) size of ore to be treated.

These factors are considered in relation to the viscosity and stability of the various media available and to such medium characteristics as: (a) ease of recovery, (b) the availability of high densities, (c) low rates of degradation by abrasion, (d) resistance to corrosion, and (e) operating costs.

Medium recovery and cleaning

Considerable quantities of medium follow both the sink and the float products from the separating vessels and their separation from the coarser ore and waste materials is effected on screens. Particles of medium adhering to the larger solids are removed by high-pressure water sprays. The underflow is diluted and the ferrosilicon or other medium is recovered magnetically and returned to the main circuit. A wet drum magnetic separator is used, sometimes preceded by a thickener or centrifugal densifier.

After being recovered magnetically, residual magnetism in the ferrosilicon or other magnetic media particles causes them to cohere into large flocs. Settling rates are increased together with stratification in the separating vessel. Demagnetization is accomplished by passing the suspensions of media through an AC coil. Alternately, the high shear forces encountered by the medium in its passage through a dynamic separatory vessel are sufficient to break up the flocs thus rendering demagnetization unnecessary.

Some losses of medium are inevitable during all of these operations and the medium that is recovered is more or less contaminated by slime-sized particles that have escaped the washing process. Slimes must not be allowed to build up in the circuit or the proper functioning of the plant will be adversely affected due to increased medium viscosity. Normally, therefore, a portion of the medium is bled off continuously to a special cleaning circuit to remove the non-magnetic slimes particles. However, further losses from this circuit constitute a considerable proportion of the total medium losses and special attention must be paid to its design (Bath, 1978).

7.4.8 Design considerations – a modern gravity treatment plant

The point has been made consistantly in this and previous chapters that, if the lower-grade deposits of the future are to be worked economically, designers must begin to look seriously to the introduction of fresh concepts for gravity treatment-plant design and reject past inefficient methods and arrangements. For example, with few exceptions, new bucket-line dredgers are still being commissioned with treatment layouts that have not changed in 50 years and in which the only improvements are in the materials of construction and drive mechanisms. Some recovery inefficiencies that were tolerated then are still being tolerated by many operators today. On the other hand, a few designers have gone to the other extreme by incorporating into their designs extremely sophisticated electronic control equipment in circumstances where such equipment can neither be reasonably understood or maintained.

Common sense as well as vision is needed and with it a flexibility of thought that allows each new project to be studied objectively and for the selection of methods that are the most appropriate for what has to be done. A hypothetical set of data put forward by Redmond (1970, personal communication) for a floating treatment plant to process 100–120 $m^3 h^{-1}$ of tailings from a previous tin-dredging operation is used to illustrate the point. The proposal was soundly based but the concept of using Reichert Cones instead of jigs for tin recovery was too revolutionary for the times and the plant was never built. It is summarized here to demonstrate the manner in which a thoughtful designer incorporated features that provided for both metallurgical and operational efficiency. In order to round off the exercise, designs for the main slurry pumping installations are given in Chapter 5.

In the model used (Fig. 7.16) the combination of surge bin, concentrating units and tailings stackers provides good stability, ample deckspace for all purposes, economic construction, efficient tailings disposal, a convenient grouping of equipment and simplicity of reticulation for both the slurries and power. The pontoon is of modular construction for ready dismantling and re-erection when moving to another area. The overall proportions are around 15 m × 11 m × 1.8 m draught subject to detailed design for buoyancy and trim.

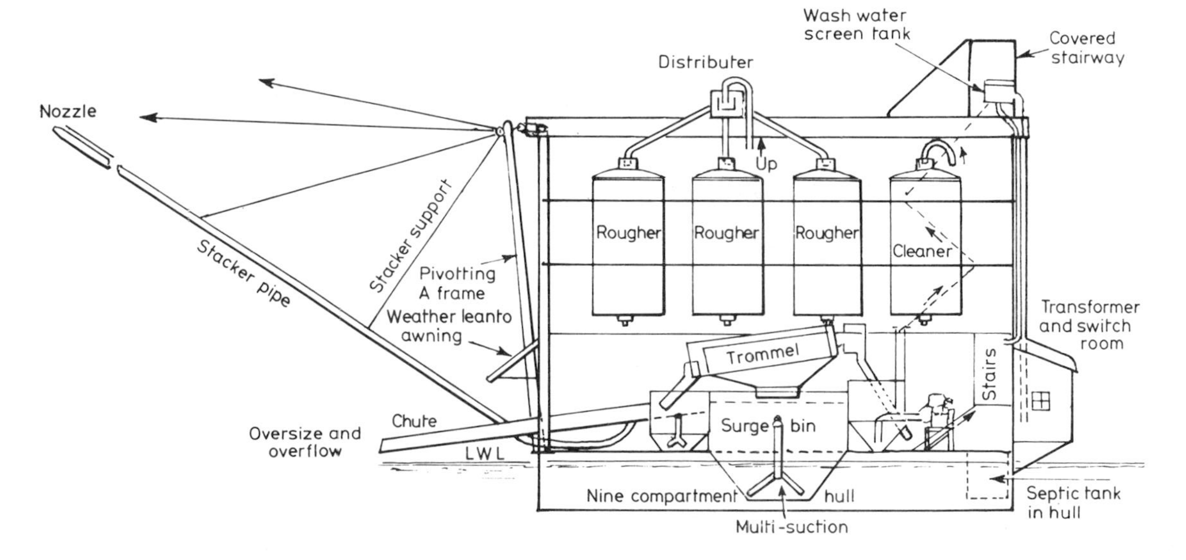

Fig. 7.16 Sectional arrangement of 200 Th^{-1} floating treatment plant

A brief description is given of the layout of process units at each level on the floating plant barge with some reasons for the various emplacements.

Lower deck

All pumps and bins except the surge bin are above deck level so that spillages will not harm the motors. The surge bin is set centrally within the pontoon to provide stability and all motors face outwards away from the launders and possible spill points. Only one make of pump is used and all bins overflow to provide steady head drowned suction conditions.

The tailings stacker is pivotted outboard to allow the A frame and stays to be luffed on the same axis without fouling the main structure. The stacker pipe is supported from the frame at several points to eliminate the need for a separate supporting truss. An important feature of the design, it provides lightness, flexibility and a low level of wind resistance as well as providing the winch with a better mechanical advantage, at flatter angles of the stacker.

An onboard workshop and storeroom is mounted on the port side with a light mono-rail circling the unit and passing over all important pumps and motors as well as projecting outboard at one point for taking on heavy stores and equipment.

The feed line from the dredger can be brought on board from either side, side connections being preferred to end ones, because they allow better 'coiling' of the feed line in the pond. The surge bin is mounted on both centrelines so that fluctuations in loading do not affect the trim. Other 'live' loads are also kept close to the centrelines and are placed symmetrically for the same reason.

The transformer and switchroom are located slightly above floor level and outboard towards the front of the pontoon to act as counter weights for the stacker boom. They are placed directly under the control panel to reduce the lengths of wiring and are protected from spillages from above. Being above the main floor level, the electrics are not affected by water splashing onto them when the deck is hosed down.

The spuds are set as far apart as the width of the pontoon permits to allow for maximum travel and to keep the sidelines (15° on either side of axis of movement) clear of the tailings. The control station for spuds, sidelines and stacker winches is situated at the rear of the plant to enable simultaneous viewing of all units. The wash-water ring main equalizes pressure to all of the discharge points.

First (operating) floor

A machinery console in the operating cabin has all control machinery except the spuds and washwater tank in full view and from a position in front of the console an operator can start or stop each motor and visually check the effects without having to move anything but his head. The console has an ammeter for each motor together with indicator lights.

In the work section of the floor, the trommel screen, cones and spirals are placed so that rougher tailings to scavengers and other fractions can pass to their respective bins with a minimum of crossovers. For this reason they are not disposed evenly around their various bins. Clearances of 0.75 m minimum around each unit give ready access for adjustment and observation.

Second floor

The recleaner stage projects through to this floor and special means of access are provided to the distributors at the top of the spirals. The operators ablutions and other amenities are situated on this deck in space that is available above the trommel.

Third floor

The distributers to the Reichert cones are accessible from this deck and pipe feeds are all below roof level except at the rougher distributer.

Roof

The rougher distributer is located centrally above the four units to be serviced. The roof has a shallow pitch and is strong enough to be walked on with safety.

7.5 MAGNETIC SEPARATION

7.5.1 Magnetic mineral types

Magnetic minerals fall generally into four classes; diamagnetic, paramagnetic, ferromagnetic and antiferromagnetic. In the broadest of terms, diamagnetic minerals have magnetic permeabilities less than unity; paramagnetic minerals have magnetic permeabilities greater than unity; ferromagnetic minerals are paramagnetic minerals with high magnetic permeabilities; antiferromagnetic minerals behave like paramagnetic minerals above a certain critical temperature. Below this temperature, the magnetism decreases with decreasing temperature.

Diamagnetic minerals

The structure of an atom involves a central nucleus with a positive charge surrounded by negatively charged electrons in orbit around it. In the presence of a magnetic field the orbits are modified in such a way that the electrons are slowed down and there is induced in diamagnetic minerals a magnetic intensity in the opposite direction to the magnetizing field and proportional to it. All minerals have some degree of diamagnetism but this may be masked by other superimposed effects such as from the presence of small quantities of iron and manganese. Common placer minerals that are dominantly diamagnetic are quartz, limestone, feldspars and gold.

Paramagnetic minerals

In addition to having orbital motions electrons also rotate about their own axes and each electron acts as a small permanent magnet. In a magnetic field they tend to align themselves in the direction of the field with a positive susceptibility that increases directly in proportion to the field strength. This effect is known as paramagnetism and it is due primarily to the content of iron and manganese.

Due to thermal agitation, the susceptibility of paramagnetic minerals decreases inversely with the temperature and at high temperatures the susceptibility is small. Figure 7.17(a) (after Strangeway, 1970) shows the relationship between magnetization and temperature for a typical paramagnetic mineral, biotite. Other common paramagnetic placer minerals are garnets, amphiboles, rutile, zircon and ilmenite.

Ferromagnetic minerals

The above two classes of minerals exhibit magnetism only in the presence of an external magnetic field and lose that magnetism when the field is removed. Ferromagnetic minerals are magnetic even in the absence of a magnetic field because the atoms have a magnetic moment and align themselves spontaneously in the same direction. Ferromagnetic substances have a positive susceptibility and high magnetic retention or remnant magnetism that is reversible when the applied field is weak. It results in a permanent magnetism that can only be reversed if the field is reversed when the applied field is strong.

Ferromagnetism disappears when heated above a certain temperature and the substance then behaves as a paramagnetic mineral (Fig. 7.17(b)). Although iron, cobalt, nickel and manganese are typically ferromagnetic, in nature none of the minerals or rocks are truly ferromagnetic but instead, minerals such as magnetite, chromite, ilmenite etc. are classed as

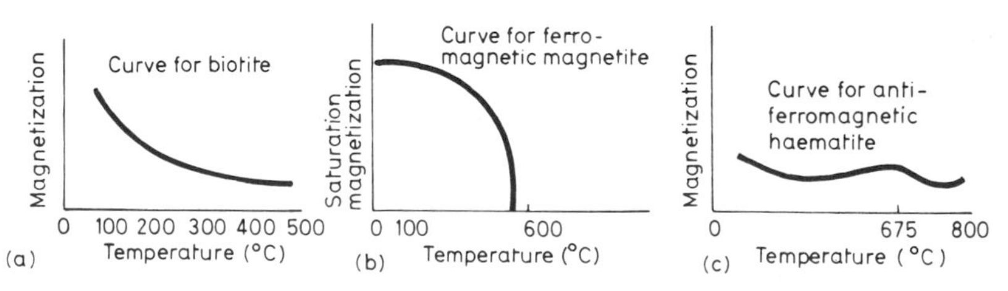

Fig. 7.17 Relationship between magnetism and temperature for biotite, magnetite and haematite

ferromagnetic when there is a net magnetic moment in the absence of a field, or as antiferromagnetic.

Antiferromagnetic minerals

Antiferromagnetic minerals have susceptibilities similar to those of paramagnetic minerals because of both positive and negative exchange effects that cancel out so that the net magnetic moment is zero. They differ from paramagnetic materials in that, below certain critical temperatures, the magnetization decreases with decreasing temperature. Haematite, ilmenite, ulvospinel and pyroxenes are the most important antiferromagnetic minerals in placers (Fig. 7.17(c)).

Obviously, therefore, a simple classification of minerals according to the above standards is not always a reliable guide to their separating qualities; mineral particles have varying degrees of purity and in a magnetic field, small particles may behave quite differently to large particles of the same material. Paramagnetism is often a manifestation of weak ferromagnetism due to varying textural associations with magnetite or ilmenite, or to an antiferromagnetism from intergrowths or inclusions of minerals such as haematite and pyroleucite. Quartz and feldspars are typical diamagnetic materials which tend to behave as non-magnetic minerals in conventional magnetic separators. However, to some degree there is a diamagnetic effect in all minerals. According to Nessett and Finch (1980) haematite displays a form of magnetic anisotrophy called 'canted' antiferromagnetism where the susceptibility decreases with increased field strength, i.e. less than expected recovery benefits with increase in field strengths.

Table 7.5 which lists typical values for the relative magnetic attraction of some of the more important placer minerals is therefore to be taken only as a guide. In plant practice there are many instances where the expected does not happen; minerals such as magnetite, ilmenite and cassiterite have widely varying magnetic properties and, inevitably, a small proportion of cassiterite reports with the magnetics because of minor impurities of ferromagnetic minerals.

Low susceptibility paramagnetic minerals are separated from diamagnetic minerals at high-magnetic-field intensities. Consequently the high-intensity magnetic processes should be preceded by low-intensity magnetic separation to remove the ferromagnetics and tramp iron.

The design of magnetic separating machines is based upon a fundamental law of electromagnetics which states that any moving electrical charge generates a magnetic field and that the induced intensity is governed by the magnetic susceptibility of the mineral and the magnetizing field strength.

Magnetic separation thus involves the passage of mineral particles through a magnetic field with the various operating parameters selected so

Table 7.5 Relative magnetic susceptibility of typical placer minerals in descending order of susceptibilities based on iron = 100

Mineral	Chemical composition	Density	Relative magnetic suscepti-bility	Type of magnetism
Magnetite	Fe_3O_4	5.2	40	Ferromagnetic
Maghemite	Oxidized form of magnetite		36	
Ilmenite–haematite Solid solution	$FeTiO_2$ and Fe_2O_3		9	Ferromagnetic
Chromite	$Fe\ Cr_2O_4$			Ferromagnetic
Ilmenite	$FeTiO_2$	4.5–5.0	12	Antiferro-magnetic
Haematite	Fe_2O_3	4.9–5.3	5	
Pyroxene	$FeSiO_3$			
Garnet	Various Fe, Ca silicates	3.5–4.0	7	Paramagnetic
Monazite	(Ce, La, PO_4)	4.6–5.4	4	
Pyroxene	$FeSiO_3$		2	
Biotite			2	
Amphiboles			2	
Ilmenite	$FeTiO_2$	4.5–5.0	12	
Rutile	TiO_2	4.2–4.25	0.9	
Zircon	$ZrSiO_4$	4.2–4.9	0.5	Diamagnetic
Quartz	SiO_2	2.65	0.4	
Gold	Au		0.1	
Cassiterite	SnO_2	6.8–7.1	0.1	

that non-magnetic minerals are unaffected while magnetic minerals are attracted or deflected according to the means provided for separation. The selection of a magnet type for any new plant or expansion of an existing plant is influenced in all cases by the mineralogical composition of the feed, the particle sizes of the various minerals and by whether the material is to be treated wet or dry. The following brief descriptions of processes and machines apply mainly to mineral sands operations, however, they all have some applications in other fields.

7.5.2 Wet magnetic separation

Low intensity separators of the drum type are used for recovering highly magnetic minerals such as magnetite and for removing tramp iron. High intensity wet magnets are used primarily for recovering products of such minerals as ilmenite and chromite or for scavenging impure or valueless magnetics from a circuit.

Wet drum magnets

Minerals of high magnetic susceptibility such as magnetite, titano magnetite and tramp iron require only a small magnetic potential to attract them to a low-intensity magnet and the drum type provides permanent magnets fitted inside a non-magnetic shell. The magnets are located along the bottom portion of the shell and remain stationary while the drum rotates around them (Fig. 7.18). In the single concurrent feed-type described, magnetic particles attracted to the shell surface are washed off when the shell travels beyond the influence of the magnets. The non-magnetics (by ferro-magnetic standards) flow through to the bottom discharge spigot.

Drum magnets are used to recover magnetite and titanomagnetite products from alluvial and beach placers; to recover and densify ground magnetite and ferrosilicon in heavy medium and coal-washing plants and to scalp out ferromagnetic particles including tramp iron from materials that are subsequently treated using high-intensity magnetic separators. Any ferromagnetic particles entering with the feed will immediately attach themselves to the pole pieces, causing bearding and bridging across air gaps; separation will be affected adversely and may cease altogether.

Field strengths are varied according to the material and product requirements. At Bacnotan, La Union Province, Luzon Island, The Philippines, sand containing titanomagnetite is delivered to a primary drum-type magnetic separator (Fig. 7.19) equipped with two magnetic drums, one operating as a rougher at 1200 gauss, the other as a scavenger at 1000 gauss. The rougher concentrates pass to a two-stage magnetic cleaner with first and second cleaning stages set at 600 and 350 gauss variable respectively. The final product contains at least 60% Fe and approximately 6% of TiO_2.

At Nueva Gorgona, Panama, the flow sheet (Fig. 7.20) was designed to produce a very-high-grade concentrate from a simple ore by creating a large circulating load and producing finished tailings in one pass. Feed to this plant averaged about 16% of recoverable titanomagnetite and the

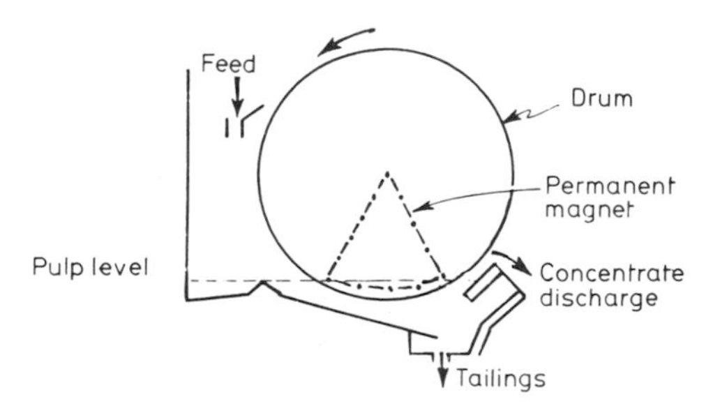

Fig. 7.18 Low-intensity wet drum magnet

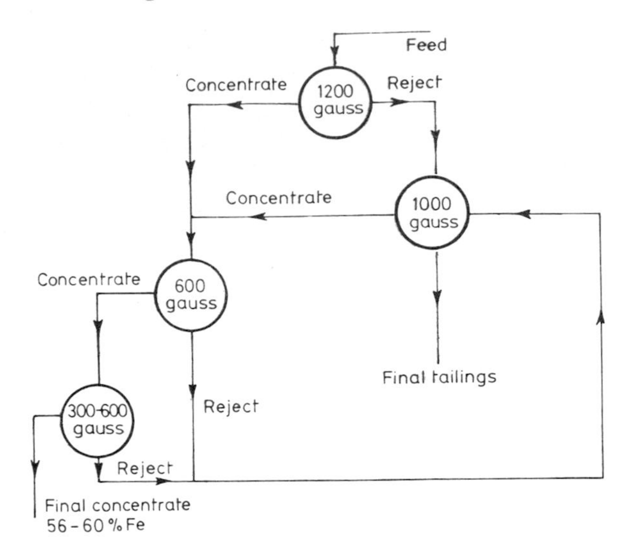

Fig. 7.19 Typical flowsheet for complex magnetite ore, Bacnotan type

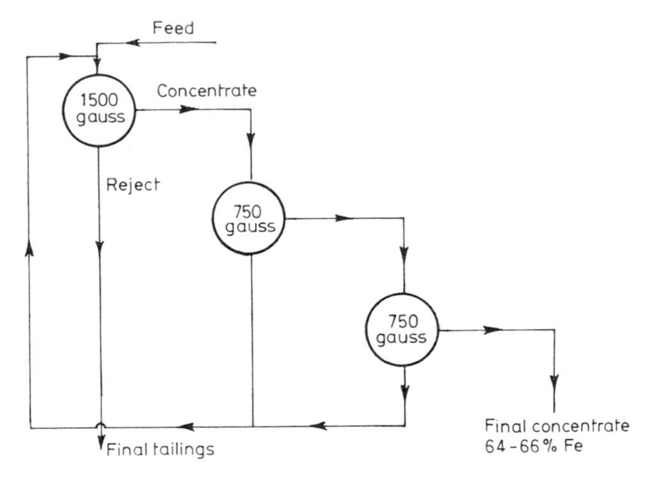

Fig. 7.20 Typical flowsheet for simple magnetite ore, Nueva Gorgona type

final product assayed 64–65 % Fe with 3–4 % TiO_2. Satisfactory recoveries at high Fe grades were possible with this simple arrangement because of the purity of the ore. Less-pure materials require more complex arrangements for treatment because of the wider range of magnetic susceptibilities. Product grades, in any case, are limited according to the composition of the recovered minerals.

Experience has shown that feed-slurry densities should not exceed $C_W = 35\%$ with a maximum particle size of 1.5 mm and that a close control is needed, also, of the rate of feed and volume of wash water. Some designers tend to minimize the importance of interposing a surge bin with densification controls between the dredger and the magnets but good feed preparation is as important in this form of processing as in any other.

Wet high-intensity magnetic separators

Wet high-intensity magnetic separators (WHIMS) operate at field strengths up to more than 20 000 gauss although 11 000 gauss is suitable for most placer operations. WHIMS are used in a variety of applications including:

(1) The production of saleable magnetic products, e.g. ilmenite production, Richards Bay, South Africa. At Richards Bay, the ilmenite fraction constitutes more than 50 % of the feed and responds well to a magnetic field. The magnetic waste materials pyroxene and garnet are measurably lower in the scale of magnetic susceptibility and a high-quality ilmenite product is made in its initial wet state, thus effecting considerable savings in cost by eliminating drying. The ore is first passed over drum magnets to remove magnetite then through two stages of WHIMS. Magnetics from the first stage are passed over a second stage to reduce entrainment of the less-magnetic minerals and the final product is very high grade.
(2) Removal of ilmenite, chromium and other paramagnetics from glass sands.
(3) Removal of unwanted magnetic minerals such as high chrome ilmenite from beach sands; garnets and pyroxene from zircon sands; ilmenite from tin concentrates.

Several types of WHIMS are available and, of the principal ones, the Jones, Gill, Carpco and Readings are well known in the placer industry.

Jones separator

The Jones, high-intensity, wet magnetic separator (Fig. 7.21) removes magnetite with a fraction of an ampere current on the magnetising coils and various other minerals at amperages up to 50 amps at which stage a magnetically saturated field is produced. The matrix consists of parallel grooved plates set perpendicular to the magnetic field direction.

It functions by means of an automatically operated and controlled cycle of frequency of 10–15 cycle min^{-1}, each cycle consisting of three separate actions, thus:

(1) The feed period: the feed valve opens and the magnetic particles adhere to the grooved plates, the non-magnetic particles passing through.
(2) The washing period: the feed valve closes and wash water is passed over the grooved plates in pulsations to free any non-magnetic

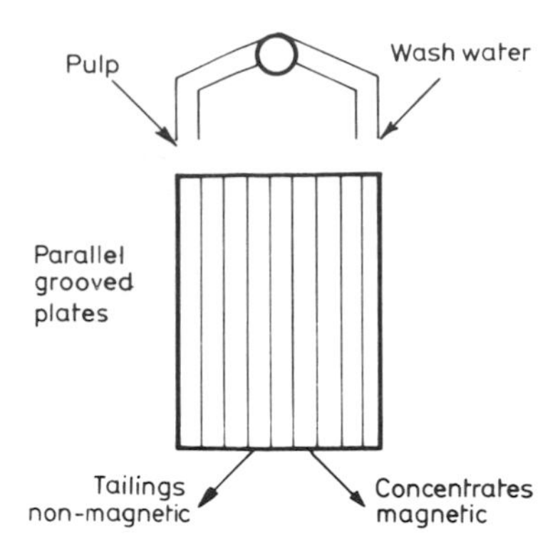

Fig. 7.21 Jones WHIMS (cyclical) operational features

particles that have become entrained with the magnetic particles. This produces a middlings product.

(3) The scour period: the scour water valve opens to wash the magnetics off the grooved plate into the magnetic product bins.

Low operating costs are claimed for the separator and, on a large machine with a saturated field, power consumption, including that of the drive and stirring motors is generally less than 3 KW $tonne^{-1}$ treated.

Gill separator

Although now superceded by the Reading WHIMS the Gill separator has been widely used in Australia for scalping high chrome ilmenite from some heavy-mineral sands and for the production of ilmenite concentrates from others. It represented a departure from the Jones type of cyclical operation by providing a grooved, laminated rotor, rotating about a vertical axis between electromagnetic pole pieces. The magnetic particles are attracted to the rotor and are washed off at intervals between the magnetizing pole pieces (Fig. 7.22). Feed to the Gill was limited to 1.5 mm maximum-sized

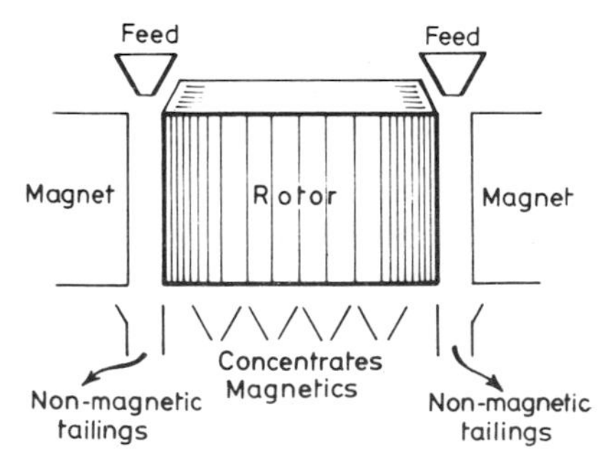

Fig. 7.22 Gill WHIMS (medium intensity) operational features

particles and required a preliminary scalping out of tramp iron, magnetite and other ferromagnetics. A maximum field strength of 11 000 gauss limited its application.

Carpco separator

This separator uses an annular ring containing a matrix composed of steel balls passing between the poles of an electromagnet. The magnetics adhere to the matrix and are flushed off in non-magnetic positions (Fig. 7.23). The Carpco Company of Florida claims that a 22 000 gauss field strength can be achieved, making it one of the most powerful of the WHIMS.

Reading separator

The Reading Separator was developed to take the place of the Gill magnet. It has a similar configuration to the Gill but incorporates some important new features including:

(1) Sixteen oil-cooled coils which provide more ampere turns than the air-cooled Gill coils.
(2) The rotor consists of an annular ring containing parallel laminated grooved plates normal to the direction of rotation.
(3) The arrangement provides eight feed positions as against four for the sixteen-pole Gill magnet.
(4) A feed capacity of up to 35 tonnes h^{-1} whereas the Gill had a maximum feed capacity of 8 tonnes h^{-1}

In operation, feed is introduced in the form of a slurry to the feed distributor and distributed evenly to stainless-steel feed boxes set above the rotor. The slurry from each feed box then passes to the rotor assembly at the ascending rate of magnetic intensity; the magnetics are attracted to serrations on the salient pole plates while the non-magnetic particles fall freely into the catch box below. A controlled flow of wash water ensures the continuity of flow of the non-magnetics and the freeing of any entrained non-magnetic particles. The magnetic particles are scoured off the salient poles by water jets under pressure in the 'null' zone located between two poles of identical polarity. Catch boxes beneath the rotor

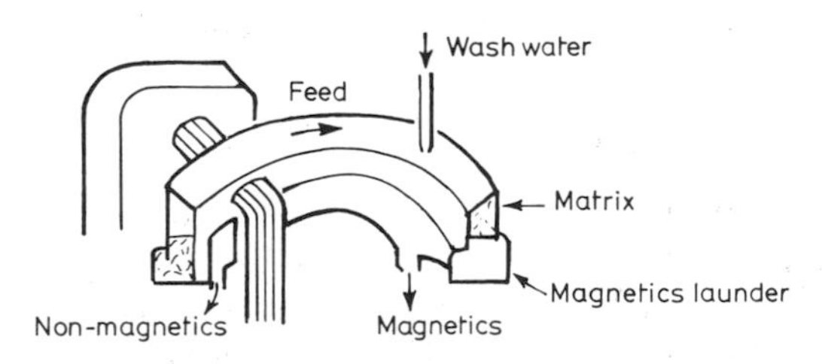

Fig. 7.23 The Capco WHIMS operational features

collect the non-magnetics, middlings and magnetics from each feed point and gravitate them through rubber hoses to common collection boxes and bins.

As with most other WHIMS types, the Reading machine operates satisfactorily only with feeds that have been properly prepared. Important requirements are: the prior removal of all oversize particles that might cause clogging; the prior removal of all ferromagnetic materials for the same reason and controlled slurry densities. For example, beach sands are usually treated at between 30 and 35% of solids by weight. Higher pulp densities result in increased production rates but only at the expense of an increased entrainment of non-magnetics with the final product. Low pulp densities reduce the entrainment of non-magnetic minerals but may give rise to losses of both coarse and very fine magnetic particles. Slurries of predominantly very fine magnetic minerals are better treated at higher pulp densities, although such materials are rarely found in placers. In all cases, excessive amounts of slimes-sized particles smaller than about 20 μm are deleterious to the operation and should be removed at the feed preparation stage.

Advantages of WHIMS

(1) The feed does not require close sizing except for the removal of oversize particles and slimes-sized particles. Good separations can be made down to a few microns.
(2) Although first cost is relatively high, power consumption is low, around 3–4 kW h^{-1}/tonne treated.
(3) They may be used at the wet concentration stage, prior to incurring the handling and drying charges attached to dry-mill separation. In some instances a final ilmenite product can be made in a single pass. In others WHIMS are used to scavenge feed to the dry mill in order to remove such troublesome contaminants as the weakly magnetic rock forming minerals garnet, pyrobole etc.

Disadvantages of WHIMS

They are less selective than some dry magnetic separators where, for example, minerals of close magnetic susceptibility such as garnet, monazite and xenotime are required to be separated from one another.

7.5.3 Dry magnetic separation

Dry magnetic separation involving, as it does, increased costs for drying and transportation is normally used in roughing service only where wet magnetic separation methods cannot be applied successfully or where a greater selectivity is needed. In the example given of ilmenite recovery at Richards Bay, the production of high-grade concentrates was possible by these means because the ilmenite was strongly magnetic and the pyroxene and garnets were weakly magnetic. WHIMS might not have been suitable if much of the ilmenite had been weakly magnetic or if the magnetic silicates had contained an appreciable proportion of more strongly magnetic particles. Greater selectivity would then have resulted from using dry methods of magnetic separation, the main types being, (a) permanent drum magnets, (b) high-intensity crossbelt magnets, (c) disc magnets, and (d) induced roll magnets.

Permanent drum magnets

Permanent drum magnets used for low-intensity, dry magnetic separation are similar in construction to wet drum magnets except that the feed is brought in around top dead centre and the metal shell rotates around an inside stationary magnetic assembly in the top right-hand quadrant (Fig. 7.24). Particles attracted to the magnet are held against the shell as it revolves and fall away from it in a zone of diminished magnetism below the drum. The non-magnetics are thrown from the drum in a normal trajectory. A splitter separates the two streams.

Applications

The main applications for such magnets are in scalping out ferromagnetics and tramp iron ahead of higher-intensity magnetic circuits.

Crossbelt magnets

Crossbelt magnets are lift-type magnets so arranged that the magnetic particles are lifted from the body of the feed and discharged to the side. The separator consists of a flat feed belt passing between two or more poles of electromagnets and crossbelts which service each set of poles and move under the upper pole pieces across the direction of the main belt. A typical arrangement is illustrated in Fig. 7.25.

The temperature of the feed material should not be more than 75° C or vary excessively if a range of magnetic products is to be made. Apart from

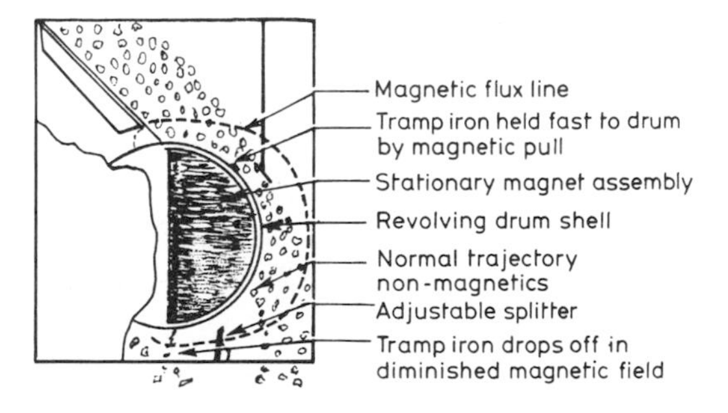

Fig. 7.24 Permanent drum scalping magnet

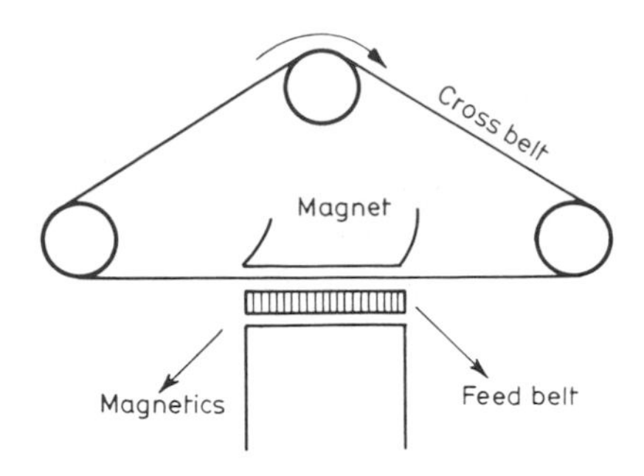

Fig. 7.25 Operational features of the crossbelt lift-type magnet

damage to belts because of the excessive heat, variations in temperature have varying effects upon the magnetic susceptibilities of different minerals. For example, in a mixture of ilmenite and garnet, a good separation may be made at 40° C but not at 70° C.

Testwork will determine the optimum temperature for separation but in all cases in which the total feed is made up of new feed plus a recirculated fraction particular care must be taken to see that all of the particles enter the magnetic field at around the same temperature. Other variables such as the main belt speed, crossbelt speed, applied current and feed rate are also determined initially under test conditions and, once fixed, seldom require adjustment unless the mineral composition of the feed alters. All machines separate more effectively if the total feed is sized into an appropriate number of fractions for individual treatment.

Feed presented in a thin layer onto the main belt passes successively between each set of pole pieces. Magnetic minerals are lifted and adhere to the under surface of each crossbelt and are carried across the faces of the poles to drop into containers beyond the influence of the magnetic field. Non-magnetic minerals pass unaffected under all poles and fall into a product box at the discharge end of the machine.

Selectivity is provided by varying the air gap at each set of poles. In the Reading Crossbelt magnet the magnetic field strength can be varied from a minimum of 9500 gauss (3/8 in air gap) on the first of six crossbelts to a maximum of 19 000 gauss (3/16 in air gap) on the last crossbelt. Readings give the following example of separation of a Malaysian feed containing 88% and 10% cassiterite:

Product	Weight %	Sn %	Sn distribution %
Crossbelt magnetics	79.0	0.02	0.3
Crossbelt middlings	2.0	0.60	0.2
Crossbelt non-magnetics	19.0	31.40	99.5

Applications

Any free-flowing materials finer than 2 mm can be fed to crossbelt magnets. Selectivity is high due to the lifting action for closely sized particles at feed rates that may vary up to 30 tonnes h^{-1} in some of the larger machines. Typical examples of their uses are found in the recovery of cassiterite from 'Amang' minerals in south-east Asia; the recovery of wolfram from colluvium in eastern Australia and the production of ilmenite from beach sands at Pulmadai in Sri Lanka.

Disc magnets

Disc magnetic separators also have a lifting action but they differ from crossbelt magnets in the use of grooved discs instead of crossbelts for picking up and removing the magnetic particles. The grooving provides a high magnetic field gradient and the air gap at each pick-up point is varied by raising and lowering the individual discs and by tilting. Minerals of close magnetic susceptibilities can be separated from one another by closely sizing the feed and then providing a long residence time for the particles.

Applications

Disc magnets are used for such separations as garnet from monazite and monazite from xenotime. They are used also for producing final monazite products from zircon-rich fractions. In such applications, high-intensity roll magnets might be used first to recover the magnetic conductor minerals; disc-type magnets would then differentiate between the monazite and other weakly magnetic minerals, garnet etc. Selectivity is high but capacities are low, of the order of 0.5 tonnes h^{-1} in most disc-type machines. Feed particle sizes should not exceed 1.2 mm, particularly in cleaning service. The action of a McLean type disc magnetic separator is illustrated in Fig. 7.26.

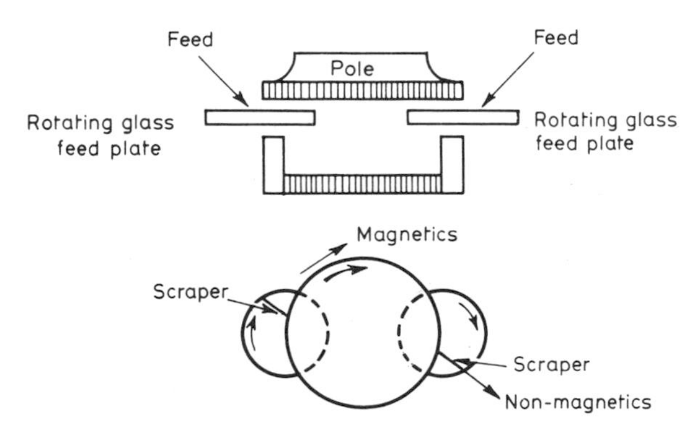

Fig. 7.26 Operational features of the disc type magnet

Induced roll magnets

Induced roll magnetic separators utilize a series of laminated rotors rotating between the poles of powerful electromagnets. Each fixed pole is shaped to follow the curve of its rotor and is positioned to induce a high-intensity field between the pole and the rotor at a point about 15° below dead centre to give maximum residence time for the particles.

Materials fed onto the rotor pass between the surface of the rotor and the adjustable pole piece. Rotors are normally fabricated with alternate strips of conducting and non-conducting materials so that the lines of force between the two poles converge onto the narrow conductor strips, thus avoiding eddy currents and increasing the magnetic intensity at the rotor surface. The magnetic particles are attracted to the centre of the rotor rather than to the pole pieces and are brushed off in a zone of low magnetism as illustrated in Fig. 7.27. The non-magnetic particles fall away under the dual influence of centrifugal force and gravity. Adjustable splitters are used to direct the products into their respective boxes.

The operating variables are feed rate, roll speed, pole strength, air gap and splitter positioning. Flux density between the poles and the rotors is continuously variable and the magnetic circuit should be capable of saturating the poles of the magnets in the separating zone. The main mineral variables are magnetic susceptibility, density, distribution and absolute and relative particle size. The feed should be dry so that the individual particles are quite discrete and do not agglomerate.

Strongly magnetic minerals are attracted to the tip of the nose piece and cause clogging if the variables of flux density, air gap and rotor speed are not suited to the characteristics of the feed material. In such cases, low intensity scalper rolls can be fitted ahead of the top high-intensity rolls to remove the tramp particles. Care must also be taken not to use a flux density greater than needed to recover the required magnetic fraction, otherwise agglomerations of both magnetic and non-magnetic particles

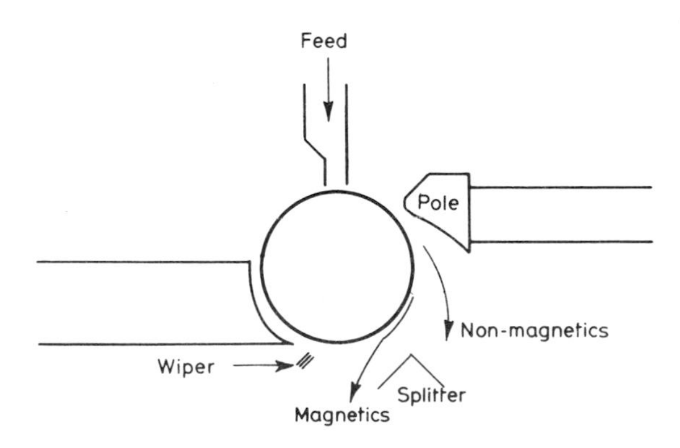

Fig. 7.27 Operational features of the induced roll magnet

will tend to bunch and choke the air gap in the same way as do strongly magnetic materials at normal flux densities.

Roll speeds are determined according to the particle size of the feed and the type of material to be treated. In mineral-sands operations the zircon cleaning-roll magnets are generally run at slow roll speeds (120–140 rpm) to ensure maximum removal of magnetic impurities. Roll speeds are higher for rutile cleaning circuits (140–200 rpm) because the conductor magnetics (e.g. ilmenite) are more magnetically susceptible. The higher roll speeds do not prejudice the recovery of ilmenite particles measurably but ensure more effective throwing of the rutile particles, thus effecting a better split.

Induced roll magnetic separators have high capacities but lack selectivity. They are used normally in a non-magnetics retreat configuration (see Fig. 7.28) with the non-magnetics from the first roll being retreated on the next roll and so on. Sizing effects are opposite to those in crossbelt magnets in that the finer particles are taken off first. The maximum flux density of the standard Reading machine is around 18 000 gauss at 3.2 mm air gap with nominal feed rates up to 3.3 tonnes h^{-1}/1.5 m of roll length. The actual capacity of a machine depends upon the properties of the material being treated and the required product grades.

Applications

Roll magnets may be used:

(1) To produce a clean non-magnetic product, e.g. removing ilmenite from rutile in a conductor circuit or ferromagnesians and silicates such as

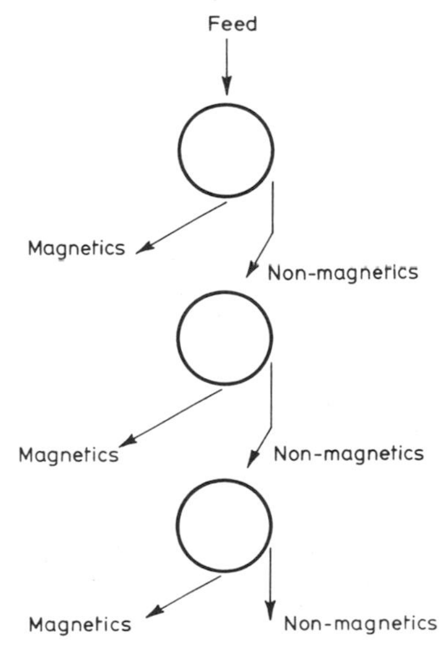

Fig. 7.28 Non-magnetics retreat configuration

garnets, hornblende, pyroxene and tourmaline from zircon in a non-conductor circuit.

(2) To remove magnetic particles from concentrations containing diamonds and sapphires. Special nose irons and rollers can be fitted to treat mineral particles as large as 4 mm.
(3) To clean refractory minerals such as chromite.
(4) To clean coarse cassiterite, tungsten and tantalum concentrates.
(5) For separating minerals such as ilmenite and some monazites that differ only slightly from one another in magnetic susceptibility, a degree of lift may be obtained using specially designed adjustable pole pieces which tend to direct the magnetics into a trajectory that is concave downwards and to steepen the trajectory of the non-magnetics.

In general, coarse magnetics are less likely to be attracted to the roll than fine magnetics because the higher momentum imparted to coarse particles causes them to be thrown outwards from the roll. Sizing the feed into coarse and fine fractions and treating each separately at different roll speeds and magnetic intensities usually provides for more efficient separation.

Isodynamic separators

Isodynamic separators of which the Franz is the best-known type, are used in preliminary studies to provide a measure of the specific susceptibility of individual mineral types to magnetic separation. The basic Franz unit comprises a vibrating chute mounted centrally between the pole pieces of an electro magnet. The unit can be tilted in any direction by virtue of its universal mounting, the most sensitive separation being obtained by using the highest practicable current and transverse slope, and the lowest practicable longitudinal slope. The electro-magnetic current is continuously adjustable from 0–1.5 amperes and with suitable calibration the separator will provide absolute measurements of the specific susceptibility of individual particles. Figure 7.29 after McAndrew (1957) depicts the forces acting upon a particle to move it down the chute of a Franz isodynamic separator. F_g is the force due to gravity; F_m the magnetic force

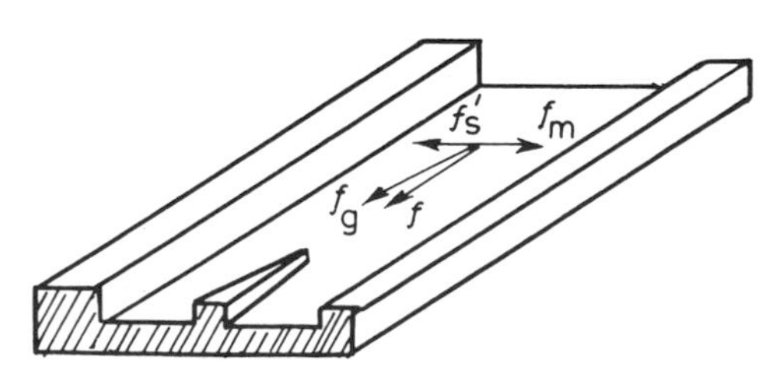

Fig. 7.29 Forces acting on a particle on the chute in a Franz isodynamic separator

acting at right angles to the longitudinal axis of the chute; the component F'_g across the chute of the gravitational force is given by

$$F'_g = mg \sin \theta \quad (7.1)$$

in consistent units. F is the resultant force on the particle in the direction of the chute length.

The action of the separator on a single particle is theoretically a function only of its magnetic susceptibility and is independent of both mass and density. In practice the separator provides cleaner products from feed materials that are closely sized because the movement downslope of mixtures of sized particles is more uniform.

7.6 ELECTROSTATIC AND HIGH-TENSION SEPARATION

Methods of electrostatic and high-tension separation utilize the fact that particles of different minerals act as either conductors or non-conductors in an electrical field and hence can be separated from one another if suitably charged. However, some basic differences distinguish the two methods and, although it is common to refer to the separation of minerals at high potential as electrostatic separation, regardless of whether ionizing or non-ionizing charges are used to exploit the conductivity differences of minerals, this leads to some misunderstandings in terminology and sometimes to a lack of appreciation of the fundamental differences in separation techniques. High-tension separators utilize differences in the surface electrical conductivity of individual mineral particles to effect separation; electrostatic separators depend upon differences in electrostatic attraction rather than upon differences in ionization and pinning.

7.6.1 Factors influencing separation

The degree to which minerals of different electrical properties are separated from one another in an electrical field is governed by how closely the following variables are controlled.

Applied voltage

Normal direct current requirements are between 25 000 volts and 32 000 volts but most rectifiers are designed with a high degree of flexibility up to 50 000 volts DC, continuously variable.

Feed rate

The capacity of a unit varies with the class of material treated and the required degree of separation. Nominal capacities are about 3 kg h^{-1}/mm of roll length for most HT separators and between 0.35 and 0.7 kg h $^{-1}$/mm of plate or screen width for electrostatic separators. The actual capacity in each case is defined by the conditions of separation and the properties of the materials being separated.

Feed temperature

Electrostatic separators operate at lower temperatures than those required for HT separation, generally between 20 and 80°C. High-tension temperatures are kept as low as possible conducive with good separation but individual units must be capable of accepting feed at temperatures as high as 150°C. Normal operating temperatures in HT rougher service are around 120° C.

Ability to make adjustments

Machine performance is influenced by how effectively and quickly adjustments can be made to feed rates, splitting etc. The electrode position is a critical factor in determining the inductance and the dimensions of the zone over which the forces act, hence the electrodes must be fully adjustable at all times in accordance with local regulations governing safety.

Humidity

Extremes of humidity affect both ionizing and non-ionizing charging and sudden changes may throw an operating plant off balance until compensating adjustments can be made. Such adjustments relate primarily to voltage regulation and electrode positioning.

Residence time

Residence time should be long enough for receptive particles to acquire their appropriate charges but not so long that other particles become wrongly charged.

Particle size

Size limitations range generally between 60 μm and 400 μm. The finer particles have lower terminal velocities in air than larger particles and are more easily deflected in an electrically charged field. Even where they do pick up large charges, the larger particles are more affected by gravitational and centrifugal forces than are smaller particles.

Configurations

Various configurations are used in order to reduce headroom, achieve a more constant feed rate and provide for ease of setting and adjustment. Compact modular arrangements provide for full interchangeability of any of the high tension and electrostatic systems.

7.7 ELECTROSTATIC SEPARATORS

An electrostatic separator consists essentially of an earthed roll or metal surface which acquires, by conductance, a charge of opposite sign to that of the charged electrode. The intensity of the charge is comparatively small compared with high-tension charging but it increases with the proximity of the electrode to the metal surface. The roll-type electrostatic separator can, in fact, be readily converted to high-tension separation by the addition of a fine wire electrode to the electrode system and the different effects can be seen in Fig. 7.30 (a), (b).

The separation of conductor particles from non-conductor particles is due to differences in electrostatic attraction and depends upon the fact that when solid particles pass freely over an earthed metal surface in a non-ionizing electrostatic field, the conductor particles acquire a charge of similar sign to that of the metal surface and are attracted away from it in the direction of the electrode. Non-conductor minerals are little affected and their trajectories, instead of being curved towards the electrode, are directed more vertically downwards due to gravity. Adjustable splitters divide the stream into two or more fractions as required.

Two closely allied types of electrostatic separator have been developed for slightly different duties. Plate electrostatic separators (Fig. 7.31) are used for predominantly conductor feeds and are more tolerant to varying operating conditions than screen plate separators (Fig. 7.32) which are used for predominantly non-conductor feeds and are more selective under optimum conditions. Roll-type electrostatic separators, as described in Fig. 7.30 are seldom used.

7.7.1 Plate separators

Plate separators consist of stationary earthed plates and splitters arranged in stages. Any number of stages is possible but the usual number is

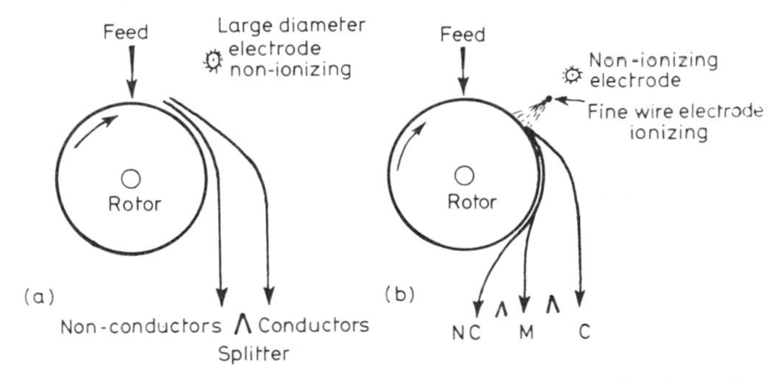

Fig. 7.30 (a) Electrostatic separation process converted to high-tension separation (b) by interposing a fine wire ionizing electrode between non-ionizing electrode and rotor

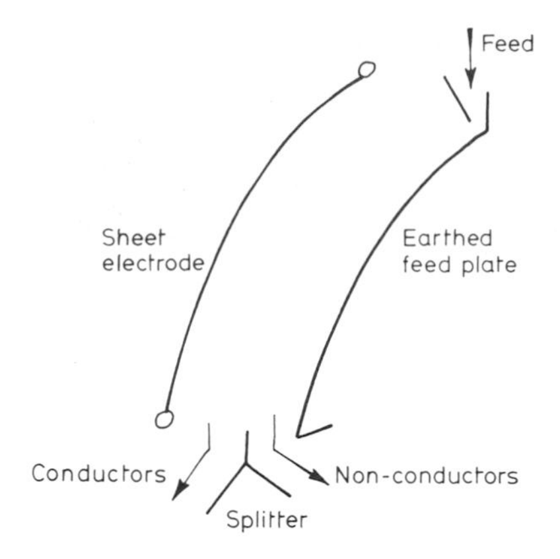

Fig. 7.31 Single stage of plate electrostatic separator

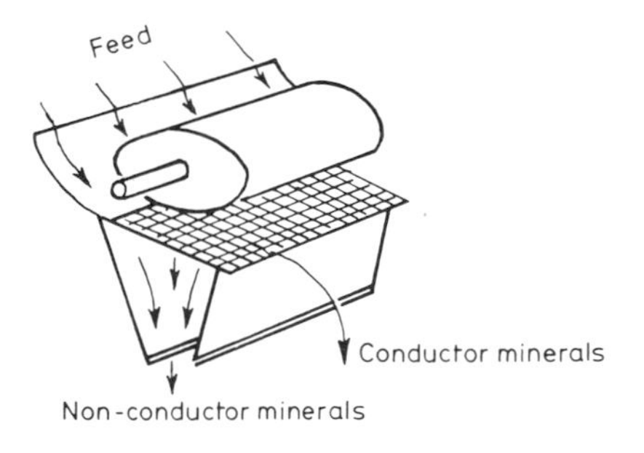

Fig. 7.32 Single stage of screen plate electrostatic separator

four. Each plate has its own divergent electrostatic field induced by a large curved electrode. The non-conductors rill down each plate in succession, remaining in virtual contact with it throughout, and are largely unaffected by the electrostatic field. The conductors acquire a charge of the same sign as that of the plate and are attracted to the electrode. Because centrifugal forces are minimal, one of the most important applications of plate separators is in the final cleaning of high-tension conductor products to remove coarse non-conductors that are too large to be pinned on high-tension rolls in an ionizing field; another is in the treatment of high-tension middlings in which the conductor particles tend to be much finer than the non-conductors.

Plate separators are usually constructed with two starts, back to back, and in a number of stages, usually four, depending upon the difficulties of separation. The main operating variables are the applied voltage and polarity, distance between plates, feed plate angle, feed temperature and rate of feed.

7.7.2 Screen-plate separators

Screen-plate separators are designed primarily for cleaning non-conductor fractions, i.e. to remove small quantities of fine conductor particles that have become entrained physically with non-conductor products from high-tension pinning. Feed gravitates onto the lower electrode which is a combination plate and metal screen cloth as illustrated. The purpose of the plate section is to allow the conductors to pick up a charge before reaching the screening section. The metal screen cloth has apertures slightly larger than the largest particle in the feed so that rejection may be entirely according to electrostatic repulsion and not by size.

Similarly as for the plate separator, the conductors pick up a charge of the same sign as that of the electrode but instead of utilizing a splitter to separate the conductors from the non-conductors, the charged metal screen is used and the conductor particles bounce along making repeated contact with the earthed electrode (screen) until finally discharged at the end. Non-conductors are unaffected and pass through the screen. Banks of screen-plate separators may be used to clean and reclean, or scavenge in the same manner as for other separator types.

7.7.3 Polarity reversal

It has been established that if the polarity of an electrostatic separator is reversed, the trajectories of some minerals remain unchanged whereas those of certain other minerals such as monazite, zircon and quartz may be changed (Macdonald, 1973). This is because particles of such minerals tend to acquire charges when rubbed one against the other. Some minerals are reversible positive, others are reversible negative and, for example, quartz particles can sometimes be induced to behave as conductors by reversing the polarity thus providing a means of separation from particles of zircon. The effects are little understood and apparently other factors are involved because reversals of polarity do not always yield the same results.

7.7.4 Machine selection

Although both plate and screen-plate separators have similar operating principles and can both be used in the same service, plate machines are more capable of handling a range of operating variables and screen-plate separators are more selective in stable conditions.

The plate machines are usually preferred when the purpose is to remove non-conductors from a conductor fraction and screen plates are selected for feeds that are predominantly non-conductor and conductors have to be removed. The two methods are complementary and modern designers now present systems in modular form for particular applications and for complementary usage with differing reactions to feed variations. Figure 7.33 presents typical configurations for plate and screen-plate electrostatic separators.

Each separation bank can be operated in its basic form and additional complementary components provide all of the facilities needed for complete functional systems. Each bank normally contains four rolls or six plate or screen-plate separators. The control module which has integrated control panels and high-voltage power sources is placed at any required point in the bank, depending upon plant layout and operational requirements. Advantages for the modular system include more efficient space utilization, pre-wired control and high-voltage systems and operational flexibility.

7.8 HIGH-TENSION SEPARATORS

A high-tension separator is essentially a roll-type electrostatic separator in which the addition of a fine wire electrode to the electrode system so increases the rate and intensity of electrical discharge that an ionizing zone is formed between the electrode and the earthed roll surface in addition to the already non-ionized electrostatic field. As a result, high voltages impressed upon the electrode system induce charges of opposite sign on the surface of the earthed roll in a cone-shaped electrostatic field between the large electrode system and the roll, and an inner cone of ionization between the thin wire electrode and the roll. Particles entering

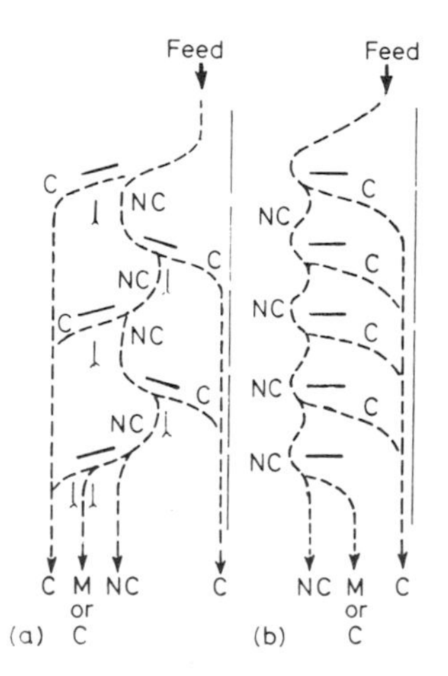

Fig. 7.33 Typical configurations: (a) plate electrostatic separator, (b) screen plate electrostatic separator

into this zone first acquire conductance charges on their surfaces and are then 'sprayed' with much larger ionizing charges. With the continued rotation of the roll, the particles pass through the ionizing zone back into a wholly conductance-charging zone where conductors are subjected to an additional lift force and beyond to where no charging of either type takes place. Conductance charges are of opposite sign to that of the electrode and surface-charged particles tend to be attracted towards the electrode and away from the earthed roll. Ionizing charges have the same sign as the electrode charges and ionized particles are repelled away from the electrode and attracted to the earthed roll. Figure 7.34 illustrates the basic features of HT separation.

7.8.1 Machine selection

Conductor minerals offer little resistance to the passage of the ionizing charges and according to Nilkuha and Hudson (1962) may lose all of their charge to the roll and retain only conductance charges attracting them towards the electrode and away from the roll. Non-conductors absorb and become saturated with ionizing charges tending to 'pin' them to the roll and on them, conductance charges have little effect. Thus, if all particles were of equal size, shape and density and could be passed over a high-tension roll without any crowding or overlapping of electrical properties, a clean separation of all of the constituent minerals could be made in one pass. But in practice this is never so and systems must be designed to handle feed materials consisting of particles of widely varying and overlapping physical and electrical properties, often complicated by grain coatings that vary individual dialectrics, and always subjected to crowding effects that result in the mechanical entrainment of particles having

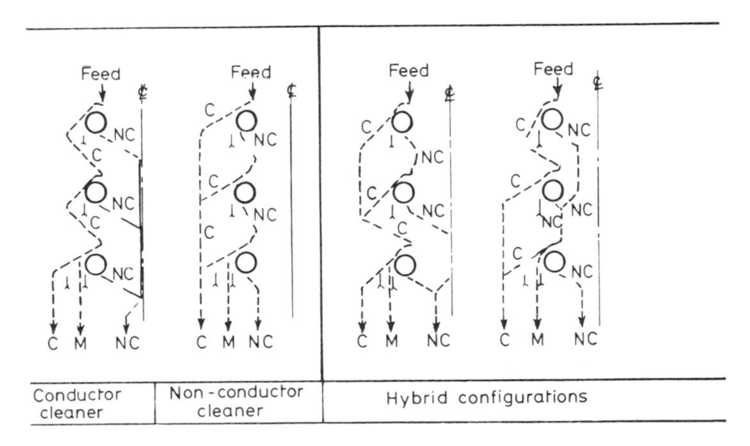

Fig. 7.34 Typical configurations of high-tension separators

different conductivities, particularly in the finer sizings. Multi-stage circuitry is needed to handle high circulating loads and, except for very simple separations, high-tension separation is only one stage in a continuing process being dependent also upon other electrostatic and gravity-separating processes for final cleaning.

Operating variables

Feed is presented to a high-tension roll at around top dead centre and the subsequent behaviour of the particles is governed by centrifugal forces which vary with the diameter and speed of the roll, and electrical body forces which depend on the configuration, positioning and charging of the electrode system. The main operating variables affecting the selection of HT units and their configurations are (a) flow rates, (b) roll speed, and (c) electrode voltage and positioning.

Flow rates

Estimates made of flow rates to be handled at each stage of processing are based either upon existing plant data or upon the results of laboratory or pilot-plant testing already discussed. The data will be more or less comprehensive depending upon their source but, in any case, a metallurgical balance is needed to predict the proportions of conductor and non-conductor minerals of all types at each stage of the process. There are four main possibilities as follows:

(1) Conductors > non-conductors.
(2) Conductors = non-conductors (approximately).
(3) Conductors < non-conductors.
(4) Middlings fractions in which a build up can occur due to such causes as wide variations in particle size, particle surface coatings and particle contamination due to intergrowths or other textural associations. Where such conditions occur the relevant fractions must be removed from the circuit for separate treatment by such means as:
   (a) attritioning to remove particle surface coatings
   (b) further stages of wet gravity and other means of separation including sizing in order to eliminate the unwanted particles. Sizing may be either by screening or sedimentation, depending upon the required cut-off point, however, consideration must be given to the effects of rheological differences in the splits due to differences in the particle sizes of solids in the various fractions. Typical rheological effects are found in the slow settling of slurries of the finer fractions in re-treatment circuits and the consequent need for larger sumps for density control; and in the tendency for fine dry particles to hang up in chutes designed for more granular materials.

Flow rates to individual process units are thus, governed by factors that may not in the first instance be capable of close definition. Factors of safety are applied to the selection of appropriate machines based upon the reliability of the data used. Readings of Lismore present a typical worked example as follows:

Assumptions.

(1) Flow rate to the stage – 30 tonnes h^{-1}
(2) A conductor, non-conductor re-treat bin configuration
(3) Machine feed rate 3 kg/mm of roll length/h
(4) Factor of safety (load factor) – 1.55.
(5) Space limitations consistant with choice of 2 × 2 roll configuration conductor, non-conductor type separator.

Calculations.

$$30 \text{ tonnes h}^{-1} = \frac{30\,000 \text{ kg h}^{-1}}{3} = 10\,000 \text{ mm of roll length}$$

$10\,000 \times 1.55 = 15\,500$ mm of roll length required
Effective length of one roll (Reading type) = 1314 mm

$$\text{Therefore number of primary rolls} = \frac{15\,500}{1314} = 11.8 \text{ primary rolls}$$

Since the number of primary rolls available on the selected 2 × 2 machine is 2 then the number of machines required for the stage = 6, and the type of machine chosen in the conductor, non-conductor re-treat bin configuration is a 2 × 2 × 270 mm × 1524 mm high-tension separator.

The same principles apply regardless of whether a new processing plant is involved or whether the intention is to modify an existing plant. The essential feature is to establish the basic criteria with care and to draw upon knowledge and experience as required.

Roll speeds

Particles remain in contact with the roll as long as the pinning forces attracting them towards the centre of the roll are greater than the centripetal forces acting tangentially away from the roll. Between 0°–90° and 270°–360°, gravity forces augment the pinning forces; from 90°–270° they subtract from them. Within the zone of electrostatic influence, electrical forces from surface conductance charges have components in the direction of the centripetal forces which vary as the square of the roll speed. Variable speed gearing is normally provided either hydraulically or through some form of mechanical gearing such as the Reeves gear so as to be able to strike a balance between the opposing forces over the full operating range.

Electrode voltage and positioning

The effectiveness of electrical charging is related both to the impressed DC voltage on the electrodes and their positioning, and to the temperature of the feed material. The ionizing effect is increased by increasing the potential or by narrowing the air gap between the electrodes and the roll; it is limited by the dialectric strength of the airspace. The break point occurs when the airspace becomes saturated electrostatically and a corona discharge (sparking) follows. The electrode voltage is usually held just below the break point which varies with the temperature in the air gap and hence with the temperature of the feed which is kept as low as possible conducive with good separation. However, temperature requirements vary and individual feed materials may need to be processed at various temperatures between 60° C and 150° C depending upon their responses to applied conditions.

The electrodes are in two parts and adjustments are provided so that each is free to move and both can be locked into position in any desired configuration relative to the feed roll. Experience has shown that the greatest ionizing effect is normally obtained with the fine wire placed parallel to the large-diameter, non-ionizing electrode and directly between that electrode and the centre line of the roll. Placed in this manner, the system develops a very strong discharge pattern concentrated in a narrow beam. Nilkuha and Hudson (1962) investigated the effects of changes in the electrode positions relative to each other and to the feed roll but concluded that there was a need for further work to study the effects of such other variables as feedrate and splitter control, load factors for a range of fluctuating plant conditions, changes in minerals characteristics and feed temperatures at every phase of the operation.

Such variables are of great concern both to the designer who must incorporate measures for ready adjustment and control and to the operator whose successful use of the machines depends upon how effectively they can be controlled.

7.8.2 Configurations

Mineral Deposits Limited has developed high-tension roll separator configurations in the form of two start by three rolls within standard modules, with a unique system of reversible chutes enabling individual units to operate in either the conductor cleaner mode, the non-conductor cleaner mode, or in a hybrid mode if desired. The configurations as illustrated in Fig. 7.34 allow any machine to be used for any function in a mill circuit with changeovers from one mode to another being achieved in less than 1 h without requiring additional components. This has opened up new possibilities for exploiting some deposits containing only small proportions of rutile, zircon and monazite for which it has been im-

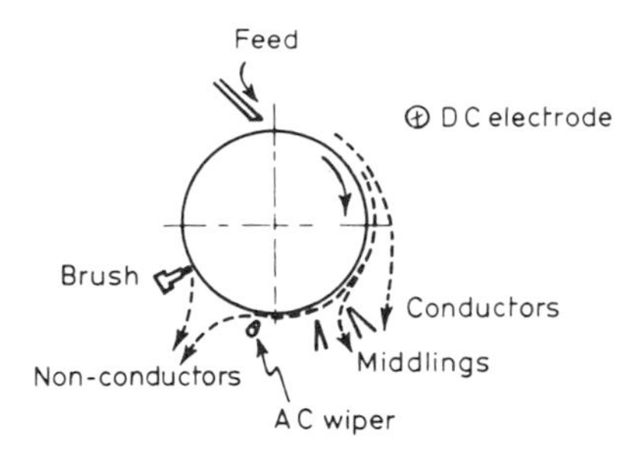

Fig. 7.35 Basic HT roll separator features

practicable to provide the necessarily complex circuitry for separation. Such deposits may now be worked economically using the rougher circuit for stockpiling separate conductor and non-conductor fractions; the one set of standard modules can then be used to treat each fraction separately at a flow rate appropriate to the size of the fraction and its content of valuable minerals.

REFERENCES

Bath, M. D. (1978) *Beneficiation of Ores by Heavy Medium Separation*, presented to Institution of Engineers, 22nd March, Perth.

Cotts, Mitchell (1970) New techniques in heavy media separation, *Australian Mining Review*, December.

Dreissen, H. (1980) *Theory, Practice and Developments of the DSM Heavy Medium Cyclone Process for Minerals*, Stanicarbon-DSM, the Netherlands.

Dunkin, H. H. (1953) Concentration of zircon rutile beach sands, in *Ore Dressing Methods in Australia and Adjacent Territories*, (ed. H. H. Dunkin) 5th Empire Mining Congress, Melbourne, 230–74.

Ferree, T. J. and Terrill, I. J. (1978) The Reichert cone, an update, gravity separation technology, in *Placer Mining and Exploration Short Course*, Mackay School of Mines, Nevada.

Harris, J. (1976) Placer mining, what options, in *Placer Exploration and Mining Short Course*, Mackay School of Mines, Nevada.

Johnson, R. E. and Thring, M. W. (1957) *Pilot Plants, Models and Scale-up Methods in Chemical Engineering*, McGraw Hill, New York.

Lawrence, L. J. (1965) The application of ore microscopy to mining geology, in *Exploration and Mining Geology, 8th Comm. Min and Met. Congress, Australia and New Zealand*, pp. 231–9.

Macdonald, E. H. (1973) *Manual of Beach Mining Practice, Exploration and Evaluation*, 2nd edn, Dept. For. Affairs, Aust. Govt. Printing Office, Canberra.

McAndrew, J. (1957) Calibration of a Franz isodynamic separator and its application to mineral separation, *Proc. Aust. I.M.M*, 181–73.

Nessett, J. E. and Finch, J. A. (1980) Determination of magnetic parameters for field dependent susceptibility minerals by Franz isodynamic separator, in *Mineral Processing and Extractive Metallurgy, Trans. Inst. Min. Met. London*, **89** December, C161–C166.

Nilkuha, C. and Hudson, S. B. (1962) The effect of roll speed, electrode voltage and electrode position in high tension separation, *Aust. IMM Proc.* 204.

Peele, R. (1927) *Mining Engineers Handbook*, John Wiley & Sons, New York.

Pullar, Stuart S. (1965) Developments in Separating Equipment in the Australian Heavy Mineral Sands Industry, *8th Commonwealth Mining and Metallurgical Congress*, Vol. 6, *Proceedings*, Melbourne. pp. 1343–57.

Rose, T. K. (1902) *The Metallurgy of Gold*, Chas Griffin & Co., London.

Strangeway, David W. (1970) *History of the Earths Magnetic Field*, McGraw Hill, New York.

Terrill, I. J. (1980) *High Capacity Gravity Separating Systems for Gold Recovery*, Mineral Deposits Ltd, Southport, Queensland.

Terrill, I. J. and Villar J. B. (1975) Elements of high capacity gravity separation, *Can. Min & Met. Bull*, May.

Wolfe, Ernest N. (1976) Small scale placer mining methods in Alaska, in *Placer Mining and Exploration Short Course*, Mackay School of Mines, Nevada.

8

Placer Valuation

8.1 INTRODUCTION

As in any other branch of mining a placer project is attractive as an investment only if, after first recovering all of the capital and operating costs and interest on borrowings, it offers substantial profits commensurate with the risks involved. Private investors and governments alike provide finance in expectation of an adequate return and both want quick returns so that the money can be re-invested for the continued expansion of their respective interests.

However, although benefits are measured traditionally in terms of money, human values are also important and economic factors relating to balance sheets and profit and loss accounts are normally considered along with such non-economic factors as employee and community welfare, operational safety, ecological safeguards and other social priorities. For example, operators in underdeveloped and heavily populated regions have a responsibility to the community to provide maximum employment opportunities and normally, only where local manpower resources are inadequate, and labour must be imported, is greater emphasis placed upon mechanization. Generally, in such cases, low-cost money is available from international agencies and a balance can be struck which will allow the operator to make a profit and still protect the interests of the developing society.

In recent years valuations have tended to become increasingly complex. There are many reasons: changing social and economic relationships have introduced new and sometimes indeterminate factors into every-day life; high inflation rates have become endemic and outbreaks of epidemic proportions occur seemingly at random; political and religious differences have divided some communities, even to the point of civil war; rising energy costs and other inflationary factors continue to force prices up, thus increasing the risk of competition from other sources and materials. Extraordinary problems are sometimes faced in obtaining rights to mine, as well as in complying with special conditions laid down for mining in ecologically sensitive areas.

Even so, the risks of failure through faulty evaluations are probably much less than in the past because great masses of supporting data are being obtained and stored and mathematical statisticians, scientists and engineers have introduced new methods and systems for sampling, and fresh interpretive concepts that give additional assurance to the prediction of future trends. The basic principles of valuation have not changed, only the means for complying with them.

8.2 THE PLACER FINANCIAL MODEL

Cost-price relationships in placer valuations depend upon, (a) the ore reserves, quantities and grades, (b) the methods and rates of mining and treatment, (c) the logistics of proposed operations and environmental safeguards, (d) possible future price trends and currency movements, and (e) the anticipated project life.

Such relationships invoke the following questions of economic and money management, (a) is there a continuing market for the product at stable prices, (b) can the product be marketed competitively, (c) what uncertainties surround the prediction of future labour and materials costings, (d) what profits can be expected from the proposed operation, (e) what are the political, legal and taxation environments, and (f) what are the risks of failure.

The placer financial model (Fig. 8.1) considers the cost structures of various possible alternatives for mining and treatment based upon data from exploration and testing (Chapters 4 and 6). Estimated costs, both capital and operating, are weighed against the anticipated returns from marketing. Surpluses are discounted over the life of the proposed undertaking to estimate its worth and to provide a basis for comparison with alternative investment opportunities. The risks are assessed and if the predictions appear to meet the required investment criteria, the reliability of the data and the conclusions reached will be finally tested in a detailed economic and engineering study called a 'feasibility study'.

8.3 ORE RESERVES

To be classified as 'ore', a placer accumulation must be capable of being mined and its products sold at a profit *now*. The term 'ore' is thus dependent upon time and present economics and it cannot be defined mathematically

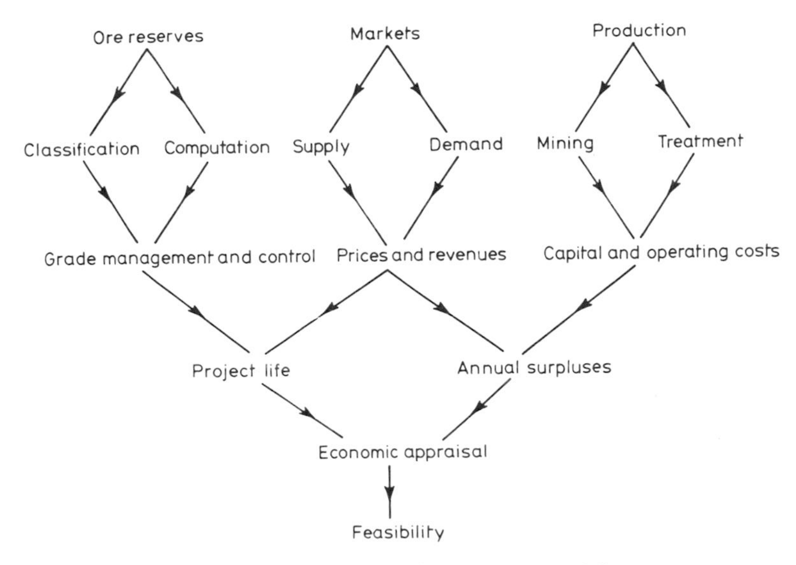

Fig. 8.1 The placer financial model

except in the present. For example, deposits that cannot be worked profitably now may later become more valuable because of lower unit production costs resulting from improved technologies or because of increased product prices due to increased demand. On the other hand some deposits, now classed as ore, may have to be set aside and the ore reserves downgraded because of falling prices due to such causes as downturns in the economy, reduced demand, government decisions to release minerals from stockpile and other 'bearish' market factors. Both types of movement were illustrated graphically by the events leading to a sudden rise in the price of zircon from US$ 49.81 in 1974 to $159.49 in 1975 and the subsequent market reaction in 1977 back to $82.18.

8.3.1 The classification of ore reserves

Ore reserves are classified according to the adequacy of the data from which they are calculated. This is a topic upon which there has been extensive debate. No one has yet been able to do more than express, in the broadest of terms, opinions held at various levels of knowledge, experience and temperament. No two deposits are exactly alike and attempts to standardize and define the terminology have had only limited success. This is largely due to the fact that each exploration team has individual thoughts on such matters as how to accumulate and process data, how to determine adequate sample intervals and densities, and how to assess likely margins of error for the particular types of equipment and the methods used. The uncertainties are reflected in the use of terms such as 'drill indicated', 'geologically inferred' and 'semi-proven' which mean something to those who use them, but are often quite confusing to others.

For example King (1955) lists the following recommendations based upon United States Bureau of Mines definitions:

(1) *Proved* reserves are those in which the ore is developed or so well known – the tonnage having been computed from dimensions revealed in outcrops, pits, trenches, workings, and/or drill holes, and the grade computed from the results of detailed sampling – that tonnage and grade estimates have a high probability of being accurate within close limits.
(2) *Measured* reserves are those for which, although tonnage is computed for dimensions revealed in outcrops, pits, trenches, workings, and/or drill holes, and the grade from the results of detailed sampling, and tonnage and grade are as well known as they will be prior to extraction, nevertheless the geological character of the ore body and the method of mining are such that tonnage and/or grade estimates do not have a high probability of being accurate within close limits.
(3) *Developed* reserves are those for which comparable information is available but which are more appropriately described as 'developed'.
(4) *Indicated* reserves are those for which tonnage and grade are computed partly from specific measurements, samples or production data and partly from projection for a reasonable distance on geologic evidence. The sites available for inspection, measurement, and sampling are too widely or otherwise inappropriately spaced to outline the ore completely or to establish its grade throughout.
(5) *Partly developed* reserves are those for which comparable information is available but which are more appropriately described as 'partly developed'.
(6) *Inferred* ore is ore for which quantitative estimates are based largely on broad knowledge of the geologic character of the deposit and for which there are few, if any, samples or measurements. The estimates are based on an assumed continuity or repetition for which there is geologic evidence. This evidence may include comparison with deposits of similar type. Bodies that are completely concealed may be included if there is specific geologic evidence of their presence. Estimates of inferred ore should include a statement of the spatial limits within which the inferred ore may lie.

Such classifications become more and more complex as they seek to cover every eventuality and, in many instances tend to defeat the purposes for which they were drawn up. As a result, the main weight of opinion now leans towards classifications that are simple, practical and generally applicable and, essentially there are three main categories. In the first category ore is reported as being proven on statistical grounds at a predetermined level of confidence and within specific fiducial limits. A second category is viewed with less confidence because of incomplete data but, nevertheless, ore reserves in this category are expected to be generally confirmed when the density of sampling is increased. The third category tends to reflect either the optimism or the knowledge and experience of the estimator and perhaps none, some or all of his predictions will eventually be found to measure up to what is hoped for. The main consideration is that a definition of the terminology used should accompany all calculations and, if this is done, the early terminology of proven, probable and possible ore reserves may be as expressive or vague as any other terms depending on how well they are defined.

Clearly, for valuation purposes, only 'proven' reserves have any real status and hence there must be a clear understanding of what proven means. For a start it means that the sample data used for estimating the quantities and grades have been obtained reliably under standard conditions using appropriate techniques for taking and analysing the samples. The data must be adequate, in that the standard deviation will not be reduced significantly by taking more samples, and any two computers given these data will arrive at similar conclusions. The term proven means, moreover, that in determining what is or is not 'ore', all relevant mining and treatment studies have been completed and a full appreciation has been made of the economics of the proposed operation in order to define the relevant parameters for each section of the deposit.

8.4 THE COMPUTATION OF ORE RESERVES

The two main systems for processing sample data and thereby calculating volumes and grades are 'geometrics' and statistics, and both give acceptable results if correctly applied. However, the geometrical methods are more tedious and normally require the taking of more samples; *geo*statistical studies have an added advantage of providing a ready means for determining risk factors and sensitivities. Furthermore, although neither system is based fully on geological grounds, geostatistics is able to apply interpretive reasoning to such geological phenomena as minerals pay streaks and 'nugget' effects and the method is now applied to some degree in the more important valuations. Regardless of the system used, however, all data must be examined critically to ensure that the physical properties and characteristics of the deposit are fairly represented by the density of sampling and that the results have been reported in an appropriately standard form.

8.4.1 Geometrics

Geometric methods of computation involve the weighting of individual samples first through the arbitrary assumption that each sample truly represents a specific volume of material surrounding it, and then by assessing the importance of each volume of influence relative to the whole. All such methods assume a steady gradation of values between adjacent sample positions and, in effect, rely upon the Law of Diminishing Returns in that when a sufficient number of unbiased samples have been taken geometrically across a deposit the results will allow a close approximation to be made of its true value. Three methods commonly used for analysing placer sampling data are, (a) the method of sections, (b) the method of triangles, and (c) the method of polygons.

Method of sections

This is the usual method applied to any long narrow placer. It involves the preparation of cross sections for each line of boreholes across the deposit (Fig. 8.2). These sections also provide essential data for planning the subsequent mining operations.

Volumes of influence are determined by multiplying the average area of each pair of cross sections by the distance between them, or by multiplying each cross-sectional area by one-half of its distance to the adjoining cross sections.

Grade calculations involve the weighting of values affecting each volume of influence according to the line and borehole spacings and sample depths. Values are considered to extend for a distance of one-half the borehole spacing beyond the last payable borehole along each margin of the deposit.

Some mineral sands placers are also concerned with a third dimension, that of density and, as demonstrated below, the values according to volume

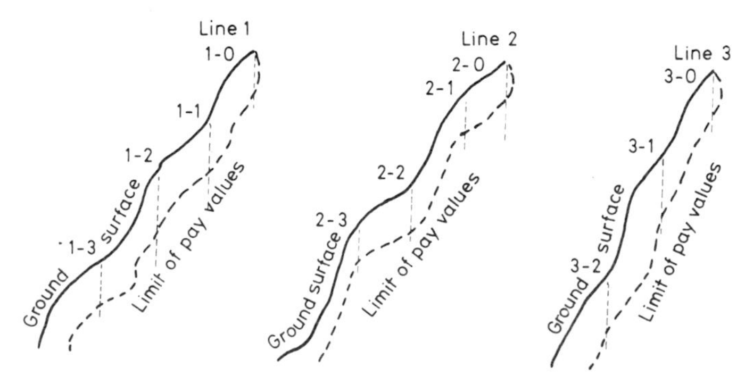

Fig. 8.2 Schematic layout of cross-sections along narrow placer showing limits of pay values from borehole samples

weighting alone can be significantly different to those that also take account of differences in mass.

Consider for example, two adjacent samples of a typical drilling program in which hole 1 averages 6.8 % of heavy minerals for a volume of influence of 104 m^3 and hole 2 averages 21.6 % of heavy minerals for a volume of influence of 96 m^3.

If mass effects due to density differences are ignored the weighted average grade for the combined volume of influence of 200 m^3 is given by

Hole 1	$104 \times 6.8 =$	707.2
Hole 2	$96 \times 21.6 =$	2073.6
Totals	$200 \times 13.9 =$	2780.8

Therefore average grade = 13.90 % heavy minerals

On the other hand if mass effects are taken into consideration using weight factors of 1.6 and 1.78 tonnes/m^3 respectively, extrapolated from Fig. 2.9 (Chapter 2) the result becomes

Hole 1	$104 \times 1.6 =$	166.4×6.8	= 1131.52
Hole 2	$96 \times 1.78 =$	170.88×21.6	= 3691.01
Totals		337.28×14.30	= 4822.53

Therefore average grade = 14.30 % heavy minerals

Such mass effects are not confined to differences in total heavy-mineral percentages, the proportion of valuable particles in an assemblage of valuable and non-valuable minerals also increases exponentially with increasing heavy-mineral grades. Hence, in terms of real values, evaluation procedures should be taken a step further and for each deposit, a similar graph to that of Fig. 2.9 would plot percentages of valuable heavy minerals versus percentages of total heavy minerals to obtain a more accurate basis for calculation. For example, if the heavies from hole 1 contained 7.8 % of valuable heavies and the heavies from hole 2 contained 14.5 % of valuable heavies which would not be an unusual order of relationship, the new calculations would give:

Hole 1	$166.4 \times [6.8 \times (7.8/100)]$	= 88.26
Hole 2	$170.88 \times [21.6 \times (14.5/100)]$	= 535.20
Totals	$337.28 \times 1.85\,\%$	= 623.46

The resulting weighted average of 1.85 % of *valuable* heavy minerals is a realistic figure. The figure of 14.30 % of *total* heavy minerals gives no indication of its valuable mineral content.

Method of triangles

Figure 8.3 demonstrates the method of triangles for computing volumes

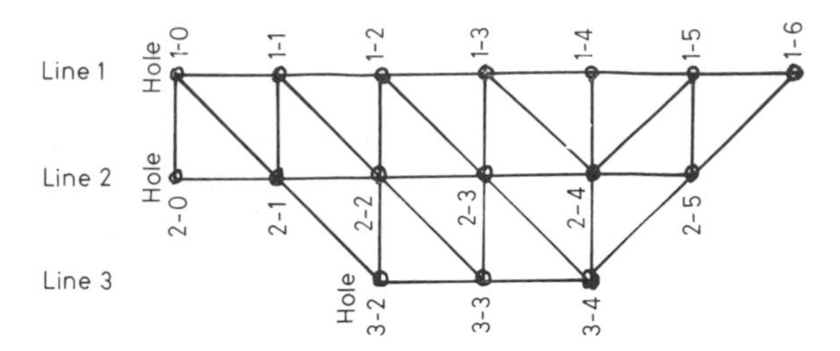

Fig. 8.3 Layout of triangular blocks

and grades. This method also assumes a linear change in grade between boreholes in direct proportion to the distance between the sample points. Triangles are constructed by drawing straight lines from each borehole to its adjacent boreholes. The volume of each triangular prism is calculated by multiplying its plane surface area by the average depth of its three boreholes. The grade of each block is the weighted average of the three borehole values. Weighting is done according to the depth, value and distance of each borehole from the centroid of the triangle and, where applicable, to differences in density of the sample material.

The outer limits of the placer are assumed to extend to the centroid of each triangle having at least one payable corner hole along the main channel. The boundaries are defined by a series of lines joining the outermost centroids.

Method of polygons

Although again an approximation based upon linear changes in grade and depth, the method of polygons is generally preferred to other geometric methods for computing volumes and grades in placers having a large aerial extent and an irregular distribution of values. Polygons are constructed around individual sampling points by erecting perpendiculars from the midpoints of lines connecting adjacent boreholes and joining the common points of intersection. Figure 8.4 shows the construction of a polygon around a central sampling point. The construction can be extended for all other sample points to cover the entire area in polygons, each one with an

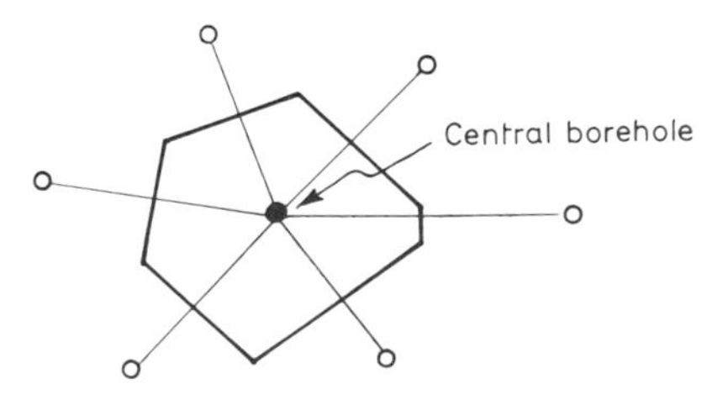

Fig. 8.4 Construction of polygonal block around central borehole from perpendiculars at mid-points between holes

area of influence for each sample reaching halfway to its adjacent drill holes.

The volume of each polygonal block is the product of its surface area and the depth of its central borehole. The grade of each polygonal block is the grade of its central borehole. The outer limits of the deposit are defined by the polygons surrounding the outer payable holes.

The above three methods and their various adaptations give generally acceptable results if applied correctly but suffer the common disadvantage that none of them provides a means of determining when a sufficient number of samples has been taken for a required level of confidence. Generally, either the sampling is inadequate and the subsequent interpretation is faulty or the deposit is oversampled and the prospecting costs are excessive. A further disadvantage is that sample weighting may vary according to the method used and often according to the judgement of the supervising engineer.

8.4.2 Geostatistics

Statisticians have tried for many years to develop a purely mathematical treatment of sample data with results that are reproducible regardless of the computer and which, if applied from the earliest stages of valuation, would ensure the taking of no fewer and no more samples than needed. Early trend surface analytical techniques offered an alternate approach to geometric methods of computation but the method fell into disfavour for a time when it was found that the trend surface movements failed to adjust quickly enough to abrupt changes in grade. Difficulties were experienced in introducing geological concepts into the calculations and it was not until the development of geostatistical models that some of the inadequacies of trend analyses were overcome.

Statistical and geostatistical computations rely, generally, upon measurements of central tendencies and, as Clark (1979) points out for geostatics,

> whereas trend analyses allow for spatial structure by assuming that each point is a sample of the same type of distribution, but that the average of the distribution can change from place to place in a deposit, geostatistics accepts the concept that each point in the deposit represents a sample from some distribution, but the distribution at any one point may differ completely from that at all other points in the deposit in its form, mean and variance.

Most importantly the 'nugget' effects that characterize most placer accumulations are accepted by geostatisticians as being essential features of any placer distribution. They assume that differences in grades must be consistant, but not constant over the deposit.

It is beyond the scope of this text to discuss the many and varied applications of geostatistics or even to suggest that geostatistics provides the answers to all of the problems of placer valuation. For a more comprehensive treatment of the subject consult the references listed at the end of this chapter. However, it is necessary to stress the great importance of having suitable procedures available when decisions are required on such matters as the optimization of sample grids and sampling densities, and when consideration is given to the possible effects on profitability of variations in ore cut-off grades, product prices and other economic factors. Errors are involved at every stage of sampling and geostatistics presently offers the best means for predicting the range of possible effects.

The following brief notes relate to the main statistical techniques used in analysing sample values for purposes of estimating quantities and grades. It is assumed that the sample data are random in that individual sample values are independent of one another and that their randomness has been established by one or more of the statistical tests that are applicable. The evaluation is carried out in two stages. The first stage investigates the physical and statistical structure of the ore body for which the estimation is being made; the second stage is the estimation process of 'Kriging' which depends upon semivariograms constructed during the first stage.

Basic concepts

Geostatistical ore reserve estimations assume the main structural components of placers to be a series of lenses, oriented in the direction of flow. It is further assumed, for each placer, that all adjacent lenses have some structural and lithological features in common and that they can be grouped on this basis into larger structures. Royle (1976b) refers to them as 'nested' structures comprising groups of lenses of small dimensions nested inside a similar one having large dimensions.

Large deposits are built up from a multiplicity of channels, each having a main orientation and each simultaneously eroding and transporting previously deposited materials. Spatial variation occurs in both main and cross channels and, according to Royle, two kinds of horizontal anisotrophy are produced. In one, the alluvium is more or less of the same type but its spatial variation is different in two directions. In the other, the relative proportions of the alluvial constituents change persistantly in a particular direction, producing a trend in this direction.

Trends are also produced vertically from causes that range from cyclical seasonal changes to long-period crustal changes and, over long periods of time, the material being deposited will change in character. Vertical trends also follow changes in grade of heavy minerals and the trend may be in either direction, up or down, in the one deposit. Disturbances and redistribution occur at all scales from local erosion to regional changes in which the sediments of an entire alluvial field may be remobilized and relaid. Particle size relationships affect both horizontal and vertical distributions and features of bedrock geometry provide the means for trapping particles of valuable heavy minerals in pockets of enrichment. Edge effects are noted in thin parts of the alluvium where sections of the bedrock rise towards the topographic surface; structures and spatial variations then differ from those of the thicker parts and are best considered separately.

Few important placer valuations are now made without at least some reference to statistics and it is essential for field personnel to understand clearly what the statistician wants so that all of the required data are collected and presented without bias. They will need to be familiar with such terms as frequency distributions, mean and variance, and with the variogram.

Frequency distributions

Although much of the mathematics requires close collaboration with a mathematical statistician, the general formulae upon which statistical and geostatistical valuations are based are readily understood; simple statistical procedures provide the background for analysis. The first of these is based upon the proposition that all sedimentary accumulations have some form of symmetry and, hence, if all of the samples in a distribution are grouped within suitable class intervals according to their grades, the frequency with which future samples will fit into each of the intervals is predictable according to the number of samples taken. A distribution normally contains between fifteen and twenty-five class intervals designed so that there is no overlap. For example, the values in a mineral sands placer may range from less than 1% up to a maximum of 17%. Suitable grade intervals would be: 0.00–0.99; 1.00–1.99; 2.00–2.99 . . . 17.00–17.99, and there can be no doubt where each sample belongs.

The preparation of frequency distribution tables also provides the starting point for calculating such other functions as arithmetic and logarithmic means, standard deviations and variances from which the statistician prepares mathematical statements that describe the dispersion of the mineralization and its controls. The variance is a measure of how much each sample influences neighbouring areas in the same deposit and is the basis of the probabilistic concept of geostatistics. Other useful functions are the median, mode and quadratic mean.

Arithmetic mean

The arithmetic mean, or mean value estimate is simply the average value of all samples in a distribution. It is determined by adding all of the sample values and dividing by the sample size N:

$$\bar{X} = \frac{\sum X}{N} \tag{8.1}$$

Using the same sample frequency distribution the arithmetic mean is approximated closely by multiplying the mid-point of each grade interval (M_p) by the frequency of occurrence in each grade interval (f) and dividing by the sample size:

$$\bar{X} = \frac{\sum f(M_p)}{N} \tag{8.2}$$

Distinctive features of the arithmetic mean are that the sum of the deviations around $M_a = 0$ and the sum of the squares of the deviation is smaller than around any other point. Its reliability depends upon both the degree of variability of the deposit characteristic and the sample size.

Median

The median is the value of the central item in a group of items arranged in order numerically. It is calculated in a frequency distribution by finding the cumulative total for each grade interval and dividing the total by 2 to determine the position of the median.

The median is an average of position and, unlike the arithmetic mean, is independent of extreme values. It is affected by the number of items but not by any abnormalities in size or grade.

Mode

This is the individual value appearing most frequently in a population. If there is no repetition of values there is no mode.

Quadratic mean

This may be expressed as the square root of the mean square of the items, so that for a frequency distribution:

$$M_q = \frac{\sum f(M_p)^2}{N} \tag{8.3}$$

It is always larger than M_a because by squaring the items it gives additional 'weight' to the larger values.

Geometric mean

The geometric mean of a number of items is the nth root of their product and, for a frequency distribution:

$$\text{Log}\, M_q = \frac{\sum f(\log M_p)}{N} \tag{8.4}$$

Standard deviation

The standard deviation of the sample frequency distribution is used as a measure of variability and is defined as the square root of the mean of the squares of the deviations of values (d) about the mean value. It can be shown that the standard deviation of the sample (S) is an estimator of the standard deviation of the population in the expression:

$$S = \left[\frac{\sum(X-\bar{X})^2}{N}\right]^{0\cdot5} \left(\frac{\sum d^2}{N}\right)^{0\cdot5} \tag{8.5}$$

and, from the sample frequency distribution, is approximated by:

$$S = \left[\frac{\sum f(M_p)^2}{N} - \frac{(\sum M_p)^2}{N}\right]^{0\cdot5} \tag{8.6}$$

Variance

Variance (δ) is the most important measure of the dispersion of values. It makes possible the estimation of whatever errors may have occurred and is expressed as:

$$\delta = \frac{1}{N}\sum(X_1)^2 - \frac{(\sum X_1)^2}{N} \tag{8.7}$$

where X_1 is the dispersion of values.

Standard error of the mean

The standard deviation estimate (S) is used to measure the reliability of the mean value estimate using the standard error of the mean ($S_{\bar{x}}$):

$$S_{\bar{x}} = \frac{S}{(N-1)^{0\cdot5}} \tag{8.8}$$

The variogram

The variance is a measure of the influence of samples over neighbouring areas within a deposit; it is the basis of the variogram and is usually represented as a graph of variance vs. the distance L between two points. The variogram groups samples according to the distances between them. In practice only a limited number of samples are available and a model is fitted to the variogram to facilitate calculations. Variograms are taken along lines of samples so as to determine spatial variability and Royle (1976a) envisages, for alluvials, a pilot sampling scheme in which samples are taken on a cross so as to recognize the structure in two perpendicular directions. Each arm of the cross might contain twenty samples initially and more samples may be taken by extending the arms of the cross, or by infill sampling, until the structures emerge.

Case Studies in Alluvial Valuation (Royle, 1976) discuss the differences in spatial variations of light heavy minerals such as rutile (RD 4.2); diamonds, for which deposits are extremely variable, the degree of variability depending upon the granulometry; and cassiterite (RD 6.8–7.1). Evidence in Malaysia that some cassiterite deposits have been relaid complicates their spatial variation.

It must be stressed, at this point, that while geostatistics is currently the best tool available for valuation it is not yet an end in itself. Most procedures are practicable, economically, only for valuing large' placers and current procedures such as that suggested by Royal for determining structure may be too costly to apply to buried placers, particularly in the deeper waters of the continental shelf. Basically, in this respect, geostatistics is on a par with geophysics in that both are supplementary to geology and neither can function as well, alone, as when directed according to the results of detailed geological observations. At the same time, as with geophysics, the application of geostatistics plays a vital role in the development of geological knowledge by indicating possible trends in the mineralized sections of a deposit and by helping to evaluate nugget effects. If the geology is good, the data are probably capable of adequate analysis. If the geology is bad the statistician can probably point to errors in observation or interpretation that are reflected in the results from sampling. If the statistician is required to develop his own understanding of the geology from sample data alone, the exercise may prove excessively costly and, perhaps result in major errors.

The economic sampling limit is achieved when the benefits to be obtained from further sampling are marginal and when any future sampling would result in costs that could not be justified by the spread of confidence limits, i.e. the fiducial interval, about the estimated mean value. Geometrical methods do not, in themselves, provide techniques for measuring degrees of confidence in mean value estimates as do geostatistical methods, however, as Mackenzie and Schwellnus (1973) point out, this limitation is not inherent in the data, rather, it occurs because geometric methods fail to fully utilize the data.

Statistical methods also provide techniques for testing the randomness of the sample data, using nearest-neighbour, equal-expectancy and poisson-distribution tests (Hazen, 1968); and for making probabilistic estimates for assessing the risk of mine development alternatives. Such estimates accept the fact that changes in grade occur either gradually or abruptly and that while they do not alter the overall parameters in a deposit, they may have a strong influence on sub-population parameters such as the individual grades of different zones within a deposit.

8.5 GRADE MANAGEMENT AND CONTROL

Ore reserve, cost and revenue estimates are made for each technically feasible alternative on the basis of estimates of tonnage and grade.

Expected profitabilities are weighed against the individual risks for each alternative and the investment decision then involves the selection of the optimum development alternative. Thomas (1977) draws attention to the finite nature of the raw material resources available for mining and points out that, 'whereas opinions differ as to the urgency for the conservation of such resources, there is a growing tide against any form of wasting of proved reserves. In mining terms this translates into a trend towards maximizing recovery of values from known mineral deposits.' In developing the concept of 'high grading' under controlled conditions to maximize profits, Thomas draws a distinction between 'picking the eyes out of a deposit' to the extent that the remaining material is rendered physically or commercially un-mineable and, under controlled-grade management, to win ore initially from the most accessible and, hence, the lowest-cost and effectively highest-grade areas to both maximize profits and resource utilization. Grade management entails:

(1) Planning the sequence in which the various deposits or sections of deposits are mined.
(2) Applying cut-off grades that are appropriate to the individual circumstances.
(3) Blending where necessary to maximize recoveries and reduce unit costs.
(4) Controlling dilution.

8.5.1 Mine sequence planning

Two alternative sequences are considered:

(1) Simultaneous mining from two or more sections of the deposits in order to maintain an average grade mill feed for the life of the mine.
(2) Mining from the highest-value ore zone first in order to minimize the pay-back period and maximize the discounted cash-flow rate of return on the investment.

Few engineers, today, would doubt the wisdom of following the second course because of the high cost of money and the general uncertainties of future money markets and national policies. Governments do not always realize the impact upon an industry's confidence of abrupt changes to the rules under which it is operating. For example, Kelsey (1977) described the following possible effects from retrospective and unfavourable changes to royalty legislation because of subsequent increased mining costs and risk:

(1) Cut-off grades may rise decreasing the mineral reserves and the life of the mine.
(2) The cost increase can make a previously viable operation uneconomic and lead to mine closure.

(3) The changing of the ground rules by adding to the risk of a project can cause the shelving of proposed new developments and expansions.

The same effects may result, also, from increased tax levies, higher interest rates, less favourable allowances for depreciation and many other factors that may be changed suddenly according to changes in government thinking and investors are naturally concerned with whatever measures can be taken to provide additional guarantees of quick returns. In the event, the uncertainties surrounding the future have led to a much greater appreciation of the importance of grade management and control and, most importantly, to a fresh look at what actually constitutes the cut-off grade for a deposit.

8.5.2 Cut-off grade

Cut-off grades are calculated from economic considerations that take into account all of the costs of production up to and including the marketing of the finished product. They consider the overall requirement of a profit level commensurate with the risks involved and, hence, the term has a much wider meaning than has generally been attributed to it. Taylor (1972) offers the following definition for the cut-off grade:

> Any grade that for any specified reason is used to separate two courses of action e.g. to mine or to leave, to mill or to dump; also any series of grades used to truncate a frequency distribution or to separate mineralized material into graded fractions.

According to this definition no two estimators would necessarily select the same cut-off grade for any one deposit except where they were both influenced by the same considerations of policy and where all estimates of costs and revenues were identical. Ramos (1977) suggests, instead, that the cut-off grade is that grade that maximizes the total resource profit but this is a matter for debate.

In placer technology, grade management control relies upon cut off grades that distinguish between the more payable and less payable sections of ore horizons and separate bodies, each of which bears different costs due to differences in texture, overburden to ore-stripping ratios, haulage distances, bedrock conditions and geometry. A frequent error in placer valuation is to fix a common cut-off grade for the deposit or deposits as a whole, based upon average operating costs and revenues. This matter has already been discussed in Chapter 6, Section 6.4 Mine planning.

Applying realistic cut-off grades to a particular deposit is a matter for individual judgement based partly on economics and partly upon the practicalities of mining in various sections of a deposit. Consider, for example, the typical section across a mineral-sands placer illustrated in Fig. 8.5. The ore boundaries, based upon an assumed cut-off grade of 5%

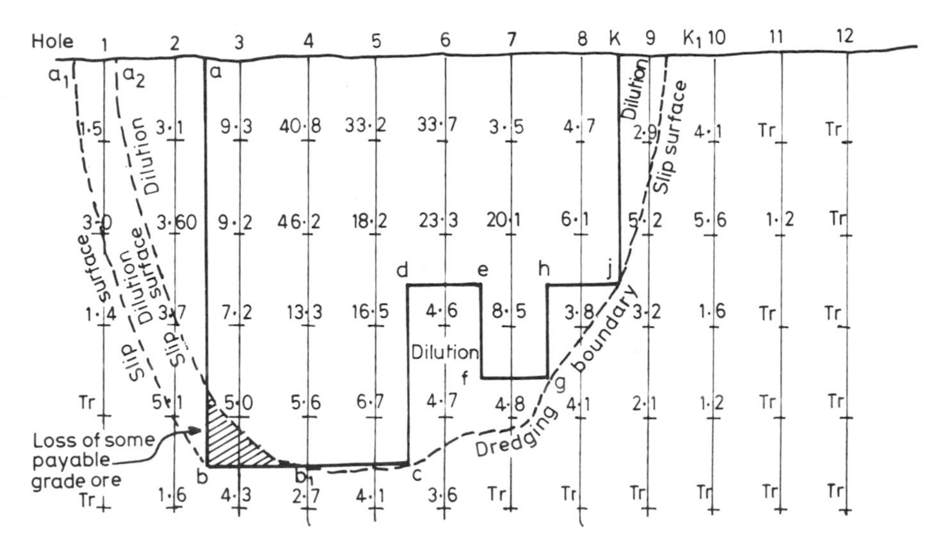

Fig. 8.5 Typical cross-section of HM placer showing possible dredging limits and dilution from fall-in and bottom scraping. The vertical scale is exaggerated
——— *limits of geological reserves for cut-off grade 5.0 % HM*
- - - - - - - - *possible dredging limits*

HM are designated by the full line *abcdefghjk*. The task of the mine planner is to determine how closely the mining limits can conform to those boundaries and the valuer must assess the resulting effects on the boundary cut-off grades. Decisions in both cases are influenced by the mining method to be used. Dry methods, being more selective, are less affected by dilution, however, dredging costs less and, for the purpose of the present discussion, it is assumed that mining is by dredging. In these circumstances:

(1) If the dredger mines down to point *b* between holes 2 and 3 there will be a fall in of low grade material as represented by the wedge shaped piece a_1ab. The planner may or may not decide to dredge out only to some other point b_1 thus sacrificing some material of cut-off grade in order to maintain the required 5 % HM along the boundary of the cut. The fall-in would be represented then by the wedge a_2b_2a.
(2) For practical dredging purposes the sub cut-off grade horizon defined by *cdefghj* will probably be included somewhat along the levels shown by the dotted line *CJ*.
(3) The 5.2 % and 5.6 % values of lines 9 and 10 are overlain by lower-grade material. The averages fall below the cut-off grade and normally they will be rejected except for the fall in represented by K_1KJ.
(4) However, a decision has to be made on whether to strip the surface material from holes 7 and 8 or to include the 3.5 % and 4.7 % material as

ore. Depending upon the logistics of what is decided upon, dredging may or may not be extended out to include hole 10 material.

Mining limits are thus selected with a view to effecting a reasonable compromise between accepting losses of valuable minerals in order to maintain the grade of cut-off and accepting an additional amount of dilution in order to maximize recoveries. Generally the most important factors involved in the decisions are those relating to the practicalities of mining, whatever means are used.

Grades for single or principal product placers are usually expressed in terms of weight per unit volume (e.g. g Au/m^3 or percent Ilmenite/m^3), ignoring the silver content in one and traces of zircon or rutile in the other. A common past practice of estimating values in terms of money per unit of volume (e.g. \$0.67/m^3) is now seldom applied because of widely fluctuating costs and prices. These fluctuations also make it most difficult to establish cut-off grades for multi-product assemblages where variations occur in both minerals distribution and product values. Estimates must be updated at frequent intervals, or whenever significant changes take place in supply and demand.

8.5.3 Blending

It is usually technically desirable for different grades of ore to be blended for treatment because most treatment units operate more efficiently when the feed materials are uniform in grade and texture (see Chapter 7). However, it is not always economically feasible to blend and, even where it is, the optimum level of blending must be determined according to the economics of the operation as a whole. In doing so, the benefits of maximizing throughputs and recoveries are weighed against the additional costs for such exercises as stockpiling and double handling and for mining more selectively, perhaps from a number of different faces simultaneously.

8.5.4 Dilution

Some dilution is inevitable if the deposit is mined to its boundaries either laterally or in the vertical plane. The extent of the problem depends upon how much care is taken to define the geological boundaries and from them to determine the economic mining limits. Specific problems relate to placer deposits located in perched horizons for which dilution occurs from all directions; and continental type placers where valuable particles find their way down into cracks and weathered portions of the bedrock. In the majority of such cases mining is extended for some distance into the bedrock but not so far as to be uneconomic.

8.6 MARKETS

The placer-mining industry operates in conditions of fluctuating world prices and demand for the various products that follow closely upon the state of the national economies in user countries. For some minerals, e.g. cassiterite and zircon, the fluctuations are dampened by the manipulation of buffer stocks held by both consumers and governments, but in boom periods there is always a tendency towards inflation and prices rise. Corrective government action is often taken late and may be harshly applied. The consequences are periods of recession, a lower consumption of raw materials and depressed price levels (Macdonald, 1973).

Other important market considerations affecting valuation predictions are changes in defence policy, accumulation of stocks or releases from stockpile, the opening up of new mineral fields and scientific discoveries that open up new uses for a particular material, or provide increased competition from some other material. A specific problem of the mineral-sands industry is that short-term financial crises may occur if much more than 25% of the output has to be negotiated in any one year. Individual producers usually rationalize their sales programmes by entering into forward contracts for the bulk of their output and disposing of the balance in spot sales.

Marketing is influenced also by differences in standards amongst individual purchasers. As a result, while the nominal prices for concentrates quoted on various metal exchanges can be used by producers as bases for negotiation they do not necessarily indicate firm prices for their own particular products. For minerals such as zircon and cassiterite there are inherent differences in composition and quality of concentrates from different localities. The ability to negotiate satisfactory forward contracts depends upon the following main considerations.

8.6.1 Continuity and uniformity of supply

A manufacturer who is forced to change his source of raw material may also be forced into making radical process changes because of differences in minerals composition and reactivity. Furthermore, a new material must be subjected to exhaustive testing before it can be accepted and utilized on a plant scale and this can be extremely costly. The manufacturer will always prefer to contract his supplies from producers who can satisfy his long-term requirements both quantitatively and in uniformity of grade.

8.6.2 Concentrate purity

Complex textural associations affect the composition of some mineral particles and, for titanium minerals, contaminating substances such as oxides of chromium, niobium, silicon and phosphorous add to the work of preparing constant quality pigments from ilmenite. Ilmenite concentrates containing impurities beyond the amounts allowed by individual purchasers may be subjected to penalties or even be rejected outright. Zircon buyers are concerned with the amount of free SiO_2 and TiO_2 in the product and some are also concerned with the level of P_2O_5. The buyers of monazite sometimes use the percentage of material insoluble in perchloric acid as a guide to the value of concentrates since the insolubles are composed largely of valueless minerals. Tantalite–Columbite minerals are valuable more for their tantalum content than for the niobium and the extreme variability of Ta–Nb concentrates has already been illustrated in Fig. 4.4.

Producers sometimes market a range of products advantageously through plant control and blending. Zircon may be marketed in both premium and standard grades instead of as a single standard product for which the price would be that of the standard-grade material. In respect of this, it is usually an advantage to aim at just above the minimum allowable grades. A typical example is found in allowing a proportion of leucoxene to report with a high-grade rutile product. In such cases the leucoxene, a lower value product, is paid for at the higher rutile prices while the rutile product, although slightly reduced in grade, still holds its original value.

8.6.3 Metal content

Cassiterite and wolframite concentrates fetch premium prices for high metal contents and invoke smelters penalties for impurities even when they are readily slagged away. The same conditions apply to most other minerals also. Occasionally, however, where the recovery of the metal content of an ore is dependent upon its solubility under conditions attained readily in the plant, a lower-grade material with a higher degree of solubility or a higher rate of solubility may be preferred to a higher-grade material that is less reactive. This is the case with some ilmenite concentrates for which the solubility depends very largely on the FeO: Fe_2O_3 ratio of the total iron content. Ilmenite concentrates with high ferrous : ferric ratios are generally the most reactive and, hence, the most in demand.

8.6.4 Particle size range applications

Recoveries in mineral processing plant are influenced by both particle sizes and ranges of particle size. This is a prime consideration in marketing most industrial placer minerals. Buyers specifications usually call for consistantly close sizings. Fineness is an advantage in some cases, as for instance where zircon is to be ground to a zircon flour, however, generally the product specifications are for substantially larger-sized particles.

Required particle-size frequency characteristics also depend largely upon whether the placer products are valuable in bulk as industrial minerals or valuable for their metal content. Industrial uses include sand and gravel for the construction industry and road building: silica sand for glass-making, sand for filtration and sand for foundry resin coating.

Sand and gravel for construction

The strength of concrete is controlled by its mixing water : cement ratio and the amount of mixing water is a function of the grading of the aggregate. It has been shown (Tyler 1963) that, for workable concrete of given materials and richness of mix, the coarser the combined aggregate, the less the amount of water required for a given consistency and hence the stronger the concrete. Upper limits for coarseness are established on the basis of workability of the mix.

The fineness modulus

Standards for sand and gravel used in the construction industry are based primarily on a measure for grading aggregates developed by Professor Duff A. Abrams at Lewis Institute, Chicago, during the period of the First World War. Called the 'fineness modulus', M_x, it is found by adding the percentages coarser than a range of sieve sizes with openings of 3 in (76.2 mm), 1.5 in (38.1 mm), 0.75 in (19.05 mm), 0.375 in (9.51 mm), 4.76 mm, 2.38 mm, 1.19 mm, 595 μm, 297 μm and 149 μm. The fineness modulus is found by adding the cumulative percentages of material held on the screens and dividing by 100. For example, a typical sample of sand from Boca Juan Diaz, Panama gave the following results shown by Table 8.1.

In general for this class of material the fineness modulus should be between 2.5 and 3.2 and it is readily apparent that the above sample fell short of this requirement because of a predominance of material in the lower sizings. Such materials should be mixed with coarser sand to raise

Table 8.1 *Fineness modulus – sand from Bay of Panama*

Sieve opening	Held g	%	Cumulative %	Passing g	%	ASTM standard % passing
4.76 mm	4.1	0.6	0.6	686.7	99.4	95–100
1.19 mm	3.6	0.5	1.1	683.1	98.9	50–85
595 μm	97.6	14.1	15.2	585.5	84.8	25–60
297 μm	524.0	75.9	91.1	61.5	8.9	10–30
149 μm	56.6	8.1	99.2	5.5	0.8	2–10
		Total	207.2			

The fineness modulus = 207.2/100 = 2.1

the fineness modulus to specified limits, otherwise excessive quantities of cement are required for equivalent strength.

American Society for Testing Materials (ASTM) standards, or similar, are usually adhered to in major works but availability is often a factor, as in Panama where much of the smaller construction work is done using whatever sand is available. In major works a fineness modulus of 2.3 may be acceptable provided that additional quantities of cement are used, however, fine aggregates for pipe manufacture and pre-stressed concrete must always have a fineness modulus of at least 2.5.

Although standards differ slightly in different countries, all are of the same order as specifications set by ASTM. The ASTM definition of sand for concrete aggregates (designation C58–55T) is: 'granular material passing the 3/8 inch (9.50 mm) sieve and almost passing the No. 4 (4.760 mm) sieve and predominantly retained on the 200 (74 sieve)'. Designation C33–67 provides that fine aggregates shall be graded within the limits shown by Table 8.2.

Road-making materials

Gravels and sand for pavement construction are in a different category and a conventional mix consists of the following ingredients:

Gravel	particles larger than 2 mm
Sand	particles smaller than 2 mm but larger than 0.2 mm
Silt	particles smaller than 0.2 mm but larger than 0.002 mm
Clay	particles smaller than 0.002 mm

The American Association of State Highway Officials Standard Specification Designation M147–65 for soil aggregate mixtures are shown by Table 8.3.

Foundry sands

Approximately 60 % of all zircon production is used in foundry-facing work where the important characteristics are grain size (predominantly

Table 8.2 ASTM definition of fine aggregates, designation C33–67

	Sieve opening	Percent passing
3/8 in	9.51 mm	100
No. 4	4.76 mm	95–100
8	2.38 mm	80–100
16	1.19 mm	50–85
30	595 μm	25–60
50	297 μm	10–30
100	149 μm	2–10

Table 8.3 *AASH designation M147–67 for soil aggregate mixtures*

	Percentage by weight passing square-mesh sieve					
	A	B	C	D	E	F
2 in (50.8 mm)	100	100				
1 in (25.4 mm)		75–95	100	100	100	100
3/8 in (9.510 mm)	30–65	40–75	50–85	60–100		
No. 4 (4.76 mm)	25–55	30–60	35–65	50–85	55–100	70–100
No. 10 (2.00 mm)	15–40	20–45	25–50	40–70	40–100	55–100
No. 40 (0.420 mm)	8–20	15–30	15–30	25–45	20–50	30–70
No. 200 (0.074 mm)	2–8	5–20	5–15	5–20	6–20	8–25

Note: Fraction passing the No. 200 sieve shall not be greater than two-thirds of the fraction passing the No. 40 sieve.

100–200 μm), refractoriness and low thermal expansion. Chromite sands have similar moulding properties to zircon and the two minerals compete in this market according to their respective costs.

Silica sands for glass making

Individual glassmakers are concerned with the purity, grain size and uniformity of the sands used and most of them set their own standards for both chemical analysis and granulometry. They are influenced greatly by economics, because silica sand is essentially a low-priced material, and local availability is often the deciding factor in what may or may not be used.

Common to all manufacturers, however, are the effects of grain size and grain-size frequency characteristics. In this regard:

(1) Uniformity of grain size allows better mixing of the ingredients and more even melting.
(2) Coarse grains create local zones of varying viscosities.
(3) Grains that are too fine may float to the surface and remain outside of the mix.

One Australian manufacturer sets the following size specification for green glass quality sands from local sources with the proviso that any evidence of +16-mesh material will place the shipment subject to rejection (Table 8.4).

Filtration sands

The nature of the flow of water or other viscous fluids through porous filter beds depends upon the range of grain sizes and shapes and the kinematic viscosity of the fluid. Generally desired characteristics are uniformity of grain size and shape, and grain size characteristics that afford a maximum

Table 8.4 *A glass sand specification (Australian source)*

Screen size	Opening (mm)	Maximum % allowed
US 20 mesh	0.841	Trace
30 mesh	0.595	4 (retained)
40 mesh	0.420	25 (retained)
40 mesh	0.420	5 (passing)

Screen time 10 min.

rate of filtering commensurate with retaining impurities within the bed. Oversize grains cause irregularity and channelling of the flow whereas fine materials smaller than about 0.15 mm cause clogging and reduce the flow rate.

Flow may be either laminar or turbulent and, according to Jacob (1949), experiments on natural and artificial sands of approximately uniform spherical grains have shown that departure from laminar flow begins at Reynolds Numbers of between 1 and 100 (see Chapter 2) depending on the range of grain sizes and shapes.

Sands are graded according to the size at which 90% by weight of the sand is retained on a screen and 10% passes through; and by the ratio of the size of the opening at which 60% will pass, to the size of the opening at which 90% will pass.

8.7 PRODUCTION

Ways and means of bringing the property into production are now examined (see Chapters 6 and 7) and tests are devised to study and compare the various possible alternatives. Laboratory and bench-scale investigations are followed by pilot-scale operations which lead to mine planning, design and costing. The scale of the proposed operation is determined and with it the life of the mine. Pilot-plant studies are sometimes omitted when the experience gained from similar-type operations is thought to be adequate for design purposes, however, it is generally prudent to deal with each new project afresh.

8.7.1 Mining

Placer deposits are laid down under a variety of conditions and over long periods of time during which the conditions of sedimentation change many times. Some occur in single channels or lenses, others occupy networks of channels in complex drainage systems. The controls are environmental and deposits display varying characteristics resulting from the effects of repeated cycles of erosion and deposition that have altered their original features and changed their orientation. The sedimentation properties of individual particles influence but do not entirely control their behaviour in fluids, hence, character trends also vary from deposit to deposit and within the same deposit.

Consequently, there is a common difficulty in setting technical and economic standards that apply equally to all deposits in a group and for each placer there may be several alternative methods of mining to choose from. The more primitive placers have specific handling problems because of bedrock features that are characteristically most irregular and sediments for which the particle size range may be extreme. Ancient terraces, deep leads and other fossil placers pose diagenetic problems affecting liberation and induration is a major source of machine wear. Difficulties are experienced, always, in recovering valuable particles that have settled out at bedrock, even though the bedrock may be evenly weathered, regular in outline and flat.

Problems relating specifically to placers are due to the constraints imposed upon mining by the environments within which they occur and by the sedimentary nature of the deposits. For example, mining costs are particularly sensitive to conditions at bedrock and to the difficulties posed by such features as pinnacles and bars of hard rock; and the presence of large quantities of slimes-sized particles on the one hand and frequent large boulders on the other. Placer ore reserves are similarly wasting assets to those of hard-rock deposits and each deposit has unique features that call for special consideration. These have to do not only with mining but also with treatment and marketing.

8.7.2 Treatment

Plant operational problems depend largely upon the nature of the material to be treated and upon the manner in which the various feed fractions are prepared and presented for treatment at each stage of processing. Investigational studies described in Chapter 7 deal with such matters as the liberation of valuable particles from puggy or cemented wash; the mechanical and chemical attritioning of affected particles to remove organic and inorganic coatings; the effects of overlapping electrical and magnetic responses by particles of different mineral types and so on. The full extent of the problems may not be known at the beginning of an evaluation exercise, however, there should be no doubt of how and where their effects will be felt if they are not identified and dealt with effectively.

Clearly for each deposit there is a need to concentrate attention upon costs in the main problem areas because such costs are subject to the widest variations. For placers, quite apart from lower than expected recoveries from mining and treatment and consequent higher costs for the products, both capital and operating costs are seriously affected by production bottlenecks, wasted time for plant alterations and other misfortunes that result from faulty evaluations. Errors in evaluation occur mainly because, through the lack of adequate data, the selected methods for mining and treatment are inappropriate to the full circumstances of the project or inadequate for what has to be achieved.

8.8 CAPITAL AND OPERATING COSTS

Implicit in all decisions to commit additional funds to the valuation exercise is the belief that the proposed expenditures will benefit the undertaking in both the short and the long term. Each such decision commits management to a specific course of action that cannot be stayed lightly. Formalized lists of headings, i.e. charts of accounts, help to eliminate guesswork and avoid the unsatisfactory practice of geologists and engineers relying upon their memories for the long list of things to be costed. Cost allocation is fundamental to all enterprises and cost predictions should not have to rely only upon the experience and judgement of those doing the valuation without adequate supporting data.

8.8.1 Charts of accounts

Hence, although final mine and mill development decisions are not made until the completion of exploration and prospecting, a comprehensive list of cost centres should preferably be drawn up from the beginning to ensure that all of the required data are obtained and presented in a standard form. At first the data are incomplete and estimates of costs that seek to predict the economics of processes that are only in a conceptual stage may not be more accurate than ±50%. Essentially, however, the process of valuation is a continuing exercise in which each development stage provides an increasing assurance of the outcome and will, eventually, supply all of the necessary information for final costings and design. By adopting a standard format from the start not only are important details unlikely to be overlooked, but strict adherence to costing according to the chart of accounts will help to isolate the most sensitive cost areas and hence the most likely sources of error at the earliest possible stage.

It is important also that the format adopted should be consistent with the accounting methods to be used when commercial operations commence and not be put aside with the implementation of the project. Systems developed for valuation purposes will be realistic only if they are equally valid when expanded to satisfy all of the future day-to-day and long-term requirements of the proposed operation. If there can not be a direct flow-on from evaluation costing to process costing it is probable that some functions have not had adequate treatment and that, hence, the framework around which the operational costing procedures are to be established may also be inadequate.

The following treatment of capital and operating costs is based largely upon Australian practice in respect of preliminary methods of assessment and the same factors may not apply elsewhere. However, the principles involved are basically the same in all communities.

8.8.2 Capital costs

Capital costs include all of the costs of setting up an operation including the costs of raising the necessary funds. They may be funded by the issue of shares, debentures or other securities or by short-term accommodation in the form of loans. The relationship between ordinary share capital and fixed income securities is dealt with in a wide range of publications on business management and economics: a good summary is given by Chambers (1966) according to Australian conditions. Other important capital-cost centres are, (a) acquisition of property or rights to mine, (b) exploration and prospecting, (c) mining plant and equipment, (d) treatment plant and equipment, (e) site preparation, access roads, transportation and materials handling, (f) services such as power, water, workshops and laboratories, (g) infrastructure, housing, etc., (h) planning, design, overheads including supervision, (i) freight and installation, (j) working capital, (k) a contingency allowance that reduces with the increasing reliability of the data used for estimating.

Capital cost estimation

Pilkington *et al.* (1966) describe various methods of estimating capital requirements (Table 8.5).

The authors, in their table, also estimate the various cost factors relating to process equipment, installation, instruments and controls, piping, electrics, buildings, utilities, land and site improvements, engineering, construction and contractors fees and working capital. In the present global climate of social upheavals and economic change such estimates must be updated frequently and cannot be relied upon except for very short periods of time.

Of the many factors that determine the adequacy of projected capital estimates, construction delays are major, and sometimes unpredictable sources of error. They lead to system bottlenecks, excessive overtime payments, double handling, reduced utilization of available machines, contract penalties and reduced efficiencies. Construction delays can be minimized but not altogether eliminated by careful planning, good management-labour relationships, and close adherence to a time and motion charted programme.

Table 8.5 Methods of estimating capital requirements

Method	Extrapolation from existing plant costs (A)	By ratio from total puchased equipment costs (B)	From unit costs (C)	From detailed bill of material and labour
Probable accuracy %	±10–50	±15–25	±10–15	±5–10
Information required	General process details, capacity, sizes and costs of existing plants of comparable size	General process details, general site information, process equipment, specifications for costing and a rough layout of process equipment	Same as (B) plus instrument flow sheets, plant layouts and engineering flow sheets	Same as (C) plus detailed designs and specifications
Contingency cost of unlisted items and unexpected price increases and plant revision	Calculated in all cases from experience factors as a percentage of the cost of the process equipment or from the uncertainty rating of the project			

Other common problems in setting up a placer-mining project concern possible increases in the costs of labour and materials; inefficiencies in purchasing; scheduling and construction; strikes, transport delays, extreme weather conditions and delays in commissioning the plant and bringing it to full design capability. All of these areas of risk must be understood and allowed for by providing in-built factors of safety and contingency allowances. Hackney (1960) uses an 'uncertainty ratio' from which he suggested contingency allowances for Australian process industries as shown by Table 8.6

Table 8.6 Contingency allowances – Australian process industries (1960)

Method of assessment	Uncertainty rating	Recommended contingency %
Equipment–ratio estimates	400 or less	±25
Unit cost estimates	200 or less	±12
Detailed bill estimates	100 or less	±6

Pilkington *et al.* (1966) from an Australian survey found that detailed bill contingency estimates ranged from a low of 2% to a high of 9%, and averaged +5%. The values might be considerably higher today, particularly if the common definition of project completion is applied. In effect, the project must have achieved its forecast production of an acceptable product within forecast cost parameters and within a specified time from the date of commencement. It follows that the prudent investor will apply stringent tests towards predicting the possible effects of all forseeable problems before they actually arise.

Effects of variations in plant size and capacity

Unit capital costs are not affected equally by variations in size and capacity. According to Battery-Limit plant costs supplied by Faith, Keyes and Clark (1957) the overall capacity factor varied at that time in America between 0.4 and 0.9 and averaged around 0.7. Such figures are useful in early studies but should be treated with caution in the later stages of valuation because of the very wide differences in unit values for individual cost centres. For example, the cost for introducing a second line of process machines may have a factor of 2. Obviously a detailed analysis is needed in which all of the unit costings are considered separately at first and then as a whole.

Working capital

Working capital is needed for starting up a project and keeping it operating until such time as the revenues adequately cover all of the outgoings. The required amounts are calculated from experience factors, detailed operation costings and marketing, and as a percentage of fixed capital. Few projects proceed exactly as planned, however, and, all too frequently, serious under-estimations in the working capital requirements lead to financial crises at the time of greatest vulnerability, that is, at the 'teething' trouble stage in the mines' life. Funds for working capital provide for the purchase of supplies, labour and administration costs and for a host of miscellaneous cash requirements. Estimates consider the overall costs of running the mine until such time as all of the functions of the outgoings and incomings are in equilibrium and a profit is being made.

In terms of projected operating costs the allowance for working capital may be almost any figure up to 100% of the operating costs for the first 4–6 months of the mine life depending upon the lapse of time between the commencement of production and realization. It varies for different placer minerals and, because gold is 100% realizable at short notice, is likely to be much less for gold placers than for mineral-sands projects. Final payments for mineral-sands shipments may not be received until consignments have been received overseas and agreement reached on quantities and grade. Progress payments are normally dependent on the completion of a batch in readiness for shipping and this may involve several months of work before the required quantities are in the bulk store.

8.8.3 Operating costs

Operating costs are all of the costs of production and marketing and each project is subject to individual charges peculiar to its circumstances. These charges comprise both fixed and variable costs and the total production cost is made up of elements of the two. Fixed charges are made for such items as building depreciation, property rates and interest on borrowings. They are called fixed charges because they are independent of production and continue to accrue regardless of whether the mine is in production or closed down. Raw materials and outward freight costs for the products are normally variable in direct proportion to the output. Costs that are variable, but not in proportion to the output, include labour charges, company overheads and some services. These costs are semi-variable and the tendency is for them to increase at a lower rate than does production itself. The variable and semi-variable costs are usually grouped for accounting purposes under such headings as: supplies – raw materials, reagents, fuels; services – electricity and water; labour and administration; repairs and maintenance; company overheads, including marketing, royalties, depreciation, insurance, taxation, etc.; freight and shipping.

The collection of data for operating-cost estimation commences, generally, with rule of thumb methods which predict individual cost levels on the basis of costs for somewhat similar operations elsewhere. Such predictions could once be made with a great deal more authority than in today's uncertain economic climate but, even then, were subject to considerable error. For example, in 1966 Buchanan and Ulherr estimated the cost of operating supplies to the process industries to vary between 5 and 20 % of the total operating costs; costs for power and drying as high as 30 % and charges for repairs and maintenance between 2.5 % and 8.0 % of fixed capital charges with an average of 4.0 %. This did not apply to vehicles and earth-moving machinery for which generally higher charges depended upon the severity of the working conditions.

The more or less informed predictions arising from such information provide considerable scope for error and, although of some use in the early stages of an evaluation exercise, should be put aside as soon as more reliable costings can be made. Placers are distributed globally in widely varying conditions with equally widely varying cost structures. For example, there are differences in time availability. A plant may operate for one, two or three shifts daily, for five, six or seven days per week: the level of religious and other holidays vary also and in Islamic countries the month of fasting, Ramadan, carries with it a measurably reduced work efficiency. At best, in a continuously operating plant the effective utilization of available work time averages only 20 h per day and between 6000 and 6500 h per annum. Plants in which the accent is on automation are usually cheaper to run because of reduced labour costs, however, the additional capital requirements for automation must be justified by the scale of savings and there must be an ability to maintain such equipment fully. This is not possible in many placer-mining environments.

A recent prefeasibility study illustrates the need for making individual assessments for each new venture. The particular project considered the mining of a placer gold deposit in a Middle East desert area and, because of the peculiar circumstances, labour and administration costs were estimated at 75% of the total operating costs in spite of the fact that other charges were also high. It was determined that what would be a barely payable grade for the deposit concerned would provide a bonanza situation in another country having lower wage levels, a less harsh climate, less rigorous standards for infrastructure, more adequate supplies of water and an availability of labour without the need for importation. Nevertheless it was noted that the deposit was located favourably compared with some other placer projects in the same region for which the operating costs would undoubtedly be much higher.

8.8.4 Process factors affecting costs

Factors affecting capital and operating costs are summarized in Tables 8.7 and 8.8 according to the various alternatives available for mining and treatment.

8.9 REVENUES

Revenues for any given treatment rate are determined by ore grades, plant recoveries, product prices and market conditions at the time of selling. Ore grade and recovery efficiencies are controlled by the operator, spot prices and market fluctuations are governed by the laws of supply and demand and are largely beyond his control. The most vulnerable undertakings in these circumstances are those for metals such as gold and tin which are generally sold 'spot'; the least vulnerable are mineral sands undertakings which operate under long-term agreements at fixed prices.

A great many uncertainties arise from social upheavals, currency fluctuations and technological changes and predictions based upon current price levels and past trends are subject to considerable error. For example in the space of a decade, gold prices have risen to as high as US$800/oz. (US$28.2/g) and fallen to as low as US$300/oz (US$10.58/g) largely on the apparent strength or weakness of the American dollar.

Table 8.7 *Process governing factors and alternatives*

Process	Governing factors	Alternatives for mining and treatment
Mining	Dimensions and volume; Sediment size range; Bedrock conditions; Required selectivity; Supply of water; Ecological requirements; Mining rate; Climate	(a) *Wet mining*: Bucket-line dredging; Suction-cutter dredging; Bucket-wheel dredging; Hydraulic sluicing; Clam-shell dredging; Dragline dredging; Drum scrapers (b) *Dry mining*: Dragline or hydraulic excavator and trucks; Bulldozers, articulated FE loaders and trucks; Bulldozers and scraper loaders; Drum scrapers
Treatment	Sediment size range; Valuable particle size range; Ease of liberation; Availability of labour; Availability of water; Rate of treatment. Treatment may be based upon, (a) one-shift operation, mine and mill, (b) daylight hours mining; two- or three-shift mill operation, (c) continuous operation, mine and mill.	(a) *Ample supply of water*: Treatment methods incorporate hydrocyclones or hydrosizers and a choice between jigs, tables, spirals, pinched sluices and sluice boxes. (b) *Limited supply of water*: Treatment methods may be restricted to dry-sizing equipment and rotating cones; in extreme cases dry gravity separation. Generally, however, there will be application for wet screening and concentration methods incorporating facilities for water recovery.

Historically, it is only in recent times that gold prices have been allowed to 'float'. According to Ridgeway (1929) the price of gold rose only gradually from 22 s an ounce in 1344 to 84 s 11½d an ounce in 1717. It remained at this figure until 1919. Similarly, during the past few years, tin prices, even with the dampening effect of buffer stock releases and purchasing, have fluctuated between boom time and depression levels. Only with forward contracts for the bulk of his sales can the producer count upon a guaranteed income within measurable limits and even then such contracts may be deferred or broken in times of industrial recession.

Table 8.8 *Summary of factors affecting placer capital and operating costs*

Cost centres	Alternative factors
Exploration and prospecting; Mine plant and equipment; Treatment plant and buildings; Services – power, water, workshops, transport, communications; Stores – spares, consumables; Communications – roads, radio telephones, airstrip; Site preparation; Freight and insurance; Design, installation and commissioning; Housing and amenities; Working capital; Establishment	(1) Permanent or semi-permanent structures depending upon projected mine life (2) Plant design oriented towards: initial high cost, low upkeep or initial low cost, high upkeep (3) Company input 100% or contribution also from outside source (4) Earth moving on operator-ownership basis, or by contract (5) Financial gearing ratios, equities, borrowings.
Mining and treatment – labour;	Balance between scale of mining and mine life. In general high production rates lead to low unit operating costs but studies must consider overall profitability and relate cost of capital to operating costs to determine optimum conditions. Operating costs are influenced most by the quality of labour and management; the avoidance of process bottlenecks and unscheduled down time.

Furthermore, while physical measurements of ore grades and volumes may usually be relied upon in all future calculations, the conditions governing supply and demand can change overnight. A government might decide to stockpile or release minerals from stockpile, reassess the value of its currency or increase its defence spending. A technological breakthrough in the re-cycling of one scrap metal could introduce an entirely new component into the competition between two different metals for the same market and so on. Uncertainties arise from all quarters and a comparatively new and little understood dimension has been added to the economic scene through the recent escalation in fuel prices.

Nevertheless, estimates must be made and increasingly sophisticated methods are being developed by economic statisticians to predict future market trends and, in doing so, to assess the risks involved and identify the most sensitive areas for detailed analysis. Of the available systems, geostatistics offers the best means at present of measuring the inherent risks in the assessment of ore reserves; a number of techniques are available to assess the value of a proposed operation, all of which rely upon accurate predictions of revenues and surpluses over a guaranteed life.

8.10 ECONOMIC APPRAISAL

From the foregoing, it is seen that the analysis of investment worth must be guided by technical, financial, economic and other studies in order to obtain the greatest economic advantage. Such analyses presuppose the selection of an appropriate method and rate of mining and primarily involve the determination of project life calculated by dividing the tonnage or volume of ore reserves by the annual rate of extraction; and estimates of annual surpluses derived from differences between the projected incomings and outgoings. A favourable means of financing must also be determined and applied sequentially according to the requirements of the mine development plan, hence the need for detailed capital-expenditure planning.

8.10.1 Capital expenditure planning

Capital expenditure is the commitment of a resource (money) made in the expectation of realizing an appropriate benefit within a forseeable period of time, and the decision to enter into such a commitment is known as an investment decision. Investment decisions should not be taken lightly; they commit an enterprise to a course or courses of action that cannot be retracted and, hence, are taken wisely only after full consideration of all of the factors involved. Decisions made without a full study of all of the project dimensions have outcomes that are less assured and may result in heavy losses or even failure.

Several aspects of capital management deserve special consideration. One is the time value of money. Another is the cost of capital. A third is the treatment of working capital. Each of these is considered in turn.

The time value of money

A sum of money has a greater value now than at any time in the future and the greater the time difference the greater is the amount by which it depreciates in value. This is basic to the cost of capital and criteria for evaluation include:

(1) The amount and timing of the investment outlay and its recoupment.
(2) The annual net cash earnings predicted for the life of the mine.
(3) The anticipated residual value of the investments at the end of the mine life.

The cost of capital

Because of the increased complexities of modern technologies, new placer projects are likely to be larger and require more financing than in the past. Thus, while simplistic calculations involving direct equity financing serve to establish broad bases for decisions to proceed with, or to abandon a project at an early stage of evaluation, such calculations are much less relevant in the final stages of an exercise when it can be assumed that the operation will probably be profitable, but there is a need to know whether it is or is not attractive as an investment. It is to be noted that cash benefits are not always the only consideration; there should also be an awareness of the social implications and, as Middleton (1965) points out, in any major conflict of interest the public interest should prevail.

The management of capital is a highly complex affair and a new enterprise has a variety of sources of capital available to it, each with a specific cost. Some of the required funds may be raised in the money market and the interest rates on such debts are less than the required yields on equity funds because of the lower risks attaching to this class of investment. Interest payments are allowable as tax deductions in most countries, whereas dividends are taxed. The balance struck between debt and equity financing is referred to as 'gearing'.

Some of the funds may also be generated internally from depreciation allowances and profits and, throughout, interest is payable at various levels. Debts are scheduled for repayment as early as practicable and fresh capital is introduced only as and when required. In some countries there is, as well as ordinary depreciation, an 'investment allowance' comparable to an additional depreciation charge. This allows an operator to obtain an additional tax deduction during the first few years of operation according to some fixed rate. One concession introduced in Australia in 1962 allowed an extra 20% depreciation in the first year of operation thus giving, over the life of the plant, a tax exemption for depreciation up to 120% of the capital cost.

8.10.2 The treatment of working capital

The investment outlay includes a certain amount of money to support the operation until it is self supporting, that is when the incomings exceed the outgoings. This investment is generally included on an accrual basis and later appears as part of the cash inflow as and when it is liquidated.

8.10.3 Cash flows

A number of techniques are applied to placer valuations and generally, the most favoured are those that allow estimates of cost and earning power to be compared with alternative investments in less-speculative fields. Early

Hoskold-type formulae have little relevance today. Debt financing has become the norm of business investment and appraisals now consider investment worth in terms of the difference between the money flowing out from an investment and that being returned. Theoretical considerations are based upon the premise that money has time value as well as face value and that, if invested, it should grow in value each year at a predetermined and acceptable rate. A positive cash flow means that the investment is profitable. The project is worthwhile if the surpluses are sufficient to satisfy the investment criteria of those providing the finance. If negative, or at a lower rate than required, the project may either be rejected or set aside for future re-appraisal.

The present worth of an investment is defined as the discounted value of the total revenue received less the total costs of operation, ownership, royalties and taxation. In this context, both the present worth of a dollar and the reliability of future market projections are reduced progressively with each succeeding year and cash-flow projections are rarely extended beyond 15 years. Generally, however, the investors will expect satisfactory rates of return over a much shorter period and a pay back time that is reduced to a practical minimum.

A hypothetical set of data describes a typical cash-flow projection in which:

(1) The net sales vary in value from year to year according to the grade of ore treated.
(2) Operating costs in real money terms are considered to remain constant at $5 million/annum. The product unit price similarly remains constant in real-money terms.
(3) Capital expenditure is at the rate of $5 million in Year 1 and $7 million in Year 2.
(4) Capital is supplied by 40% equity and 60% borrowings at 14% simple interest per annum repayable in equal instalments in Years 3, 4, 5 and 6.
(5) Total interest on borrowings is considered to be provided on bank overdraft in Years 1 and 2 at the same interest rate and repayable in full in Year 3.
(6) Redemption is at the rate shown.

The discounting of cash flows

Using the discounted cash-flow (DCF) technique, the 'present worth' of an investment is the algebraic sum of the annual cash flows discounted to the present time at a rate that is generally equal to the cost of the capital involved (Sinclair, 1966). The number of years selected for discounting varies according to such factors as mine life, investment policy, political and economic stability, and upon how much confidence is placed in

Table 8.9 *Cash flow (000s)*

Year	1	2	3	4	5	6	7	8	9	10
Net sales			12 529	10 964	10 119	9 727	7 914	7 914	6 628	6 570
Operating costs			(5 000)	(5 000)	(5 000)	(5 000)	(5 000)	(5 000)	(5 000)	(5 000)
Loan interest	(420)	(1 067)	(1 216)	(756)	(504)	(252)				
Redemption			(1 827)	(1 453)	(1 453)	(1 453)	(1 453)	(1 453)	(1 453)	(1 453)
Net operating profit	(420)	(1 067)	4 486	3 755	3 162	3 022	3 022	1 461	165	125
Profit after tax			3 287	2 253	1 897	1 813	877	877	99	75
Redemption add back			1 827	1 453	1 453	1 453	1 453	1 453	1 453	1 453
Gross cash flow			5 114	3 706	3 350	3 266	2 330	2 330	1 552	1 520
Investment	(5 000)	(7 000)								
Loan	3 000	4 200								
Repayment loan			(1 800)	(1 800)	(1 800)	(1 800)				
Loan for interest	420	1 067								
Repayment loan for interest			(1 487)							
Cash flow (equity)	(2 000)	(2 800)	1 827	1 906	1 550	1 466	2 300	2 300	1 552	1520

Notes: 1 The tax on profit in year 3 is calculated after deducting losses in years 1 and 2.
2 The pay-back period is shown to be 2.7 years after the commencement of operations.

forward market projections. The ultimate purpose of establishing 'present worth' by this method is to compare the proposed investment with alternative investments, thus determining its value in the open market. The basic data are supplied from the cash-flow statement and the more that is known of the project the more precise the data will be.

It must be stressed, however, that the true significance of DCF estimates of worth relate only to the amount of money to be invested and not necessarily to the value of the property as a whole. Because of the risks, investors require a commensurate rate of return on their investment for the number of years considered appropriate to the study. The prospect of a much longer life will not justify a greater capital expenditure than the cash flow can support and because the present worth of the dollar becomes almost negligible after 15 years. This factor has led to a misconception in the purchasers favour that the DCF values the property as well as the investment. It does not, as the following example will show.

> The investment worth of a property having reserves for 30 years is very little more than for a similar property with reserves of only 15 years. This suggests that the first property has no more value to the purchaser than the second. The reasoning is false, of course, because if the cash flow for the 30 years property is discounted to zero at 15 years and paid for accordingly, the investor then owns a similar 15-year property for which he has made no investment and for which he might only need to have paid the normal costs for maintenance and replacement (already discounted) in order to continue producing. (Macdonald, 1973)

Assuming that the cost of capital is between 16 and 18% the present worth of the investment is found by discounting the annual cash flows in table 8.9 to zero at each of the two rates (Table 8.10).

Table 8.10 *Present worth of an investment (000s)*

Year	Cash flow	DCF 16% factor	Present worth	DCF 18% factor	Present worth
1	(2000)	0.8621	(1724)	0.8475	(1695)
2	(2800)	0.7432	(2081)	0.7182	(2011)
3	1827	0.6407	1171	0.6086	1112
4	1906	0.5523	1053	0.5160	983
5	1550	0.4761	738	0.4371	678
6	1466	0.4104	602	0.3074	451
7	2330	0.3538	824	0.3139	721
8	2330	0.3050	711	0.2660	620
9	1552	0.2630	408	0.2255	350
10	1520	0.2267	345	0.1911	290
Total			+2047		+1509

This means that if the cost of money to the company is 16% the present worth of the proposal is $2.047 million whereas if the cost of the money is 18% then the present worth is only $1.509 million. If on the other hand the cost of the money is 35% the investment can only be made at a loss and its present worth would be −$407 000.

Alternatively, the annual cash flows can be discounted to zero at the present date again using annuity tables to give the present value of $1 due in n years at various rates percent (see Appendix). Using the cash-flow data already described, the annual figures are discounted by trial and error until the zero discount rate falls on some point between two of the rates. In the case being considered, it falls at around 29.5% for the equity capital of $4800 as shown in Table 8.11.

By lineal interpolation the DCF = 29.5% on the equity capital involved. This is otherwise known as the 'Internal Rate of Return'.

Normally, for small investments, a pay-back period is aimed at between 3 and 4 years and rarely more than 6–8 years for the larger ones. The shorter the period the safer is the investment because of the uncertainties of continuing political attitudes and markets. Very high profits are usually looked for in the first few operational years even to the extent, sometimes, of 'high-grading' the deposit and shortening the project life. Some governments exacerbate the situation by levying excessive royalties and taxes, thereby encouraging work only in the richer sections of deposits.

Notes to accompany a DCF analysis

The bare essentials of a discounted-cash-flow analysis are made more intelligible through the use of supplementary notes describing the

Table 8.11 Discounted cash flow (000s)

Year	Cash flow	DCF 25% factor	Amount	DCF 30% factor	Amount
1	(2000)	0.8000	(1600)	0.7692	(1538)
2	(2800)	0.6400	(1792)	0.5917	(1657)
3	1827	0.5120	935	0.4552	832
4	1906	0.4096	781	0.3501	667
5	1550	0.3277	508	0.2693	417
6	1466	0.2621	384	0.2072	304
7	2330	0.2097	489	0.1594	371
8	2330	0.1678	391	0.1226	286
9	1552	0.1342	208	0.0943	146
10	1520	0.1074	163	0.0725	110
Total			+467		−62

background for the various figures and measurements used. Typically, in a detailed analysis they will include such descriptions as:

(1) The bases for estimating the capital costs and escalation factors used to allow for inflation if the expenditure is to be phased in over a number of years.
(2) Details of resource purchases and the terms of payment including commissions.
(3) A breakdown of infrastructure costs into the respective categories of housing, mine buildings, services etc.
(4) The bases for calculating revenues from sales, commissions etc. Much of the credibility of the analysis might depend upon what proportions of the output are to be disposed of under contract and what proportion by spot selling.
(5) The basis for estimating mining and treatment costs and the allowances made for contingencies. Credibility obviously increases with each succeeding stage of valuation.
(6) The amounts and rates of payments for government levies and royalties to land owners.
(7) Estimated freight rates and port charges if applicable.
(8) The bases for calculating depreciation and/or amortization allowances.
(9) Provisions of the taxation laws as they apply to the proposed operation.
(10) Concise definitions of the various terms, net cash flow, equity cash flow etc. used in the statement of analysis.
(11) The scope of the valuation exercise. For example, what alternative investment rates have been considered.
(12) Details of the proposed debt servicing.
(13) The reliability of the exercise and the checks made, if any, by independent consultants and valuers.

8.10.4 Allowing for risk

An element of risk attaches to every stage of valuation from the initial taking of samples to the final assessment of market opportunities and costings. Individual errors may be minimized by careful attention to detail, however, the cumulative effects are usually significant and must be allowed for.

The highest risks are associated with the production of minerals for which there is no assured market, either because of inferior product quality or because the market requirements are already being satisfied from other sources. The decision to break into such a market demands strong evidence that the buyers can be persuaded of benefits that will more than offset any possible increases in risk. The venture itself is at considerable risk until all of the assurances given have been accepted. Other risks involve machinery obsolescence, currency fluctuations, political upheavals and global recessions; risk may sometimes be considered only intuitively and then, only in terms of ultimate success or failure.

Risk must not be confused with uncertainty. Risk relates to situations in which a probability can be assigned to each of the possible outcomes. Uncertainties are not measurable and cannot be assigned a factor for correction. One method of allowing for risk is to aim at a higher discount rate and in early studies where the risks are greater a DCF rate of, say, 25% might be required whereas in the final studies 15% might be considered quite a favourable level. One problem in this approach is to set arbitrary levels of discount rates, for risks that have not been evaluated mathematically.

If a study has been carefully done, a contingency factor can usually be applied realistically to every component of a venture and, in such cases, the risks can be computed and numerical values obtained that are finite. Statistical methods offer theoretical means of measuring risk factors; an approach developed by Hertz (1964) involves the use of simulation techniques to derive a probability distribution of rates of return, recognizing that a number of factors influence the cash forecast. A computer is used to carry out the trials and Middleton (1965) refers to one evaluation in which 3600 DCF calculations, each based upon a selection of nine input variables were run in 2 min. The resulting rate-of-return variables were read out immediately and graphed.

The risk in not completing a project because of lack of capital is one of the greatest hazards faced by new ventures and, in a highly geared operation where the borrower has only 10–20% of the equity, it is the lender who takes most of the risk. Although it is easy to say that this problem should be taken care of before it arises, cost overruns are all too common in the present highly inflationary environment. The risk can be minimized only by having additional resources available if required, either contractually with the parties to the agreement, or with some other lending agency. In any case, it is always prudent to incorporate a completion undertaking in any agreement entered into.

8.10.5 Sensitivity

One element of risk that should not be overlooked is the sensitivity of a project's cash flow to small changes in the economic conditions. The purpose of a sensitivity study is to identify the most critical of the factors involved so that special attention can be given to their control. Such studies examine the effects on DCF factors of small changes in one variable at a time while holding all of the others constant.

Sensitivity analyses are most important in marginal situations which highlight the common difficulty of setting standards for a wide range of placer types, each having varying sensitivities to changing socio-economic relationships. At the other end of the scale, bonanza-type situations may lead to a complacent acceptance of the sensitivity of some factors with consequent sloppy management and reduced profit margins. Table 8.12 summarizes the main points to be considered in the economic assessment of placer projects.

8.11 FEASIBILITY STUDIES

Feasibility studies are multi-disciplinary exercises designed to test the validity of data at all levels and to provide the justification for major investment decisions and courses of action which may be extremely far reaching. Such decisions cannot be taken lightly, nor are they concerned only with the flow of capital and expectation of profits; they are also

Table 8.12 Notes on economic assessment and profitability

	Main factors	Alternatives
Revenues	Throughputs; Ore grades; Recoveries; Gold prices; Treatment rates.	Based upon a certain treatment rate in any plant: (1) Increased throughputs may give higher yields but only at expense of lower recovery rates (2) Higher cut-off grades and hence, increased feed grades, give higher yields and, within capacity of treatment plant, higher unit recoveries (3) Lower ore grades give lower yields and lower unit recoveries in all circumstances. Revenues vary directly with changes in the price of the products sold.
Internal rate of return	Method of financing gearing ratios; Interest rates and repayments; Operating costs; Mine path; Royalties; Taxes; Company overheads; Life of mine; Revenues.	Basic choice between: (1) Maximum total profits (2) Minimizing pay back period and risk. Best situation can be assessed from the discounting of cash flows when all risk and sensitivity factors have been determined. Fundamental aspects of studies revolve around mining, treatment, marketing and socio-economic relationships.

directed towards the environment in which the studies are taking place and to all of the social and political implications arising out of them. Feasibility studies appropriate to placer mining in any environment can be considered in four main stages, thus.

8.11.1 Stage I. Orientation study

This study includes reviews and checks on:

(1) Mining operations past and present.
(2) Ore-reserve computations followed, if necessary, by independent check drilling and calculations to investigate apparent anomalies.
(3) The geological findings of the exploration programme and how they fit in with the accepted pattern of ore reserves and proposed mine developments.
(4) The validity of both field and laboratory procedures.
(5) Pilot-scale and other mineral-processing investigations.
(6) All legislation pertaining to mining operations in the proposed area.
(7) The amount and character of any additional work which may be needed before commencing Stage II of the exercise which involves the development of conceptual plans, and an interim report.

8.11.2 Stage II. Development of conceptual plans

Plans involve the preparation of:

(1) Mining and materials handling systems.
(2) Treatment plant flowsheets.
(3) Preliminary maintenance workshop, power and water reticulation studies.
(4) Preliminary infrastructure studies including housing, harbour and ship-loading facilities, access roads and communications.
(5) Organization charts and estimates of staff and labour requirements.
(6) Provisional plant and equipment selection and costings.
(7) An environmental impact study.
(8) Preliminary capital and operating-cost estimates based upon similar models and equipment selection for the proposed scale of operations.
(9) Detailed market studies and projections.
(10) An interim economic assessment of project viability to justify the third stage.

Engineering note

Report writing is seldom wasted if its prime purpose is to document critical analyses of what has gone before. Items overlooked in mental processing may not be apparent in reviews; gaps in written statements of fact are readily detected, allowing corrections to be made before critical decisions are arrived at.

8.11.3 Stage III. Engineering

Includes:

(1) The preparation of detailed engineering-design drawings and specifications for all aspects of the mining–processing complex and materials handling; and for township planning and construction if applicable.
(2) The determination of optimum production rates and cut-off grades.
(3) The design of dredge or mining paths to optimize reserve utilization.
(4) The preparation of tender documents; the advertising of tenders and the processing of bids from interested contracting firms and supply organizations.
(5) A final project costing which should be both detailed and accurate so that final feasibility can be determined in Stage IV, Feasibility.

8.11.4 Stage IV. Feasibility

This entails:

(1) A consideration of alternative methods of financing.
(2) The preparation of cash-flow projections combined with sensitivity and risk analyses for the various alternatives.
(3) Measuring the results obtained against the original project requirements. Does the project satisfy these requirements, and, if so, does it appear to be better or worse than some alternative project.
(4) Determining what national interests may be involved and how they may be evaluated.
(5) The preparation of a final feasibility report summarizing all of the relevant data and stating both the facts and the conclusions reached for each aspect of the study. Positive recommendations on whether and how to proceed to the construction and commissioning stage.

REFERENCES

Buchanan, R. H. and Uhlherr, P. H. T. (1966) Operating cost estimation, in *Costs and Economics of the Australian Process Industries* (eds. Buchanan, R. H. & Sinclair C. G.), West Publishing Co., Sydney.
Clark, Isobel (1979) The semi variogram, *Geostatistics Part 2, E & MJ*, July, 90–4.
Chambers, R. J. (1966) Company Structure and Capitalization, in *Costs and Economics of the Australian Process Industries* (eds. Buchanan, R. H. and Sinclair, C. G.), West Publishing Company, Sydney.
Faith, W. L., Keyes, O. B. and Clark, R. L. (1957) *Industrial Chemicals*, John Wiley, New York.

Hackney, J. W. (1960) Estimate production costs quickly, *Chem Eng.* March 7, 113–30; April 4, 119–34.
Hazen, Scott W. (1968) Ore reserve calculations, in *Ore Reserve Estimation and Grade Control*, Can Inst. Min. and Met. Special Volume 9, pp. 11–32.
Hertz, David B. (1964) and Jacob, C. E. (1949) Risk analysis in capital investment, *Harvard Business Review*, Jan–Feb. in *Flow of Ground Water in Engineering Hydraulics*, (ed. Hunter Rouse), Institute of Hydraulic Research, Iowa.
Kelsey, R. D. (1977) Royalty Systems and Their Impact on Mining Investments, *Mining Economics Symposium*, UNI, New South Wales.
King, Haddon F. (1955) Classification and Nomenclature of Ore Reserves, *Proc. AIMM*, 174, p. 21.
Macdonald, Eoin H. (1973) *Manual of Beach Mining Practice, Exploration and Evaluation*, Australian Government Printing Office, Canberra.
Mackenzie, B. W. and Schwellnus, J. E. G. (1973) *Tonnage Grade Estimation for Mineral Deposits and Assessment of Ore Reserves.* DP/UN/INT-72-064, United Nations. pp. 4–34.
Middleton, K. A. (1965) *The Economics of Capital Expenditure.* The Australian Society of Accountants, Sydney.
Pilkington, J. T., Hercock, D. W., Buchanan, R. H., Uhlherr, P. T. H. and Balak, I. *Capital Cost Estimation – Costs and Economics of the Australian Process Industries.* West Pub. Corp. Sydney.
Ramos, H. C. (1977) Impact of Grade Management on Project Economics, *Mining Economics Symposium*, UNI, New South Wales.
Ridgeway, Robert H. (1929) Summarized Data of Gold Production, *U.S. Bur. Mines Economic Paper 6*.
Royle, A. J. (1976a) Structure and Valuation in Alluvial Deposits, in *Placer Mining and Exploration Short Course*, Mackay School of Mines, Nevada.
Royle, A. J. (1976b) *Case Studies in Alluvial Valuation.* Department of Mining and Mineral Sciences, Leeds University, England.
Sinclair, C. G. (1966) Alternate investment, in *Costs and Economics of the Australian Process Industries*, West Publishing, Sydney.
Taylor, H. K. (1972) General Theory of cut-off grades, *Inst. Min. Met. Trans.*, July edition, 160–179.
Thomas, E. G. (1977) Justification of high grading, at the feasibility study stage, *Min. Ecm. Symposium*, UNI, New South Wales.
Tyler, W. S. (1963) *Testing Sieves and their Uses.* Handbook 53, Tyler Company, Cleveland, Ohio.

APPENDIX 8.1 Present value of $1 in *n* years hence at various rates of compound interest (assuming that the $1 is received in a single payment of the last day of the year)

Years hence	Percent compound interest					
	5	10	15	20	25	30
1	0.952 38	0.909 09	0.869 57	0.833 33	0.800 00	0.769 23
2	0.907 03	0.826 45	0.756 14	0.694 44	0.640 00	0.591 72
3	0.863 84	0.751 31	0.657 52	0.578 70	0.512 00	0.455 17
4	0.822 70	0.683 01	0.571 75	0.482 25	0.409 60	0.350 13
5	0.783 53	0.620 92	0.497 18	0.401 88	0.327 68	0.269 33
6	0.746 22	0.564 47	0.432 33	0.334 90	0.262 14	0.207 18
7	0.710 68	0.513 16	0.375 94	0.279 08	0.209 72	0.159 37
8	0.676 84	0.466 51	0.326 90	0.232 57	0.167 77	0.122 59
9	0.644 61	0.424 10	0.284 26	0.193 81	0.134 22	0.094 30
10	0.613 91	0.385 54	0.247 18	0.161 51	0.107 37	0.072 54
11	0.584 68	0.350 49	0.214 94	0.134 59	0.085 90	0.055 80
12	0.556 84	0.318 63	0.186 91	0.112 16	0.068 72	0.042 92
13	0.530 32	0.289 66	0.162 53	0.093 46	0.054 98	0.033 02
14	0.505 07	0.263 33	0.141 33	0.077 89	0.043 98	0.025 40
15	0.481 02	0.239 39	0.122 89	0.064 91	0.035 18	0.019 54

Appendix

Conversion Tables and Factors

Note: Units of mass (e.g. pounds, grams etc.) are distinguished from units of force which are termed pound weight, gram weight and so on. Conversions from gravitational to absolute units are based upon the international standard acceleration due to gravity:

$$g = 32.1740\,\text{ft s}^{-2} = 980.665\,\text{cm s}^{-2}$$

All dimensions of quantities are given in terms of mass M, time T and length L.

A.1 CONVERSION TABLES

A.1.1 Length (L)

1 in = 25.40 mm
1 ft = 12 in = 30.48 cm
1 yd = 3 ft = 0.9144 m
1 mile = 1760 yd = 1.609 km
1 chain = 22 yd = 20.12 metres
1 nautical mile = 1.151 miles = 1.853 km
1 fathom = 1.8288 m

1 mm = 0.039 37 in
1 centimetre = 10 mm = 0.032 81 ft
1 metre = 100 cm = 39.37 in
1 kilometre = 1000 m = 0.6214 mile

A.1.2 Volume (L^3)

1 Imperial gallon = 1.201 US gallons = 0.1605 ft^3 = 4.546 l
1 US gallon = 0.8327 Imperial gallon = 0.1337 ft^3 = 3.785 l
1 ft^3 = 28.32 l = 0.02832 m^3
1 m^3 = 1000 l = 35.32 ft^3

A.1.3 Area (L^2)

1 ft^2 = 144 in^2 = 929 cm^2
1 cm^2 = 0.1550 in^2
1 m^2 = 10.76 ft^2 = 1.196 yd^2
1 ac = 40.47 are = 0.4047 ha = 10 $chain^2$ = 43 560 ft^2
ha = 2.471 ac = 0.003861 $mile^2$ = 100 acres = 10 000 m^2
$mile^2$ = 640 ac = 259.0 ha

A.1.4 Mass (M)

Avoirdupois

1 pound = 16 oz = 7000 grains
1 ton = 2240 lbs = 20 cwt = 4 quarters
1 grain = 0.0648 g

1 oz = 28.35 g = 437.5 grains = 0.9115 oz troy
1 g = 0.035 27 oz = 15.43 grains
1 kg = 2.2046 lb
1 lb = 0.456 kg = 7000 grains
1 ton = 1.016 tonnes = 2240 lb
1 tonne = 0.9842 ton = 2204.6 lb
1 tonne (metric ton) = 10 quintals = 1000 kg
1 US long ton = 2240 lb = 1 English ton
1 US short ton = 2000 lb = 0.8929 English ton = 0.9072 tonne

Troy

1 grain troy = 1 grain avoirdupois = 1 grain apothecaries
1 oz troy = 480 grains = 31.1 g = 1.0971 oz avoirdupois
1 penny weight (dwt) = 24 grains = 1.5552 g = 0.05 oz troy

A.1.5 Linear velocity (LT^{-1})

1 mph = 88 ft min^{-1} = 0.4470 m s^{-1}
1 m s^{-1} = 2.237 mph
1 cm s^{-1} = 1.967 ft s^{-1}

A.1.6 Density (ML^{-3})

g cm^{-3}	*lb in^{-3}*	*lb ft^{-3}*	*slug ft^{-3}*
1.000 0	0.036 13	62.43	1.94
27.68	1.000 00	1728.00	53.71
0.016 02	0.000 58	1.00	0.031
0.514 8	0.018 62	32.17	1.000

A.1.7 Power (ML^2T^{-3})

Note: pound means pound weight and kilogram means kilogram weight.

1 HP = 0.7457 kW = 550 ft lb s $^{-1}$ = 76.04 kg m s $^{-1}$
1 kW = 1.341 HP = 737.6 ft lb s $^{-1}$ = 102.00 kg m s $^{-1}$
Btu h $^{-1}$ = 0.000 392 9 HP = 0.000 293 kW

A.1.8 Work (ML^2T^{-2})

1 Btu = 0.252 kg calorie = 778 ft lb = 1055 joule = 0.000 293 kW h
1 therm = 100 000 Btu
1 joule = 1 watt-second = 10^7 erg
1 erg = 1 dyne-centimetre

A.1.9 Pressure (MLT^{-2})

lb m $^{-2}$	*kg cm* $^{-2}$	*atmospheres*	*in Hg*	*mm Hg*	*ft water*
1.000 00	0.070 31	0.068 04	2.036	51.712	2.306 7
14.233	1.000 000	0.967 6	28.96	735.31	32.81
14.696	1.033 3	1.000 00	29.921	760.00	33.90
0.491 2	0.034 53	0.033 42	1.000 00	25.4	1.132 9
0.019 34	0.001 359	0.001 316	0.039 37	1.000	0.044 61
0.433 5	0.030 48	0.029 49	0.882 6	22.419	1.000 00

A.1.10 Miscellaneous

1 g (metric) = 5 international metric carats
1 carat (international metric) = 200 mg = 3.086 grains (troy)
1 picul = 133 1/3 lb (avoir) 1 miners inch = 1.5 ft^3 min $^{-1}$
1 electrical unit = 1000 Wh
1 knot = 6080 ft h $^{-1}$ = 1853 m h $^{-1}$
Velocity of light = 186 330 mile s $^{-1}$ = 299 860 km s $^{-1}$
Velocity of sound in air (0°C) = 1087 ft s $^{-1}$ = 331.3 m s $^{-1}$
Velocity of sound in water (15°C) = 4714 ft s $^{-1}$ = 1437 m s $^{-1}$

A.2 CONVERSION FACTORS – PUMP CALCULATIONS

Although it is recommended to work in SI Pump Units, there are occasions when it will be necessary to convert known British, US or old metric units to the proper SI Unit.

The SI pumping units shown are in accordance with AS1686–1974.

To obtain the SI value, multiply the other value by the conversion factor. If the result of a metric calculation requires checking divide by the conversion factor.

For ease of reference physical quantities are listed in alphabetical order.

Physical quantity	Various units		SI pumping units		Conversion Factor
	Name	Unit	Name	Unit	
Area	square inch	in^2	square millimetre	mm^2	6.452×10^2
	square foot	ft^2	square metre	m^2	9.290×10^{-2}
	acre	ac	square metre	m^2	4.047×10^3
	acre	ac	hectare	ha	4.047×10^{-1}
	hectare	ha	square metre	m^2	1.000×10^5
Concentration	ounce per British gallon	$oz\,gal^{-1}$	miligram per litre	$mg\,l^{-1}$	6.236×10^3
	part per million	ppm	miligram per litre	$mg\,l^{-1}$	1.000
Density	pound per cubic foot	$lb\,ft^{-3}$	kilogram per cubic metre	$kg\,m^{-3}$	1.602×10
	pound per British gallon	$lb\,gal^{-1}$	kilogram per cubic metre	$kg\,m^{-3}$	9.978×10
	pound per US gallon	lb/US gal	kilogram per cubic metre	$kg\,m^{-3}$	1.198×10^2
Electrical input energy per volume pumped	kilowatt hour per 1000 British gallons	$kW\,h/10^3$ gal	kilowatt hour per kilolitre	$kW\,h\,kl^{-1}$	2.200×10^{-1}
Energy (work and heat)	foot pound-force	ftlbf	joule	J	1.356
	kilogram-force metre	kgfm	joule	J	9.807
	British thermal unit	Btu	joule	J	1.055×10^3
	therm	therm	megajoule	MJ	1.055×10^2
	kilocalorie	kcal	megajoule	MJ	4.187×10^{-3}
	kilowatt hour	kW h	megajoule	MJ	3.6
Force	pound-force	lbf	newton	N	4.448
	kilogram-force	kgf	newton	N	9.807
	ton-force	tonf	newton	N	9.964×10^3
Head of liquid and NPSH	foot	ft	metre	m	3.048×10^{-1}

For centrifugal and axial flow pumps head should be in metres total head of liquid pumped, since this is a fundamental performance parameter.

Physical quantity	Various units		SI pumping units		Conversion Factor
	Name	Unit	Name	Unit	
Length	inch	in	millimetre	mm	2.540×10
	foot	ft	metre	m	3.048×10^{-1}
	yard	yd	metre	m	9.144×10^{-1}
	mile	—	kilometre	km	1.609
Mass	pound	lb	kilogram	kg	4.536×10^{-1}
	long ton (2240 pounds)	ton	tonne	t	1.016
	tonne	t	kilogram	kg	1.000×10^{3}
Mass flow rate	pound per hour	$lb\,h^{-1}$	kilogram per second	$kg\,s^{-1}$	1.260×10^{-4}
	long ton per hour	$ton\,h^{-1}$	kilogram per second	$kg\,s^{-1}$	2.822×10^{-1}
	tonne per hour	t/h	kilogram per second	$kg\,s^{-1}$	2.778×10^{-1}
Power	horsepower (electric)	Hp	kilowatt	kW	7.456×10^{-1}
	foot pound-force per second	$ft\,lbf\,s^{-1}$	kilowatt	kW	1.356×10^{-3}
	kilogram-force metre per second	$kgf\,m\,s^{-1}$	kilowatt	kW	9.807×10^{-3}
	kilogram-force metre per second	$kgf\,m\,s^{-1}$	kilowatt	kW	9.807×10^{-3}
	British thermal unit per hour	$Btu\,h^{-1}$	kilowatt	kW	2.931×10^{-4}
	Long ton of refrigeration	—	kilowatt	kW	3.517
Pressure	pound-force per square inch	lbf/in^2	kilopascal	kPa	6.895
	kilogram-force per square centimetre	kgf/cm^2	kilopascal	kPa	9.807×10
	bar	b	kilopascal	kPa	1.000×10^{2}
	millibar	mb	kilopascal	kPa	1.000×10^{-1}
	standard atmosphere	atm	kilopascal	kPa	1.013×10^{2}
	millimetre of mercury	mmHg	kilopascal	kPa	1.333×10^{-1}
	torr	torr	kilopascal	kPa	1.333×10^{-1}
	inch of mercury	in Hg	kilopascal	kPa	3.386

When specifying pressure, the density of the liquid must be stated for conversion to head of liquid pumped.

Torque	pound-force inch	lbfin	newton metre	N m	1.130×10^{-1}
	pound-force foot	lbfft	newton metre	N m	1.356
	kilogram-force metre	kgfm	newton metre	N m	9.807
Velocity	foot per minute	ft/min	metre per second	$m\,s^{-1}$	5.080×10^{-3}
	foot per second	ft/s	metre per second	$m\,s^{-1}$	3.048×10^{-1}
Viscosity					
Dynamic	centipoise	cP	millipascal second	mPa s	1.000
Kinematic	centistoke	cSt	square millimetre per second	$mm^2\,s^{-1}$	1.000
Volume	cubic inch	in^3	cubic millimetre	mm^3	1.639×10^4
	cubic foot	ft^3	cubic metre	m^3	2.832×10^{-2}
	cubic foot	ft^3	litre	l	2.832×10
	British gallon	gal	litre	l	4.546
	US gallon	US gal	litre	l	3.785
Volumetric Flow rate	cubic metre per hour	$m^3\,h^{-1}$	litre per second	$l\,s^{-1}$	2.778×10^{-1}
	British gallon per minute	gpm	litre per second	$l\,s^{-1}$	7.577×10^{-2}
	million British gallons per day	mgd	litre per second	$l\,s^{-1}$	5.262×10
	cubic foot per second (cusec)	$ft^3\,s^{-1}$	litre per second	$l\,s^{-1}$	2.832×10
	acre foot per hour	$ac\,ft\,h^{-1}$	litre per second	$l\,s^{-1}$	3.426×10^2
	acre foot per day	$ac\,ft\,d^{-1}$	litre per second	$l\,s^{-1}$	1.428×10
	US gallon per minute	US gpm	litre per second	$l\,s^{-1}$	6.309×10^{-2}
	barrel (42 US gal) per day	—	litre per second	$l\,s^{-1}$	1.840×10^{-3}

Index